Broadband Networking: ATM, SDH, and SONET

For a complete listing of the *Artech House Telecommunciations Library,*
turn to the back of this book.

Broadband Networking:
ATM, SDH, and SONET

Mike Sexton
Andy Reid

Artech House
Boston • London

Library of Congress Cataloging-in-Publication Data
Sexton, Mike
  Broadband networking : ATM, SDH, and SONET / Michael Sexton, Andrew Reid
      p.   cm. — (Artech House telecommunications library)
    Includes bibliographical references and index.
    ISBN 0-89006-578-0 (alk. paper)
      1.Broadband communication systems.   2. Asynchronous transfer mode.
3. SONET (Data transmission)   I. Reid, Andy, 1961-   . II. Series.
TK5103.4S49   1997
    621.382'1—dc21   97-27618
                                                          CIP

British Library Cataloguing in Publication Data
Sexton, Mike
  Broadband networking : ATM, SDH, and SONET
  1. Broadband communications systems
  I. Title. II. Reid, Andy, 1961-
  621.3'821

  ISBN 0-89006-578-0

Cover design by Jennifer L. Stuart

© 1997 ARTECH HOUSE, INC.
685 Canton Street
Norwood, MA 02062

International Standard Book Number: 0-89006-578-0
Library of Congress Catalog Card Number: 97-27618

10 9 8 7 6 5 4 3

Contents

Preface

Network bandwidth continues to grow apace. Much of the core network is already served by optical fiber while silicon digital signal processors are opening up the previously band-limited copper and radio access segments with high capacity compression and transmission technologies. The synchronous digital hierarchy (SDH) has fulfilled much of its early promise as the standard for optical transmission in the multioperator environment with extensions to 10 Gbit/s and now, wavelength division multiplexing as well, holding out the promise of unlimited bandwidth for the future. The asynchronous transfer mode (ATM), with its almost infinite flexibility, is still the strongest contender for service bandwidth management in this new broadband world, but there is still some way to go before it can be said that its potential has been fully realized in the context of a modern, full-service broadband network. At the same time, the exponential growth in the Internet makes it likely to be the largest single user of broadband networks.

This book started out as essentially a second edition to *Transmission Networking: SONET and the Synchronous Digital Hierarchy* (Artech House - 1992), and indeed a large part of the book builds on the first edition with a substantial update of developments in the field of SDH and SONET. But we have also taken the opportunity to expand our scope considerably in this second book to include other closely associated, contemporary developments. In particular, we have devoted considerable space to describing ATM and its developing role in the multilayer, multiservice network. As before, we have set great store by functional analysis as the key to understanding network architecture and it has been gratifying to note that the generic principles we set out in the first edition in an SDH context have now been widely accepted in the ATM community also.

Our story is still very much the development of technical standards for international interworking, although the focus in the second book has shifted away from the International Telecommunications Union (ITU) and the regional standards bodies that laid down the basic frameworks, towards a number of important industry interest groups who have done much to promote practical solutions within the agreed standards framework. Born mainly out of frustration with the established fora, these groups have often found out for themselves that the development of detailed agreements on complex technical issues in a quasi democratic context is never easy. Nevertheless groups such as the Network Management Forum and the ATM Forum have produced substantial added value, and we can look forward with some confidence to the rapid appearance of real multivendor, multioperator interworking systems based on these expanded recommendations.

We have in general avoided being very specific about vendor implementations and operator deployment plans. Apart from the real risk of contentiousness, these are very often quite ephemeral. We have concentrated instead on what we believe to be the basic truths of our subject, which we believe will outlive the implementations that give them effect. Although we have in the main constrained ourselves to an objective presentation of the facts as we see them, it would not be natural in such volatile circumstances to avoid expressing opinions about probable or preferred outcomes. In these cases, the opinions expressed are our own and not necessarily those of our companies. We are indebted, as before, to our many friends from manufacturing and operating companies across the world who have collaborated under the banner of international standards to produce an impressive body of work from which we can all benefit. We also thank our companies Alcatel and BT for giving us the opportunity to participate in these activities and build up a body of knowledge, which it is now our privilege to place before you. Again, we must thank our wives Maureen and Pippa who despite the trauma of the first edition nevertheless acquiesced in the present enterprise and showed commendable forbearance throughout. Finally, we would like to dedicate this work to our parents, three of whom are sadly no longer with us; they would nevertheless have taken some pleasure in seeing their sons' names appended to a creditable if somewhat esoteric publication.

Evolution of Telecommunications Services

The original Greek word *architect* referred to the chief builder. Today, architecture is generally understood to be the art and science of building design. According to well-established practice, structures are analyzed to gain an understanding of their component parts, the functions of those parts, and their relationships with one another. The knowledge obtained is then used to design bigger or better buildings. The same process is used in the study and construction of telecommunication networks. Thus, network architecture is a description of a network's component parts, their function, and the relationships between them.

However, there is an important difference in scale between these two examples of architecture. A great building generally has a more or less well-defined scope and function and generally is conceived and executed according to a single plan. The public network on the other hand, covers the globe like a web. Its construction proceeds in a fragmentary manner by the more or less cooperative activity of a large number of independent agencies, and yet, when viewed from a wide enough perspective, it is undeniably a single entity providing near instantaneous communications between individuals and organizations throughout the world.

The building metaphor is extended if the architect can similarly widen his or her perspective to include the cities of the world, major and minor, together with the network of land, sea, and air links between them. Indeed, the comparison now becomes somewhat more than metaphoric. Many commentators see in the international telecommunication network a modern counterpart to this physical transportation network established over many centuries that has enabled the present high levels of economic activity between nations, cities, and all the smaller communities of interest into which mankind is divided.

This extended architectural perspective reveals further similarities between telecommunications networks and terrestrial transportation networks. Each new project, be it a road bridge or a submarine transmission system, may be viewed by the builder or systems engineer as a self-contained entity of well-defined scope, defined by its intended function and its relationship with the environment in which it will be placed. But from our widened architectural perspective, each represents a small incremental growth in a network that, by its very nature, is never complete. Similarly, the process of renewal, whereby redundant or obsolete facilities are abandoned or destroyed to be replaced as circumstances demand, is characteristic of each.

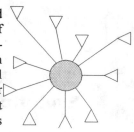

Such human artifacts studied on this scale display an apparently random complexity and intricacy that is more characteristic of the natural world than the well-ordered world of engineering. Indeed, it would be surprising if it were otherwise because, as has already been stated, these structures are not the result of some grand design but rather the separate activities of a very large number of independent agencies acting largely out of local self interest and a desire to integrate and harmonize with their surroundings. As the process is cumulative, each new extension is restrained by the legacy of the past and the need for compatibility with its neighbors and its environment. Recommendations for the construction, interconnection, and operation of telecommunication networks have been provided by the International Telecommunications Union (ITU-T, formerly CCITT) since its

inauguration in 1932, but these only give a very partial insight into those aspects of network architecture that are the subject of this chapter.

It is therefore appropriate in this introductory chapter to approach the study of network architecture in the manner of the natural scientist or the archeologist, analyzing the existing transport network as we find it; tracing its evolution, classifying components according to form and function, and studying their interrelationships with one another. It is impractical and (fortunately for the present purpose) unnecessary to acquire a detailed understanding of the underlying technologies in each component. It is our intention only to provide sufficient commentary on current and past practice to provide a sound intuitive basis for the general propositions that follow.

1.1   THE BEGINNINGS OF THE TELEPHONE SERVICE

1.1.1   Architecture of the Early Telephone Network

The first public telephone services appeared in the latter half of the 19th century, serving small communities of interest in well-developed commercial areas and fashionable residential areas. A few hundred subscribers situated within a radius of less than 1 km from the exchange building was typical of a well-developed early example. One can pass over the pioneering work of Bell, Edison, and others whose inventions made this possible, but

pause to note some features of these early networks that have significance for the transport network architecture seen today. This small group of initial subscribers will have been drawn from a community of interest that already formed part of a network of communicating individuals accustomed to forming associations for the purpose of communication among themselves, whether for business or pleasure. They will visit one another's houses or offices or arrange to meet in public places for this purpose. When their communication is complete, the association is broken and the individuals part company. The telephone network provides an alternative way of forming such associations for communication purposes, requiring no mobility on the part of individual subscribers. Each subscriber possesses a telephone instrument that converts between the normal mode of voice transmission through the air and the electromagnetic mode of transmission along copper wires. Each instrument is linked via a copper connection to the exchange where an operator is able to connect together two such connections, an association between a pair of subscribers for communication purposes.

Already in this very simple case we can see examples of the key architectural components of any transport network: the adaptation between the mode of information transfer within one established network and the mode of information transfer in another providing a service to the first, and a topology whereby the (copper) links of the new (transmission) network connect the adapted information sources and links (the telephones) to a switching node (the switchboard) where they can be flexibly associated for the purpose of transparent information transfer across the end-to-end circuit.

1.1.2   Evolution in the Management of Connectivity

Management of these early networks was based on a similarly simple control architecture in which the basic features recur in more sophisticated form to this day. A request to set up an association originates from one of the terminations (subscribers) on the network. This request is communicated verbally to the network manager (the operator), who will have been alerted by means of an access protocol relying on the manually driven magneto and bell. The network manager (the operator) makes a routing decision based on an analysis of the original request and a knowledge of all the possibilities within the network. The destination termination (the called subscriber) is alerted, the association is made, and the end-to-end circuit established. The opening exchange between the terminations (the calling and the called subscribers) conventionally serves to confirm to the terminations (the subscribers) and the network manager (the operator) the validity (or otherwise) of the association. Performance is continuously monitored by the terminations (the subscribers) for the duration of the association 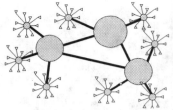 and reported to the network manager (the operator) if discovered to be below expectations.

1.1.3   Effect of Growth on the Topology and Control of Connectivity

Further technical developments around the turn of the century provided significant elaborations of this basic architecture. Long transmission lines meant that each switchboard could be connected to its neighbors. Two subscribers in different towns could communicate directly and the simple star network topology evolved to a more complex meshed network. The growing number of subscribers and increasing level of long-distance calls resulted in many of the originally small switchboards becoming so large they had to be housed in huge switch rooms in order to handle the thousands of subscribers and many hundreds of "trunks," as the interexchange lines were called.

Network management now involved a growing army of operators and was fast

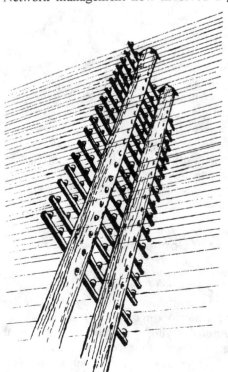

becoming the main limitation to further growth. In 1889 the step-by-step mechanism was invented by Armon B. Strowger that enabled subscribers to set up their own call using a 10-digit dial on their telephone. Local calls were the first to be handled this way, leaving the operators free to deal with the more complex task of trunk routing.

The first 40 years of this century were dominated by the further elaboration of Strowger's invention and the associated, mechanical call-processing technology. This, together with the standardization of the international numbering plan, enabled more and more of the *public switched telephone networks* (PSTNs) to become automated, resulting in a significant elaboration of the control architecture of the PSTN and a corresponding vast increase in scale of the transport architecture. That is to say, switches got bigger and more numerous, numbers

of subscribers increased, and more and more transmission links were provided, but the basic architectural principles remained the same. The *network node interface* (NNI) between the switching components and the transmission components of the network and the *user to network interface* (UNI) quickly stabilized, albeit with significant regional variants. This allowed the transmission and switching aspects of network design to develop independently with well-defined roles for many years.

Throughout this early period of network development, the analog voice frequency signals generated and interpreted in the subscriber's telephone instrument were transmitted between network nodes by a single, dedicated pair of copper wires. The large number of copper pairs entering a telephone exchange were connected onto a distribution frame. The ports on the switching equipment were also connected to this frame so that each port could then be connected to an incoming line. The frame provided a flexibility point between the external line plant and the internal switching equipment. Subscribers' lines terminated on the main distribution frame while interoffice trunks terminated on the trunk distribution frame. Such flexibility was essential for maintenance purposes and to deal with the changing requirements of what we can now recognize as the PSTN.

1.1.4   Transmission, Multiplexing, and the Resulting Stratification

The copper pairs used for transmission between network nodes were first strung from poles. At a later stage, groups of copper pairs were assembled into cables and buried in ducts under ground. Amplifiers, equalizers, loading coils, and many other innovations served to increase the feasible distance and improve the quality of transmission. However, none of these technical improvements resulted in any significant architectural change in what can now be recognized as the transport network, whose subservient role had become well defined. This was the faithful reproduction of voiceband signals presented at one network node to another network node remote from the first.

Humble though it might seem compared to the vast elaborations of Strowger's original concept, this role has proved to be a very demanding one as distances and traffic volume have increased. These drivers were then responsible for the next significant development in transport network architecture with the introduction of multiplexing in the 1930s. This enabled analog voiceband signals to be assembled into 12-channel groups that were adapted for transmission over media capable of supporting the higher bandwidth. Point-to-point radio systems and new high-quality coaxial cable transmission systems were developed to supplement the limited capabilities of the existing copper pairs, and the trunk distribution frames appeared to provide flexibility for this new 12-channel group layer.

It was not long before another stage of multiplexing was introduced whereby five of the original 12-channel groups were assembled together to form a 60-channel supergroup. At first, this new capability was regarded merely as a means of providing a larger number of trunks between two PSTN nodes at lower cost than before. However, this transmission method became so widespread that it was convenient to route some supergroups through a transit node or building. This was equivalent to the introduction of a new supergroup layer to the trunk distribution frame. Also around this time, the through-group filter was introduced, which allowed one or more groups to be dropped or added from a supergroup transiting the node.

It is now possible to observe one of the most striking architectural features of the evolving transport network, namely the expansion mechanism that spawns new, subservient transport layers, each providing simple point-to-point transport between the nodes of its predecessor. Each new layer is seen to possess separately all the components of a network. The network nodes manifested as distribution frames are capable of supporting flexible associations between links. The nodes themselves are connected to one another by links provided by the point-to-point transport capability of another, higher capacity layer to which they were adapted in these early analog networks by the *frequency division multiplex* (FDM) technique. Monitoring at network terminations was provided by a system of supervisory tones and pilots that monitored performance and transport integrity at the layer boundaries.

1.1.5   Non-PSTN Services

Although public telephony was by far the greatest driver, it was by no means the only telecommunications service offered in those early days. The infrastructure provided by the telephone network could be used to provide transport service for a variety of other teleservices.

1.1.5.1   *Special Circuits and Leased Lines*

From the earliest days, there has been a demand for "fixed" private lines connecting two subscribers in a semipermanent manner. The public telephone companies achieved this by "patching" across the distribution frame between the appropriate terminating lines, thus establishing the semipermanent association. This procedure was closer to that of the early switchboard operators than to the automatic switching machines of the PSTN. Thus, even at this early stage we can observe two distinct networks. The PSTN and the private circuit network were both supported on a single infrastructure of ducts, cables, and buildings that housed the switches and distribution frames.

1.1.5.2 Data Transmission

The original telegraph predated telephony by several decades, but its mid-20th century successor, in the form of telex, had grown to a network of wide international extent which (although small by PSTN standards) provided an early demonstration of the power of entirely digital communications. It provided a public switched data service between teleprinters in users' premises with its own international numbering plan. Architecturally, it contained the same generic components as the other network layers, including links provided by the same well-established analog transport network that was supporting the PSTN. Digital switching and stored program control were first exploited in the telex network before adoption in the PSTN.

1.1.6 Service Layer Growth and Fragmentation

This pattern of development continued well into the 1960s. A steady growth in PSTN subscribers fueled an even greater increase in traffic levels. This forced the pace of automation in the PSTN, which in turn enabled still more growth. Quite extensive private networks emerged using scaled-down PSTN technology to provide switching and control with internodal links provided by leased capacity on the transport network. New communities of more specialist users appeared, leasing capacity on the transport network in wider bandwidth channels. Broadcasters were typical of this class of user, with their requirements for transporting wide-bandwidth broadcast material between studios and transmitters. Thus although based on some common infrastructure the network was effectively partitioned between largely independent service user communities.

1.2 THE DIGITAL REVOLUTION

The first evidence of digital operation in the transport network can be traced to the early 1960s, when digital transmission was introduced in the interoffice or junction networks in several countries. Adaptation of the analog voice information to the digital mode of transport exploited the principle of *pulse code modulation* (PCM). This requires that the analog voice signal be sampled at a rate greater than twice its maximum frequency component. Each sample is measured and transmitted as an equivalent number, and a set of such samples derived from many different channels can be transmitted sequentially over a wideband medium; a process described as *time division multiplexing* (TDM).

During the mid-1960s, a series of standards were developed in ITU-T to define the adaptation process, the structure and size of the multiplexing groups, and the characteristics of the NNI for this new transport network. There was universal agreement on a sampling rate of 8 kHz and a 64 kbit/s channel rate, the minimum requirement for switching, but standards for the supporting transport layers in Europe and North America differed in

almost every other important respect. Japan aligned roughly with North America, but diverged at the higher bit rates. This set of standards has retrospectively been called the *plesiochronous digital hierarchy* (PDH), in contrast with the *synchronous digital hierarchy* (SDH), which is one of the main subjects of this book. The ITU-T Recommendation G.702 defines the hierarchical levels of the PDH. A sequence of subsequent recommendations in the G series defines the detailed frame structures and the functional requirements of equipment.

At this stage in the evolution of the network, operators and planners had not identified a pressing need for transport integration or management automation. Operations relied exclusively on manually connected *digital distribution frames* (DDF) and paper records, while operations support systems used simple extensions of the electromechanical relay, lamp, and alarm bell technology established in the early days.

The evolution towards digital technology in the network was mainly driven by technical considerations. The major operators all operated as regional monopolies, which meant that (although long-term cost benefits and market need were cited as part justification) deployment was not markedly driven by market forces. The power of the emerging semiconductor technologies was well-matched to the digital TDM technique as very large switches could be constructed using combinations of time and space switching. This in turn was well-matched to the new processor control technologies. In the transport network, digital technology held out the possibility of high-quality, high-capacity transmission independent of distance and the possibility to bypass the trunk DDF by terminating digitally multiplexed groups directly on the switch.

From the viewpoint of architectural evolution, the new digital technology enabled a degree of circuit layer integration as it dealt equitably with voice and data but also introduced new, digital transport layers parallel to the well-established analog transport network. The characteristic features of circuit layer fragmentation and transport layer stratification were still very evident in the digital environment.

The succeeding quarter century has seen the widespread deployment of digital transmission and switching in the industrially developed countries. Packetized data multiplexing and switching allowed a statistically, more efficient handling of data traffic. Optical fiber is driving down the cost of transmission. Processor control and internationally standardized signaling protocols are enabling new standards of PSTN service, and the size and number of special networks continue to increase. During the same period, digital technology, developed in the data-processing industry, has been deployed in administrative and operations support roles, adding new dimensions of architectural complexity.

1.2.1   Integrated Services

The main technical objective that drove integration in the digital revolution was the possibility to handle any traffic, no matter the bandwidth or information characteristics, in a single network fabric. Circuit- and packet-switching fabrics on a common digital transport platform was considered an acceptable compromise. The basic channel rate 64 kbit/s/s channel and its multiples was considered to be a sufficiently flexible unit of connectivity for all foreseeable needs, and packet multiplexing provided sufficient flexibility for bursty data traffic. These assumptions were contradicted even before the integrated service digital network (ISDN) standards were complete. There are many theories to explain the comparative lack of penetration of ISDN even in the mid-1990s, but the lack of cost-efficient adaptability to services of different rates and traffic statistics other than the fixed, continuous rate ISDN channel is a very fundamental one.

Encouraged by the success of packet networks, the concept of fast packet switching based on small, fixed-length packets was tested during the mid-1980s. This has since been termed the asynchronous transfer mode (ATM) and is now the focus of current efforts to realize a true ISDN. Traffic rates are not constrained to the channel rates of the supporting fabric and very low rate or sporadic data can be accepted just as readily as multimegabit data bursts from unpredicted and unpredictable new services.

1.2.2   The Rise of the Internet

The consolidation of the digital network with its inherent integrated digital service (ISDN) capability promised increasing synergy between the data-processing and telecommunications industries. Confident predictions about the convergence of computing and communications were commonplace in the middle and late 1980s, and few doubted that ISDN would be the key enabler. In fact, almost unnoticed in the telecommunications world was an experiment in information sharing between computer networks started in the United States that initially connected together a small set of defense establishments and academic research centers with the support of the National Science Foundation (NSF). Shared media local area networking (LAN) appeared soon afterwards, allowing physically collocated computers to share peripherals and exchange information. The NSF experiment generated a set of high-level protocols enabling effective data transfer between LANs. These were collectively termed internetworking protocols, and the federation of independent local networks became known as the Internet. It soon grew to support a range of services including file transfer and electronic mail.

Today, the Internet has matured from an academic novelty into a large commercial reality with many millions of users worldwide, both business and residential. It is a real live example of the sort of diversity that can be sustained at the service level on a ubiquitous transport platform, achieving ever greater economies of scale shared across services. It is also interesting in that it is based on the connectionless mode of operation in which complete datagrams with destination address information are launched into the network without first creating a connection and committing resources. Routing is automatic, according to routing information embedded in each network node. The Internet is gradually acquiring more capabilities for multimedia, interactive applications, and other real-time services. These are currently charged at a flat distance-independent, service-independent tariff, although guaranteed service quality is a problem. The big question today is whether the Internet will develop into a true broadband integrated service network, making obsolete the connection-oriented architectures developed over the past century, or will it join with the developing ISDN as part of a wider Internet federation of networks. Either way, it will no doubt continue to draw on the services of a global broadband transport platform.

At the time of writing, there is much heated discussion between Internet advocates and ATM advocates. If the target is a single, universal, multiservice network which meets all requirements, then only the Internet Protocol (IP) or only ATM can be that answer. There is, however, a growing consensus that such a utopian solution can never exist. Both Internet and ATM have great strengths, and practical solutions will use both. Indeed, since Internet only defines the higher *inter*networking protocols, it must use a transport network, and ATM provides an increasingly cost effective answer. This is the basic working assumption we adopt in the book. Irrespective of the protocol used by broadband applications, an underlying broadband transport network is required. It seems likely that the Internet will be the largest single user of this broadband network, however, it is very unlikely to be the only user.

1.2.3   Multiservice Traffic Growth

The growth in total traffic levels from all sources was reflected in increasing demand on the transport network from all the client networks that it serves. The DDFs had now become so big that they formed a major barrier to growth and a limiting factor in transport network quality, in much the same way as the large, multioperator switch rooms were limiting PSTN development in earlier years. This pressure has stimulated the deployment in recent years of distributed bandwidth management and electronic crossconnect systems controlled from remote network management systems in a manner more or less integrated with the processor-based operations support systems. It was against this background that the SDH standards were conceived and developed.

The potential for increased variety of services with a wide range of traffic characteristics is only now being seen in the developing Internet. Already, the limitations of both the narrowband service-switching infrastructure and the present connectionless packet-routing networks are being strained. The provision of a true service-independent switched infrastructure has been the corresponding objective behind the development of the ATM standards. The service layer generally can be expected to continue to fragment under the combined pressures of competition and regulation. The service-independent transport layer serves to lower the threshold of viability of any new service.

1.2.4   Transport Network Transition Strategies

During the transition period from analog to digital, a special set of elements emerged whose function was to ease the coexistence of the two parallel technologies. Supergroup and hypergroup codes could be found carrying high-capacity analog signals across digital islands. Conversely, high-capacity digital modems were developed to carry digital groups between digital islands across the high-capacity analog network. And, most impressive of all, transmultiplexers appeared that used a combination of high-resolution analog-to-digital conversion and complex digital processing to convert analog channels, which had been multiplexed in one technology directly to their multiplexed equivalent in the other. All of this equipment provided interlayer adaptation functions directly between analog and digital transport layers without exposing the circuit layer. As we shall see later, the transition to SDH and ATM is generating similar adaptation functions, with the main purpose being to ease the transition from PDH and narrowband ISDN, respectively.

1.2.5   Globalization

The early stages of the evolutionary process described above took place more or less independently in the different regions of the world. Even when voice frequency transmission standards were developed to allow stable communication over long distances without echo, international telephony was still a precarious affair with each call requiring extensive preplanning. Introduction of subscriber trunk dialing, and more recently common channel signaling, is now providing international telephone service of extraordinarily high quality. Translation at international boundaries is still required to take account of the different regional transport standards that have developed. This is costly, although tolerable, for the PSTN where connections have to be made at the level of the individual voice channel, but for the transport layers that are increasingly required to arbitrate between competing services in a multiservice environment, incompatibility at regional boundaries has been a significant barrier to growth.

Almost certainly the new generation of optical cross-border links will use SDH and ATM, extending the scope of management and control automation to these high-capacity global highways. Open interfaces allow the automation of bandwidth management between and among cooperating network operators. Thus, transport paths originating with one operator, transiting the network of a second operator, and terminating with a third may be set up, restored on failure, and taken down when no longer required, entirely by automatic processes with no manual intervention. International traffic is the fastest growing and the highest earning of all sectors. It is also fast becoming the most competitive. The pressures are very strong, primarily from the international business community, to improve capability and reduce the cost of global networking.

The preceding discussion on network evolution has served to informally introduce the main components required for a functional analysis of the transport network and will serve as an introduction to the formal description that follows. In the following, we briefly consider some of the nontechnical factors that are also influencing the future evolution of the network architecture.

1.3   THE REGULATORY ENVIRONMENT

The picture of network evolution presented so far is primarily one of organic growth driven by demand. This has been the factor determining the technical developments resulting in the high level of functional integration, service layer fragmentation, transport stratification, and increasingly distributed control that we now see. Most of these developments have taken place while the major networks were under the control of large, though generally benign regional monopolies. Not surprisingly, these monopolies have attracted the attention of legislators in most of the industrialized world. This intervention has taken different forms in different regions, but in all cases has led to some form of liberalization accompanied by some compensating regulation. Regulation in all its different forms will have a major effect on the future evolution of the network.

Legislators and regulators have not found their task easy. The architectural complexities, which have been outlined above and will be considered in more depth later, have proved to be a considerable challenge to political and legal minds, and there can be few who have been entirely satisfied with their work. Their task is to balance the conflicting interests of the ordinary subscriber (including those in remote or disadvantaged groups), the business community, the network operator, and the equipment supplier in a regulatory framework that is seen to be fair and open. At one extreme is the view of telecommunications as a natural monopoly that must be strictly regulated. At the other extreme is the view that it is a service industry like any other and can be expected to function best in a pluralist, competitive environment with minimal regulatory constraints. Most solutions have fallen somewhere between these two extremes.

1.3.1   Divestiture in the United States

The first significant regulatory intervention was the restructuring of United States telecommunications in 1984, whereby the almost universal provider, *American Telephone and Telegraph* (AT&T), was replaced by seven Regional Bell Operating Companies (RBOCs) acting as *local exchange carriers* (LEC), and one long-haul carrier. In this case, the regulators took the view that the LECs should remain as regulated local monopolies, while open competition was encouraged in the long-haul market. These broad categorizations were refined by a series of subsequent judgments that clearly marked out the scope of all the licensed players. The LECs were further divided into *local access and transit areas* (LATAs), with inter-LATA transport reserved for the long-haul carriers. The LECs were required to present such traffic at a *point of presence* (POP) on the long-haul operators network. This, together with a matching commitment to a multivendor open market, has been one of the major factors in the development of high-capacity, standardized transport interfaces.

The enforced separation between local access and long-haul segments of the market played a large part in the explosive development of corporate private networking based on digital leased lines at the primary rate. The long-haul carriers were allowed to sell this service directly to corporate users and bypass the LECs who, on their own, could not provide end-to-end service. Broadly the same customer base has been the consumer of wide area data network services for interconnecting LANs. Such leased lines form the Internet backbone. Dr. Peter Huber, in his report to Congress in 1988 on the results of divestiture four years earlier, described a vision of what he called the geodesic network. Its main feature was the prevalence of distributed intelligence in a highly meshed network, with inexpensive universal transport as an alternative to the fixed hierarchical structure that had developed over the preceding century. The Internet and the private networks based on PBXs (private branch exchanges) are both examples of this principle.

In the access segment, the reservation of voice services for exclusive operation by the RBOCs was balanced by a prohibition on their participation in local cable TV (CATV) services. This made it unattractive for the RBOCs to invest in local fiber distribution systems, which could seldom be justified for telephony alone. Pressure for a review has built up over several years and the Telecommunications Act, signed into law by President Clinton in February 1996, makes sweeping changes that have the effect of opening up all areas for open competition.

1.3.1.1   *The Telecommunications Act of 1996*

This far-reaching act affects all sectors including local, telephone, CATV, long-distance, and manufacturing alike. The primary effect is to generate new competition in the local loop. The way is open for new local access providers to offer a complete range of services including telephony and CATV. The RBOCs are no longer prohibited from CATV provision in their local markets, and public utility holding companies are no longer banned from offering telephone service. The RBOCs may also enter the long-distance business previously prohibited to them and also provide information services. In return for these new freedoms, the RBOCs must demonstrate their effectiveness in fostering competition in their local markets. In particular, they will be judged by their success in enabling open interconnect to other carriers on a fair and equitable basis. In other words, their licenses in the new areas are conditional on satisfying the Federal Communications Commission (FCC) on a number of open networking issues:

- *Resale*—Unreasonable or discriminatory conditions will not be allowed. Incumbents must offer at wholesale prices any service provided at retail rates to customers who are not telecommunication carriers.
- *Number portability*—Incumbent telcos must provide facilities allowing customers to retain their phone numbers when switching to another phone company.
- *Dialing parity*—Must be provided to all competing providers, allowing nondiscriminatory access to all destinations without unreasonable dialing delay.

- *Access to rights of way*—Incumbent must provide reasonable access to poles, ducts, cable ways, and so forth at fair and reasonable prices.
- *Reciprocal compensation*—Incumbents and new entrants must set up agreements covering the origination and termination of calls.
- *Interconnection*—The incumbent must provide interconnection facilities at any technically feasible point in its network equivalent to that provided internally.
- *Unbundled access*—The incumbent must provide access to local loop plants on a fair and equitable basis to competing operators.
- *Access to certain service platforms*—The incumbent must provide nondiscriminatory access to directory services, call completion services, and so forth.

To allow a period for adjustment, the long-distance operators are not allowed to offer bundled (i.e., long-distance component and local completion components) for three years. With a similar time delay, the CATV companies are exempted from rate regulation on their upper tier services to allow them to adjust to the new competition.

This has precipitated a phase of feverish activity where potential partners seek new alliances, failed alliances seek disentanglement, and telcos together with their traditional suppliers try to come to terms with the Internet. The effect of all this legislative activity and corporate restructuring is not easy to predict, but we can be certain it will have a far-reaching and dominant effect on the evolving architecture.

1.3.2   Liberalization in Europe

In the past, telecommunication networks in Europe were administered on national lines, with the *Committee European de Post et Telegraph* (CEPT) providing a technical framework of base standards and a commercial framework to supervise accounting adjustments. Interoperator communications were agreed bilaterally, and the regulatory provisions varied from country to country. The development of the *European Union (EU)*, and the intention to operate an open market from 1992, has been a major factor in the gradual liberalization of telecommunications in Europe. The United Kingdom has enacted the most extreme changes so far by licensing many new operators. Similar distinctions are evident in the roles these new operators are expected to play in the evolving network, like those between LECs and international exchange carriers (IECs) in the United States. The principle of open network provision (ONP) mandates equitable access to transmission capacity (long and short haul). Dominant telcos are being required to operate a transparent and cost-based tariff structure. It is clear that transport is seen generally as a business of selling transport services wholesale to the service networks, who subsequently act as retailers, selling telecommunication services to end users. Therefore, from a user viewpoint, monopoly is being gradually transformed into pluralism.

In 1993, the Bangemann report emphasized the need for efficient and effective communication infrastructure and urged member states to accelerate the process of liberalization of the telecommunications sector by opening up to competition infrastructures and services still in the monopoly area. Subsequently, in October 1994, the European Commission adopted Part I of a Green Paper on the liberalization of telecommunications infrastructure and CATV networks. In that document, the Commission set out general principles and a timetable for action forming the basis for consultation on the development of a common approach to infrastructure in the EU. Part II of the Green Paper subsequently made detailed proposals, most of which have been adopted and since become directives:

- Removal of special and exclusive rights over use of infrastructure for the delivery of voice telephony services to the general public from January 1998 (in particular, CATV);
- The safeguarding of the universal service obligation (USO) and the recommendation of a framework for financing that avoids access charges;
- Rights and obligations of public telecommunication infrastructure providers with regard to interconnection, including the establishment of dispute resolution mechanisms;
- Common principles for interconnection charges and recovery of USO based on net cost calculations;
- Application of ONP and competition rules to ensure open access to adequate transmission capacity with transparent accounting and uniform standards;
- Nondiscriminatory approach to infrastructure authorization and licensing, but service provision not subject to individual licensing;
- Ensure fair competition in initial period by screening interconnection agreements, monitoring joint ventures, assessing the financing of USO, ensuring open access to rights of way, and reviewing effects of different networks and joint provision arrangements. This last requires transparent accounting between the infrastructure components and the services component;
- Provision of arrangements to ensure equal access to numbers and provision of a European numbering space for special services;
- Incumbents must provide open access to directory services and subscriber data.

The European legislation is broadly comparable to that enacted in the United States. One notable exception is the lack of measures to force unbundling of the local loop. Nevertheless, the combined provisions for network access and ONP represent a significant liberalization. Certain countries have negotiated a derogation beyond the date of January 1998, but others are already enacting certain aspects under pressure from their own local legislators.

The structural changes in networks set in motion by liberalization and competition-enhancing regulation is driving the network to become more fragmented at the operating level, with many new interoperator boundaries to administer, while the end user must still perceive a homogeneous, integrated service. National operators that had been vertically integrated are now obliged to deal equitably across layer boundaries with other operators on either side of the boundary. All those issues—operational, commercial, or regulatory—that were previously relevant at international gateways now become relevant at boundaries between different operators in the same layer and in the same territory. But just as fragmentation is progressing under liberalization pressures, new alignments are developing (particularly in the direction of horizontal integration). It is too early to judge where these processes will find their equilibrium. There are many gray areas in the regulations governing interoperator relations and the regulators are allowed considerable discretion in resolving conflicts, but the die is cast and the industry can ever be the same again.

1.3.3   International Standards

Maintaining homogeneous services to end users in such an internationally fragmented situation relies heavily on powerful, high-quality interworking standards, and the roles of the various standards bodies, both regional and international, as well as the industry fora, become critical to the success of the whole concept. The operators are under ever greater pressure to reduce costs and find it difficult to provide the level of standards support they had previously managed. A much greater part of the responsibility to maintain adequate research and standards development activity now falls on the component supplier community.

Much of the remainder of this book tells the story of standards development for the operating scenarios outlined above. Standards development, far from being the dull, bureaucratic task of popular imagination, is placed in the forefront of network development. It has become in many ways the top-level system design task for the largest system imaginable: the global telecommunication network.

References

The following describe the strategic development of telecommunications in the United States:

Huber, P. W., *The Geodesic Network — 1987 Report on Competition in the Telephone Industry*. (The report is to Congress on the effects of divestiture. It is a good overview of the benefits and the shortcomings of divestiture in the US which has been a major input to later phases of liberalization, and was remarkably prescient about many recent developments.)

U.S. Legislation, *The Telecommunications Act of 1996*.

Computer Systems Policy Project Report, *Perspectives on the Global Information Infrastructure*, 1995.

The following describe the strategic development of telecommunications in Europe:

European Commission, *Green Paper on the Liberalisation of Telecommunications Infrastructure and Cable Televison Networks*.

European Commission, *Europe and the Global Information Society — Recommendations to the European Council* (known as the Bangemann report after the commissioner who headed the study, it gives framework and rationale for network liberalization in Europe).

ETSI, *Report of the Sixth Strategic Review Committee on European Information Infractructure*, (SRC6).

Architectural Framework 2

In the previous chapter, we described the almost organic process of evolution in telecommunications up to the huge size and complexity we see today. We described in an informal way how architectural concepts are used to master the inherent complexity. The increasing level of automation and the introduction of a wide range of new broadband services have made the formulation of a clear, comprehensive, and comprehensible architectural framework even more essential. Most of this book is concerned with elaborating on the architecture that will be implemented to support the full-service broadband network. In this chapter, we try to build on the informal review of the introductory chapter a more formal framework; the rules, guidelines, methods, and procedures for developing a comprehensive network architecture, for this, we believe, is the key to mastering complexity and ensuring consistency of design, management, and operation.

This chapter, along with chapters 3 and 4, is principally concerned with the functional modeling of broadband networks; in other words, the functional architecture. It is based mainly on the work of the International Telecommunications Union (ITU) on transport network modeling as defined by ITU-T in Recommendations G.805 (generic), G.803 (SDH specific), I.326 (ATM specific), and Draft Recommendation G.otn (WDM specific), but also draws on the current work on global information infrastructure (GII) in ITU-T, International Organization for Standardization (ISO), regional bodies, and forums such as the Digital Audio Visual Council (DAVIC). In Chapter 3, we develop the functional architecture using the framework described here to describe the transport aspects of broadband networks, while in Chapter 4 we specialize this further in order to describe the control and management aspects of broadband networks.

2.1   BACKGROUND AND REQUIREMENTS

The functional architecture must describe the overall conception of a broadband network and give the framework within which detailed design can take place. It should:

- Be comprehensive in scope;
- Be comprehensible to a wide community of network designers, planners, builders, and operators;
- Be acceptable within the international multioperator, multivendor telecommunications networking community;
- Support tools and methods for consistently expressing concepts and specific capabilities;
- Allow an appropriate level of specification that is precise enough to ensure compatibility between components and successful operation of the overall network, while at the same time not placing unnecessary restrictions on the way an implementer can realize particular components of the network.

Meeting all these requirements is not simple, and this is further complicated by the fact that telecommunications professionals have become accustomed to a particular set of architectural concepts manifest in the traditional network, which has evolved in an ad hoc manner over the preceding half century, and are not always aware how strongly these presumptions color their thinking.

2.1.1   A Historical Perspective

The architecture of the existing network developed over several decades divides the network into the four primary components of access, switching, transmission, and management. This has been extended more recently by the concept of the intelligent network (IN). This architecture has developed in an essentially homogeneous network service environment, namely the public switched telephone network (PSTN), offering the single service: telephony (now often referred to as plain old telephony service — POTS).

The components with which the networks were constructed were constrained mainly by technological limitations to be relatively simple, and even an abstract logical representation expressed through the conventional "box and line" drawings of the systems engineer mirrored the physical implementation. The complexity increase in the network and service environment is paralleled by the complexity of high levels of functional integration made possible by technology developments in recent years. New, more powerful abstractions and more expressive analytical and specification tools are now considered necessary. The advent of the synchronous digital hierarchy (SDH) around 1987 brought this situation to a head. This coincided with an unprecedented level of industry restructuring and a pressing need for better and more open standards.

At first, the reader who is accustomed to the existing network architecture may find the functional architecture abstract and removed from the "real" network. However, we hope that by the time we have presented the other aspects of broadband networks, the great value, indeed the necessity, of this approach will be apparent.

2.2   GENERAL PRINCIPLES OF FUNCTIONAL ARCHITECTURE

The functional architecture relies on a small number of very general principles. We shall see these principles recur many times and in many forms throughout this book.

2.2.1   Definition of a Function

In general terms, a function is a part of a wider system that carries out a set of well-defined actions on the basis of inputs to the function from the wider system and outputs to the wider system. In the context of broadband networks, the inputs and the outputs to the functions are information, while inside a function, information is used to initiate and control the actions of the function, and can also be stored. A generic function is illustrated in Figure 2.1.

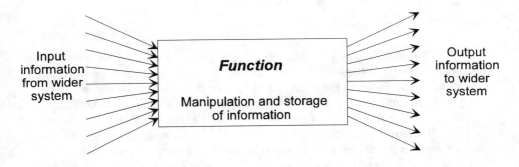

Figure 2.1   A function.

There are other terms that broadly fit with this definition each with a differing emphasis:

- *Process* emphasizes the dynamic activity within a function which manipulates information.
- *System* emphasizes the objectives of a group of cooperating functions.
- *Component* implies a function that is one of a collection of functions that are normally implemented together.
- *Object* emphasizes the encapsulation of functionality, which may only be accessed through declared interface operations; a feature that tends to be inherent in hardware implementations but must be consciously enforced in software implementations.
- *Agent* implies an object that acts independently or on behalf of another function.

Finally, as we are dealing with broadband networks, which are concerned with the transport of information, most of the functions of concern are informational (and not physical). This is an important point that means that functional definitions may be all the more effectively abstracted from their physical realization.

2.2.2   Definition of a Logical Interface

All our systems and processes are made up of functions that exchange information with one another. They converse across a logical interface, which is the formalization of the information that passes between them together with the way in which the passage of information is handled by both functions. This is illustrated in Figure 2.2.

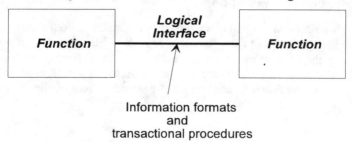

Figure 2.2   A logical interface between two functions.

Logical interfaces come in several forms. They can be two-way, in that information passes in both directions between the two functions, or one-way, in that information only passes from one function to the other but not in the reverse direction. In addition, the passing of the information can be in the form of *discrete* or *continuous* messages.

As with the term "function" there are many close analogues with the term "logical interface." The term "protocol" originally was used in a context of messages passing between functions to achieve a "handshake" between the functions; however, it is now increasingly used in a way that is synonymous with our definition of logical interface. The term "interface," on its own, often implies some form of physical connection as well as the logical one intended here. We purposely exclude this implication in general and do not want to presume any physical aspects in defining a logical interface. When a physical interface is intended, we say so explicitly. Many current usages are already in line with this; notably application programming interfaces (APIs), discussed in Chapter 4, which are certainly not physical. We try to use the term interface with caution to avoid confusion, and always try to use it in a clear context (e.g., as in API). Terminology is always a difficult subject when dealing with new concepts. Our aim is to be precise and self-consistent without being pedantic. We try to use generic terms in a sense close to their natural meaning and indicate where a formal definition is intended. Thus we hope to avoid misleading the reader.

2.2.3   General Principles of Functional Analysis and Design

In analysis, we decompose, classify, and abstract functions to obtain clearer understanding. In design, we aggregate, encapsulate, and distribute functions to form components of a coherent system, balancing the often conflicting requirements of integrity, reliability, and efficiency. We aim to identify the functions and logical interfaces between them in a way that is at once simple, parsimonious, and complete. There is a small number of basic principles:

- Identify functions at the correct scale for the level of design under consideration;
- Look for common functions where the design, specification, or implementation can be used many times;
- Specify functions to be complete stand-alone entities that may also be used outside the context of the immediate system;
- Specify logical interfaces in a common way so it is easy for other, unconnected design teams to use a function through its logical interface without understanding the full working of the function;
- Design the functions of a system so the system can be expanded and evolved by adding new functions and not by altering old functions;
- Design systems that are resilient to failure in components;
- Hide the internal design of a function so other functions in the system may operate without such knowledge.

Many of these requirements can be fulfilled by using properties of functions described in the following sections.

2.2.4   Encapsulation

This is a general principle of being able to hide the inner working of a function from the outside users of a function. To the outside, the function has well-defined logical interfaces and responds according to specification. Internally, the function can be composed of a number of smaller functions that need not be visible. This is illustrated in Figure 2.3.

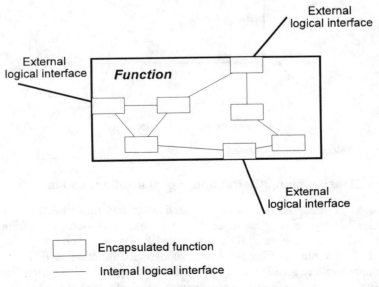

| Figure 2.3 | Encapsulation of smaller functions within a larger function. |

Encapsulation aids modularity. A design or specification may be broken down into manageable-sized pieces with a mechanism by which the pieces can be related. This can follow through to development and testing so that each component of a system can be developed and tested without the need for the whole system. Of particular importance is the ability to separate the role of a component builder concerned with the encapsulated functions from that of a system builder concerned with aggregation of encapsulated components into larger systems. Encapsulation also provides a mechanism to define commercial boundaries between network operators, between network operators and service providers, and between equipment vendors and operators.

2.2.5   Recursion

Recursion is a powerful technique that allows reuse of the same function at several levels of encapsulation. If, on looking inside an encapsulated function, there are smaller functions of the same type as the larger function, the encapsulation is recursive. As we shall see in Chapter 3, by using recursion it is possible to reduce the specification of very complex systems, specifically broadband networks, to a relatively small number of basic functions. Recursion is illustrated in Figure 2.4.

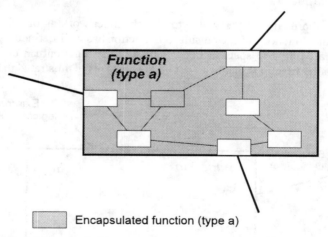

Encapsulated function (type a)

Figure 2.4   Recursive encapsulation.

2.2.6   Classification, Abstraction, Specialization, and Inheritance

Classification of functions makes their specification and implementation a great deal easier. Often, a situation of great apparent variety may be rendered comprehensible, and thereby manageable and more easily implemented, by classification according to common features. For example, in Chapter 4, the subnetwork function is defined with some precision applicable to a wide variety of subnetwork types. The specification is *abstract* in that it includes only the parts that are common to all subnetworks. It is then *specialized* to particular subnetwork types by adding a rather small increment of specialized definition and specification. For example, a VC-4 subnetwork is a subnetwork that connects SDH

VC-4s. The specialization is done in this case by a process called *inheritance*, whereby the specialized function inherits all the properties of the abstract function and adds the specialized features. This is a "top-down" approach to classification.

In a bottom-up approach, functions can be grouped into classes such that all the members of the same class have the same basic characteristics. Classes can then be grouped together to form more general classes. For example, all animals and plants in the world are classified according to their species. Species with common evolutionary inheritance are then grouped together into genera. This grouping into more general classes is called *abstraction*.

Both approaches are valid and, in general, both may be used to arrive at a good functional design. The concept of classification and classes is illustrated in Figure 2.5, and as the concept of inheritance does not restrict the number of superclasses to which an object can belong, a subclass can inherit its properties from more than one superclass. This concept is called *multiple inheritance,* and is illustrated in Figure 2.6.

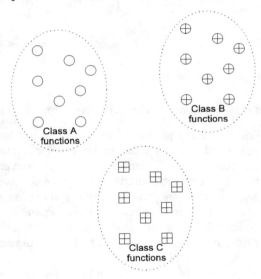

Figure 2.5   Object classes.

2.2.7   Separation of Functions From Their Implementation

With the large increase in functional integration that is now possible using silicon hardware and object oriented software technology, it is no longer appropriate to assume that one piece of telecommunications equipment carries out one well-defined function. A piece of equipment in a broadband network will usually carry out many functions and, with the extensive use of software in equipment, the number of functions may increase or change over time.

In this step of the development of the functional architecture, we separate the functions of the system from the equipment, fiber, software modules, and so forth that implement these functions.

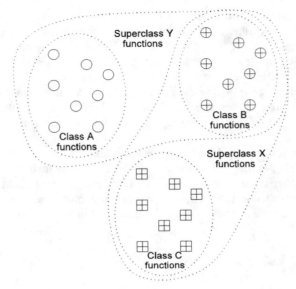

Figure 2.6   Superclasses and subclasses.

This method has many advantages. The most noteworthy at this point are as follows:

- The basic specification of equipment can be written in purely functional terms, allowing a supplier full freedom in the implementation, including the selection of the most advantageous level of integration in their product environment.
- The specifications of the physical interfaces to a piece of equipment can be flexible and self-defining, as they are built up from aggregation of logical interfaces supported by the functions implemented by the equipment.
- Operation and control of the network to provide service is effectively made independent of the maintenance and repair of the equipment, a more physical activity.
- The design and planning of a network can take place without prior assumptions about the way the network will be implemented by an individual network operator or manufactured by an individual vendor.

There are two aspects to the separation of functions from their implementation. The first derives from the basic and profound fact that information is abstract, while the second reflects the choice of an appropriate level of encapsulation to achieve the design objectives of cost efficiency and effectiveness. The first aspect of this separation is illustrated in Figure 2.7. The second aspect, which also introduces the concept of an implementational interface, is illustrated in Figure 2.8.

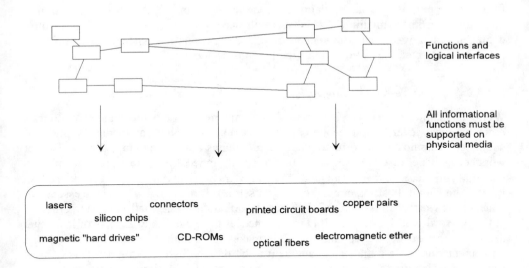

Figure 2.7   Support of functions by physical media.

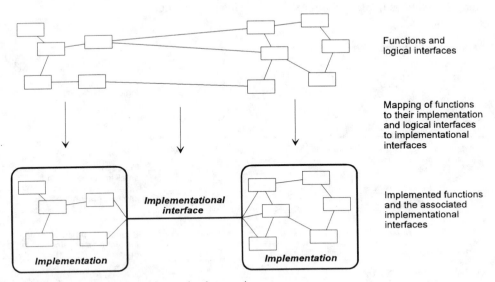

Figure 2.8   Mapping functions to implementations.

2.2.7.1   Supporting Implementation and Physical Media

The analytical goal of separation between function and implementation is not restricted to telecommunications; however, the abstract nature of the telecommunications world makes this split much more natural and straightforward than in many other branches of engineering. Information is abstract by nature, and while it must always be expressed in physical form for the purposes of manipulation, storage or transmission, the abstract function is quite distinct from the medium in which it is expressed. When the implementation is faulty, degraded, or removed, it is no longer able to support functions; the existence of the function itself is not affected, it is simply unavailable to the system.

2.2.7.2   *Implementational Design*

While the distinction between information and implementation is real and profound, at a more practical level "implementation" is more than the physical media. A piece of equipment may not work correctly because it contains a damaged chip; it may also not work correctly because a chip has been incorrectly designed. Likewise, a piece of software may not work because there is an area of damaged memory; it may also not work because it has a "bug" in it. In strict terms, the design itself is informational. This is the essence of much of the work of Boltzman and Shannon. Information is essentially linked to ordering in the physical world; however, it is more practical to regard many aspects of low-level design as part of the implementation. This makes the boundary between function and implementation more a matter of judgment to be defined by the designer. This is most easily viewed in terms of the encapsulation principle. Many of the implementational design aspects are still functional in the strict sense, however, they are encapsulated into the most basic functions that are important to the system design. Incorrect design at this level can then be viewed as implementational failure. This is illustrated in Figure 2.9.

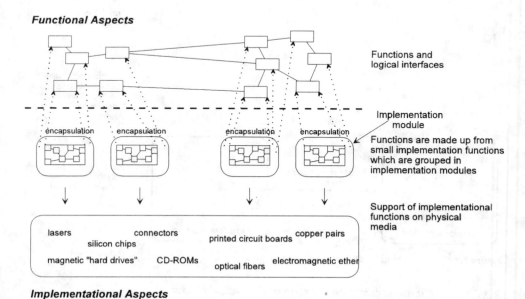

Figure 2.9   Encapsulation of low-level functions in implementational modules.

2.2.7.3   Definition of the Terms "Equipment" and "Software Module"

A piece of equipment is a physical entity that implements a set of functions, which the manufacturer sells and the operator buys. It may become faulty and need repair. In short, while the functions are what we want, the equipment is what we must buy and maintain.

In popular usage, the distinction between software and hardware is inextricably linked to the medium of storage. Code stored on a hard or floppy disk is seen as software, while the same code stored in a silicon ROM is seen as hardware. No distinction is normally made on the function of the stored code. Certainly, today, the distinction between hardware and software is diminishing. Application-specific integrated circuits (ASICs), for example, are designed and tested as software; however, the finished product has the software "burned" into a silicon chip. We would like to distinguish between software that is implementational and software that is functional regardless of the manner in which it is implemented (stored or burned in).

We define a *software module* as implementational. It is like a piece of equipment's physical replaceable unit except that it may be changed remotely without physically changing the equipment. A software module is a part of the implementational design and must be hosted on a piece of equipment. Equipment and software modules are illustrated in Figure 2.10.

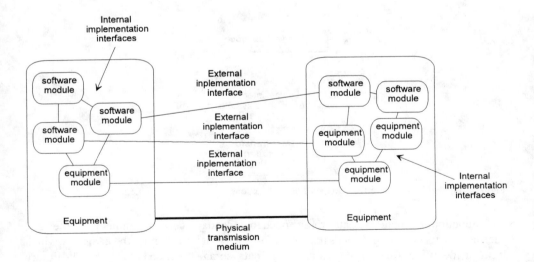

Figure 2.10   Equipment and software modules

2.2.8   Categorization of Functions for Broadband Networks

Information systems are concerned principally with the manipulation, storage, or transfer of information, and in this world of information, telecommunication networks perform the transfer role.

2.2.8.1   *Data Storage and Data Transfer Functions*

There are three elements, any or all of which may be found in information systems. First, there is the understanding and manipulation of information, which leads to decisions within the system; second, the storage of information for later retrieval; and third, the transfer of information from one function within the system to other functions. Neither the storage nor the transfer functions are concerned with understanding the information they are storing or transferring, but are simply holding or carrying it and then faithfully and accurately reproducing it as required. We use the term *data* to define information not understood by a function but that may be held or transferred by a function. Data storage and data transfer functions must, however, understand messages about the data to be able to carry out the storage or transfer. Logical interfaces to these functions will therefore comprise data flows that the function will look after but not understand and information flows that it will understand, interpret, and act upon. This is illustrated in Figure 2.11.

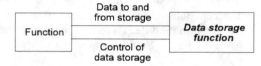

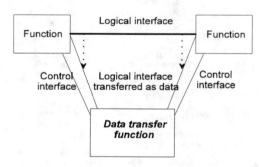

Figure 2.11   Logical interfaces to data storage functions and data transfer functions.

Functions dedicated to data storage need to understand certain parameters of the data, such as its size, how long it needs to be held for, and most importantly, the names by which it can be identified. An everyday example of a data storage function is a library. The library holds books; however, the library itself is not concerned with the content of the books. The library can hold books (and now other media such as CDs, tapes, and vinyl records) and allocate space to them. At the heart of the library is an indexing system that gives the user of the library a way to search for and find a book. The logical interface for using the library includes not just borrowing and returning books, but also providing information on the book title, subject, author, publisher, and so forth. This in turn enables the user to find the book and the librarians to classify and place the book on the shelves in such a way they can be found.

Functions dedicated to data transfer need to understand other aspects of the data, such as whether the data will be transferred in discrete amounts—packets, large files, or continuous streams. In addition, the data transfer function must understand to where the data must be delivered by using some form of identification scheme of possible destinations. All telecommunications networks are basically data transfer functions set up between information sources and destinations that are identified using a numbering and addressing scheme. The data is transferred from source to destination.

2.2.8.2 The Role of the Data Transfer Function

The primary role of a data transfer function is to support the logical interfaces between functions that are physically separated. When the functions are operated by humans, the logical interfaces usually take the form of conversations, meetings, exchanges of diagrams, and so forth, and these can be carried across the PSTN using telephony service and facsimile (fax) services. Indeed, this is one important way of understanding the marketplace for telephony and fax; they implement the logical interfaces between human-operated functions at a distance. This may sound like a rather mundane description of the way many of us spend our lives; however, this insight provides a strong link to the similar behaviors of communicating software systems where data networks implement the logical interfaces between them.

There are three basic information transfer scenarios that may use such a logical interface: person to person, person to computer, and computer to computer, as illustrated in Figure 2.12. In addition, we can identify a fourth type of communication that may be required, which is a link between two or more components for data transfer functions.

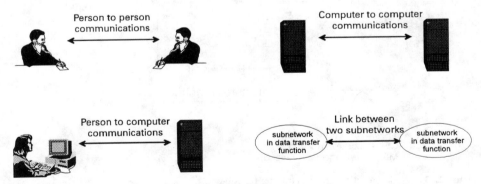

Figure 2.12   Common types of information transfer.

This unifying view of telecommunications networks helps to explain the need for multimedia, multiservice networks, as we can understand the drivers for different forms of data and the services required to transfer them. Moreover, it gives us a common framework in which to understand network management. Many network management functions are currently operated by people; however, these functions are being rapidly automated. Our functional architecture offers a unified approach to these concerns. We can move from people-operated functions to software-operated functions with minimal impact on many aspects of system specification. This theme will be developed further in Chapter 4.

2.2.8.3   *Types of Functions Within Data Transfer Functions*

The data transfer function (which is the essence of the telecommunications network) itself comprises three basic types of component function, namely:

- *Transport functions*—Associated with the transport of the information between remote locations using a shared transport network;
- *Control functions*—Associated with the identification and routing to end points within the transport network and that allow a user of the transport network to set up and take down connections within the context allowed by management functions;
- *Management functions*—Associated with planning and design of the data transfer function, establishment of the context within which users can use the control functions (i.e., service management), maintenance of services, maintenance of the equipment in the network, control of network inventory, and so forth.

These are illustrated in Figure 2.13.

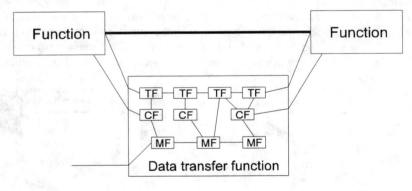

TF - transport function
CF - control function
MF - management function

Figure 2.13   Composition of a data transfer function.

2.3   TYPES OF DATA TRANSFER SERVICE

Before considering the design of the data transfer function in more detail, we must consider the type of services that must be provided to adequately meet the requirements of supporting logical interfaces in the four forms illustrated in Figure 2.12.

2.3.1 Types of Data to be Transported

Broadband networks call for a much wider range of transport requirements, especially compared to the PSTN/ISDN, which is oriented around the transport of interactive voice communication between people (which it does very well). Much of the alleged difficulty in using the ISDN for computer communications is that its characteristics are optimized around interactive voice communication and are perhaps not quite so well adapted for other types of communication.

To understand the requirements of broadband networks, we must first introduce four types of data transfer styles: messages, files, nonreal-time streams, and real-time streams. This is not a rigorous classification, but nevertheless acts as an introduction to the wide range of transport service classes that a broadband network may be required to support.

2.3.1.1 *Messages*

A message is generally a short amount of information passed from one function to another as part of a transaction protocol. The message will normally be fully understood by both the sending and receiving functions, and acted upon, rather than stored or forwarded to another function. This means that messages are principally, but not exclusively, computer-to-computer communications. Signaling messages are a good example. They pass between connection control functions and are directly acted upon within the connection control functions, which fully understand each signaling message they send and receive. The messages themselves are relatively short and form part of a transaction protocol, which is the logical interface between the connection control functions. Many electronic document interchange (EDI) applications, such as those used in banking, are message-based applications where every "document" is a message from one banking transaction processor to another.

The requirements for the transfer of messages is normally quick, accurate, and reliable. A term that is often used in this context is "latency." Low-latency transport is where the time between the sending function initiating the sending of the message and the receiving function receiving and acting on the message is low. The message is "hidden" from the communicating functions for a short period of time while in transit. Latency is affected by the length of the message, the bit rate of the transport, and the transport delay across the network.

$$latency = \frac{message\ length}{transport\ bit\ rate} + transport\ delay$$

2.3.1.2 *Files*

A file is generally longer than a message. Normally, at least one of the functions involved in sending or receiving the file will not understand its contents. One or other of the functions may be a data storage function. A file can be a collection of messages, but more commonly consists of multimedia content for display at a human-computer interface. This content is therefore human-readable rather than computer-readable. A file can range in size from, for example, a short electronic mail of a few hundred bytes to a file holding a full-length feature film, possibly many tens of gigabytes; the upper limit is growing all the time. Each file has a defined size of data, and this is characteristic of the file.

The requirements for the transfer of files are normally to be complete and accurate. Transferring 90% of a file is normally totally inadequate, as is corrupting even one byte. On the other hand, the time taken to transfer a file is normally not so important. Completeness and accuracy are absolute requirements, while the transfer time is a subjective measurement of quality.

$$transfer\ time\ (quality)\ =\ \frac{file\ size\ in\ bytes\ (fixed)}{bandwidth\ (network\ parameter)}$$

2.3.1.3   Real-Time Streams

A stream is a continuous flow of information between functions. A real-time stream is one where the time position of each piece of information in the stream is significant. When audio or video is sent directly to a human interface and not to a data storage function, the time position of each audio sample or video picture frame is essential to the successful rendering of the sound or picture to a human. If the audio or video stream is being played one-way, for example, from a data storage device, the absolute transfer time will not normally be important. Therefore, a certain amount of buffering can be built into the human interface to smooth out any impairments to the real-time information of the stream, which may be introduced by the data transfer function. However, if the stream is two-way and interactive, any absolute delay becomes a major impairment to the quality of the stream transfer.

Telephony is a two-way, interactive audio stream service and the effects of absolute delay in the transfer of the stream are well understood. Any delay in excess of a few tens of milliseconds can affect people's conversations when they use the service.

There are, therefore, two delay parameters that are important in real-time stream transfer: the absolute transfer delay and the variation in transfer delay. The latter must either be buffered and so produce more absolute transfer delay, or cause fluctuations and/or breaks in the recovered stream. In addition, loss or corruption in the transfer of the stream may be tolerable in that it will cause a momentary break in the stream; however, it is a quite distinct impairment and contributes significantly to the perceived quality of service.

The transfer of real-time audio and video streams has almost the inverse requirements to those of file transfer. The number of bytes transferred in an audio or video stream will affect the quality of the sound or picture—the more the bytes, the better the quality. However, the time duration (the play time) is fixed, thus:

$$data\ quantity\ (quality)\ =\ \frac{stream\ length\ in\ time\ (fixed)}{bandwidth\ (network\ parameter)}$$

In addition to audio and video streams, we can also identify the links of networks that carry audio and video streams as themselves real-time streams. Thus PDH and SDH both support real-time streams.

2.3.1.4   Nonreal-Time Streams

Links in networks that transfer only messages or files are themselves streams in that they are generally carrying continuous flows of data consisting of multiple sequential files or messages. However, they do not have the real-time requirements of audio or video streams. We call these *nonreal-time streams*.

2.3.2   Types of Broadband Services

Between the current PSTN/ISDN, PDH networks, SDH networks, data networks, and now ATM networks, a wide variety of data transfer services have already been defined to meet the needs of the various classes of data transfer. These are summarized below in Table 2.1.

Table 2.1
Data Transfer Services Defined to Meet Real Data Transfer Requirements

Types of Use	Data Types	Connection Type
Person-person Links	rt-streams	Circuit switched
Computer-computer Person-computer	Messages Files	Connectionless packet
Person-person Links	rt-streams	Constant bit rate (CBR)
Person-computer	rt-streams	Real-time variable bit rate (rt-VBR)
Person-computer nrt-links	Files nrt-streams	Nonreal-time variable bit rate (nrt-VBR)
Person-computer nrt-links	Files nrt-streams rt-streams	Available bit rate (ABR)
Computer-computer Person-computer nrt-links	Messages Files nrt-streams	Unspecified bit rate (UBR)
Person-computer Computer-computer	Files	ATM block transfer - delayed transmission (ABT-DT)
Person-computer Computer-computer	Files	ATM block transfer - immediate transmission (ABT-IT)

2.3.2.1   Connectionless Datagram Service

A connectionless service is ideally suited to the transfer of messages, especially when the function needs to send messages to many other functions. In the connectionless mode, the full destination address and other control information required by the network is attached, as a header, to the message data to form a variable-length packet, or datagram. This is launched into the network and the header address is used directly for routing. An implication of connectionless services is that the connectionless routers must buffer the packet until it has looked up its route table and determined where to send the packet. In addition, if there is no capacity available to route the packet, the router has no option but to buffer the packet or dump it if its buffers overflow. Examples of connectionless layer networks are IP networks, SMDS networks, and LANs based on media access control (MAC) addressing.

The most notable of these is the IP network of the Internet. However, there is currently considerable effort among the Internet community to upgrade the capability to support a quasi connection-oriented mode, which would reduce the incidence of dropped packets to the extent that it may be used for real-time services. While IP and connectionless services are generally outside the scope of this book, a connectionless network requires link transport, which will use real-time or nonreal-time stream service to support its links. This is discussed further in the following chapters.

2.3.2.2   Circuit-Switched Services

Circuit-switched services are connection-oriented in that a control negotiation takes place between the client function and the data transfer function prior to any data being transferred. The data transfer function has a "right of refusal" if it deems that accepting the request would adversely affect the network.

Circuit-switched services include those offered by the PSTN/ISDN along with the PDH and the SDH path layers. The capacity allocated to each connection is predefined (or is an integral multiple of the predefined amount), the capacity allocation takes place during the connection setup phase, and once the capacity has been allocated to that connection, it cannot be used by any other. In this way, blocking can occur at connection setup; however, if the connection setup is successful, the network will not lose data due to congestion.

2.3.2.3   Constant Bit Rate (CBR) Services

Constant bit rate (CBR) services can be offered by ATM networks and are similar to circuit-switched services, except that the bit rate is not predetermined. The actual bit rate can be specified at connection setup and, if granted by the network, can be changed during the connection. CBR is the term used by ATM Forum; ITU-T use the term deterministic bit rate (DBR) to describe the same service type.

2.3.2.4 Real-Time Variable Bit Rate (rt-VBR) Services

Variable bit rate (VBR) services can be offered by ATM networks and are similar to CBR services except that the user can vary the bit rate, within pre-agreed limits, at will during the connection. Three parameters are used to characterize a VBR service bit rate, and they are negotiated at connection setup. These are the peak cell rate (PCR); the sustained cell rate (SCR), which is the cell rate averaged over a longer period of time; and the maximum burst size, which is the maximum amount of data the user will send at the PCR at one time. Using these parameters, it is possible for the ATM network to statistically multiplex a number of VBR connections together on a single link. This allows the link to be filled to a level between the total of the PCRs and the total of the SCRs of all the connections. The statistics of VBR services are discussed in detail in Chapter 9.

In addition, the real-time VBR service allows the user to specify two delay parameters, the absolute cell transfer delay (CTD) and the cell delay variation (CDV). In ITU-T, real-time VBR service is called real-time statistical bit rate (rt-SBR) service.

2.3.2.5 Nonreal-Time Variable Bit Rate (nrt-VBR) Services

Nonreal-time VBR services are basically the same as real-time VBR services except that the user cannot specify any delay parameters. This allows the network to use bigger buffers in the statistical multiplexing and, potentially, achieve a higher statistical gain. In ITU-T, real-time VBR service is called nonreal-time statistical bit rate (nrt-SBR) service.

2.3.2.6 Available Bit Rate (ABR) Services

Available bit rate (ABR) services can be offered by ATM networks. In some ways, the ABR service is similar to the VBR service in that the bit rate available to the user changes during the connection. However, in this case, as well as negotiating a PCR and an MCR at connection setup, the user must negotiate with the network before changing the bit rate. The user can request to send at a higher rate; however, the network will respond with the maximum rate that the network can tolerate at that moment. To do this, a resource management (RM) control protocol is required during the connection as well as the normal connection setup and takedown control protocol.

ABR services are useful for a number of requirements as they have many of the advantages of both CBR services and VBR services. However, this has to be set against the cost of the RM protocol. Both ATM Forum and ITU-T are currently considering distinguishing between real-time and nonreal-time ABR services.

2.3.2.7 Unspecified Bit Rate (UBR) Services

Unspecified bit rate (UBR) service is often referred to as "best-effort" service. The basic principle is that the user can send data at will and the network will make its best effort to transfer the data. This service builds on the idea of the TCP/IP service of the Internet but is based on connection-oriented transport rather than connectionless. The user requests connection at connection setup but only negotiates a PCR, and thereafter the user is free to send at any rate up to this PCR.

This is basically the same, other than the preestablishment of the routing, as the connectionless service of IP networks. Traffic management in connectionless IP networks is effectively done by the TCP protocol that runs end to end across the IP network. Put simply, if there is congestion in the IP network, the router drops a packet, and the TCP protocol will respond by slowing down the rate at which it sends packets. If an ATM network is mainly carrying IP, UBR may be of interest. However, there are a number of reasons for believing the simple use of UBR to carry IP may cause significant traffic management problems.

As a result of the operation of IP and TCP, the current Internet is based on a single, universal quality of service, which is the lowest common denominator of all users' requests. It is currently not really possible to "buy" a better quality of service. In addition, the traffic stability of the Internet depends critically on the good behavior of the TCP functions in each user's terminal, and the network has no mechanisms of its own with which to control traffic stability. The aim of ATM is that quality of service can be negotiated between the user and the network with, presumably, the user paying a premium for a high quality of service. In this environment, it seems less likely that all users will be quite so "well-behaved," and the traffic management across more than a single, universal quality of service is considerably more difficult. Many operators are nervous about trusting the traffic stability of their networks to the successful operation of software in the users' terminals.

Two proposals are currently under discussion. The first is a hardware-based traffic control mechanism in every switch handling UBR service called weighted fair queuing (WFQ). With this scheme, the probability that a cell is dropped is actively biased towards dropping cells from users sending at higher rates. The second alternative is to include the specification of a minimum cell rate (MCR) and maximum burst size (MBS) at connection setup and to run the service as a form of VBR service, but with fewer guarantees. WFQ is potentially expensive but would make some sense if ATM were to be dedicated to IP. If this is not to be the case, and this is the current view of both authors, the second proposition offers a better solution.

2.3.2.8   *ATM Block Transfer (ABT) Services*

ATM block transfer (ABT) services are currently only specified by ITU-T and are designed for file transfer. ITU-T has specified two types of ABT service. One can give a cell loss guarantee but requires the user to buffer the file until given permission to send by the network; this is ABT delayed transmission (ABT-DT). The other type does not give a cell loss guarantee but allows the user to send immediately; this is called ABT immediate transmission (ABT-IT).

2.3.3   Distinction Between Switched Services and Leased Services

The normal distinction between a "switched service" and a "leased service" is that connection of a switched service is established using a signaling control protocol while the connection of a leased service is established by administrative procedures. From a technology viewpoint, this distinction is now largely irrelevant. There are a number of factors that ensure the distinction persists, but we should anticipate these will eventually disappear.

The first is organizational. Leased circuits have been the domain of the "transmission" part of a network operator's organization, while switched services have been the domain of the "switch" part of a network operator's organization. Organizational inertia on the part of the network operators has meant that few operators have fully considered the rationale for maintaining the distinction.

Another is the control protocol used. Switched service is normally activated by the end user using signaling protocols, while leased circuits are activated through a service management capability and use network management protocols. The protocol differences are rather minor, and we can expect some convergence in time.

A major factor is the predominant tariff structures evident in today's operating scenario. Most services offered by incumbent "telcos" tend to be based on a general tariff formula of distance × bandwidth × time. Switched services generally charge linearly for the time element, often measured with one-second accuracy. A leased circuit, on the other hand, typically charges on a week by week or month by month basis and the connection can be used continuously during that time. Leased circuits will persist as long as users see a clear price advantage in using them, and this will be the case as long as switched services are tariffed linearly with time, on a second by second basis.

This tariff structure is under threat from two directions. First the general cost structure of today's switched services simply does not reflect the distance × bandwidth × time formula. The primary cost components are now access, which is normally dedicated to a user, and the use of the processor to effect the connection setup. Commercial, competitive pressure will eventually force a tariff structure that more accurately reflects costs. Second the Internet, which has a totally different tariff structure (mainly subscription with no distance component and little or no time component), is starting to offer CBR services (in particular, voice telephony). This, if nothing else, will force telcos to review these long-standing anomalies in tariff structures.

In summary, there is no real justification for persisting with the distinction between switched services and leased services. This is even more apparent since the ATM Forum and ITU-T have defined the idea of a "soft" private virtual channel (S-PVC). This has the properties of a leased connection but is set up using the signaling protocol.

It has been disappointing to see the SDH community in general restricting its ambition to meeting the quasi static, leased connection requirement by implementing slow response connection control architecture in SDH NEs when faster response could have been achieved at similar cost. With slow connection control response, it is not only difficult to offer services with greater churn and shorter holding times—which would be very attractive to users but would also significantly improve network utilization (see Chapter 13)—but it also limits the effectiveness of service restoration of SDH paths following a node failure. The opportunity does now exist to use the ATM Forum P-NNI signaling protocol (which supports S-PVC) for network-level management of SDH networks.

References

The following standards set out a general framework including enterprise models, service models, as well as functional and implementational models.

ITU-T Draft Recommendation GII.PFA, *Principles and Framework Architecture of the Global Information Infrastructure*, Study Group 13 Report R04, 1997.

European Programme on Information Infrastructure Coordination Committee (EPIC), Report of Project 4.1, *Principles and Framework Architecture*, available from ETSI, Sophia Antipolis, France.

Computer System Policy Project (CSPP) Report, *Perspectives on the National Information Infrastructure: Ensuring Interoperability*, 1994.

The following describes general modeling language. UML has been submitted to the OMG for its adoption.

Booch, G., *Object Oriented Analysis and Design*, Second Edition, Benjamin/Cummings Publishing, 1994.

The Unified Modeling Language (UML) Specification 1.0, Rational Software Corporation, 1997.

Transport Network Architecture 3

The study of transport network functional architecture was initiated in the International Telecommunications Union — Telecommunications (ITU-T) at a time when asynchronous transfer mode (ATM) and synchronous digital hierarchy (SDH) format standards were already well advanced. Architectural recommendations were considered necessary to provide a sound basis for the development of equipment and network management standards. The subject is introduced here, in advance of a detailed consideration of bit rates and frame formats, in order to understand the architectural principles that were applied, albeit intuitively, in the development of the interface standards. By emphasizing the generic aspects of functional architecture and the continuity with earlier stages of network evolution, the transmission specialist can find justification for the apparently liberal overhead provision. The nonspecialist can find a broader, more intuitive base from which to develop an understanding of transport networks and the role that ATM and SDH will play in transport network evolution. However, many readers may find the concepts in this chapter and their importance becomes clearer if the reader revisits them after reading other chapters.

Transport functionality is conventionally described in functional block diagrams, where each block represents a transport function defined by the manner in which information is processed between a set of inputs and a set of outputs, as illustrated in Figure 3.1.

Figure 3.1   Generalized transport function.

A network is itself such a function. It can be described as a connecting function with many inputs and many outputs. At any point in time, the function is in a connectivity state in which information presented at a set of inputs is passed transparently to a set of outputs. Information presented at an input is reproduced, subject to allowable degradations, more or less faithfully at the connected outputs. The set of input-to-output connections is commonly expressed as a connection matrix, listing the connections as associations between inputs and outputs.

Such a connecting function is invariably associated with a management or control function that determines its connectivity state, mediating between the requirements of external controlling agencies and the capabilities of the connecting function itself. We will return to management and control functions later when we have considered the transport architecture in more detail.

The international public switched telephone network (PSTN) is a well-developed example of a connecting function. It is a single functional block with many inputs and many outputs where each input is associated with a single output in the telephone of an individual subscriber. The connection matrix then represents the association of pairs of telephones interconnected by means of bidirectional connections across the network.

3.1   LAYER NETWORKS

From a high level of abstraction there is no difference between the PSTN example and a network based on the 2048 kbit/s primary level of the European plesiochronous digital hierarchy (PDH) or, indeed, the digital level 3 (DS3) of the North American PDH. In both these cases the inputs and outputs to the connecting function may currently be found in transport terminal equipment, while the network derives its connectivity from DDFs and crossconnects rather than telephone switching equipment. Existing connections can be described by the connection matrix in each case, while the ability to support new connections depends on the available topology and the spare capacity in the network.

Networks such as these are called layer networks, as they exist within a single layer of a multilayer transport network. They are defined by the set of compatible inputs and outputs that may be interconnected and characterized by the characteristic information that is transported. The inputs and outputs are regarded as access points (APs) on the layer network. The functions that represent these components of layer networks are abstract in the sense that they represent an abstraction of the underlying, network structure.

3.1.1   Partitioning of a Layer Network

The first important property of a layer network is that it can be progressively decomposed to show more and more detail, finally revealing atomic functions below which (from an architectural viewpoint) further detail is of no interest. The process of decomposition within a single layer is referred to as partitioning, and is discussed in relation to connectivity and topology. Partitioning of a layer network is an example of encapsulation as described in Chapter 2.

Functional partitioning of a layer network provides an understanding of the detailed structure within a single layer. It forms the basis for the definition of administrative boundaries between operators and the definition of independent routing domains for the management of connectivity within the layer. It also enables the identification of boundaries within the layer network for the apportionment of performance objectives between the various elements and subsystems of which it is composed.

3.1.2   Partitioning With Respect to Connectivity

A layer network supports a multiplicity of network connections, where a network connection is defined as the transport function capable of transferring information transparently across the layer network from input to output. This definition describes a unidirectional network connection with the user's information flowing in only one direction. A bidirectional network connection naturally consists of a contradirectional pair of unidirectional network connections. A point-to-multipoint network connection is the transport function capable of transferring information transparently between one input and several outputs of the layer network.

The layer network also has inputs and outputs which, while not currently supporting network connections, may be associated at some future time via the transport resources of the layer network to form network connections. Network connections are setup and released as part of the layer connection management process, thereby providing a degree of flexibility in operating the network.

A layer network is illustrated conceptually in Figure 3.2. Characteristic information is shown entering the network at its inputs and leaving the network at its outputs. The dotted lines indicate the associations between inputs and outputs that we refer to as network connections. Two modalities of association are illustrated—point-to-point and point-to-multipoint—and two directionalities—unidirectional and bidirectional.

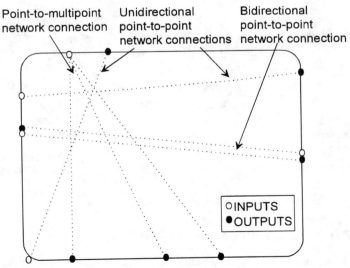

Figure 3.2   Top level abstraction of a layer network.

A unidirectional network connection is an encapsulation of smaller transport functions. It is constructed by the concatenation of a number of more basic connection types corresponding to the components from which the network is constructed. These connection types naturally share the same capacity for transparent transfer of characteristic information as the network connection. Closer examination will reveal that they are of two distinct types. The first is flexible, like the network connection itself, in that it may be set up and released as part of the layer process in setting up and releasing the network connection of which it forms a part. The second is fixed and inflexible as far as the layer management process is concerned.

The set of points that may be flexibly associated as part of the layer management process are the inputs and outputs of an abstract function defined as a *subnetwork* and the point associations that it supports are called *subnetwork connections*. When two subnetworks are topologically adjacent, the interconnections between the two subnetworks contain matching subsets of points and the subset of points in one subnetwork has a fixed association with the corresponding subset of points in the other subnetwork. These fixed or inflexible associations are termed *link connections*. A full set of such link connections connecting two topologically adjacent subnetworks is called a link. The decomposition of a network connection into subnetwork connections and link connections is illustrated in Figure 3.3.

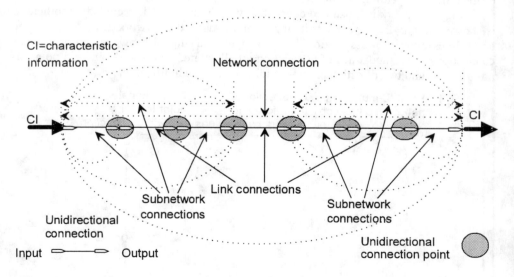

Figure 3.3   Decomposition of a unidirectional network connection.

3.1.3   Partitioning With Respect to Topology

While the connectivity of the layer network describes the actual connections of each type that exist and is expressed as the association between particular points, the topology of the layer network describes the connections that could exist and is expressed as the relationships between sets of points. Thus, a subnetwork is a set of points in the same layer network that may be interconnected by operation of the layer connection control process, and a link is an expression of the fixed relationship between a set of points in one subnetwork and a corresponding set of points in another subnetwork topologically adjacent to the first.

A layer network is therefore an encapsulation and may be decomposed into subnetworks and the links between them. Each subnetwork is also an encapsulation and may be further decomposed recursively into smaller subnetworks interconnected by links until the desired level of detail is revealed. This will generally be when the subnetwork is equivalent to a single switch, frame, or crossconnect at one location and the level of detail becomes implementational as described in Chapter 2. In many cases, the switch or crossconnect may itself be decomposed in the same way to reveal further, linked subnetworks, however, the internal implementation architecture of a network element (NE) is of no concern within a network-level chapter; however, it can be useful to view an interconnected cluster of collocated NEs as a single nodal matrix, in which case the further decomposition of the node into its components will certainly be relevant.

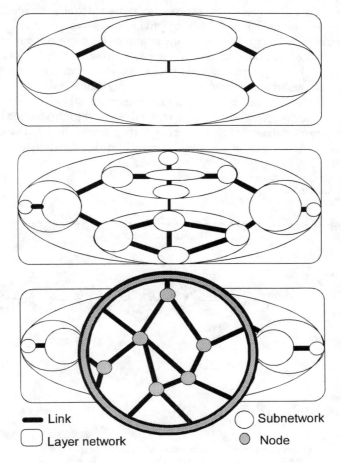

Figure 3.4   Topological decomposition of a layer network into subnetworks and links.

The progressive decomposition of the layer network into subnetworks and links is illustrated in Figure 3.4. The lowest level of decomposition that reveals the individual nodes (or nodal subnetworks) is here viewed metaphorically through a magnifier. We shall return to the subject of topology later in this chapter when we consider specific network structures in more depth.

3.1.4   Integrity of Information Transfer in a Trail

As a lesson learnt by its general omission in 64 kbit/s and PDH layer networks, it is increasingly necessary to provide, along with the transport capability, some positive measure of the quality of the information transfer and the validity of the supporting connection. This can be achieved in a limited way in some layers by monitoring specific structural properties of the characteristic information at the output of the network connection. However, this has generally proved insufficient for modern needs, and it is now normal to introduce additional intralayer processes designed to achieve specific operations, administration, and maintenance (OAM) objectives. This gives rise to additional overhead information transmitted from the input to the output of the network connection.

Such a validated information flow has been termed a *trail* in ITU-T. The term was introduced to express the generic characteristics of a class of transport entities that includes transmission paths and transmission sections—transport entities with the ability to supervise the integrity, validity, and/or quality of information transfer across a network connection. The network architecture is thereby further refined by the introduction of a trail which is the encapsulation of the network connection with a termination source function at the input to the network connection and a trail termination sink function at the output of the

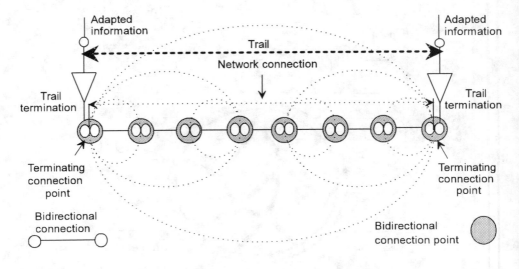

Figure 3.5   A bidirectional trail in a layer network.

network connection. These functions process the information entering and leaving the layer network to achieve the desired OAM objectives. The concept of a trail is illustrated in Figure 3.5.

Although the concepts of validated information and in-service performance monitoring are important features of trails in ATM and SDH layers, not all network layers have such capabilities and the 64 kbit/s transport layer is one notable example. Even within SDH, it has been useful to model functions such as protection and tandem path monitoring using sublayers with no specific intralayer validation processes. The concept of the trail has now been generalized so that integrity, quality, or validity may be inferred directly by intralayer monitoring, indirectly from other information in the network, or not at all.

3.2   INTERLAYER RELATIONSHIPS AND ADAPTATION

The link connections in a layer network provide connectivity between topologically adjacent subnetworks and are provided by the services of a trail in another layer. This layer is called the server layer, and the layer in which link connections are provided is called the client layer. The two layers are said to participate in a client-server relationship. The complete set of link connections between two subnetworks defines the link between the two subnetworks. Requirements to change the topology of a layer network in response to growth or churn in a client layer result in requests from the client layer management process on one of its servers for trails providing more or alternative link connections in the client layer. The mechanisms by which this can happen are described in Chapters 4 and 13 and can rely on automated or manual mechanisms.

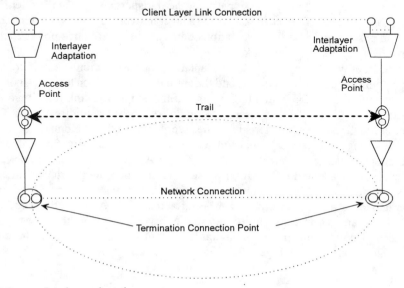

Figure 3.6   Interlayer adaptation.

The nature of the interlayer relationship is much the same whether the management processes involved are human or autonomous. Links are fixed as far as the client intralayer process is concerned and may only be changed by a request on the server layer process. A change in client layer connection service requirement forecasts will generally result in a need to enhance or otherwise modify the client topology, which can only be achieved by connection change requests on the server.

Client layer information must be adapted for transmission in the server layer. The nature of the adaptation process naturally depends on the characteristic information in each layer, but typically involves rate changing, multiplexing, aligning, coding, or some combination of these. For example, adaptation between the analog VF (voice frequency) layer and the digital 64 kbit/s layer is based on the pulse code modulation (PCM) principle whereby the analog signal is sampled and quantized. With ATM and SDH, adaptation requires many different mechanisms in order to carry the wide variety of voice, video, and data signals, as well as the links of client layer networks. The generic interlayer adaptation function is illustrated in Figure 3.6.

3.3   LAYER NETWORK REFERENCE POINTS

The point at which the output of one atomic connection is bound to the input of another is called a connection point (CP). Similarly, the point at which a trail termination source output is bound to the input of the network connection and the point at which a trail termination sink input is bound to the output of a network connection are both called terminating connection points (TCPs). Of course, CPs and TCPs can be bidirectional or unidirectional according to the directionality of the transport entity that they delimit. The CPs and TCPs serve as reference points within a layer network, across which characteristic information passes, and it is the associations between these reference points that define the connectivity of the layer network.

The point at which the output of the adaptation source function is bound to the input of the trail termination source function and the similar point at which the output of the trail termination sink function is bound to the input of the adaptation sink function are called the access points (APs). The APs can also be bidirectional or unidirectional. The trail termination functions are characteristic of the layer to which they belong. In other words, they are client layer (or service) independent. However, an adaptation function will exist for each client-server layer pair that is defined. The AP acts as a reference point across which server layer adapted information passes and at which the interlayer client-server relationship is defined. All the adaptation functions belonging to a server layer network are constrained so that the information passing across their APs has the same adapted format, which can be transferred uniformly in the server layer. Both the access point and the connection point are formed by binding the input of one function to the output of another. This is illustrated in Figure 3.3, Figure 3.5, and Figure 3.6.

3.3.1 Associations Between Reference Points: Topology and Connectivity.

The AP derives its architectural significance from marking the functional boundary between transport network layers, where defined interlayer information is transferred and where the server layer trail is delimited. From the server viewpoint, it is a routing destination that may support a trail. From the client viewpoint, it represents a point at which it is possible to procure link capacity. This places the adaptation function in between layers. However, this and the associated APs are conventionally considered to "belong" to the server layer from a management and control viewpoint, thus placing the layer boundary for many practical purposes somewhere between the adaptation function and the client layer CPs, as illustrated in Figure 3.7.

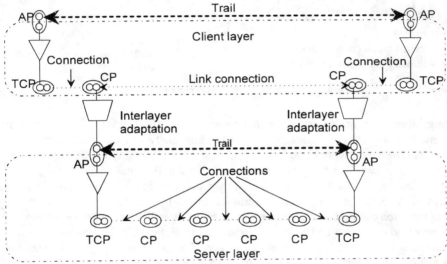

Figure 3.7   The client server connectivity relationship between layer networks.

The set of APs serving a single client entity is termed an access group (AG). Thus if a subnetwork is defined as a connectable set of CPs and/or APs acting as an intermediate routing destination, the AG is a set of APs defined as allocated to a specific client entity to provide connection services and acting as a final routing destination in its layer. Subnetworks, links, and access groups are topological entities whose relationships define the *topology* of the layer network. The AG may be a subset of the connectable set in a subnetwork or may be connectable by virtue of their attachment via an access link to a subnetwork.

The topology of a layer network at any particular level of abstraction is defined by the complete set of associations between the subnetworks and AGs at that level of abstraction; that is to say, by the links between them. The set of CPs that are connected via link connections to a corresponding set of CPs in a topologically adjacent subnetwork is termed a transit group (TG). A link is therefore delimited by a pair of TGs. Topological relationships are illustrated in Figure 3.8. Here, the internal transit groups (TGi) are

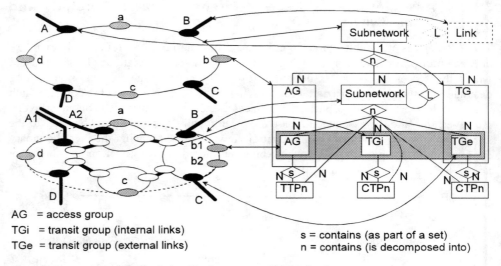

AG = access group
TGi = transit group (internal links)
TGe = transit group (external links)

s = contains (as part of a set)
n = contains (is decomposed into)

Figure 3.8   Topological relationships between partitioning levels.

distinguished from the external transit groups (TGe). The former, like the subordinate subnetworks and links, are encapsulated and not normally visible to the client, while the latter mark the subnetwork boundary and are visible to the external process (whether it be a client layer network or a superior partitioning level in the same layer).

A network connection is represented by an association between TCPs in the same layer, and the complete set of network connections defines the connectivity of that layer. The association between a client layer CP, a server layer TCP, and the AP through which it is expressed represents an instance of an interlayer relationship. The complete set of such associations defines the access that the client layer has to the server layer. The set of server APs available to a particular client is termed an access group (AG). The topology of each client layer is defined by the connectivity in all its server layers and the access that it has to those server layers.

The associations between CPs (including TCPs) in the same layer represent the link connections and subnetwork connections into which the network connection may be decomposed. Therefore, the layer network connectivity can be described to any desired level of detail by expressing each network connection association as a list of its constituent link and subnetwork connection associations. The resulting list of associated points represents the concatenation of connections of both types that make up the network connection.

3.3.2   Tandem Connections

Any serially connected set of connections, regardless of whether they form a subnetwork connection or not, is referred to as a tandem connection, as illustrated in Figure 3.9. Note that an n node tandem connection actually includes $n(n - 1)/2$ tandem connections, including the individual connections and the defining tandem connection. In the four-node example of Figure 3.9 this amounts to $4 \times 3/2 = 6$.

A special case of the tandem connection that is of special interest is the tandem link connection, also illustrated in Figure 3.9. In this case, all the constituent connections are link connections. The tandem link connection is equivalent to a link connection in its inflexibility and in its role as part of a link describing the topological relationship between two subnetworks.

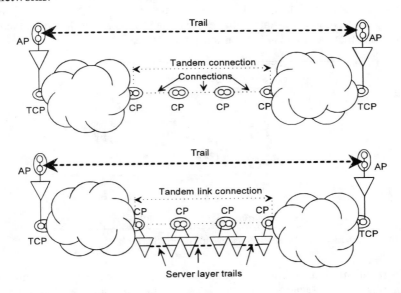

Figure 3.9   Tandem connection and tandem link connection.

3.3.3   Different Perspectives on the Layered Network Model

Layer networks can be visualized as parallel planes where all the CPs belonging to a single layer network lie on a single plane. The connection associations between CPs within the layer network are represented by lines on the plane connecting the associated CPs. Client-server associations between CPs in a client layer and TCPs in a server layer are represented by interconnecting lines between planes that pass through the APs that are directly related to the TCPs.

The information processes acting on the client layer information as it passes into a server layer effectively add information. (It is interesting to note that in the terms of information theory, this is equivalent to reducing entropy as bits that were previously undefined are now assigned to useful information. The height of the planes in the geometric model described above can therefore be interpreted as indicating the energy of the layer's characteristic information. In the adaptation and trail termination functions, the processes of alignment, multiplexing, and overhead generation all serve to reduce entropy.)

The layered model is very powerful as a conceptual tool, but it is not usually possible to illustrate in more than two dimensions at a level of detail that is representative of the real world. Figure 3.10 and Figure 3.11 illustrate aspects of the layered model from several different perspectives. In Figure 3.10, a small part of three layer networks ("k," "l," and "m") is illustrated, sufficient to show relationships between points within planes and between planes. In this example, a single link connection is illustrated between two CPs in layer "m." This is provided together with three similar link connections by a trail in layer "l." This trail consists of a network connection in layer "l" that is made up of two link connections and a subnetwork connection. The two link connections are each provided by trails in layer "k," and the subnetwork connection is provided by a layer "l" switching matrix within an NE. This example makes no assumptions about the components of the layer "k" trails that are not visible from this perspective.

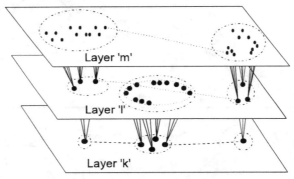

Figure 3.10   Inter and intralayer relationships in a layered network.

Where the building architect conventionally represents a three-dimensional world in terms of plan view, elevations, and sections, the network architect represents all the multidimensional complexities of a transport network in a related set of two-dimensional views using the principles of partitioning and layering.

3.3.4   Local and Network-Level Relationships Between Reference Points

The tasks of planning and managing integrated layer networks revolve around identifying, defining, and manipulating the relationships between points and sets of points. The example of Figure 3.10 identifies one set of points that is characterized by the ability to support subnetwork connections between one member of the set and another. This set of points defines a subnetwork and is enclosed by a dotted ellipse. Connectability within and between sets of CPs is used to describe topology.

Figure 3.10 also identifies sets of points that exist in a single layer within a single NE. These are shown enclosed by dashed ellipses. The dashed ellipses are conic sections of a conical surface that encloses all the CPs, TCPs, and APs contained within a single NE. From this perspective, the individual client-server relationships and the associated trail termination and adaptation functions are viewed as NE functions being contained wholly within a single NE.

Figure 3.11 illustrates in more detail the intersection between the NE view and the network as a whole. In this example, a single unidirectional connection between G.703

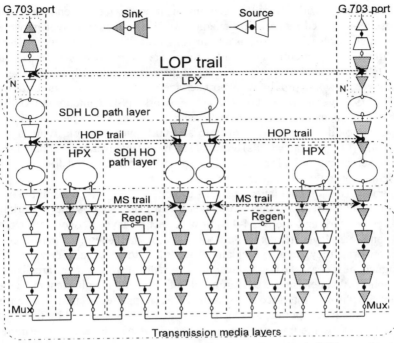

Figure 3.11   A single unidirectional connection: cross section through an SDH network.

ports is observed as it is transported across an SDH network. Access multiplexers interfacing to the PDH network, path layer digital crossconnect systems, and simple regenerators are all visible from this perspective. (The SDH layer network terminology is introduced later.) As the example is symmetrical, it is possible to fold it about the center line so that source and sink functions become superimposed. The set of source-sink pairs can be associated as a set of bidirectional reference points, thus creating a bidirectional example. It will be left to the reader's imagination to visualize extensions to this example by, for instance, adding more connections, more ports on crossconnects, and more nodes on the network. Although the complexity increases without bound in terms of individual functions and relationships, the power of the abstraction provided by layering and partitioning makes it possible to comprehend the structure on different scales and from different perspectives, and thus to analyze and compare particular features independently of one another.

3.4   CLASSIFICATION OF NETWORK LAYERS

The layering and partitioning principles described above are believed to be generic, applying to all networking technologies past, present, and future. Indeed, in developing the principles, many illustrative examples have been drawn from existing practice. ITU-T has classified network layers into three broad categories: circuit layer networks, path layer networks, and transmission media layer networks.

Circuit layer networks provide telecommunication services for end users. The PSTN and the packet-switched network are examples of historic circuit layer networks while ATM virtual channels will be used as a basis for a broadband circuit layer network in the future. Circuits (i.e., circuit layer trails) terminate in the user's premises and their connectivity is controlled by a process invoked directly or indirectly by the user.

Path layer networks provide transport services to the circuit layers (or to other path layers). The DS3 layer network, the VC4 layer network of SDH, and the ATM virtual path layer network are all examples of path layer networks. A leased line network is a path layer network that provides transport services to a circuit layer network, typically a private network. Paths (i.e., path layer trails) terminate at the circuit layer or other path layer access points, and their connectivity is controlled by a process that is invoked directly or indirectly by the client layer trail management processes.

The transmission media layer networks provide transport services to the path layers or, less commonly, directly to the circuit layers. The 139264 kbit/s coded mark inversion (CMI) layer network and the STM-4 multiplex section layer network are examples of transmission media layer networks. Although sharing the generic properties common to all layer networks, they are also specialized according to the media for which they have been designed, be it coaxial cable, optical fiber, or radio. Two sublayers are distinguished: the section layer that determines the information format and the physical media layer that determines the physical interface characteristics. Sections (i.e., section layer trails) terminate at path layer access points, and their connectivity is determined by a process that is invoked indirectly by the transport requirements of the path layers as determined by the path layer management processes.

Figure 3.12 is an entity-relationship (E-R) diagram illustrating the main transport layers and the client-server relationships between them. A line between two transport layer entities represents the existence of a client-server relationship, with the small circle designating the client end. A fully comprehensive representation would be overly complex; therefore, some simplifications are made for clarity.

The legacy PDH layers are referenced only by their bit rates. In fact, the PDH has developed a richly varied internal structure with different path layer formats and interlayer capabilities. The use of a primary rate path to carry synchronization reference information also distinguishes it from those that do not. These distinctions have not been shown. Many well-established PDH clients, particularly special services designed to exploit the PDH network, have also been omitted for the same reason.

The 64 kbit/s (DS0 in the North American hierarchy) layer and the ATM VCh layer are shown as transport layers. As such, they may be used to provide semipermanent connections as a transport or bearer service for the digital PSTN, the ISDN, or the broadband ISDN.

3.4.1   ATM in a Multilayer Network

ATM has been organized essentially as two administrative layers. The ATM virtual channel (ATM VCh) provides VCh connections acting as a bearer service for B-ISDN in support of a wide variety of applications requiring data transport. In addition, ATM virtual channels can be used to provide link transport for other networks, notably data networks

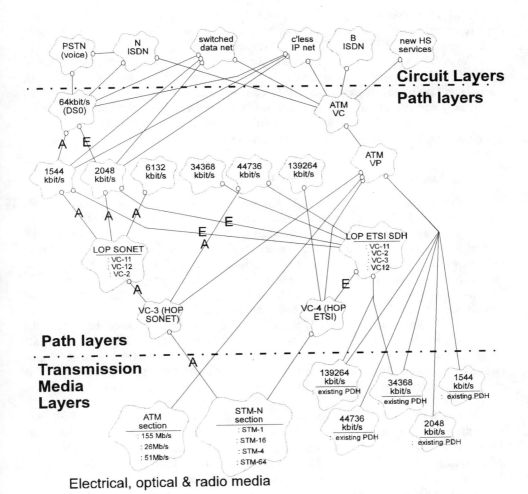

Figure 3.12   Transport layer networks and their client server relationships.
Note: A = Only used in SONET (i.e., ANSI SDH), E = Only used in ETSI SDH.

including networks based on the Internet Protocol (IP). It is in turn served by a path layer
based on the bearer service capability of an ATM virtual path (ATM VP). The ATM VP
is in turn served by a wide range of possible transport path layers, both within the SDH
and from the legacy PDH network, collectively described in the ATM community as the
physical layer, although they may well be multilayer transport networks in their own right.

Cell-based section layers are also defined for ATM transport that operate at a constant,
media-determined bit rate, having transmission overheads similar to those in an SDH
section. Currently defined as UNI options, they may well become more widely used in the
future as ATM services become more ubiquitous.

ATM VChs may also support PDH or SDH connections in circuit mode using the CBR (constant bit rate) mapping. These mappings are for evolution or other specialist applications and in the interests of clarity, they are not illustrated.

The ATM VCh layer could emerge as a provider of the universal bearer service in the multiservice networks of the future. It provides transport for the well-established, narrow band circuit layers as well as B-ISDN and a whole range of potential new services with widely varying traffic characteristics, many of which do not exist today. However, its relationships with IP networks as a provider of universal bearer service is still very much debated. This is discussed in more detail in Chapter 13.

3.4.2   Layer Structure of the SDH

A prime objective in developing SDH has been to remedy some of the deficiencies associated with PDH. The main feature has been a reduction in the number of path layers to two, with maximum transparency between them and between each of them and the section layer. Clients of the SDH path layers are first multiplexed into one of the lower order path (LOP) layers (the use of "lower" originated from PDH multiplexing and can be confusing as it is actually "higher" when shown on client/server layer diagrams). The LOP layers are then mutiplexed into a higher order path (HOP) layer. In turn, the HOP layer is carried on SDH transmission media layers.

This has sometimes been loosely described as "one-step multiplexing" because of the "observability" of an individual 64 kbit/s channel transported within a 2.5 Gbit/s SDH data stream. A complete specification for the transmission media layers has also been developed to enable high-capacity open interconnection between network node equipment. This capability is often loosely described as "midfiber meet" or more formally as section layer transverse compatibility.

The multiplexing principle and the structure of signals at the NNI are introduced in 5, but the nomenclature used to describe this structure is used in Figure 3.12 to designate the different SDH layer networks. For the present purpose, these can be taken as arbitrary layer network designators. A list of layer designators along with summary descriptions of the layer characteristic information is given in Appendix B.

The significance of the E-R diagram of Figure 3.12 is in the specific client-server relationships supported between the designated layer network entities. The SDH as defined by the ITU-T is the superset of which the ANSI SONET and the ETSI SDH are subsets. Figure 3.12 is marked to indicate those relationships that only occur in SONET and those that only occur in the ETSI version. All unmarked relationships are common to both. This convention has not been extended to areas that, though region-specific, are nevertheless uncontentious. DS3 signals, for example, are seldom encountered in Europe, but where they are, the standard ANSI mapping is used.

Finally, evolutionary procedures exist to support SDH paths on plesiochronous servers. These are also left unillustrated.

3.4.3 SONET and ETSI SDH

The most striking difference between SONET and ETSI SDH can be seen at the higher order path layer. In SONET, this is based on the VC-3 layer — about 50 Mbit/s, while in ETSI SDH it is based on the VC-4 layer — about 150 Mbit/s. This differentiation reflects a degree of administrative convenience in matching the new SDH transport rates to the comparable PDH transport rates already well-established in North American and European networks. Interworking at intercontinental boundaries, however, will use the higher order path layer based on VC-4.

Within the transmission media layers, four rates are currently defined for optical networks and these are common to both SONET and ETSI SDH. They are each derivatives of the basic synchronous transport module (STM) and provide transmission between transport nodes at rates of 155, 622, 2488, and 9953 Mbit/s.

In ETSI SDH, VC-3 is a lower order path layer alongside VC-12 and VC-2, and is primarily intended to transport the third levels of the PDH legacy network, namely 44736 kbit/s and 34368 kbit/s. In SONET, VC-4 is transported using the capacity of three VC-3 connections in parallel and the bit sequence integrity is maintained in this arrangement by a process known as concatenation described in Chapter 5. The VC-4 has been defined as the transport layer to support ATM cell streams, i.e. the ATM VP layer, at the NNI in B-ISDN. Thus, SONET and ETSI SDH are identical when supporting ATM layer networks such as B-ISDN.

Finally, ETSI SDH supports the VC-11 layer by transporting it on VC-12 layer connections using a special adaptation, thus allowing ETSI SDH crossconnect equipment to be better matched to the prevalent 2048 kbit/s usage in Europe. Support of VC-11 and VC-12 simultaneously in a single link or network element is discussed in Chapter 5 and

Table 3.1

SDH Layers

	SONET	*ETSI SDH*
Lower Order Path Layers	VC-11 (1664 kbit/s) VC-12 (2240 kbit/s) VC-2 (6848 kbit/s) VC-2-nc (n × 6848 kbit/s)	VC-11 (1664 kbit/s) * VC-12 (2240 kbit/s) VC-2 (6848 kbit/s) VC-2-nc (n × 6848 kbit/s) VC-3 (48960 kbit/s)
Higher Order Path Layers	VC-3 (48960 kbit/s) VC-4 (150336 kbit/s) VC-4-nc (n × 150336 kbit/s)	VC-4 (150336 kbit/s) VC-4-nc (n × 150336 kbit/s)
Transmission Media Layers	STM-0 (51840 kbit/s) STM-1 (155520 kbit/s) STM-4 (622080 kbit/s) STM-16 (2488320 kbit/s) STM-64 (9953280 kbit/s)	STM-1 (155520 kbit/s) STM-4 (622080 kbit/s) STM-16 (2488320 kbit/s) STM-64 (9953280 kbit/s)

* carried in the capacity of a VC-12

from an implementation perspective in Chapter 12. All the SDH layers are summarized in Table 3.1.

Despite these important differences, there is sufficient commonality between SONET and ETSI SDH to greatly simplify interworking between operators across intercontinental boundaries and also to go some way towards promoting greater equipment design reusability in different markets.

3.4.4   Link Diversity

A link is considered, according to G.805, to contain all the LCs between its referenced terminators (subnetworks or access groups). But if the topology is configured so that LCs may be allocated to two or more disjointly routed subsets as a result of the connectivity configured in the server, the subsets may be regarded as subsidiary links of the original link in which they are aggregated. The aggregate link therefore contains a set of disjoint subsidiary links. As the subnetworks are decomposed, there is a matching decomposition in the links at each level in the recursion. This is illustrated in Figure 3.13.

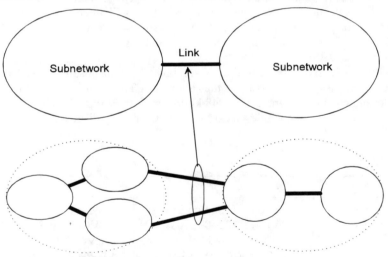

Figure 3.13   Topological decomposition of links.

Where two disjoint links connect two disjoint node pairs (as in a dual-node intercon-nect), then at the next partitioning level where each node pair is aggregated into a different superior subnetwork, the two original disjoint links form a set of disjoint subsidiary links between the two newly generated superior subnetworks. This provides the basis of a simple and consistent abstraction of dual-node behavior.

The concept of disjoint link sets may be extended to cover a larger number of links in a mesh network. Where a set of physical links has been selected to support a ring, for example, or simply to support load sharing or 1 + 1 SNC protection, then they may be declared as members of a disjoint set, and this information is enough to support disjoint routing within the client domain without explicitly declaring the server connectivity provided to achieve it. The declaration of a disjoint link set in this way is merely a formalization of the routing constraint implicit in any subnetwork (whether a ring or any other topology where diversity is exploited to improve availability).

3.4.4.1 Management and Control of Disjointness

The management and control of disjointness is one of the more challenging aspects of transport network operation particularly in a multioperator environment. In the past, when a client required disjointness between a set of leased facilities, he would request figuratively to "walk down the route." In other words, the server was required to explicitly declare what facilities were used and the client then made a judgment on the acceptability of the level of diversity required. This is completely at odds with the concepts of management automation that are developing around ATM and SDH where underlying physical relationships are encapsulated and therefore hidden from view. The concept of managed subnetworks constructed from disjoint link sets interconnected by a reliable dual-node interconnect strategy provides a simple and workable solution for small subnetworks. The operating mechanisms of self-healing subnetworks are discussed more fully in Chapter 10.

This simple proposition does not scale well. In larger subnetworks that grow over time, it is often the case that while not completely disjoint, a useful high level of disjointness is nevertheless achievable. In general, therefore, a more quantitative measure of the probability of common mode failure amongst a set of nodes and links is required so that the best, or at least an adequate, choice is made for the components of a protected connection.

The server topology must be analyzed for probability and impact of failures. Each individual, low-level failure mechanism is quantified and allocated a unique identifier. When a resource is allocated by a server to a particular client service role, the set of uniquely identifiable failure events to which it is subjected is also declared by the server. The client does not need to know the nature or location of the failure mechanism, only its unique identifier, its probability of occurrence, and the impacted resource provided by the server. The client can then estimate the probability of common mode failure between any set of offered resources and thereby assess the acceptability of the offer. When the topology of an intermediate path layer is constructed by the server, its failure data is built up from an aggregation of all its servers in turn. Such a common mode failure topology may be consulted during the connection setup process along with the subnetwork topology in searching out acceptable routings for a particular connection request.

3.4.4.2 Prefabricated Link Connections

In the examples above, a disjoint topology is provided by a specific routing policy in the server, which is naturally devised to meet disjointness commitments to its clients. It can also be useful for the client to generate diversity by prefabricating links from tandem connections, an extension to the role of resource manager. It has already been noted that a link at a particular level of decomposition may consist of a string of shorter links. Such links have been called tandem links. The link connections in a tandem link are similarly called tandem link connections. A link connection that cannot be further decomposed into smaller link connections can be considered atomic. These properties of links and link connections are illustrated in Figure 3.9.

The essential properties of a link connection from the viewpoint of an SNC setup process are its characteristic information and the fact that it has one end in each of the two linked subnetworks (or access groups). Therefore, a tandem link connection (a tandem connection consisting only of elementary link connections and no subnetwork connections) may be considered a subclass of a LC as it has the same connectivity properties as the elementary LC. The link connection of G.805 is such an elementary link connection.

We may also define another LC subclass, called prefabricated link connection (PLC), constructed on request from a resource management process and consisting of a sequence LC-TC-LC where TC represents any valid tandem connection. LCs of whatever subclass may all be allocated to links to form disjoint link sets. This extended view of link connection is illustrated in Figure 3.14. Although the PLC may be constructed by the same set of mechanisms used for setting up subnetwork connections, once constructed and allocated to a link it is regarded as a fixed feature of the topology, equivalent to an LC provided by a server trail and available to be selected as a component of a new subnetwork-to-network connection.

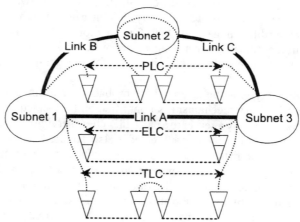

Figure 3.14   Extensions of the definition of link connection to take in prefabricated diverse routings.

3.4.5   Further Considerations on Layering

The layering concept as described so far has been developed mainly along technological lines. In other words, new layers have developed in response to technical pressures for higher transmission bandwidth or to support different modes of communication. Certain of these layers have also been accorded some administrative significance and play a role in management of connectivity and topology. There are also examples of layers that have been created expressly for administrative purposes, with no other functional or technical raison d'être. Trunk groups, for example, and the organization of pairs in large cables both predated multiplexing, while the "bundling" of channels into units of capacity unrelated to the hierarchical levels for routing through the network as a single entity is a service feature planned by some operators in an SDH environment. The main classifications of circuit, path, and transmission media, with which the subject of layering was introduced, were subsequently decomposed to reveal the layers that have been illustrated in detail in Figure 3.12. Features as diverse as protection, tandem path monitoring, and bundling appear to generate new layers. What, then, is the nature of a layer in the transport network? What rules govern the generation of new layers, and how should such layers be modeled?

The functional network architecture developed in this chapter has enabled us to describe very complex networks to whatever level of detail is required in terms of three generic functions. They are connection and termination, which are defined for each layer, and interlayer adaptation, which is defined for each pair of layers that participate in an interlayer client-server relationship. New layers can be generated for various purposes in three ways by further decomposition of any one of these three functions. The generation of new layers by functional decomposition is illustrated in Figure 3.15.

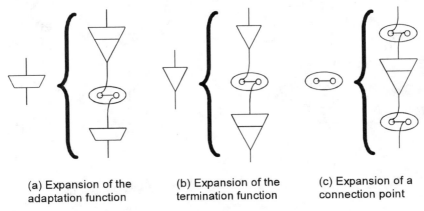

(a) Expansion of the adaptation function	(b) Expansion of the termination function	(c) Expansion of a connection point

Figure 3.15   Generation of new network layers.

3.4.6   Generation of New Administrative Path Layers

New path layers can be added above, below, or in between the existing path layers by decomposition of the serving adaptation function as described in Figure 3.15(a). The SDH path layers already defined and, indeed, the decomposition of the section layer into multiplex section (MS) and regenerator section (RS) have in effect been generated in this way. The latter is illustrated in Figure 3.16(a). The process is reversible, and there are several examples in recent years where this has been done. A notable example is the C-bit parity mapping of DS-1s directly into DS-3s. This was done to release overhead previously dedicated to DS-2 justification to improve the OAM capabilities of the DS-3 signal in DS-1 transport applications. This is illustrated in Figure 3.16(b).

The same modeling procedure would be used to create administrative levels for handling groups or bundles of paths with or without concatenation. Such administrative bundling is a feature of present networking practice and can certainly be expected in broadband networks based on SDH and ATM. It is equally applicable to administrative groups formed from higher order paths within a high-capacity section or lower order paths within a higher order path.

Creation of administrative domains for various purposes, such as the application of performance management or the operation of protection schemes, are also modeled by this mechanism. Examples of the use of this layer-splitting mechanism applied to both of these applications are given later in Chapters 8 and 10, respectively, which deal with these topics.

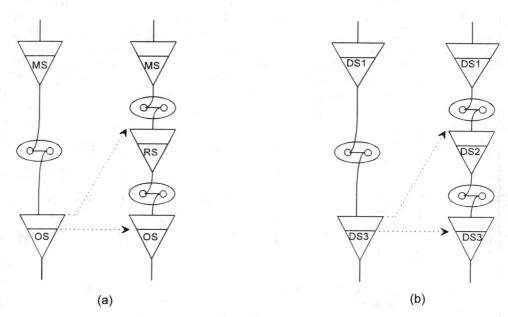

(a)   (b)

Figure 3.16   Examples of layer generation by expansion of the adaptation function.

3.4.7   Further Decomposition of Trails and Termination Processes

The ability to monitor the integrity, validity, and quality of transport entities in service has become an essential requirement of the modern operating environment. Indeed, improvement in monitoring capability has been one of the main drivers for the replacement of PDH by SDH.

The concept of the trail was introduced earlier in this chapter as a generalization, of which circuits, paths, and sections were specialized types. It was defined as a type of connection between layer boundaries with processes to safeguard integrity, validity, and quality. In ATM and SDH these objectives are generally achieved by embedded processes that we describe collectively as "termination process." PDH trails had rather variable provision, ranging from nothing (in the case of DS0 or 2048 kbit/s clear channel) to something (in the case of DS3 or G.704 primary rate):

- *Integrity*—The integrity of a trail is defined by its current availability for service. The criteria for availability are defined in terms of loss of signal, loss of frame, far end receiver fail (FERF), alarm indication signal (AIS), and so forth.

- *Validity*—A trail is valid if its trail termination point identifiers (TTPId) and labels match those planned prior to its establishment.
- *Quality*—The performance of a trail is defined in terms of the parameters in G.826, which are computed from block error-detecting codes specified in each layer.

The embedded termination process is generally made up of several quite independent subprocesses directed to these major objectives. They each contain at least information sources and information sinks, located in the termination functions at the layer boundary, and a dedicated communication channel from each source to each sink provided by the same server as the link connections to which they refer. This is illustrated in Figure 3.17 for a typical unidirectional trail having error-monitoring, trail validation, and signal label-matching processes. Although the aggregated definition of G.805 is good enough for many purposes, it is often important to distinguish further between trails of nominally the same type but differing in the exact termination processes supported.

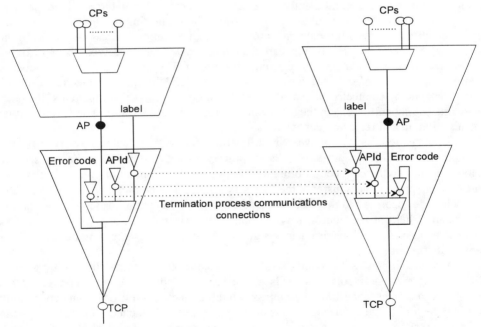

Figure 3.17   Trail termination processes.

3.4.8   Connectionless and Connection-Oriented Data Transfer Modes

The transport model described in this chapter was developed very much from a connection-oriented (CO) perspective. In a real sense, the concept of a connection between reference points in a layer network is fundamental to the whole concept of the model. However, the connectionless transfer mode was developed in the context of data communication because:

- It was more natural to a large class of intra and intercomputer communications based on transactions and interprocess messaging via a shared medium.
- The overhead of a separate connection control process for the creation of committed resource connections is not justified for such short data transfers.

If the duration of the message (which depends on both its size and the latency of the data transfer function) is shorter than the transfer delay, then the overhead of connection control is generally likely to be prohibitive. While many distinctions have been suggested, we use this to define the difference between connectionless and connection oriented modes.

3.4.8.1   *Characteristics of Connectionless Layer Networks*

The connectionless mode of data transfer involves forming the data to be transferred into a datagram or packet whose header includes both the destination address and the source address. Even data that is inherently streamed is buffered and transferred as independent datagrams across a connectionless layer network. The supporting network layer then has the capability of forwarding datagrams towards the destination address contained in the datagram header. The destination can distinguish datagrams received from different sources by the source address, also carried in the header. The connectionless layer network has the same topological structure of subnetworks and links as connection oriented networks, but the connectivity-related behaviors of the topological components (the layer network, the subnetwork, and the link) work by routing and forwarding datagrams rather than setting up and releasing connections.

A connectionless layer network is still defined by the set of data source APs and the complementary set of data sink APs towards which datagrams may be transferred. However, unlike the connection-oriented case, the route taken is not prenegotiated with the network prior to the transfer of the user's data but each datagram is routed dynamically at each node according to the destination address in the datagram header and the routing information held at the node. As with connection oriented networks, a group of APs collocated at one client host is an access group (AG) and acts as a routing destination for any of its contained APs.

A connectionless subnetwork is still a topological component delimited by a set of links to other subnetworks (and AGs at the layer boundary). In most connectionless networks, however, each link has only one link connection with one connection point joining the link to the subnetwork. The datagram is merely launched into the link, taking whatever resources it needs, and is distinguished at the other end of the link by the source and destination information in the header rather than by a link connection channel number known to be part of a predetermined connection. (Links in connectionless networks with more than one link connection are possible and may well emerge as a way of managing differential quality of service.)

The topological information supporting the routing process normally takes the form of route tables that contain a list of route choices (either next hop or full source to destination link lists) for all destination groups in the subnetwork routing domain. Broadly the same range of automatic, semiautomatic, and manual approaches to routing management and information exchange are available to both classes of network. The additional diagrammatic adornments illustrated in Figure 3.18 are introduced to indicate the connectionless and connection oriented specializations of the generic topological components.

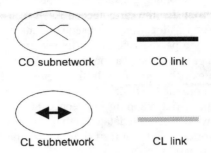

Figure 3.18   Conventions to distinguish connection oriented and connectionless topological components.

3.4.8.2   Resource Reservation

The main behavioral difference between connectionless and connection oriented networks is the prior negotiation and acceptance by the network of the implicit commitment of a connection as opposed to the probabilistic expectation that a properly addressed datagram will be delivered. There are two independent aspects to the commitment implied in a connection:

- A route exists and has been marked (without regard to sufficiency of resources);
- Resources have been reserved to support expected traffic needs on the selected route.

When a connection is set up in a connection oriented network, a routing is established and link connection resources identified and reserved on each link dedicated to the support of the connection on the basis of the traffic requirements of the data transfer service type. The connection setup process involves procuring link resources on each link on the recommended route. If any of the transit links is faulty or full and cannot provide resources, then the connection is refused.

A connectionless subnetwork, on the other hand, will successfully forward a datagram up to the point where a congested link is unable to provide the requested resource. The datagram may then be stored until resources become available, or ultimately discarded if either there is insufficient buffer storage or the queuing time is considered too long for the class of service. This is the characteristic of the connectionless router that distinguishes it from a connection oriented switching matrix.

3.4.8.3   Datagrams as Short Connections

The pros and cons of connectionless and connection oriented networks are currently debated frequently and passionately, and it is unfortunate that little effort has been devoted to an architecture which can satisfactorily encompass both. In this section we present architectural definitions which we believe can set a basis for such an architecture, however, it is unlikely that standards bodies and fora will reach any agreements in this area in the near future.

In the connectionless environment, each transit group normally contains a single CP that is allocated to a datagram for the time needed for it to traverse the link. It is then immediately released for future allocation to subsequent datagrams. This directly implies that the datagram is, from the perspective of the transit group, a transitory, short-lived connection. As a datagram passed through a connectionless network, the sequence of these short-lived connections is exactly equivalent to the sequence of links used for routing a connection in a connection oriented network.

In order to achieve the strict mapping between G.805 functional architecture and connectionless networks, we define each datagram to be an independent network connection. As its duration is shorter than the transit time across the network, it is set up and released as it traverses each part of the networks.

3.4.8.4   Flows

The above definition allows a formal and rigorous common architecture, however, at a practical level, connectionless networks frequently transfer many datagrams associated by a common routing. There is currently no definition in G.805 to cover this concept and we choose to define the term *flow* for this concept. This term is used in several areas of telecommunications in a general way, however, we believe our more formal definition here is consistent with most common usages of the term.

As each datagram is fully independent in the connectionless network, a flow is a very general concept. In some cases a flow can look like a connection, however, it has properties that a connection does not have.

The trail was defined in 3.1.4 as an end-to-end network connection formed by the association between APs at the layer boundary together with a termination function at each end. This definition applies equally well to the association formed between communicating clients of a connectionless data network and the TCP protocol of Internet provides this functionality. However, this trail is supported by a flow and not a network connection. Each datagram in the flow is routed separately and so each could take a different routing. In addition, the network does not correlate the probability of successful delivery of one datagram in the flow with any other. Neither of these is true of a network connection.

This connectionless trail then is simply the association between APs, but it "blindly" entrusts the data-forwarding task to the resource allocation procedures of the intralayer process in the connectionless subnetwork. The equivalent connection oriented trail requires a negotiation with the connection-oriented subnetwork to at least assign a valid, explicit, and persistent routing to the association. (Note that the final resourcing of the chosen connection is often implied but is, in fact, a separate process. Connections may be resourced at setup as in the PSTN or in SDH; resourcing may be delayed until some later time when the connection is needed, as in some ATM protection scenarios as discussed in Chapter 10. Alternatively, resourcing may be delegated to a separate intralayer process such as the ATM ABR process described in Chapter 6 or the flow control and windowing mechanisms used in X.25 networks.)

The concept of a flow is applicable to connection oriented networks as well as connectionless networks and we use the following as its definition. A flow is a set of connections all of which are routed towards the same transit group or access group.

This concept is important in shared media subnetworks which are described in the next section. Without the concept of a flow, these networks would need to describe multiplexing to a new layer network every time another connection is added to the flow.

This points to another property of a flow which is not true of a connection, which is that it can support multiplexing of connections (including datagrams) within the same flow. This allows multipoint-to-point flows where connections from different sources routing towards the same sink can be directly aggregated into the same flow and treated as a single traffic entity.

3.4.9   Shared Media Subnetworks

This is an important class of subnetwork because of its formative role in the development of the architecture of data networks. In fact the term "internetworking" and its abbreviated derivative "Internet" first appeared in connection with integrating such basic subnetworks.

In the shared media subnetwork, data sources and sinks share access to a single medium. Within the environment of IEEE 803 series LANs, for example, the sources effectively broadcast data to the whole shared media subnetwork, but an attached sink only accepts information addressed to itself. We can regard this as a connectionless subnetwork in that there is no prior connection establishment and the data transfer mechanisms, including contention resolution, are not visible to the client. But we can also note that such networks depend on an address resolution process to establish which hosts are attached. In response to a broadcast query, each host will respond with its media access control (MAC) layer address and host identity. In this way, each host builds up a set of addresses of all its LAN neighbors. This is equivalent to creating a set of point-to-multipoint "connected neighbor" associations or simply connections according to the definition above. The connections are, however, unresourced and some discipline must be applied to constrain host sources from transmitting simultaneously. The discipline is embodied in the resource management capability of the subnetwork and is achieved in the MAC layer using contention resolution techniques such as token passing or collision detection.

There are other interesting classes of shared media subnetworks commonly used for multiservice access to telecommunications networks, most notably those based on passive optical and radio technologies.

A passive optical access network can be regarded as containing a fixed set of media layer connections, determined by its structure, which are effectively setup when the physical network is built according to the disposition of splitters. A 1: N passive optical splitter supports connections between a root node and each of the branch nodes as illustrated in Figure 3-19, but not between the branch nodes themselves. The dual-node architecture implies that branch node sources can broadcast to both root node sinks. Such unresourced connections are then subsequently resourced by allocating shared media capacity amongst the sources by time or wavelength division multiplexing among them. Time division multiplexing is generally supported by a ranging process by which remote sources automatically make allowances for their relative transmission delays.

Mobile radio access uses a hybrid process whereby a user "goes off-hook" by signaling to the base station on a signaling channel shared with all other mobiles in the

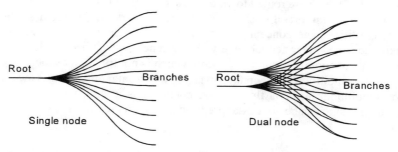

Figure 3-19   Connectivity constraints in PON splitters.

cell, which is hence subject to corruption by collision of coincident off-hook messages. Contention resolution on this channel is similar to the LAN environment. The base station responds with a channel allocation selected from the scarce spectrum resource available. If none is left, then the call is blocked.

We therefore prefer to regard all the shared media subnetworks in this unified way as essentially networks in which point-to-multipoint connections exist between a source and the set of sinks which can be reached by broadcast. The differentiation is then according to the manner in which resource allocation to connections is managed. It is interesting to note that data network architects, starting from their different viewpoint, have regarded the shared media LAN as the generic subnetwork and have regarded the simple subnetwork described in Section 3.1.3 as a specialization of this, which they then term a nonbroadcast multiple access subnetwork (NBMA).

3.4.10   The Layered Protocol Model

We have described the functional decomposition of transport networks in terms of layering and partitioning. In the domain of data communications, the International Organization for Standardization (ISO) has provided a different perspective on functional decomposition in terms of the seven-layer protocol model. In view of the widespread adoption of both models, it becomes important to consider their relative compatibility. Some discussion has taken place in ITU-T on this subject, but it is still some way from resolution. In this section, we attempt to position the ISO model with respect to the ITU-T transport model and incorporate aspects of the ISO viewpoint to provide a unified view of the two approaches. The ISO seven layer model is illustrated in Table 3.2.

The first point to note is that the layered protocol model was conceived within or around a single transport layer in the G.805 sense and took no particular account of partitioning. It was concerned instead about decomposing the different functions involved in transferring data across a layer network with a single partition. We find that layer 3 and 4 have a quite close alignment with the subnetwork connection and trail processes, respectively, and there are similarities and overlaps also at other layers. We do not propose

Table 3.2

The ISO Seven-Layer Protocol Model

Layer 7	Application layer
Layer 6	Presentation layer
Layer 5	Session layer
Layer 4	Transport layer
Layer 3	Network layer
Layer 2	Link layer
Layer 1	Physical layer

to provide a detailed explanation of the ISO model, but instead to apply a similar procedure to the transport model and note the relationships to the ISO model where appropriate.

Thus, the data transfer process is decomposed into subprocesses operating independently in each of the protocol layers, which we shall henceforth refer to as bands (to avoid confusion between the two usages of the term layer). In general, a band manages transport resources and uses them to provide services to higher bands as illustrated in Figure 3.20.

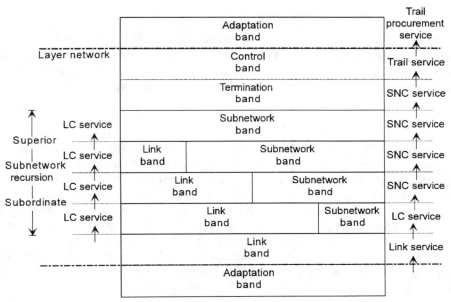

Figure 3.20   Banding within a layer network.

3.4.10.1   The Link Band

The link band manages link resources and offers a link resource service to the subnetwork band. Link resources are capacity and link connections. Link connection resources without capacity allocation are just channel numbers reflecting the granularity of capacity allocation. The capacity of a link connection is defined by its bandwidth, its bit rate, or, in the case of an ATM VCh link connection, its traffic parameter values. In the case of an ATM link band, the VCIs and VPIs may be allocated separately without capacity allocation. SDH VC link connections imply a fixed, predetermined capacity in terms of bit rate available. The link band procures link resources by using the services of the adaptation bands of one or more serving layers.

3.4.10.2   The Subnetwork Band

The subnetwork band offers a subnetwork connection service. This service may be offered to the termination band, in which case the subnetwork connection is also a network connection, or the service may be offered to a superior subnetwork band in the same layer network according to the partitioning rules. The subnetwork uses the services of the link band and also the services of subordinate subnetwork bands. The resources it uses are access groups and transit groups, and the access points and connection points contained. The subnetwork band provides similar functionality to the ISO network layer, with the additional ability to deal with the recursion inherent in network partitioning.

3.4.10.3   The Termination Band

The termination band offers a trail service to the adaptation band. The trail service is the basic service of the layer network, providing connectivity between client layer access points. The trail service may also include embedded processes to monitor validity, integrity, and quality. In this sense, it is analogous to the ISO transport layer.

3.4.10.4   The Control Band

The control band contains functionalities to do with procuring connectivity from the termination band to support the traffic demand of the client layers expressed through the adaptation band. We prefer to model this as an intralayer process, but typical current SDH implementations introduce a manual operator interface here so that the control band acts as an agent in a management process procuring connectivity in the layer network. When applied at a circuit layer providing end-user service, the control agent provides a signaling interface in what is there referred to as the control plane. In our target architecture, the control band provides the control interface directly to the client layer and is essential for interlayer automation in the future. This is roughly equivalent to the OSI session layer which, when applied to the PSTN model, takes on the functions of call control.

3.4.10.5   The Adaptation Band

The adaptation band offers an appropriately adapted data transfer service to an application. When the application is the support of a link in a client layer network, the adaptation band offers a link service to a link band in a client layer. It may provide access to part of the server capacity, and the capacity may be offered in channelized form as link connections. An adaptation band is specific to a particular client-server combination. It uses trail services provided by the termination band and partitions the capacity available for the use of the client layer link band. The adaptation band performs a role somewhat analogous to the ISO presentation layer. The OSI link layer includes the basic link resource management function of the link band above, but also much of the termination band and adaptation band functionalities for the serving layer below it. The OSI model does not recognize these functions as recursions in a different network layer.

3.4.10.6   Application and Physical Protocol Layers

In the recursive transport model, the physical layer is effectively the service supplied by a server layer. Ultimately, at the bottom of the recursion, we find a transmission media layer, which is closely equivalent to the ISO physical layer. Similarly, at the top of the stack we can say that the transport application is, in fact, another client layer network. In a similar way, this may ultimately be a user application like telephony on the 64 kbit/s circuit layer or videoconference on an ATM VCh layer.

3.4.11   Practical Application of Architectural Principles

In this chapter, we have introduced the main architectural concepts by which telecommunication transport networks are defined:

- The layer network defined by the characteristic information structure at its reference points, allowing subnetwork and link connections to be supported between reference points;
- The client-server relationship that exists between network layers and the superior-subordinate relationship that exists between partitions in the same layer;
- Connectivity as the relationships between subnetwork and link connection components concatenated to provide trails within a layer network;
- Topology as the relationships between sets of reference points that describe a layer network's connectability.

In doing so, we have remained faithful to the principles elucidated in ITU-T Recommendation G.805, extending them only slightly to expose further issues relevant to transport network management. Although, like G.805 itself, the main objective has been to provide an abstract description of the architectural features of a static network, we have attempted to set this in the context of a dynamic network process, recognizing that the various entities described have a lifetime in which creation, deletion, and change of state in operational service are ever-present factors. Indeed, we have shown that connectionless datagrams are instances of very short-lived connections. Termination processes or subnetwork protection processes, for example, are embedded within elements of this transport architecture, while these same elements are themselves components that may be manipulated in other larger processes such as layer management, service restoration, or connection control. Some of these are discussed more fully in later chapters.

When planning, designing, or installing a real network extension, the operational staff concerned will have to make decisions and choices as to how new network resources are allocated to the network. Some judgment is required in the choice of subnetworks and access groups as management entities, and subsequent allocation of resources to them. The choice is mainly driven by administrative and operational concerns such as maintenance domains and routing efficiency, but must also take account of such restrictions as commonly exist in the infrastructure. Close coupling of components in a protection group or nonuniformities in feature penetration, for example, force administrative choices that can conflict with other objectives. We will examine several such cases in more detail in later chapters.

The level of abstraction of the functional model may be considered too high for many purposes. We make no apology because we consider that a rigorous abstract definition that commands wide international support is a major prerequisite for the successful development of operation and control systems for the large multioperator, multivendor networks that we expect to characterize network evolution at the turn of the century.

References

The following form the main standards base for the architectural discussion in this chapter:

ITU Recommendation G.805, *The Generic Functional Architecture of Transport Networks"*.

ITU Recommendation G.803, *Architecture of Transport Networks Based on the Synchronous Digital Hierarchy*.

ITU Recommendation I.326, *Architecture of Transport Networks Based on Asynchronous Transfer Mode*.

Control and Management Architecture   4

The architectural framework presented in Chapter 2 is based on the concepts of functions and logical interfaces. We noted a particular function, called the data transfer function, which can transfer information between two functions that are geographically separated and we noted that a data transfer function would be made up of transport functions, control functions, and management functions. In Chapter 3, we presented the transport functions, the functions through which the client information will pass. They structure the client's information into a data format, according to the precise needs of the interlayer and intralayer processes, and transfer the information across the layer network between the desired end points.

In this chapter, we consider the control and management functions within the data transfer function. These are required to control and manage a layer network and allow client functions to request service of the layer network. We can see from Figure 2.13 that the composite interface to the data transfer function comprises a data interface and a control interface. The functions behind the data interface were described in Chapter 3, while the functions behind the control interface will be described in this chapter.

While there are still some differences of history and approach between the synchronous digital hierarchy (SDH) and asynchronous transfer mode (ATM) communities, the transport functions are now reasonably well-established in standards and there is a high degree of confidence that the transport functions described in Chapter 3 do form a complete and consistent set. However, the same cannot be said of the control and management functions. There is a growing consensus on the general shape that these processes should take; however, there are still a number of alternatives at almost every level of architectural detail. This is due in large measure to the differences in approach between four different communities, namely, the telecommunications signaling community, the network management open system interconnection (OSI) community, the distributed processing interests of the computer industry, and the Internet community. Work in each of these areas has generally been quite disconnected. There is today an increasing awareness of the inherent similarity of the problems facing each of these groups.

The scope of control and management is very broad, and many different approaches have been tried. The approach we present here draws on a number of these; however, the primary emphasis is on the development of a functional architecture for control and management that is consistent across the four different communities. In the same way that the transport functional architecture is based on a set of implementation-independent transport functions, so control and management functional architecture is based on a set of implementation-independent control and management functions. The set of functions presented in this chapter is necessarily incomplete as the scope is very wide. Emphasis is placed on the control and management functions required for the control and maintenance of connections. This reflects the central objectives of this book and its orientation towards functions for which interworking standards are necessary. Other interesting areas such as service control and IN functionality are not discussed.

4.1   HISTORICAL PERSPECTIVE

Over the last 30 years, computers have taken over several business processes like double entry accounting, customer sales records and invoicing, engineering design calculations and system simulations, production process control, and so forth. In general, these are processes that are carried out at the "head office" or at one location, such as a factory site. They were normally implemented in large, centralized, proprietary mainframe computers managing a single monolithic database, and while the information could be accessed from almost anywhere, all the data and processing of the data would be done at a single site. This computer architecture, shown in Figure 4.1, is called Von Neuman architecture after its inventor and has dominated the information-processing industry.

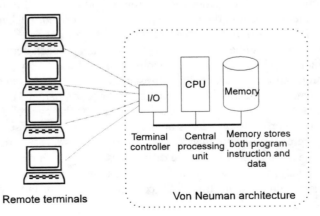

Figure 4.1   Typical Von Neuman computer architecture.

Von Neuman architecture is based on the idea that data is manipulated by a set of instructions, one instruction at a time, called a program. Both the data and the program are stored in memory. To increase the capabilities of a system based on this architecture, the memory must be made bigger and the processor made to run faster. This had been the major preoccupation of the computer industry for many years.

However, over the last ten years, there have been some serious challenges to the dominance of simple Von Neuman architecture, not necessarily in the way in which an individual computer is designed, but in the way applications are designed to run on computers and the way in which computers can be linked together to support these applications. Today, computers are typically linked to other computers to form a network, so applications can be designed in such a way that they can share the distributed resources of the computer network and are so designed to be split into several different, distributed components. This has led to the development of a new branch of computer science called "distributed computing."

More recently, the term "global information infrastructure" (GII) has been used in connection with this emergence of a general technology that supports distributed applications. It is based on a fusion of computing, telecommunications, and consumer electronics technologies, and is attracting widespread interest and support. Laying down an acceptable framework for this activity is not at all straightforward as the gap between entrenched positions in these different communities is quite large. The standards fora involved range from fully accredited international standardization bodies such as the International Telecommunications Union (ITU-T), the International Organization for Standardization (ISO), and international exchange carriers (IEC), to industry special interest groups such as the Open Management Group (OMG) and DAVIC, and also bodies such as the Internet Engineering Task Force (IETF), which has recently made the transition from an academic research group under the U.S. Department of Defense (DoD) to internationally recognized guardian of the Internet. In addition to these more or less open and democratic fora, there are the "developers groups" run by commercial companies such as Microsoft and Sun Microsystems who canvas support from interested users.

4.2 LEVELS OF CONTROL AND MANAGEMENT

ITU-T Recommendation M.3010 forms a useful starting point in partitioning the management problem, although it does only consider "management" processes and not control processes. Four management levels are defined: business level, service level, network level, and network element level.

Another attempt, with a wider scope, is defined in ITU-T Draft Recommendation GII.PFA, and this gives four models: an enterprise model, a structural model (concerned with services), a functional model, and an implementational model. These models are fundamentally different from one another; nevertheless, as we are concerned only with functional aspects in this chapter, they give a good insight into some of the basic activities involved in operating a broadband network. The major difference is that the GII.PFA models reflect the distinction between functions and implementation described in Chapter 2, while M.3010 makes a distinction between activities within a network element and activities beyond the scope of a single network element. The experience of ITU-T SG15 in developing the management interface specification for synchronous digital hierarchy (SDH) was that while both distinctions have their uses, the distinction between functions and implementation is the more important. The model we present in this section reflects this and is based on the following levels of control and management:

- *Business-level control and management*—Which is concerned with the business planning, forecasting, and financial control that ensures the ongoing viability of network operator's business;
- *Service-level control and management*—Which is concerned with the services that are offered to clients, their quality of service, and the account and billing for the services;
- *Network-level control and management*—Which is concerned with controlling the main functionality of a layer network to provide connectivity for services at a requested level and restore the connectivity if needed;
- *Implementation-level control and management*—Which is concerned with the installation, commissioning, and repair of equipment.

These control and management levels relate to layer networks in a particular way that reflects the client-server relationship between layer networks. Each layer network (or indeed, each partition of each layer network) is treated as a separate enterprise, effectively a separate business activity. Each layer network, therefore, has its own business-level control and management that designs and plans both the services offered by the layer network and the functional design of the layer network itself. To provide the link capacity for the layer network, the business-level planning process must request the services of a server layer. The basic interlayer client-server relationship described in Chapter 5 can be viewed as a fully commercial relationship between commercial enterprises as illustrated in Figure 4.2. In addition, the layer network must also procure equipment on its own behalf to provide node functionality. Some equipment may be shared between layer networks if

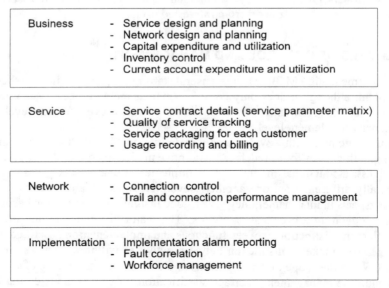

Business	- Service design and planning
	- Network design and planning
	- Capital expenditure and utilization
	- Inventory control
	- Current account expenditure and utilization

Service	- Service contract details (service parameter matrix)
	- Quality of service tracking
	- Service packaging for each customer
	- Usage recording and billing

| Network | - Connection control |
| | - Trail and connection performance management |

Implementation	- Implementation alarm reporting
	- Fault correlation
	- Workforce management

Figure 4.2   Telecommunication management and control processes.

they can be configured to provide sources to various layers. In this case, the costs may be shared across the layer networks. This is discussed in Chapter 13.

4.2.1   Business-Level Control and Management

Within the context outlined above, business management is the collection of processes that control and manage the commercial aspects of a partition of a layer network. General areas of control and management associated with the business level are

- Service design;
- Traffic forecasting;
- Promotion and advertising;
- Network planning and design;
- Inventory and capital control;
- Management accounting.

We do not consider these in this chapter, as they are normally proprietary. However, network planning and design along with aspects of forecasting are discussed in Chapter 13, and some of the functions required to support network planning and design are described in the functional architecture.

The network design and planning process for the layer network could be largely automated by dynamically monitoring the level of spare capacity in the layer network and automatically allocating more node and link capacity to the layer network as required. This can be carried out in real time, so the capacity allocated to a layer network could vary with time. For example, most public switched telephone network (PSTN) capacity is not required overnight and could be released to another layer network when not in use. It is only when the physically installed switch, crossconnect capacity, or line system capacity is nearly exhausted that human intervention is required to install more capacity, and even that could easily be triggered by the automatic planning process.

4.2.2   Service-Level Control and Management

The service management level is concerned with establishing and coordinating a client's requirements and then coordinating with the network control and management level to give the required level of service to the user. General areas of control and management associated with the service level are

- Service creation management;
- Customer account management;
- Service/session control;
- Enhanced service control (e.g., IN features);
- Customer billing.

Again, most of these are not covered directly in the functional architecture of control and management described here, as they go beyond our scope. However, the functional architecture does take account of the existence of these processes and the requirement for the functions described in this chapter to link with them.

The supply of a transport service can be broken down into a number of phases: client attachment to one or more layer networks, creating a number of access points; connection across a layer network to a distant access point; transfer of the data; disconnection from the distant access point; and detachment from the network. For each phase, the features of the transport service, its notional price, and its expected quality can be defined as illustrated in Table 4.1.

Table 4.1

Example of a Service Definition Matrix

	Price	*Speed*	*Accuracy*	*Dependability*
Attachment	Amount per attachment	Time to make new attachment	Attachment of correct AP	Attachment at planned time/date
Connection setup	Amount per connection setup	Connection setup delay	Connection between correct endpoints	Connection setup not blocked
Information transfer	Amount per unit volume or per unit time	Transmission delay	Error performance	Availability in presence of failures
Connection release	Amount per connection release	Connection release delay	Correct connection released	Release of resources complete
Detachment	Amount per detachment	Time to make detachment	Detachment of correct AP	Detachment as planned
Accounting	Discounts for scale	Delay between usage and billing	Correct amount billed	Correct bill remitted on time

A portion of this matrix is defined in ITU-T to describe the quality of service of the 64 kbit/s integrated services digital network (ISDN) bearer service in Recommendation I.350. It is unclear to what extent service management processes will ever be standardized, as network operators feel that this may reduce or eliminate the scope for competitive advantage. However, some level of standardization of services is essential.

The requirement to carry more than simple voice channels on broadband networks means that they must support a large number of different forms of connectivity. Asynchronous transfer mode (ATM) makes possible a large number of different connection types, and it must be possible to define these precisely when a service instance is requested to ensure that the network control and management processes allocate network capacity sufficient for the service. The ATM Forum and ITU-T have specified a range of connection types, and these were described in Chapter 2.

4.2.3 Network-Level Control and Management

The network control and management level is concerned with the control and management of layer networks and subnetworks, and their network performance. General areas of control and management associated with the network level are:

- Naming, addressing, and routing management;
- Traffic management and capacity allocation;
- Connection control;
- Trail performance management.

These are all covered by the functional architecture described in this chapter, and in some cases these functions provide linkage to areas associated with business and service levels. We have already mentioned the different approaches to routing, capacity allocation, and connection control emanating from the four different communities of networking expertise. Broadband networking based on ATM and SDH demands a single, consistent approach. In this chapter, we try to take the best from these approaches to present a unified solution.

There is a close coupling between the related activities of provisioning and restoration of connectivity, and there is also a close coupling between the commissioning and repair of equipment, while the couplings between other combinations of these activities is correspondingly loose. This leads us to accord less weight to the coupling expressed in M.3010 between fault correlation and network connectivity. This aligns closely with the principles developed in SDH and is quite central to the models in GII.PFA that propose that the control and management of the desired functionality of a network should be well separated from the maintenance of the implementation. This principle is essential to implementation-independent control and management. As a result, fault correlation is considered part of the implementation-level control and management, and not part of the network element management level of M.3010.

4.2.4 Implementation-Level Control and Management

The implementation-level control and management is concerned with the commissioning and repair of equipment and maintaining build control of software mounted on equipment. General areas of control and management associated with implementations are

- Equipment commissioning management and software build control;
- Fault correlation and identification;
- Equipment repair management;
- Workforce management.

At the implementation level, much will necessarily be proprietary. In addition, many of the processes will be manual, as they involve the installation and repair of equipment and physical media.

In principle, broadband equipment ought to be able to understand its own state and be able to report failure alarms. However, there will certainly be circumstances when a failure is not detected locally, when it will be necessary to use information derived from trails that pass through equipment to localize a failure. Similarly, if the network-level process has not restored a connection, the service level may need to ascertain the progress on equipment repair to report to affected customers on likely down time for connections. While the degree of coupling between the implementation level and the network and service levels of control and management are relatively weaker, they are nevertheless coupled (albeit loosely) to satisfy this sort of requirement.

4.2.5   Developing the Generic Functions of the Functional Architecture

Having looked at the different levels of control and management, which are summarized in Figure 4.2, the next step is to develop and define the generic functions in the control and management functional architecture.

Developing these functions is an exercise that should in many ways be independent of the analysis of the control into management levels described in this section. These give a reasonably comprehensive description of the "problem space;" however, the functional definitions should also be oriented towards more general desirable properties of a functional design, as described in Chapter 2. These include such characteristics as reusability, extensibility, and disambiguity.

4.3   FUNCTIONS ASSOCIATED WITH VISIBILITY OF RELATIONSHIPS

In Chapter 3, we discussed the functional components of the transport architecture and the relationships between them. Transport components were defined in terms of relationships between reference points and topological components in terms of relationships between sets of reference points. The management and control of a transport network is concerned mainly with the manipulation of these relationships. The relationship information is recorded and presented by a *relationship function*. There are three basic classes of relationships that recur with different semantic context throughout the model. They are listed below:

- A one-to-one relationship exists directly between a function and only one other.
- A one-to-many (or many-to-one) relationship is not symmetric. The one function may see the many functions to which it is related while each of the many can each only see the one to which it is related.
- A many-to-many relationship is also asymmetric, as each function may see the many functions to which it is related. Such relationships may become quite complex.

Any significant relationship between functions in the architectural model must be visible to the management and control functions or their agents. As a matter of principle, we show each direction of such reciprocal relationships independently. However, depending on the functional requirement, we may show one or the other direction of the relationship, or both. For example, it may be that in a one-to-many relationship, we may need to frequently navigate from each of the many to the one, but seldom from the one to the many. The three types of relationships mentioned above are illustrated in Figure 4.3 together with the way

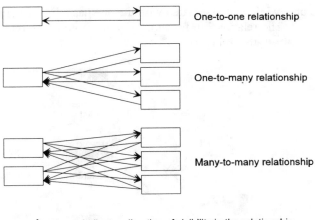

Arrow indicates direction of visibility in the relationship

Figure 4.3 Visibility of relationships.

the relationship is shown in the relationship function. Some of the more important relationships for control and management are listed in Table 4.2.

Table 4.2

Examples of Relationships

Relationship	Type of Visibility	Directionality
"contains"	one to many	bidirectional
"names"	one to many	bidirectional
"is connected to" (sink to source)	one to one	unidirectional
"is connected to" (source to sink)	one to many	unidirectional
"is connected to" (bidirectional)	one to one	bidirectional
"affects"	one to many	unidirectional
"is supported by"	one to many	unidirectional
"manages"	one to many	bidirectional
"is managed by"	one to many	bidirectional

The relationship is shown within each of the participant functions by means of a "pointer" or a list of pointers. A pointer is an attribute which can hold the name of another function. In the case of the one-to-one relationship, each relationship function has a single pointer that holds the name of the other function. In the case of the one-to-many relationship, each relationship function associated with the many has a single pointer with the name of the one, while that associated with the one has a list of pointers with all the names of the many. The relationship functions in a many-to-many relationship all have lists of pointers as illustrated in Figure 4.4.

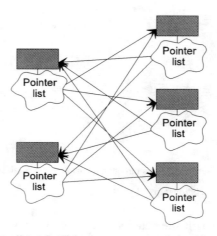

Figure 4.4   Use of pointer lists to model a many-to-many relationship.

This general way of modeling relationships by means of relationship functions that may (but need not) be encapsulated within the participant function to which it applies allows for some extensibility. As the application matures, functions may be called upon to participate in new relationships, and these may be added without affecting the earlier relationship functions defined. Of course, the general classes of relationship listed above are specialized in practice by the particular semantics attached to the relationship. Two important relationships illustrate this: the "is connected to" relationship between connection termination point functions, and the "affects" between implementation agents and the functions they support.

4.3.1   The "Is Connected to" Relationship

The "is connected to" relationship represents the connectivity between connection points participating in a connection. The context of the "is connected to" relationship derives from the transport network architecture, from which we note that a data source may be connected to many data sinks, but a sink may only be connected to one source. This means that the type of "is connected to" relationship that the connection point is allowed to participate in depends on whether the connection point is a sink, a source, or bidirectional.

To make the relationship visible to control and management, we must consider the sink and source separately so the bidirectional connection point is viewed as a combination of a sink and a source. Basic connectivity is represented by a single pointer relationship function associated with the connection point sink that holds the name of the source, as illustrated in Figure 4.5. The bidirectional connection is automatically visible, as each

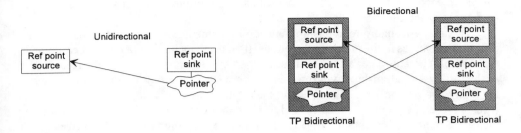

Figure 4.5   Modeling the unidirectional sink to source "is connected to" relationship and the bidirectional "is connected to" relationship.

bidirectional connection point has a sink part that has an associated pointer holding the name of the other end.

The case not covered by this is the point-to-multipoint connection. In this case, the relationship from the sink to the source is visible using the single pointer relationship function described above; however, the source does not have the names of all the sinks. This accurately reflects the real-world situation in that any sink can know to whom it is connected by the signal it is receiving; however, the source cannot know all the sinks that might to listening to the signal the source is sending, whether legally or illegally. If it is necessary in an application for the source to "know" the sinks to which it should be connected, this can be done by adding a "list of pointers" relationship function for the purpose, as illustrated in Figure 4.6.

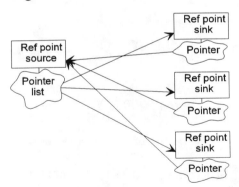

Figure 4.6   Modeling the unidirectional sink to source "is connected to" relationship and the unidirectional source to sink "is connected to" relationship.

4.3.2   The "Affects" and "Is Supported by" Relationships

An implementation is represented to control and management by a closely coupled equipment agent. If the implementation fails, many functions may be affected. This can be made visible by a relationship function encapsulated in the implementation agent that lists all the functions that may be affected if the implementation fails. On the other hand, the successful operation of a function may require the successful operation of several implementation parts, and this too can be made visible by a relationship function encapsulated in each function that lists the names of all the implementations that are involved in supporting the function.

This many-to-many relationship between implementations and functions again has different semantics in the different directions. If an implementation (a piece of equipment, for example) is faulty, not all the trails supported by the equipment may be affected, and so the "affects" relationship gives information on which trails *may* require rerouting while the faulty equipment is replaced. Ideally, the implementation agent should give an alarm when a piece of equipment is faulty, and this should be the only information necessary to locate the fault and indicate the repair action (i.e., replace the faulty unit). However, if a trail is detected as failed and no corresponding equipment failure is independently indicated as being the probable cause, special action will be required to diagnose the problem and determine the probable cause. In general, it becomes necessary to trace all the equipment with "is supported by" relationships to the affected trail. In a well-partitioned design, this procedure should be rare; therefore, we can say that the participant objects may be loosely coupled. Thus, we may decide that the "affects" relationships may be held closely coupled to the participants, while the "is supported by" relationships need only be computed if and when they are required. These are illustrated in Figure 4.7

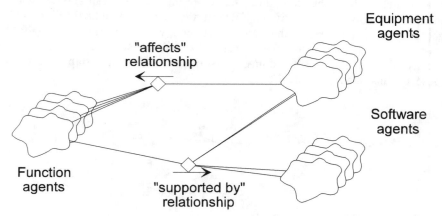

Figure 4.7   Representation of the relationships between implementations and the functions they support to control and management.

4.4   TRANSPORT FUNCTIONS AND THEIR AGENTS

The basic transport functions were described in Chapter 3 in a manner that made no reference to their control and management. In the control and management context, we consider each transport function to have a closely coupled agent that embodies the role it is required to play for the purposes of control and management. The transport function control agents interact with other functions, participating in the control and management enterprise through interfaces and presenting information or executing operations as required by the process. In the traditional signaling systems used in telecommunications, these interfaces were often proprietary. In the context of SDH management, this group of functions is fully defined with open interfaces. We refer to such functions generically as "agents," emphasizing their similarity whether they are found in the control plane or the management plane.

In SDH, the manager-agent principle was adopted as illustrated in Figure 4.8 to define the view presented by network resources to the manager, and this has now been carried through to both ATM and WDM networks. The transport function holds and records its own internal state and communicates through proprietary local interfaces with its associated agent, thus providing a standard logical interface to other control and management process agents.

The transport function will also create relationships between inputs and outputs in response to external requests and reflecting its internal connectivity. The connectivity information must be presented to the control and management system through the agent interface. As illustrated in Figure 4.8, transport function agents are associated with reference points on the transport function boundary, so they are also used to carry information relating to the state of the transport function. The connectivity within the transport function is shown by the values in the relationship pointers between the transport agents.

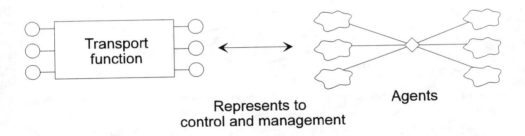

Figure 4.8   Internal connectivity visible through pointers in termination points.

4.4.1   Topology and Transport Function Agents

The two transport functions associated with topology are the subnetwork and the link, and these are delimited by access groups and/or transit groups. The link is presented to control and management via two agents associated with the groups of reference points by which it is delimited, one for each end of the link. They will either be an *AG agent* or a *TG agent*. Similarly, the subnetwork is presented to control and management, at least for the purposes of topology, by a function for each link that attaches to the subnetwork and, again, these will be either AG agents or TG agents. This is illustrated in Figure 4.9. The topology itself is represented by the relationships expressed in the relationship functions in the control agents. As an example, the properties of the TG agent should include

- Answer name of subnetwork (belongs to relationship);
- Answer address of TG (own identity);
- Answer list of names of TG or AG (subnetwork relationship);

- Answer name of TG or AG (link relationship);
- Answer list of names of CTP* (member link connections relationship);
- Answer of name of resource manager*;
- Answer list of names of implementations (supported by relationship).

* These are discussed in later sections.

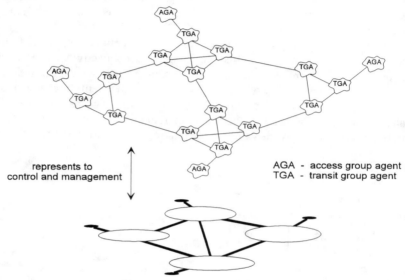

Figure 4.9   Topological agents carrying relationship information relating to the transport topology.

The properties of the AG are very similar; however, it will not have both the subnetwork relationship and the link relationship. It will have one or the other.

4.4.2   Connectivity Transport Mediation Functions

Adaptation, termination, and connection transport functions are represented to the control and management community with two classes of transport agents: *trail termination point (TTP)* and *connection termination point (CTP)*. The relationships between these transport agents are used to present the configuration of the adaptation function as well as which connection points have been connected to form connections and trails, as illustrated in Figure 4.10. The number and the nature of the relationships between these functions is restricted by the general rules of the transport network architecture.

As the transport functional architecture is fundamentally unidirectional and a connection point is a bound group of unidirectional ports, the transport control agents must also reflect this. Depending on the application, a TTP could be a TTP source (TTPSo), a TTP sink (TTPSi), or a TTP bidirectional (TTPBi), which is the encapsulation of both the TTPSo and TTPSi. There is the similar set for the CTP; that is, the CTP source (CTPSo), CTP sink (CTPSi), and CTP bidirectional (CTPBi).

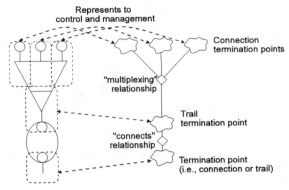

Figure 4.10   The manner in which TP relationships are used to model transport functions

An example of a very simple VC-4 crossconnect equipment is shown in Figure 4.11 with both STM-4 and 139264 kbit/s interfaces. The connectivity agents for the crossconnect are illustrated in Figure 4.12.

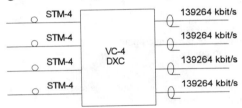

Figure 4.11   A simple VC-4 crossconnect with STM-4 and 140 Mbit/s ports.

All these connectivity agents in whatever layer network have similar properties and therefore are all members of a more abstract class of transport agents called a *termination point (TP)*, which is therefore a superclass for these subclasses. When the sink and source aspects are distinguished, then there is a complex multiply inherited set of classes set up as illustrated in Figure 4.13.

The following gives an indication of the properties that the TTPSo, TTPSi, CTPSo, and CTPSi should have. A TTPSo control function should have the following capabilities:

- Answer name of layer network (belongs to relationship);
- Answer address of access group (belongs to relationship);
- Answer address of TTPSo (own identity);
- Set and answer path trace name to be sent;
- Answer list of names of far end TTPSi (trail relationship);
- Answer name of CTPSo or GTPSo (connection relationship);
- Answer list of names of client layer CTPSo (client-server relationship);
- Answer list of names of implementations (supported by relationship).

A TTPSi control function should have the following abilities:

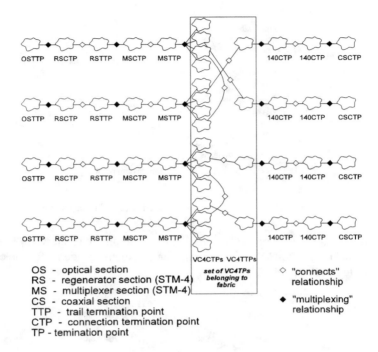

OS - optical section
RS - regenerator section (STM-4)
MS - multiplexer section (STM-4)
CS - coaxial section
TTP - trail termination point
CTP - connection termination point
TP - temination point

◇ "connects" relationship

◆ "multiplexing" relationship

Figure 4.12   Transport control agents required to manage and control a VC-4 crossconnect with STM-4 and 140Mbit/s ports.

- Answer name of layer network (belongs to relationship);
- Answer address of access group (belongs to relationship);
- Answer address of TTPSi (own identity);
- Report health of received trail;
- Report match of path trace name received;
- Answer name of far end TTPSo (trail relationship);
- Answer name of CTPSi or GTPSi (connection relationship);
- Answer list of names of client layer CTPSi (client-server relationship);
- Answer list of names of implementations (supported by relationship).

A CTPSo control function should have the following abilities:

- Answer name of subnetwork (belongs to relationship);
- Answer address of transit group (belongs to relationship);
- Answer address of CTPSo (own identity);
- Answer list of names east end CTPSi, GTPSi, or TTPSi (east connection relationship);
- Answer name of west end CTPSi, GTPSi, or TTPSi (west connection relationship);
- Answer name of server layer TTPSo (server-client relationship);
- Answer list of names of implementations (supported by relationship).

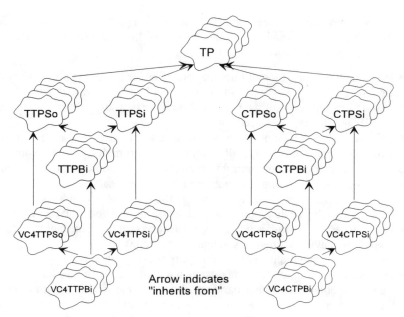

Figure 4.13   Multiple inheritance relationships between termination point classes.

A CTPSi control function should have the following abilities:

- Answer name of subnetwork (belongs to relationship);
- Answer address of transit group (belongs to relationship);
- Answer address of CTPSi;
- Report health of passing connection;
- Answer name of west end CTPSo, GTPSo, or TTPSo (west connection relationship);
- Answer list of names of east end CTPSo, GTPSo, or TTPSo (east connection relationship);
- Answer name of server layer TTPSi (server-client relationship);
- Answer list of names of implementations (supported by relationship).

There is an additional complication with trails that use "inverse multiplexing" where one trail is carried on several parallel connections, as is the case with SDH concatenation. When these parallel connections are established, it is important to be able to identify the connections that are bundled together to form a part of the single trail. This is done using an additional agent that controls a group of parallel connection points that are used for the group of connections. This is called a *group termination point.*

4.5   NAMING, ADDRESSING, AND ROUTING FUNCTIONS

Routing within a subnetwork depends on the existence of an identification scheme for the access points and connection points on its boundary. Every access and connection point in a layer network needs to be unambiguously identifiable so the routing process can find and route to it. To pursue this subject, we introduce the following terminology (which now enjoys fairly widespread support):

- A *global address* is a globally unique identifier for an access point, a connection point, an access group, or a transit group that is structured in such a way that it can be used directly for routing using no more than route tables (in the existing PSTN/ISDN, this is also called a *number*).
- A *relative address* or simply *address* unambiguously identifies any point in a layer network, but may not be globally unique. It is normally derived from a global address (the telephony numbering plan allows shorter addresses for local calls and adds access codes for long-distance and international calls).
- A *route table* is a function that can, on the basis of an address (or number), determine which transit group or access group on the boundary of a subnetwork should be used for onward routing of a connection (simple route tables in the PSTN, for example, have a subnetwork scope of one node).
- A *name* is a unique identifier for some purpose *other* than routing.
- A *directory* is a function that can relate a name in one naming scheme to the name of the same object in another naming scheme. In particular, we are interested in relating names for access points, connection points, access groups, and transit groups to their addresses.

According to these definitions, the normal "number" for a mobile telephone is not a number or an address at all, but is in fact a name. The control system must use considerably more than a route table to route successfully to the mobile and track any connection, as the mobile may move. The same is also true for the 800 type of service where the 800 "number" is not a number or an address, but a name. Again, the control system must use considerably more than route tables to route to the 800 "number." In both cases, a full translation is carried out that uses the services of a directory function to find the correct fixed network address associated with the mobile name or the 800 name. The routing system can then use this fixed network address to route successfully using only route tables. This is illustrated in Figure 4.14.

4.5.1   Addressing Schemes

An addressing scheme is likely to be long-lived, since it is a logical property of a layer network and very difficult to change once in service. It will persist as long as the layer network itself, and much longer than any technology implementing it. While there are many addressing schemes currently in use, the two of greatest importance are the ISDN numbering and addressing scheme described in ITU-T Recommendation E.164 and the Internet addressing scheme; however, both of these are currently experiencing difficulties, as their use has expanded well beyond that originally envisaged. An addressing scheme must balance a number of important factors including the following:

- Administration (the way in which individual addresses are allocated);
- Alignment with the topology of the network to minimize route table complexity;

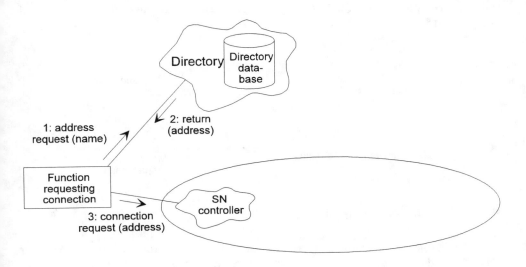

Figure 4.14   Use of a directory to translate from a name to an address

- Comprehensibility by human users;
- Relationships with other addressing schemes present or envisaged in future.

Addressing schemes are all derived from the hierarchical tree in which branches from a common root are uniquely identified in relation to their root node so that the address may be simply expressed as a sequence of root root-branch identifiers. The hierarchy may reflect the administration and allocation structure, the topology of the layer network itself, or, more generally, both. It is the balance between the conflicting requirements of administration and topology that determines the size of the route tables.

In addition, two "pseudo addressing" schemes should be mentioned that are significant for broadband networks. These are the OSI network service access point (NSAP) addresses and the media access control (MAC) addresses used in IEEE local area networks (LANs). After much discussion, ITU-T has endorsed a difficult compromise loosely based on E.164 for use in SDH networks, and this is also supported by many regional standards bodies.

4.5.1.1   E.164 Addressing Schemes

ITU-T Recommendation E.164, entitled "Numbering Plan for the ISDN Era," is intended to cover numbering and addressing for the PSTN, the N-ISDN, and the B-ISDN. The format for the numbers is simple in that it is a string of up to 15 decimal digits and starts with a country code. The digits after the country code are a matter for the administration in each country to structure and allocate. The format also includes a "subaddress," which is a further field of up to 20 digits that can be appended to the 15 digits and can be structured and allocated by a customer for routing across a private network. The structure is illustrated in Figure 4.15.

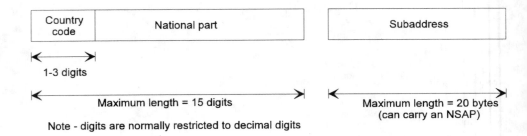

Figure 4.15   E.164 addressing scheme

The E.164 address structure has been agreed as viable for both the ATM layer networks and SDH path layer networks. Addresses can be reused in different layer networks, as the identity of the layer network itself taken together with the address provides unique identification.

4.5.1.2   Internet Addressing Schemes

The existing Internet addressing scheme in IP version 4 is a 4-byte binary number. Because IP is connectionless, the address on every packet must be globally unique so there is no distinction between global addresses and relative addresses. The basic structure of the 4 bytes is in two fields: the first field is a network identifier and the second is a host identifier. Three classes of IPv4 addresses were identified, as shown in Figure 4.16.

Class B addresses became rapidly in short supply; to overcome this, the IETF defined "subnetworks," which effectively allows a fully flexible divide between the network identifier and the host identifier. Even so, this has been a temporary fix, and ultimately the 4-byte address is insufficient. IETF has now agreed to a completely new version of the Internet protocol, IPv6, which is based on a 16-byte binary address.

4.5.1.3   NSAP Format

An NSAP is not so much an address but more a holder for an address. It has three fields, as illustrated in Figure 4.17. The first simply identifies the addressing scheme, which is used in the other two fields and is called the authority and format indicator (AFI) field. The second, the initial domain identifier (IDI) field, is the "public" part of the NSAP in that it contains a public address of the format indicated by the AFI field. Together, the AFI and IDI form the initial domain part (IDP), and as well as indicating an address, the IDP must implicitly indicate the authority that is responsible for allocating addresses in the domain-specific part (DSP) of the NSAP. The DSP is like the subaddress in E.164. The DSP is carried across a public network and used with a private network for final routing to the end point.

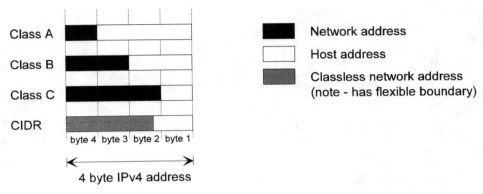

Figure 4.16   Internet Addressing (version 4).

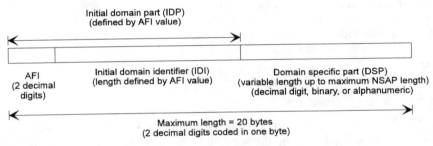

Figure 4.17   Network Service Access Point (NSAP) address format.

There is much discussion in the ATM Forum over the use of NSAP for points in ATM layer networks. They are attractive from the point of view that they are simple, have a single format, and are therefore quick and easy to specify as a solution to the problem of agreeing to a standard. However, as the NSAP is not so much an address but a container for an address, specifying an NSAP does not really tackle the issue surrounding the structure of the address to be helpful for routing. An NSAP can carry an E.164 address with the main address in the IDI and the subaddress in the DSP, as illustrated in Figure 4.18.

4.5.1.4   ISO Addressing Schemes

ISO has also specified in ISO 3166 a set of data country codes (DCCs) that are, for addressing purposes, similar to E.164 country codes. However, its origins are different. The emphasis with DCCs is on the allocation authorities for the addresses rather than a concern for efficient routing. The DCCs essentially specify a national standards organization that is responsible for allocating addresses. When this is used in an NSAP, only this country code is contained in the IDI and the address as allocated by the national standards body is contained in the DSP.

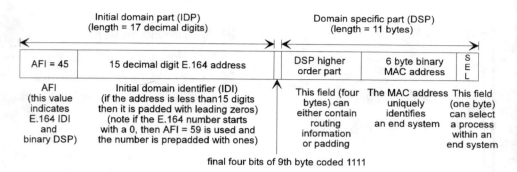

Figure 4.18   NSAP carrying an E.164 address.

In addition, ISO has also specified a set of international code designators (ICDs) for identifying international organizations, including companies that are authorities for allocating addresses. When ICDs are used in NSAPs, only the ICD is carried in the IDI, and this indicates the authority that has allocated the DSP.

The ATM Forum has endorsed both of these addressing schemes for ATM layer networks.

4.5.1.5   MAC Addressing Schemes

Shared media local area networks (LANs) achieve data transfer by broadcasting onto the media. While every end point on the network receives the data, only the designated end point will collect the data in order to deliver it to its client. A basic requirement is that the endpoint identifiers must be unique within the subnetwork formed by the shared media. To achieve this with minimum administrative burden for users, the IEEE defined the media access control address, which is, in fact, globally unique although it only needs to be unique within its local area. Obviously, if an address is globally unique amongst the superset of all such addresses, then it must also be unique amongst any subset.

The MAC address format is defined in IEEE 802.1 and address allocation is administered by the Xerox corporation on behalf of the IEEE. MAC layer addresses are allocated in blocks for a small administrative charge to any organization that wants them. They are most frequently "burned" into Ethernet chips by the manufacturers when the chip is made, so when a new node is added to the shared media subnetwork, no special procedure is necessary to guarantee it has a unique address. MAC addresses are more akin to channel labels for the distributed packet multiplexer, which is the LAN.

MAC layer addresses have no network significance and are therefore no use for routing globally. Every end point on a LAN must also have a network address that is globally unique but carries topological location information in its structure. To map the network layer address to the MAC layer address, an address resolution protocol is defined. A query may be broadcast to all nodes on the LAN, which is recognized and understood by every end system. Each end system responds with its MAC layer address and its network layer address. The same mechanism may be used to allocate network addresses initially. In this way, a "directory" may be built up at the LAN "gateway," relating network layer addresses to MAC layer addresses. Messages arriving at the gateway directed to one of the network layer addresses can be delivered over the MAC layer by looking up the corresponding MAC layer address.

4.5.1.6   SDH Layer Network Addressing Schemes

While ITU-T agreed early on to the need for an identification scheme for the access points, connections points, access groups, and transit groups for the SDH layer networks, it only recently agreed to a format for these addresses. The agreed upon format is 15 bytes that starts either with an E.164 country code or an ISO data country code. The format after the country code is a matter for national administration, but may include an operator field, as illustrated in Figure 4.19.

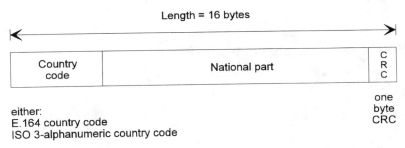

Figure 4.19   SDH access point identifier.

In the past, ports on transmission equipment tended to be identified in relation to their implementation, with perhaps a structure of region, operator, building name, equipment number, or port number. Furthermore, there was no coordination between such naming schemes within different operator domains. Such an approach is no longer sufficient in international managed networks. To identify these important network functions by means of the equipment that implements them implies that if the equipment is changed, replaced, or upgraded, its network location information changes even though the function remains unchanged and is still represented in route tables far and wide. The address should be decoupled from the implementation of the network, but coupling the address closely to the topological structure is also problematic because the topology must change from time to time, and this should not induce unnecessary changes in addressing. A degree of coupling to the topology is desirable, but not too much. This is discussed in more detail in the next section.

4.5.2   Route Table and Directory Functions

The distinction between the route tables based on addresses and the directories based on names is not fundamental, but is really a distinction of scale. The scale is determined by the degree of linkage between the naming or addressing scheme and the topology of the network. One extreme would be an addressing scheme that is exactly tied to the topology of the network and that would not require any route tables, since the address would directly identify each link in the connection. In other words, the address would be identical to a source route. This is clearly impractical, as addresses would need to be changed every time the topology of the network is altered. Therefore, some level of decoupling between the addressing scheme and the network topology is important. The translation necessitated by this decoupling is done using the route tables. This is illustrated in Figure 4.20. A directory could be viewed as an extreme of decoupling between the address scheme and the network topology. However, there is an exponential relationship between the extent of the decoupling and the size of the route tables whereby the size of the route tables grow exponentially for a linear increase in the degree of the decoupling. Most of the PSTN is currently based on route tables with up to around 1,000 entries, although this could easily be increased with new technology. In the Internet, the route tables in IP routers have grown very large as the decoupling between the network topology and the addressing scheme has also grown in recent years. These route tables can have 10,000 to 100,000 entries.

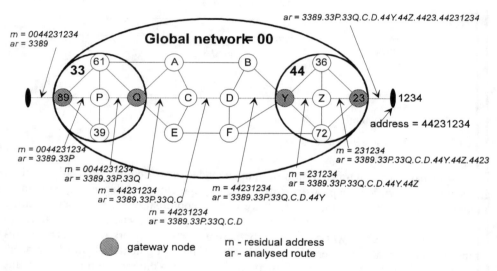

Figure 4.20   Translation from address to route

The simple route table at a subnetwork node contains a list of reachable destinations in the subnetwork, and for each destination, the recommended exit link. When a route is requested for a destination address, the route table responds with an exit link on the selected route. A basic route table can be thought of as a matrix whose rows are groups of destination addresses and whose columns are the transit groups (or access groups) which bound the subnetwork node. For each destination address, there will be one or more transit groups that can be used to route towards it.

One common refinement of this simple concept is the method known as source routing, whereby the whole route to the destination is specified as a list of transits that must be traversed. This is important where it is necessary to specify two alternative disjointed routes. The complete alternatives must be precomputed so that they may be compared for relative disjointedness when a routing decision is made. Route tables may also contain weighting parameters such as link state parameters related to efficiency (implied cost) or scarcity (to minimize blocking).

4.5.2.1   Alternate Routing

If only one link could be used for a given destination, then each row of the route table would have one element that indicates "use this link" and all the other elements indicating "do not use this link." However, many routing algorithms exploit alternate routing in which a number of links could be used to route to a single destination. In some algorithms, the links are prioritized, so elements in a row of the route table would indicate "1st choice," "2nd choice," "3rd choice," and so forth. Again, some algorithms require that once a link has been selected, that should continue to be chosen (the sticky random principle) until it becomes congested and so would be marked "current choice" while others could be marked "alternate choice." A combination of all of these methods may be used. Some flexibility should be designed into route table implementation to allow operators to optimize their automatic routing procedures to best suit their own situation. A route table with alternate routing is illustrated in Figure 4.21.

Route table for node C

Next address field	Route 1	Route 2	Route 3
.	.	.	.
.	.	.	.
44	D	A	E
33	33Q	A	E
.	.	.	.
.	.	.	.

Figure 4.21   Route table with alternate routes

4.5.2.2   Source Routing

Source routing allows more control over the choice of route than simple step-by-step routing. This is important when, for example, the disjointedness of a chosen route from one or more alternatives must be guaranteed. However, it requires that considerably more information be held in the route table. A large subnetwork using source routing requires, for every possible source/destination pair between transit groups (or access groups) across the subnetwork, a full route through the topology where the subnetwork is held. A source route for a particular address comprises a list of transit groups through which a connection should be routed, as illustrated in Figure 4.22. In addition, a connection setup protocol for this method must also provide a syntax for carrying the list information forward with the connection setup request. The construction of the information in the route tables for source routing is particularly amenable to the automated algorithms and routing exchange protocols described in the next section.

Source route table for node 33Q

Next address field	Route 1	Route 2	Route 3
.	.	.	.
.			
44	C.D.44Y	E.F.4472	A.B.44Y
(33)61	3361	33P.3361	-
(33)39	3339	33P.3339	-
(33)89	33P.3389	3361.3389	3339.3389
.	.	.	.
.	.	.	.

Figure 4.22   Route table with source routes

4.5.2.3   *Route Table Update Function and Routing Exchange Protocols*

It is still quite common in the PSTN to compute routing information for route tables manually. This information is then entered into the route table functions so that it is available to the routing process. When the topology of networks is reasonably static and the route tables are not too large, this can work well, and a skilled network operator can make changes to the route tables to respond to adverse network loading conditions without disrupting other parts of the network.

However, the possibility of automatically generated and maintained route tables distributed throughout the network has been known and used for some time, being first described by Dijkstra in the 1960s. In his algorithm, a *route table updating function* relays the contents of the local route table for a subnetwork to all its immediate neighbors. This may also include the status of transit links. A route table updating function receives the route tables from its immediate neighbors and updates its own local route table according to an updating algorithm to reflect all the destinations it can reach via its neighbors. This automatic method has several advantages. First, it is less likely than a manual method to generate errors in the route tables, and if errors do occur because the network topology changes, the error will be quickly eliminated as the algorithm updates. Second, if there is a failure in the network, this is perceived as a topology change and the updating process will find new route tables that will route traffic according to the new conditions in the network. To this extent, automatic route table maintenance makes an important contribution to survivability. The procedures and messages used between the route table updating functions make up the routing information exchange protocol as illustrated in Figure 4.23.

Automatically generated route tables have not normally been used in public telecommunications networks, such as the PSTN/ISDN, as network operators suffer from the perception that they are somehow deprived of control over the flow of traffic in the network. However, automatic routing information exchange protocols are the basis for routing in the Internet and have proven their value. They are, however, limited in scalability. Above a certain size, the complexity grows too large. In the Internet, automatic routing domains are limited administratively and so-called policy routing is used between domains. The border gateway protocol is used to exchange routing information at gate-

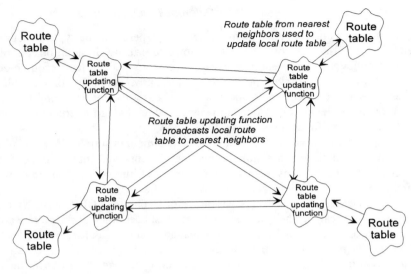

Figure 4.23   Route table updating function

ways. More recently, the ATM Forum has agreed upon a routing exchange protocol as part of the private network node interface (P-NNI) that can be used within private ATM networks. This powerful new NNI protocol introduces a hierarchical structure within the routing domain to overcome the scaling limitations of the simple Internet architectures. This seems to be the most complete and comprehensive proposal yet, and is attracting interest from network operators for use in a wider context than just private ATM networks.

4.6   TRAFFIC MANAGEMENT AND RESOURCE ALLOCATION FUNCTIONS

Traffic management and resource allocation functions are responsible for keeping track of capacity in transit groups and access groups and allocating capacity to connections in response to requests from the connection setup process. They act as agents for transit groups and access groups, performing the roles of resource manager function and the connection admission control function. Other agents act for their associated CTPs (or TTPs), including usage parameter control and traffic shaping functions.

4.6.1   Resource Manager Function

The role of the *resource manager function*, (or simply resource manager) is to keep track of the way link resources are allocated to connections. The two primary resources that are held by the resource manager are capacity and connection identifiers. The resource manager keeps track of the capacity as seen by the TG or AG and controls the allocation of capacity to connections when requested as part of the connection setup process.

4.6.1.1   *Resource Management for Different Types of Connection*

The detail of the resource management model will depend on the type of connections involved and the policies in force for reservation and priority. In addition, the resource management policy will represent a balance between simpler but less resource-efficient procedures and resource-efficient but more complex procedures. In connectionless networks, resource management takes place on a packet-by-packet basis and the resource manager effectively defaults to a queue manager, which has no alternative but to drop packets if there is insufficient capacity on the link.

In the case of circuit switching, resource management is simple since the allocation of capacity is automatic when a link connection is connected into a connection. The resource management involves recording the number of link connections in use and the total number of link connections available for use in the link.

Connection-oriented virtual services are more complex. CBR services are only slightly more complicated. The resource manager must keep track of the bit rate allocated to each connection and the total capacity left available on the link for new connections.

The capacity required by VBR connections can vary dynamically and without any direct indication from a resource manager. The resource manager must keep track of the VBR service parameters that describe the characteristics of the connection and the limits within which the connection bandwidth is allowed to vary. These parameters allow the connection admission control process to make statistical assessment on the current state of the link as to whether it can support a new connection without inducing unacceptable cell loss in the aggregate cell stream.

An ABR connection can be thought of as a special form of CBR service where the cell rate is dynamically changed by the resource management protocol described below during the course of the connection. The resource manager must keep track of the actual capacity currently allocated to each connection (called the explicit cell rate). In addition, the resource manager can keep track of other parameters such as an agreed maximum cell rate and minimum cell rate. The primary difference with ABR services is that the resource manager must also dynamically control the amount of resources allocated to the connections rather than simply keeping track of the allocated capacity.

With ABR connections, a resource management protocol is required that operates dynamically between the resource managers at each end of a link, allowing them to dynamically adjust the capacity allocated to each connection. With all the other types of connection, resource management is either nonexistent such as with UBR (equivalent to best effort) or is exercised at connection admission and subsequently policed or monitored to check that the negotiated resource allocation is not exceeded (CBR, VBR, etc.). The formal information model templates for expressing these parameters in ATM termination points are described in Chapter 7.

4.6.2   Connection Admission Control (CAC) Function

By definition, a *connection admission control function* (CAC) can only exist in a connection-oriented network. Connectionless networks such as the Internet and SMDS networks have no CAC role. However, in any connection-oriented network, CAC is a major part of the strategy for controlling congestion, and in the case of circuit switching and CBR services, it is the only mechanism.

In principle, in a connection-oriented network, there is a CAC associated with every resource manager and the role of the CAC is to decide whether there is sufficient free resource on the link to allow a new connection. If there is, the CAC may allow a connection request to proceed; if not, the CAC will indicate this to allow either the connection controller to choose a different route or, if none is available, notify the originator of the connection request that the request has been refused.

From the wider viewpoint of connection control protocols, the exact nature of the CAC is not important. It is only necessary to know whether a new connection with a given set of parameters can be admitted or not. The CAC becomes a fully encapsulated process that gives a simple response to a request to admit a new connection.

The strategy employed by a CAC will vary according to the service type and in the cases of VBR, ABR, and UBR services, becomes a quite complex issue and still the subject of intensive research.

The policy for each connection type depends on the stochastic nature of the service. The following gives a brief indication of policies for each connection type, but a full account of the statistical choices is given in Chapter 9.

4.6.2.1 Circuit-Switched Connection

In this case, the request for a connection, if granted, results in a link connection being allocated to the connection. In principle, the resource manager can be completely nondiscriminatory in its allocation of free capacity to connection requests as they will all have the same effect on the state of the resource. There may, however, be some circumstances where a CAC may refuse a new connection even when there is sufficient capacity for the connection. For example, circuits could be designated as low priority or high priority, and the CAC may choose to accept a high-priority connection while refusing a lower priority. In refusing the lower priority connection, it is effectively reserving the remaining capacity on the resource for high-priority connections. Prioritization becomes important in service restoration after a failure where the resources become scarce and some connections must be dropped. Higher priority services can reasonably attract a premium tariff.

4.6.2.2 CBR Connections

While the principles of CAC are basically similar to those for circuit-switched connections, the CAC is more likely to be discriminatory. As the connections can have different bandwidths, they will have different impacts on the remaining capacity within the link. For example, a CAC for CBR connections could employ a strategy of only accepting connections that take less than half of the remaining capacity. With such a strategy, some connections will be refused even if there is capacity available to resource the connection. This makes good commercial sense if several smaller connections are worth more than one large connection of the same capacity.

4.6.2.3   VBR Connections

With VBR connections, the CAC is not only making a judgment about the nature of future connection requests, but is also having to make a judgment on the future behavior of the connections it has already accepted onto the link based on the VBR service parameters. This is based on a statistical assessment as to whether accepting the new connection will take the likelihood of cell loss for the current traffic beyond acceptable limits.

4.6.2.4   ABR Connections

The role of CAC in ABR services depends on the ABR strategy employed. The use of the resource management protocol allows for the use of CAC; however, it is quite possible to use a policy of accepting all connections and simply regulating the rate of all the connections on the resource to ensure the resource does not become overloaded.

4.6.2.5   UBR Connections

The role of CAC in basic UBR connections is similar to that in VBR connections in that it must decide according to the requested peak cell rate on whether to admit the connection. If the parameters of minimum cell rate and maximum burst size are added to the UBR specification, then the role of CAC in this case is essentially the same as for VBR connections.

4.6.3   Usage Parameter Control (UPC) Function and Network Parameter Control (NPC) Function

The *usage parameter control function* (UPC) and the *network parameter control function* (NPC), often referred to as policing functions, are used to check that a CBR, VBR, ABR, or UBR connection is sending traffic according to the parameters agreed to at connection setup. If a connection sends cell streams with characteristics not complying with the agreed upon parameters, particularly if the connection is sending more cells in a given time than has been agreed to, then the usage parameter controller can discard cells immediately or mark them as lower priority (CLP = 1), to be discarded subsequently if congestion is encountered. If a queue in a switch starts overflowing, marked cells will be dropped first. UPC occurs at the UNI and NPC is used at the NNI, typically at interoperator administrative boundaries.

4.6.4   Traffic-Shaping Function

A traffic-shaping function can be used on a connection, close to the traffic source, to ensure that the characteristics of the traffic will conform to the parameters agreed upon at connection setup. In an ATM network, the traffic shaper can ensure that cells will not get marked as low priority by UPC/NPC functions further into the network. A traffic shaper will include a buffer, and it will seek to absorb bursts that have a higher bit rate than that allowed according to the agreed upon parameters for the connection and feed out the buffered data such that bursts fit within the agreed upon parameters. If the amount of data being sent is at a higher rate and for longer than the traffic shaper's buffer can absorb, then the traffic shaper must decide whether to launch the traffic into the network anyway and risk that some data may be discarded within the network, or it must discard data itself.

There is a difficult design judgment to make concerning the division of responsibility for traffic shaping and policing between the network and the subscriber terminal. Because this is generally a commercial interface also, it is likely that functionality will be duplicated in the two domains.

4.7   SUBNETWORK AND CONNECTION CONTROL FUNCTIONS

Signaling standards often refer to call control; however, this is a loose use of the terminology, and normally it is connection control that is meant. A "call" should be regarded as a specific example of the more general concept of a session and may involve one or more connections that may also be added to, subtracted from, or otherwise modified in the course of a single session. We can refer to a "three-way call," which has three connections and a voice bridge. Subnetwork control and connection control are only concerned with the setting up and taking down of connections.

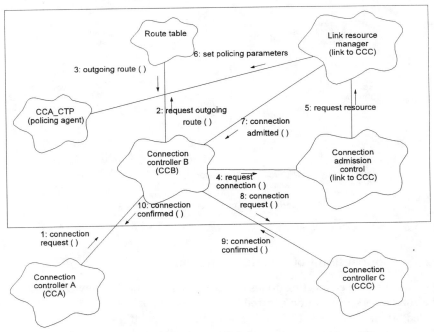

Figure 4.24   Functions inside the subnetwork controller

4.7.1   Subnetwork Control Function

The *subnetwork control function,* or subnetwork controller, has the role of executing the received connection request messages. It encapsulates a number of the functions already described, such as route tables, resource managers, connection controllers, and so forth. One possible encapsulation is illustrated in Figure 4.24. A subnetwork controller may be centralized, distributed, or copied across several locations depending on the implementation. The relationship between a subnetwork controller and the subnetwork that it is controlling will follow the partitioning decomposition, as illustrated in Figure 4.25; however, ultimately, there will be a proprietary interface between the subnetwork controller of a node and the matrix function. A connection request is given the semantics of *from* a CP *to* any member of a destination transit or access group as illustrated in Figure 4.26.

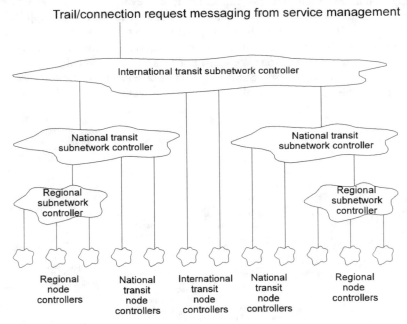

Figure 4.25   Subnetwork control and partitioning.

4.7.2   Connection Point Status Function

The status of all the connection points on the boundary of the subnetwork must be visible to the subnetwork connection controller. The *connection point status function* must be provided by the CTP (or a TTP) agent. A connection point's status hold the state of the connection point as it proceeds through connection setup and release. The connection point will answer with its status when requested, according to its current state in the connection setup and release process. Using the terminology of signaling, the status of the connection point could be

- Idle (i.e., no connection);

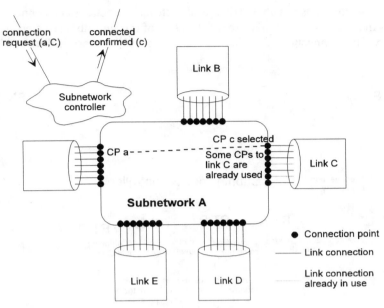

Figure 4.26   Semantics of the connection request message -
"from a CP (in small letters) to a transit group (in capital letters)"

- Seized (a connection setup is being attempted);
- Alerting (awaiting completion of the connection setup);
- Busy (a connection is using the CP and data transfer in progress);
- Clearing forward (a cessation of the connection requested by the originator of the connection);
- Clearing back (a cessation of the connection requested by the called party)
- Blocked (the connection point has been made unavailable for use);
- Blocking request (where the connection point is currently busy but when the connection is cleared, the connection point should enter the blocked state and not return to idle).

Although there is some variety in real systems, the state models of connection control generally have a broadly similar set of states to these.

4.7.3   Connection Controller Function

The connection controller is the function within the subnetwork controller that manages and supervises connection setups, releases, or modifications. It must, in response to connection requests, coordinate within the subnetwork controller the interrogation of the route table, requests to CAC, and updating of the status of connection points. Its exact operation, however, will depend on the exact type of the connection and the routing algorithm. We recognize three distinct categories of routing algorithm: hierarchical routing, source routing, and step-by-step routing. While these are all different with quite different properties, it is possible to define all route table functions and connection control functions in such a way that they are independent of the algorithm selected. The generic operation of the connection controller for routing in a subnetwork is illustrated in Figure 4.27.

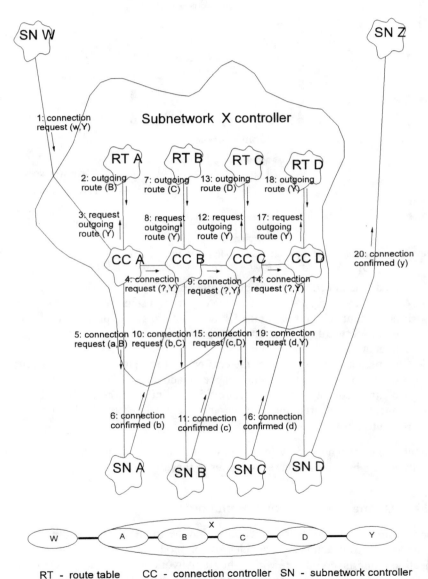

RT - route table   CC - connection controller SN - subnetwork controller

Figure 4.27   Semantics of connection requests in context of partitioning.

4.7.3.1   Hierarchical Routing

Hierarchical routing uses the decomposition of a layer network into a hierarchy of subnetworks. Each subnetwork has its own routing process that holds information about the topology within its subnetwork but has no knowledge of the topology beyond its subnetwork. The subnetwork routing process can determine an optimum route within the subnetwork to the specified exiting transit group (or access group) of the subnetwork. Hierarchical routing is illustrated in Figure 4.28 and the way functions are encapsulated into the local subnetwork controllers is illustrated in Figure 4-29.

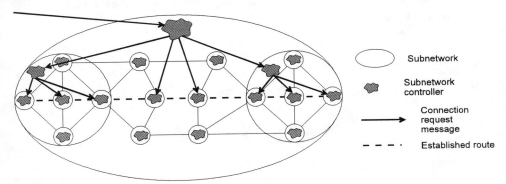

Figure 4.28   Connection setup using hierarhierarchical routing

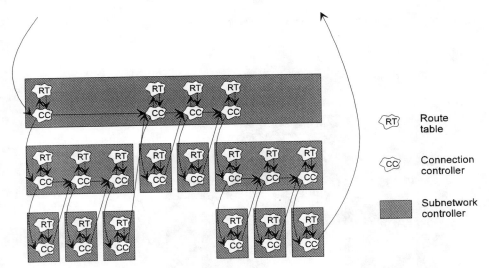

Figure 4-29   Messages present in hierarchical routing

4.7.3.2   Source Routing

Source routing is similar to hierarchical routing except that the overall routing process is divided into a number of distributed processes. There is a process for each entering transit group (or access group) of the subnetwork, and this can work out a route from that entering transit group (or access group) to any exiting transit group (or access group), but each process only knows a portion of the topology of the subnetwork. For any one connection setup, source routing and hierarchical routing are logically identical; however, source routing will normally be both faster and more reliable. Source routing is illustrated in Figure 4.30 and the way functions are encapsulated into the local subnetwork controllers is illustrated in Figure 4.31.

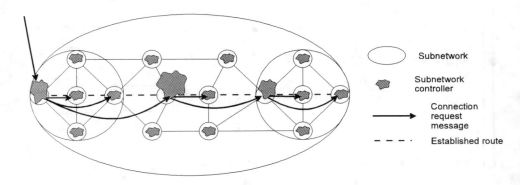

Figure 4.30   Connection setup using source routing

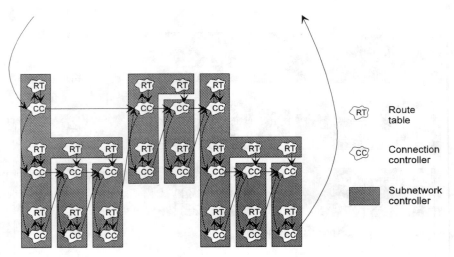

Figure 4.31   Messages present with source routing

4.7.3.3   *Step-by-Step Routing*

Step-by-step routing is a further decomposition of source routing into smaller subprocesses, but the decomposition places restrictions on the way in which the routing across the subnetwork can be determined, and hierarchical routing and source routing can support more flexible routing algorithms. Step-by-step routing has a subprocess for every entering transit group (or access group) and one for every entering transit group (or access group) of the next subnetwork decomposition. The routing through the subnetwork is calculated dynamically by a route table in each subprocess that translates the name of the exiting TP pool into the name of the next decomposed subnetwork, as illustrated in Figure 4.32. The way functions are encapsulated into the local subnetwork controllers is illustrated in Figure 4.33. Step-by-step routing is very fast and reliable, and is used on all current signaling systems, but is best suited to highly meshed subnetworks.

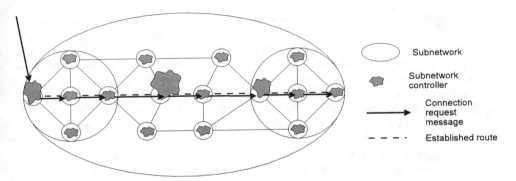

Figure 4.32   Connection setup using step by step routing

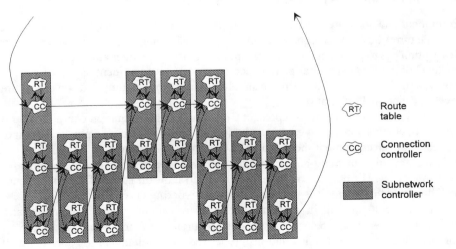

Figure 4.33   Messages present with step by step routing

4.8   TRAIL PERFORMANCE MANAGEMENT FUNCTIONS

Every ATM or SDH trail has an embedded performance monitoring process from which a number of standard parameters are calculated within the network element. These are defined in ITU-T Recommendation G.826 and are described in more detail in Chapter 10. The management system needs these parameters for several reasons, including the following:

- Verification of contracted performance of trail and connections with clients;
- Verification of the performance of manufacturer's equipment over its lifetime;
- Identification showing degrading performance to prompt remedial maintenance action;
- Provision of black spot analysis information for network quality improvement programs.

The requirement on the control and management system is to make the performance information available in a form suitable for these applications. A *performance monitoring function*, which is associated with an access point or possibly a connection point, records and reports performance information of the trail (or connection if performance monitoring is possible). The quantity of performance information collected by a network operator is potentially very large as, for example, there are, according to ITU-T Recommendation G.783, potentially six parameters to be stored for every 15 minutes for every SDH trail. The performance monitoring function has to log and record this information in such a way that it can be readily accessed by any control and management application that requires it. Different applications may require different subsets of the information and that it be presented in different formats, and the discriminator function described below can carry out this role.

4.9   IMPLEMENTATION AGENT FUNCTIONS

Equipment and software modules are built to implement functions, but the way different manufacturers design their equipment may vary considerably. To be general, the functional description has to work equally well irrespective of the implementation. Over recent years, the level of functional integration in telecommunications equipment has increased significantly, and the functional description should not make any assumptions on the nature and level of functional integration.

Implementation functions represent a particular implementation component to wider aspects of control and management, and this could be representing a physical component, a circuit card, an equipment shelf, or a piece of software. This suggests a basic classification of implementation mediation functions between *equipment agent functions* and *software agent functions*. The purpose of these functions is to show the status of the implementation, including whether or not it is in commission and working correctly.

Equipment can contain software, but software cannot contain equipment, and this relationship is encapsulated by a "contains" relationship function in the equipment and software agent functions. The relationship between an implementation and the functions that would become unavailable if the implementation were to fail is encapsulated in the implementation as an "affects" relationship function. This is illustrated in Figure 4.34. The properties of the equipment agent function should include the following:

- Answer parent equipment name (contains relationship);
- Answer equipment name (own identity);
- Set and answer administrative state (commissioned/not commissioned);
- Report operational state (working/not working);
- Answer list of implementation names (contains relationship);
- Answer list of function names (affects relationship).

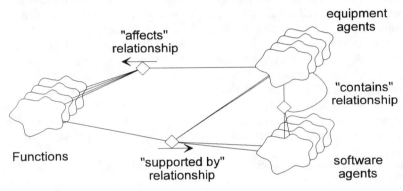

Figure 4.34   Implementation agent functions.

4.9.1   The Network Element Agent Function

A network element is an equipment which is

- Made by one manufacturer;
- Located in one place;
- Has one address (NSAP) for centralized management messages.

M.3010 places importance on the scope of management imposed by a network element and distinguishes between aspects of management that are wholly local to a network element and those where the scope requires visibility between network elements. For example, crossconnection is wholly within the scope of network element while connection across a subnetwork requires visibility across all the network elements in the subnetwork. The *network element agent function* is a specific type of equipment function that reflects these special properties of a network element.

4.10   EVENT MANAGEMENT FUNCTIONS

Many of the control and management functions, particularly the implementation agent function and termination point agent functions, can show an alarm state in response to detected defects, and this triggers an alarm notification message. We describe this as an alarm event. The notification also contains information regarding the severity of the event and the probable cause. These are used by the management system to activate a local alerting system and as information for diagnostic analysis.

The aim of event management is to direct and control a maintenance workforce to locate and repair failures, and to do this efficiently, it needs to accurately record and correlate adverse happenings in a network. For example, if a component of a piece of equipment fails, the equipment agent function should show an alarm state. In addition, all the trails that pass through that component will be affected and all their trail termination agent functions will also show an alarm state. However, these trail terminations may be distant from the equipment. The event management system reports all these individual events and may then attempt to correlate them to identify the common source of the event, and, most importantly, reduce the amount of information presented on a screen in the event management center. This correlation can be done by either a knowledge of the interrelationships in the network like a database recording that trails are supported by which equipment or by correlating the time when the events occur.

To support this second means of correlation, the time when events occur must be accurately recorded. The timestamp has to be in absolute time as related events may be distant, even in different network operator's domains. The accuracy required for the absolute timestamp has been discussed at length in ITU-T and there is still no firm agreement, with suggestions from within 1 second of absolute time to within 30 seconds of absolute time.

Under normal circumstances, the failed component should report that it has failed, and so the fault correlation system is simply used to reduce the amount of reported information. The maintenance workforce can be directed to the faulty component on the basis of its own locally generated alarm. However, there will be a small number of occasions when the local alarm detection mechanism will itself fail, and then the correlated events can be used to infer which component is faulty.

4.10.1   Record, Event Log, Discriminator, and Correlator Functions

The details of the event are stored as a *record function* that holds the information on an individual event such as the name of the function that produced the event, the nature of the event, and the time the event occurred. A *log function* is used to group together a number of record functions into an ordered list that may be examined, searched, and archived as necessary.

A *discriminator* is a function that can control which managing processes are notified about which events. Since there may be more than one management process that is interested in a particular class of events, the discriminator can control which managers are notified and the circumstances under which they are notified. For example, one management process may be interested in all events from a function while another may only be interested in events from alarms that directly affect service. A discriminator discriminating between two managing processes on behalf of a group of notification issuing functions is illustrated in Figure 4.35.

These functions are not sufficient by themselves to describe the event management, and one or more extra function types are required to describe the correlation process itself, which is normally proprietary in its implementation. We can, however, regard correlation

Objects issuing broadcast messages   **Objects receiving broadcast messages**

Figure 4.35   Use of a discriminator to control selection and onward transfer of messages.

as a single encapsulated function called a *correlation function*, which can correlate common attributes across a large number of event records

References

There is, as yet, no coherent set of standards covering this area; however, the work of a number of organizations is slowly coming together. The following are general references on software architecture:

ITU-T Draft Recommendation X.901, *Open Distributed Processing Reference Model—Part 1: Overview*.

ITU-T Recommendation X.902, *Open Distributed Processing Reference Model—Part 2: Foundations*.

ITU-T Recommendation X.903, *Open Distributed Processing Reference Model—Part 3: Architecture*.

ITU-T Draft Recommendation X.904, *Open Distributed Processing Reference Model — Part 4: Architectural Semantics*.

Object Management Group, *Object Management Architecture Guide*, Third Edition, John Wiley & Sons, 1995.

Object Management Group, "*Common Object Request Broker: Architecture and Specification, Revision 2.0 (CORBA 2.0)*, 1995.

Distributed Component Object Model (DCOM) Technical Overview, Microsoft White Paper, 1996.

JavaBeans 1.0 API Specification, JavaSoft, Sun Microsystem Publication, 1996.

The following are references specific to routing algorithms:

Dijkstra, E. W., "A Note on Two Problems in Connexions with Graphs," *Numerische Math*, Vol. 1, 1959, pp 269—271,

Thulasiraman, K. and Swamy, M. N. S., *Graphs: Theory and Algorithms*, Wiley-Interscience, 1992.

ISO/IEC Standard 10747, *Information Processing Systems—Telecommunications and Information Exchange between Systems—Protocol for Exchange of Inter-domain Routeing Information among Intermediate Systems to Support Forwarding of ISO 8473 PDUs*.

The following are references specific to addressing:

ITU-T Recommendation E.164, *Numbering for the ISDN Era.*

ITU-T Recommendation X.213, *Information Technology—Network Service Definition for Open Systems Interconnection, Annex A, Network Layer Addressing.*

ISO Standard 3166, *Codes for the Representation of Names of Countries and Their Subdivisions—Part 1: Country Codes.*

ISO/IEC Standard 6523, *Structures for the Identification of Organizations.*

Fuller, V., Li, T., Yu, J., and Varadhan, K., *Classless Inter-Domain Routing (CIDR): an Address Assignment and Aggregation Strategy*, RFC 1519, September 1993.

ITU-T Recommendation G.831, *Management Capabilities of Transport Networks Based on the Synchronous Digital Hierarchy (SDH).*

Transport Interfaces 5

The most fundamental parameters of digital telecommunications, established internationally at a very early stage for telephony, are the sampling rate of 8 kHz and the allocation of 8 bits per pulse code modulation (PCM)-coded sample. This results in a channel sampling period for digital transmission of 125 μsec which, at 8 bits per channel, is equivalent to a channel rate of 64 kbit/s. Other important characteristics of the basic rate channel, such as coding law and channel-associated signaling, were standardized differently in Europe and North America, but the basic 64 kbit/s channel became the basic unit on which the digital network was built. Other digitized services (other than voice) have had to be adapted to this regime, resulting in channel rates and structures that were sometimes less than ideal.

The information technology world developing in parallel generated its own schemes for organizing data. The 8-bit byte emerged quite early as a standard unit of data, owing much to the predominantly 8-bit data buses of the early computer architectures and the early definition of the ASCII character set. Beyond this basic unit, data is organized in files with a wide range of sizes and formats. Early serial data interfaces were low-speed and included overheads to identify octet boundaries. The digital multiplexing standards developed to meet the evolving transport needs in this environment. The plesiochronous digital hierarchy (PDH) was largely established during the 1960s and early 1970s, with only minor refinements thereafter. Before considering synchronous digital hierarchy (SDH) and asynchronous transfer mode (ATM) structures in detail, we review the basic digital transmission and multiplexing problem and the distinguishing characteristics of the plesiochronous, synchronous, and asynchronous solutions.

5.1   DIGITAL TRANSMISSION AND MULTIPLEXING

A digital signal is defined as a signal that takes one of a limited set of discrete values represented by symbols at discrete instants in time. Thus its characteristic information is defined by the symbol values allowed and the instants at which they are valid. Secondary features include the relative frequency of the symbols and correlation between them, which largely determine spectral energy distribution, an important limiting factor for transmission on physical media. Thus a binary information stream has two symbol values, but it is incomplete without its associated clock stream identifying the discrete instants in time at which the signal is valid.

If the information rate is low and the distances short, transmission of clock and data presents no problem. To transport high-rate digital information over long distances, the clock information must be combined with the data stream itself, transmitted together and separated at the receiver for the purpose of regeneration. A large part of transmission engineering has been devoted to the design of transmission media adaptation schemes (channel coding, modulation, etc.) that combine clock and data in such a way that they can share the same noise- and bandwidth-limited transmission media channel. The receiver must recover the clock before the data can be regenerated and reinterpreted.

Digital multiplexing is based primarily on the time division principle. Time is divided into small intervals. In each interval, a basic unit of information (initially the byte or octet) pertaining to one channel is transmitted. Successive time intervals carry information from successive channels. The shorter the time interval, the greater the number of channels that can be accommodated.

Multiplexing separate channels in this way implies a structuring of the bitstream. Additional information defining the structural details must be conveyed to the receiver to identify which bits belong to which channels. Much of what follows concerns the strategies for structuring data and the means for communicating the structural reference information along with the multiplexed data. The plesiochronous, synchronous, and asynchronous multiplexing paradigms, as their names suggest, differ mainly in the way that they handle these three components of the characteristic information, data, timing, and frame of reference within a multiplexed group.

- *Synchronous multiplexing—*
  Requires that the data streams to be multiplexed all have rates that are derived as integer submultiples of the aggregate rate and that they all derive their structure from that of the aggregate. Framing and timing information is transmitted once in the aggregate and difference information only is transmitted with each component. Bit synchronization information, which must be the same for all the multiplexed components, is delivered uniformly to all sources. Although many specialized variants exist, the same 125-sec frame and 64 kbit/s channel originally derived for digital voice still dominates synchronous multiplex structures today.
- *Plesiochronous multiplexing—*
  Requires that the data streams to be multiplexed are constrained to have nominally the same rate within specified limits, but no limitation is placed on their structure. The data streams are "synchronized" and interleaved into a new aggregate struc-

ture. The synchronization information coded into each stream is recovered at the receiver to restore the original rate.

- *Asynchronous multiplexing—*
  Places no explicit constraint on the rate or structure of the data streams to be multiplexed. Instead, the aggregate structure consists of a data stream structured in contiguous blocks of data. Each block is preceded by a header containing a block boundary delineation overhead and a label that uniquely identifies the block as belonging to a specific channel in the aggregate. Information from the input streams is queued and allocated to cells in the aggregate stream as required. When no information is presented, specially labeled zero blocks are inserted with no information content to maintain the rate of the serving channel.

5.1.1 The Primary Rates

One of the earliest applications of digital transmission and multiplexing using PDH was in so-called pair gain systems in the interoffice network. A number of basic rate, PCM-coded channels were multiplexed into what came to be called the primary rate digital signal This is called DS1 in North America and by analogy the European equivalent is often called E1. Different regional priorities led to a choice of 24 channels in 1544 kbit/s for North America and 30 channels in 2048 kbit/s for Europe, with the rest of the world lining up behind one or other of these. The manner by which frame synchronization and transmission path overhead is provided is very different in each case. The two primary rates have become the main administrative units at which transport capacity is managed in modern networks. This administrative focus has in turn made these attractive capacities for private leasing of clear channels, where they now dominate in the provision of digital private networks.

There was further rationalization in the context of integrated services digital network (ISDN) to yield the primary channel groups 30B+D and 23B+D, respectively, according to the combination of basic (B) 64 kbit/s channels and 64 kbit/s signaling (D) channels provided. Both these primary rate formats are generated by the synchronous interleaving of 8-bit basic channel octets and the subsequent addition of frame alignment and operational overheads. This is possible because the basic channel sampling rate and the primary aggregate rate are both derived from the same local clock source. The octet structure of the original channels is therefore preserved in the aggregate signal.

To drive down the cost of transmission, primary channel groups were multiplexed together to form higher capacity signals for transmission as a single entity. It was not possible to use the simple process of byte interleaving because, at the very least, it would have required the universal synchronization of all primary sources. The reference distribution network this would have implied was outside the scope of the early transmission deployments that were, as always, justified on the basis of minimum introductory cost. Network synchronization traceable to high-stability atomic standards has since become a reality justified by the needs of digital switching. Nevertheless, a central problem for the digital transmission pioneers was to maintain integrity between data and clock phase information despite the hazards of the transmission and multiplexing to enable exploitation of new high-speed technologies.

5.1.2   Plesiochronous Multiplexing

The technique developed for multiplexing in the PDH relies on increasing the rate of the tributary signal to be multiplexed to a rate that is synchronously derived from the new aggregate rate. Simple bit interleaving of these synchronized channels with the addition of frame alignment information and operation, administration, and maintenance (OAM) overhead was then all that was required to form the multiplexed aggregate signal.

The process whereby the tributary rate is adapted to the synchronous channel rate is termed justification or, more graphically, "stuffing," as it depends on stuffing extra bits into the channel at defined justification or stuffing opportunities in the frame. This is achieved under the control of an adaptive process that is driven by a comparison of the aggregate and tributary derived signal phases. The justification and bit interleaving process is illustrated in Figure 5.1 for the case of four-into-one multiplexing, but is otherwise representative of all the levels in each of the PDHs in common use throughout the world (i.e., in the United States, Europe, and Japan).

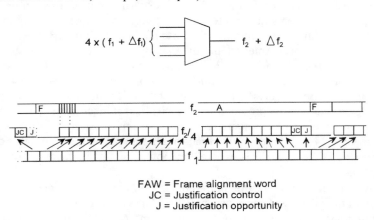

FAW = Frame alignment word
JC = Justification control
J = Justification opportunity

Figure 5.1   The justification process.

5.1.2.1   Justification

The aggregate rate in a PDH multiplexer must be a little greater than the sum of the tributary rates even when they are at their maximum permitted value to allow for the multiplexing overheads. The aggregate format is built up by choosing a frame rate and allocating some capacity for frame alignment and OAM purposes. The remaining capacity is then divided equally among the tributaries (four in the example of Figure 5.1). Each of these justified channels, which have been derived from the same clock as the aggregate signal, are further structured to form a 1 bit per frame or multiframe justification opportunity J, and a justification control channel (JC) that signals to the far end whether or not the particular frame or multiframe has been justified or not. A justified frame or multiframe carries no information in the J-bit, while an unjustified frame or multiframe carries genuine tributary data. The lower the tributary rate, the more frames on average are justified, and vice versa.

The decision to justify or not is made in each frame according to the relative phases of the tributary and the justified channel rate. The result is to maintain the average phase of the tributary data in the justified channel equivalent to the phase of the tributary signal itself so that no data is lost through overflow or underflow of data buffers. The justification process samples the tributary phase at the justification frame rate, quantizes and encodes it using 1-bit differential PCM, and multiplexes this information in the justification control channel along with the data itself so that it can be recovered at the receiver. Thus, the linear phase ramp generated by the frequency offset of the tributary signal together with any other incidental phase variations can be accurately reproduced from the composite signal.

The JC channel is generally designed to be resilient to errors because of the severity of the impairment that would otherwise be caused as a result of justification errors. Much of the science of designing such justification frames lies in achieving a satisfactory balance between efficiency, framing performance, error tolerance, and synchronizing or mapping jitter.

At the demultiplexer, the justification information is recovered from the JC channel and the recovered clock gapped accordingly. The discontinuous clock thus recovered is smoothed in a timing recovery filter, often based on phase locked loop techniques to produce a continuous signal equivalent to the original tributary input. The main impairment in such recovered signals is the timing phase error resulting from the residue of the discontinuities that will have been imperfectly suppressed, which is termed multiplexing jitter. The nature and control of timing impairments and their impact on synchronization strategy are discussed in Chapter 11.

Using this method, new multiplexed layers were generated independently in the three main operating regions (North America, Europe, and Japan), resulting in three completely different PDH hierarchies due to the relative commercial isolation of these three areas from one another. A fourth PDH has since been generated as a hybrid of the European and North American PDHs, when it eventually became necessary to interwork digitally across the Atlantic. All four hierarchies are listed in Table 5.1.

Table 5.1

International Plesiochronous Digital Hierarchies.

Level	North America	Europe	Japan	Transatlantic
0	64	64	64	64
1	1,544	2,048	1,544	2,048
2	6,312	8,448	6,312	6,312
3	44,736	34,368	32,064	44,736
4	139,264	139,264	97,728	139,264

5.1.3   Network Node Interfaces for PDH (NNI)

The G.703 series of section layer interfaces for electrical media (copper pair and coaxial cable) were standardized for each of the PDH rates in G.702 without making any restriction on the path layer structure. This standard was agreed upon at an early stage and is now used exclusively as the basis of network node interfaces (NNIs) between systems within a telecommunication complex or network node. The digital distribution frames and patch panels that use them have become the main elements of flexibility in the digital transmission network. The transmission systems between network nodes, however, whether they be twisted-pair or coaxial line, point-to-point radio or optical fiber, had resisted standardization until more recently, and provided more fertile ground for the ingenuity of the transmission engineer.

5.1.3.1   *Intraoffice NNI*

Transmission physical interfaces must deal with the problems of signal balance on the transmission medium and maintaining adequate transition density to recover timing information independent of transported information. In addition, it is desirable to add redundancy for error-monitoring purposes. These requirements have led to the development of a family of ternary (i.e., three-level, as opposed to binary or two-level) line codes, which were simple to implement and satisfied all of the above criteria. The simplest of these was alternate mark inversion (AMI), in which the marks of the binary signal are alternately transmitted as positive pulses and negative pulses thus both reducing low frequency component and increasing transition density.

This succeeded on two counts, but could not achieve adequate transition density during long sequences of zero. This limitation was overcome in the zeros substitution derivatives in which a mark was substituted for a zero after a zero sequence of specified length, the substitution being signaled by a violation of the normal alternation rule. The most common examples of such line codes are HDB3, B3ZS, and B8ZS.

One or other of these has been recommended for use at each of the NNI rates specified in G.702 except at 139264 kbit/s (DS4 in North America), where coded mark inversion (CMI) is used. This is a binary code operating at twice the channel rate. As well as achieving signal balance and adequate transition density, each of these codes is sufficiently redundant that transmission errors result in detectable code violations at the receiver, thus providing an inherent performance monitoring capability.

5.1.3.2   *Interoffice NNI*

The basic transmission requirements are the same as those described above for the intraoffice interface, but the transmission distances are greater and the impairments and environmental effects more significant. The OAM requirements derived from the remote operation of outside plant, protection switching, and coordination between the two ends of a link bring extra requirements that have never been standardized on any large scale.

For these reasons, transmission systems in the PDH were implemented in a more or less proprietary way, interfacing with the other components of the network by way of one of the NNI physical interfaces defined in G.703. Line codes, OAM overheads, and many other features could be and were optimized in the quest for higher performance and competitive advantage. The resultant lack of transverse compatibility between line systems, even where they carried nominally the same G.702 rate, meant that it was not possible to interface between equipment from different sources at the line level. This lack of standardization has only become a problem in recent years as high levels of functional integration absorb the line terminating function within the terminal equipment and much of the G.703 station interconnect structure within the flexible connectability of the nodal equipment.

5.1.3.3   Alarm Indication Signal (AIS)

When a transmission link fails in such a way that the signal is interrupted, the failure is detected at the end of the section by a detector monitoring the received signal. The G.702 path layer information stream, however, may be connected across several network nodes. Depending on the nodal function, alarms may therefore be raised at any of the downstream nodes. The alarm indication signal (AIS) was introduced in Recommendation M.20 to signal to all the downstream nodes that a server layer failure has been detected. This can be interpreted downstream as a loss of service on the particular G.702 data stream resulting from an upstream failure that has already been detected and reported. This enabled a minimum level of automatic fault management and hence cooperation between different maintenance groups in a network without the need for consultation or liaison. Maintenance staff near the actual failure can concentrate on diagnostics and repair, while other affected parties detecting the AIS signal remotely can consider service impact, such as the necessity for service restoration or not, without needing to be concerned about the location or nature of the actual failure.

AIS is signaled downstream from the failure by sending an all "1s" stream in place of the corrupted or interrupted data stream. Long sequences of "1s" are not excluded from the valid possibilities in a G.702 data stream; therefore, it was necessary to specify clearly the criteria for declaring AIS in terms of the persistence of the condition. Note that the associated loss of justification information means that the rate of the regenerated AIS signal is traceable to a local reference at the point of generation. Downstream timing recovery filters therefore experience a step change in rate at the beginning and end of a period of AIS. The basic AIS capability has been retained at all the PDH levels in all hierarchies.

5.1.3.4   The PDH Primary Path Layers

With the advent of ISDN, the PDH path layer began to take on a new significance. The specific frame structures for ISDN primary channel groups are specified in G.704. These were derived from the frames already used for PCM voice. The primary rate frame is aligned at the receiver by searching for a frame repetitive pattern inserted at the transmitter for this purpose. Framing patterns are sought that optimize frame recovery time and robustness in the presence of errors.

In Europe, the 30-channel group is carried in a 32-octet stream with 1 octet (in timeslot #16) reserved for signaling and 1 octet (in timeslot #0) used for frame alignment. In the United States, a 23-channel group is carried in a 24-octet plus 1-bit frame (i.e., 193 bits) with 1 octet reserved for signaling and the 193rd bit (termed the F bit) used for framing and network overheads. The inclusion of signaling channels in the composite structure is characteristic of the circuit layer. The signaling message set used between network nodes is outside our present scope. These formats are referred to as 30B+D and 23B+D respectively.

In both the United States and European cases, the original frame alignment structures have been enhanced to improve resilience to repetitive or imitative patterns in the user data such as may be expected from data traffic, for example, or even malicious users. This is achieved by adding a cyclic redundancy check (CRC) mechanism, which will detect frames misaligned to a false pattern and initiate a new search.

The remaining overheads reserved for OAM purposes were left open for operators to define for themselves. Belated attempts to recover this situation have met with only limited success. A 4 kbit/s data channel within the primary path overhead, for example, is now standardized for some applications.

5.1.4   User-to-Network Interface (UNI)

The UNI in an ISDN has all the same basic transport requirements as the NNI. The basic need to transfer bit and octet timing along with the information stream, to provide robust frame alignment, and network status and error-detection mechanisms for OAM purposes must be met. At the primary rate, there is no difference to the NNI in this respect.

The basic rate access format was also introduced to provide cost-efficient digital access to existing copper line subscribers. This is the same in the United States and Europe, and it uses a 2B+D channel structure. The B channels are 64 kbit/s octets and the D channel is a 16 kbit/s frame-oriented structure. The section layer coding uses a four-level transmission code that achieves all the spectrum limitation, timing transparency, and error-detection capability required. In addition, the UNI transmission media layer provides for activation/deactivation and line testing mechanisms.

The primary rate signals are assumed to be active all the time, and therefore the generic activation/deactivation procedures do not apply. The UNI presents particular maintenance issues derived from the fact that the terminal equipment and network terminations are generally on a customer's premises and therefore in a different administrative domain. The most important generic feature introduced was the remote alarm indication (RAI). This alerts the network side to a section layer failure in the direction towards the user. In the United States, a 4 kbit/s channel provided by the so-called m bits in the F-bit multiframe are used. In Europe, bit 3 of alternate TS0s provides a 4 kbit/s channel for the same purpose.

The user-to-network interface provides the means by which the user signals to the network his intentions to initiate, terminate, or otherwise modify a call and the network signals to the user the status of a call in progress. ITU-T Recommendation Q.931 defines the messages that pass between the two state machines, the one in the user terminal and the other in the local exchange. The D-channels for the basic rate interface (BRI) and the primary rate interface (PRI) provide frame-mode bearer connection for this signaling exchange.

Despite the success of digitalization during the 1970s and 1980s, the PDH, and the ITU-T standards that gave it effect, are now found to be ill-suited for the next stage in network evolution. As explained in Chapter 1, the forces of transport automation and functional integration have set the stage for the demise of the PDH. Network automation requires a multiplex structure better suited to crossconnection and management while the functional integration of high-capacity optical interfaces demands transverse compatibility at all layers, including the physical interfaces. These have been the drivers for SDH. At the same time, new switched services do not fit easily in the rigid, fixed-rate channels derived from 64 kbit/s. ATM provides a universal paradigm for multiplexing and switching combinations of signals of different and irregular rates ideally suited to the mixed rate and multimedia requirements of the evolving service scenario.

5.2 SYNCHRONOUS MULTIPLEXING

In describing the digitalization process and the concurrent development of the PDH, we have already noted that the digital network, as we know it, is based on a synchronous frame of 125 sec derived from the sampling rate used for the pulse code modulation of voice signals. This has formed the basis of the ISDN and a broad range of services delivered on frame-synchronous 64 kbit/s channels. We have also noted that the PDH primary rate signals use byte-interleaved frame structures derived from this same basic 125-sec frame. The SDH extends this same principle by defining new administrative layers derived by direct byte interleaving of 125-sec frame-synchronous signals. Such formats enable efficient use of time-space-time (TST) switching and add/drop multiplexing at higher capacity administrative layers, and are well-suited to the transport of 64 kbit/s-based service signals and their $n \times 64$ kbit/s derivatives.

5.2.1 Principles of Synchronous Multiplexing

The principles of synchronous multiplexing are illustrated in Figure 5.2. A synchronous transport module (STM) is formed based on the well-established 125-sec frame size. This forms the basis of the section layer transmission format. It consists of client layer information to be transported, which is termed its payload, and section overhead (SOH),

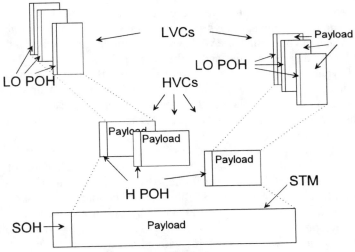

Figure 5.2   The principles of synchronous multiplexing.

which includes an alignment word to identify the frame start and additional information required for management of the section layer and adaptation to the path layers.

The section layer payload capacity is divided among a number of higher order virtual containers (HVCs), each consisting of overhead and payload. The HVC payload is further divided into lower order virtual containers (LVCs) that also consist of overhead and payload. They are termed "virtual" because they are logical entities that only exist within an STM, and "containers" because they contain client layer information in their payload. The descriptions "higher order" and "lower order" are relative and reflect the client-server relationship between the higher order and lower order path layer networks. An HVC layer network may act as a server to an LVC client layer network, but not vice versa. All virtual containers (VCs) are individually formatted according to the same recursive principles to form payload and path overhead (POH). Formats have been defined to serve all existing and planned layer networks in one or other of the VC payloads.

5.2.1.1   Diagrammatic Conventions for Illustrating Synchronous Frame Structures

The synchronous signal structure at the NNI consists of a serial bitstream that has been assembled by byte interleaving a complex combination of lower capacity streams. Although complex, the aggregate bitstream is strictly ordered so that it is possible to trace any particular component with reference to the STM frame by following the frame offset pointers that are embedded in the structure. Simple one-dimensional diagrams of the type often used to illustrate simple frame structures are inadequate to describe this complexity in any detail. The underlying structure is more clearly represented in two- and three-dimensional diagrams. Even three dimensions is not enough to represent all the component structures on a single diagram. Therefore, the dimensions of the two- and three-dimensional diagrams are selected to clearly demonstrate only a limited number of features simultaneously.

When representing the virtual structures, the gapping produced by the underlying server layers can be ignored and the virtual structure represented as though all bits were contiguous, which of course they are when viewed from the perspective of the layer being represented whose data is only defined at the gapped clock transitions. These numerical properties of synchronous frame structures provide powerful insights into the underlying structural principles. Two- and three-dimensional representations of structured serial data streams have become commonplace in describing the SDH. Some of the diagram conventions are illustrated in the examples of Figure 5.3 in which simple numbers have been chosen for the sake of clarity.

Figure 5.3(a) illustrates one frame of a synchronously structured bitstream generated by byte interleaving two channels of 5 bytes per frame, three channels of 10 bytes per frame, and one channel of 20 bytes per frame. Also visible are 3 bytes of overhead (OH) from the server layer. The phase offsets of the payload components from one another and from the server frame are an integer number of bytes but are not fixed. Figure 5.3(b) illustrates the gapping information for the server layer overhead. When considering the client layer, only those bytes that coincide with a logical one in the gapping information are of interest.

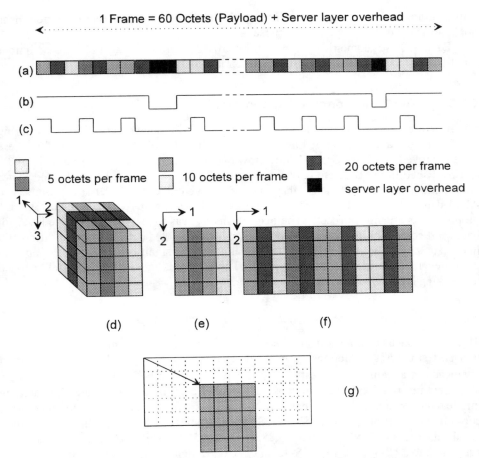

Figure 5.3   Diagrammatic conventions for illustrating synchronous frame structures.

Figure 5.3(d) shows a three-dimensional representation of this serial stream that is scanned, ignoring the gaps, in the order indicated (i.e., from front to back, from left to right, and then from top to bottom). This representation reveals more clearly that the frame consists of three groups comprising one channel of 10 bytes per frame and two channels of 5 bytes per frame in group #1, one channel of 20 bytes per frame in group #2, and two channels of 10 bytes per frame in group #3. Figure 5.3(c) represents gating information that selects the first of the three channel groups and this can be represented in the two-dimensional form of Figure 5.3(e). Finally, the whole frame may be represented in the two-dimensional format shown in Figure 5.3(f).

These representations illustrate the relationships between all the components in the aggregate structure but the frame phase of any individual component cannot easily be shown in this way. Suppose that the frame start of the 20-byte per frame channel in group #2 is at byte #29 in the 60-byte frame. This can be illustrated by superimposing a Figure 5.3(e) representation on top of a Figure 5.3(f) representation as illustrated in Figure 5.3(g).

This diagram has been drawn to a uniform scale, but, in general, such representations need not be to scale, depending on the purpose.

Each of these geometric transformations is used at different stages of the descriptions that follow according to which is most appropriate to the purpose at hand.

5.2.2   The Synchronous Transport Module

The structure of the STM and the VCs that it supports was developed in the ITU-T during 1987 and 1988 and is described in Recommendations G.707, G.708, and G.709. These were approved by Study Group XVIII (now SG13) at its plenary in Melbourne in 1988, and published in the Blue Book of that year. The SDH was derived from United States contributions describing SONET that was under development in T1X1 at that time as an ANSI standard. The evolution of the international SDH standard from the original SONET proposal was a painful process. The main difficulty was to reconcile adequate levels of backward compatibility with the prior PDH, which (as noted before) evolved differently in United States, Europe, and Japan. There was also some significant difference of opinion on the level of commonality that was justified between this and the emerging ATM-based networking layer.

5.2.2.1   *Structure of the Synchronous Transport Module*

The STM is the basic transport structure that is used to generate a range of higher capacity transport structures. It defines an ordered arrangement of payload data and transport overheads in a frame that repeats every 125 µsec and can be identified by some fixed alignment information in the overhead. The higher capacity structures are generated by interleaving, on a byte-by-byte basis, N of these basic STMs. Four such structures have currently been defined, namely STM-1, STM-4, STM-16, and STM-64, at corresponding serial data rates of 155520, 622080, 2488320, and 9953280 kbit/s. These are, respectively, equivalent to STS-3, STS-12, STS-48, and STS-192 in the SONET standard.

5.2.2.2   *Extension to Higher Rates*

Although not currently recommended by ITU-T, other rates corresponding to intermediate values of N can be generated in the same way. In principle, the structure may be expanded to still higher rates, but it is still too early to say whether the technology to achieve this will be forthcoming or when wavelength division technology will take over as the dominant technique to increase the capacity of fiber systems.

Most of the overhead is identical for each of the higher capacity structures irrespective of data rate, but certain features (notably, frame alignment and multiplex section error monitoring) are allocated proportionally more capacity in the higher capacity structures. This results in a relatively large unused capacity within the overhead structure of the higher capacity signals. Such apparent profligacy often offends the instincts of the transmission engineer, for whom transmission efficiency has long been a prime objective in an environment dominated by the high-frequency attenuation limitations of copper cable or spectrum scarcity in radio systems. However, it reflects the fact that in high-capacity optical networks, cost is dominated by terminal processing while the cost of channel capacity of the media itself is fast becoming negligible. Simplification of structure rather than optimization of bandwidth efficiency is a recurrent theme throughout the development of both SDH and ATM.

It is a common misunderstanding that STM-Ns are formed by multiplexing STM-1s. This is not the case. STM-1s, STM-4s, STM-16s, and STM-64s terminating on a network node are disassembled to recover the section overheads and the VCs that they contain. Outgoing STM-Ns are reconstructed with new overheads and new multiplexed path layer VC assemblies.

5.2.2.3   Classification of Overheads

Overheads defined in the SDH recommendations fall into four categories as follows:

- Payload-specific overheads introduced as part of an adaptation function and therefore characteristic of a particular client-server relationship. Tributary unit (TU) pointers, justification indicators, and multiframe indicator bytes are examples.
- Payload-independent overheads introduced as part of the trail termination function, and therefore characteristic of the layer itself and independent of any particular client-server relationship. Examples are bit-interleaved parity (BIP) error-monitoring bytes, and trail trace and signal label bytes.
- Auxiliary layer overheads that provide link connections in an auxiliary layer network. Examples are data communication channel (DCC) bytes and engineering order wire (EOW) bytes.
- Unallocated overheads that are reserved in the frame structure but have not yet been allocated for any specific purpose. Examples are growth bytes and reserved bytes. When allocated in the future, they are likely to fall in one or other of the categories already mentioned.

5.2.3   Signal Structure of the Transmission Media Layers

Figure 5.4 illustrates the STM and the manner in which the STM-N structures are generated from it. These signal formats are characteristic of the transmission media layer in an SDH network, and are functionally equivalent to the proprietary line system formats of the PDH. They are used to transport payload assemblies of path layer VCs between nodes of a path layer network and provide, in a standardized manner, the mechanisms such as error monitoring, protection switching, engineering order wire, and end-to-end data communications required to operate and maintain the transmission system. Figure 5.4(a) illustrates the overhead structure that has been defined for optical-fiber transmission. Other media such as point-to-point radio or satellite may require some of the currently unused part of the overhead for media-specific purposes such as forward error correction (FEC). The shading represents the section layer payload area. Each STM provides payload area for one administrative unit group (AUG). Administrative units and administrative unit groups are described below.

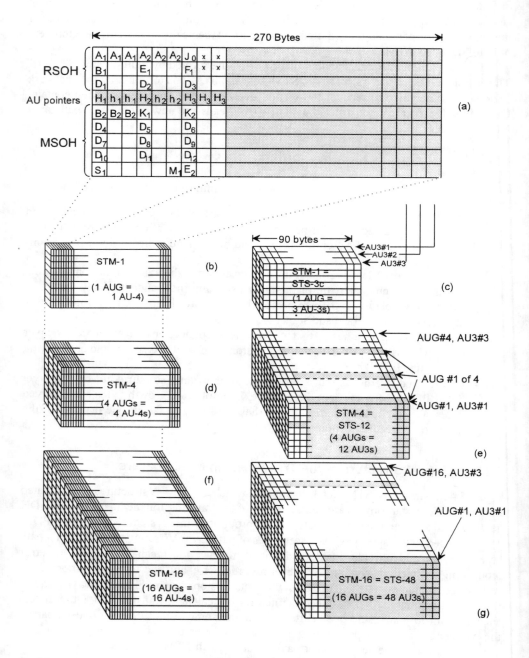

Figure 5.4   Structure of the STM and assembly of STM-N

Figure 5.4(b, d, and f) represent STM-1, STM-4, and STM-16 signal structures, respectively. The STM from which they are each generated is clearly visible in this representation. Figure 5.4(c, e, and g) represent exactly the same signals, but the representation in these cases has been transformed to emphasize the relationship to the basic 90-column STS module of SONET. In Figure 5.4, the first bit of the first A1 byte is the first bit of the frame and is located in the top left-hand corner of the STM nearest the front. The second bit is immediately behind this scanning into the page and so on until $8N$ bits have been scanned. The $(8N + 1)$th bit is at the front of the top byte of the second column in the first STM, where N represents the aggregate signal order 1, 4, and 16 in Figure 5.4 (b, d, and f), respectively, but can also be taken to represent the SONET equivalents 1, 12, and 48, respectively, in Figure 5.4 (c, e, and g).

The reader is advised to dwell a few moments on these representations to develop a feel for the relationships between the three section layer formats and to verify the exact equivalence of the two representations in each case. It will be convenient to use both of these representations later in this chapter when we consider the different HO path layer structures that have been adopted in Europe and North America.

5.2.4   Section Overheads

ITU-T Recommendation G.708 distinguishes between regenerator section overhead (RSOH) and multiplexer section overhead (MSOH). These classifications are indicated in Figure 5.5. However, with the benefit of new insights derived from the more recent architectural studies described in Chapter 3, it is possible to relate specific SOH bytes more explicitly to the architectural components that generate and interpret them. Figure 5.5 illustrates the way in which the SOH is built up as it passes through the transmission media layer. The functional representations are specific instances of the trail termination and adaptation functions described generically in Chapter 3.

The diagrammatic convention for the information format uses a two-dimensional representation of the leading STM. Some of the OH bytes are replicated in each of the N STMs, while others only appear in the leading STM. The complete outline represents the characteristic information at the reference point indicated in the associated functional representation. The dark shaded portion indicates the part of the SOH that has been contributed by the adjacent process above the reference point.

5.2.4.1   Framing (Bytes A1 and A2)

The two bytes A1 and A2 are defined for framing purposes. They are required to take the values of 11110110 and 00101000, respectively, and are replicated in each of the N interleaved STMs. The 16-bit frame alignment word (FAW) formed by the last A1 byte and the adjacent first A2 byte in the transmitted sequence uniquely defines the frame reference for each of the signal rates. The reframe algorithm is not specified, contrary to established practice in the PDH. Instead, reframe performance targets are specified in terms of time to detect out of frame (OOF) condition, mean time between false OOF detection in presence of defined error conditions, maximum recovery time from after a good signal

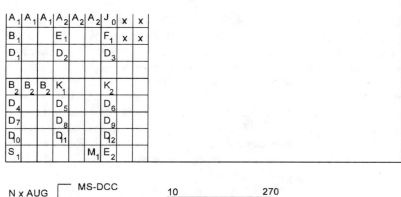

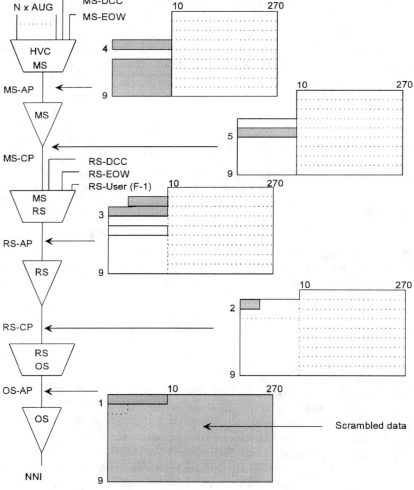

Figure 5.5   Assembling the section overhead (SOH).

becomes available, and maximum probability of false reframing on a good signal. This specification appears in G.783.

There is a significant difference between the SDH and the PDH as to the best strategy for reframing. The likely continuing availability of a high-quality clock reference at an SDH node, even when an input signal has been lost during a short break, means that (providing the local frame reference is maintained) the local and remote frame references have a high probability of still being in alignment when the input signal is restored.

5.2.4.2   STM Identifier Byte C1

The single-byte C1 is used to uniquely identify each of the interleaved STMs in an STM-N signal. It takes the binary value equivalent to the position in the interleave. Thus C1 in STM #1 takes the value 0000 0001, while C1 in STM #16 takes the value 0001 0000. This may be used to assist in finding alignment. In fact, this byte is a relic of an earlier phase in the development of the standard. When the higher rates were formed by bit (instead of byte) interleaving, it was possible to derive some benefit either in performance or implementation cost from a two-stage alignment process whereby the N interleaved bitstreams could be searched simultaneously for their alignment word. When the FAW was located, the single correct phase out of N possible phases could be quickly ascertained by reading the C1 byte. Since the adoption of byte interleaving, even to high bit rates, this capability has become somewhat redundant. The two bytes labeled "x" immediately following C1 in the first row of the STM are reserved for national use in each of the interleaves.

5.2.4.3   Regenerator Section to Physical Media Layer Adaptation

The functions of generating and regenerating the FAW at a transmitter and recovering frame alignment from an incoming signal at a receiver are considered part of the RS-OS layer adaptation function. Also part of the RS-OS layer adaptation function are the scrambling and descrambling processes and the return to zero (RZ) half-width pulse shaping specified for optical media. The equivalent RS-ES adaptation function for STM-1 electrical interfaces is the same as RS-OS, except that CMI coding is specified instead of RZ optical.

Scrambling is applied to the whole STM-N signal except for the first row of the SOH to maintain adequate transition density and DC balance. Framing information must not be scrambled because the scrambler derives its frame synchronization from the STM frame, which must therefore be recovered before the rest of the signal can be descrambled. The data to be scrambled is indicated at the bottom of Figure 5.5 by the darker shading.

The specified scrambler is a seven-stage frame-synchronous scrambler of generating polynomial $1 + x^6 + x^7$ and sequence length 127. It is set to 1111 1111 on the first bit of the first byte of the payload in every frame. This bit and all subsequent bits to be scrambled are added modulo 2 to the output of the scrambler, which is required to run continuously throughout the remainder of the frame. Scramblers of this type have been demonstrated to provide an adequate mechanism for maintaining mark density and DC balance in the transmitted signal. Control of these two parameters is essential for transmission on many media, including optical fiber.

The two bytes after the C1 byte that are reserved for national use are not scrambled as explained above. For this reason the recommendation includes a caveat restricting their use to the effect that DC balance and mark density must be preserved. What this means in practice is left to the judgment of the regional bodies themselves.

OS in Figure 5.5 represents the termination process in the optical layer, which is defined by the optical characteristics, power budget, signal-to-noise characteristics, and spectral shaping of the transmitted data, discussed later in this chapter in Section 5.6.

5.2.4.4   Regenerator Section Error Monitoring (Byte B1)

Only 1 byte in every frame is allocated for monitoring the regenerator section. In STM-4 and STM-16, only the leading STM carries a valid B1 byte. The mechanism used is termed bit-interleaved parity (BIP). The number of marks generated in the bit n position in each byte is counted modulo 2 over the whole frame. The result, 1 for an odd number and 0 for an even number, is placed in the bit n position of the B1 byte in the next frame. Thus, n independent parity calculations are made in n adjacent bit positions. In this case, the depth of the interleaving is 8 (from the 8 bits in each B1 byte) and the code is therefore termed BIP8. Parity is recalculated at the receiver and any discrepancy between the calculated and the received value is interpreted as evidence of an error block.

BIP is simple to implement and is relatively independent of error distribution. The function of inserting and reading parity is part of the RS trail termination function. There are no criteria specified for detection of RS trail failure from within the RS layer itself. RS trail failure can be inferred from loss of signal, loss of frame alignment, or AIS detected in the associated physical media layer function. Detection of either of these conditions will result in AIS (all 1s) being sent towards the client layers.

5.2.4.5   RS Engineering Order Wire and User Channel (Byte E1 and F1)

The regenerator section engineering order wire (RS-EOW) provides a channel for voice contact between maintenance personnel at terminal and/or regenerator sites. Such facilities have been a well-established feature of line transmission operations for a long time. ITU-T has made no recommendation beyond allocating the E1 byte in the SOH. The E1 byte is intended to be accessible at regenerators, where it may be connected to PCM coding and voice equipment. The EOW facilities are still an essential feature of long-haul repeater line systems where regenerators are in remote, unattended locations, but the ubiquity of mobile communication service seems likely to displace the traditional EOW in many other applications.

The user channel is in a similar class to the EOW, being a well-established feature of traditional line system operations. It is typically used by operators for remoting various physical alarms but, as might be expected with such a facility, all manner of ingenious applications have been found. Again, the ITU-T has not specified how this channel should be used but has allocated a single byte (F1) in the SOH for the function.

5.2.4.6 The RS Data Communication Channel (Bytes D1, D2, D3)

The regenerator section data communication channel (RS-DCC) provides a channel of capacity 192 kbit/s for messaging to and between regenerators. This would typically be used for intrasystem communications for management and supervision of regenerated systems. Not all RSs require a DCC. Sections in a protection group, for example, do not all carry RS-DCCs. It is therefore important that the DCC be capable of being enabled or disabled under management control. The communication protocols are described in Chapter 7.

5.2.4.7 RS Auxiliary Layers

The DCC, EOW, and user channel are examples of auxiliary layer networks served by the RS layer. As such, they are generally rather primitive examples of layer networks being not generally of very large extent, but confined within a single line system. This need not be the case, however, as there is no reason in principle why such auxiliary networks should not extend across several line systems, interconnected at nodes in the auxiliary layer, a conference bridge in the EOW, or a message switch in the DCC. Failures in the RS layer result in AIS being transmitted to each of the auxiliary client layers. The process of combining the auxiliary layers and the data from the multiplex section into the STM, and the complementary recovery process, take place in the adaptation function between the RS layer and its client layers.

5.2.4.8 Multiplex Section Error Monitoring (B2 Byte)

Three bytes labeled B2 are allocated to the trail termination function in each STM of the aggregate signal. These are organized as a BIP24 error-monitoring code for STM-1, BIP96 for STM-4, and BIP384 for STM-16, respectively. The bit-interleaved error-detection mechanism is identical to that described for BIP8 above. Parity is calculated over the MS layer characteristic information, as indicated at MS-CP in Figure 5.5. That is to say, the first three rows of the SOH are omitted from the computation but the whole of the MS section overhead, including K1 and K2 in row 5, are included.

First-generation specifications required that the trail termination function record the total number of detected parity violations. However, interpretation of such a parameter is problematic. Today, all performance parameters are derived from error blocks (in this case, errored 125-sec frames). Thus, the detail of the parity violation count will be irrelevant except in so far as it reduces to negligible proportions the probability of an undetected error block. The principles of performance monitoring in high-capacity systems are discussed in more detail in Chapter 8.

5.2.4.9 Automatic Protection Switching

The two bytes K1 and K2 are allocated in the leading STM mainly to the function of coordinating protection switching across a set of multiplex sections organized as a protection group. Automatic protection switching (APS) is a moderately complex process to coordinate across an open interface. It has for this reason been the subject of considerable controversy during the development of the standard. Protection switching and the messaging required over these channels are discussed in Chapter 10.

Apart from their role in implementing protection protocol messages, the K bytes are also used to implement some more basic layer processes. Bits 6, 7, and 8 of the K2 byte are used to signal MS-FERF (110). This indicates far end receiver failure (FERF) and is transmitted to the other end when a failure is detected at the associated local receiver. Receipt of MS-AIS (111) on these bits is taken as an indication of an upstream regenerator failure and is used to suppress local alarms.

5.2.4.10   Multiplex Section Auxiliary Layers

The multiplex section supports two auxiliary layers: a data communication channel and an engineering order wire. The MS-DCC (bytes D4 to D12) provides a 576 kbit/s channel for messaging between MS trail terminations in adjacent network nodes. This may be used for communication between management entities as part of a TMN (telecommunications management network). The MS-EOW (byte E2) is a single 64 kbit/s channel provided for voice communication between MS trail terminations in network nodes. The ITU-T has not specified the manner in which this channel should be used. Both are examples of adaptation into the MS layer.

5.2.4.11   Distribution of Synchronization References in the MS Layer

In SDH networks, synchronization reference information is transmitted over the section layer in the timing of the STM-N signal. In the recommended synchronization network architecture, it is necessary to transfer synchronization references across a number of SDH NEs (network elements) in tandem. A detailed analysis of SDH synchronization is contained in Chapter 11. The synchronization reference distribution layer is therefore considered a client of the MS layer. In recent times, the synchronization status byte has been introduced to provide an embedded layer termination process for the synchronization reference distribution layer. Essentially, the synchronization status byte carries a label indicating the class of reference source from which it originates. These are indicated in Table 5.2. The QL6 indicator plays a significant role in distribution because it is used to prevent timing loops, which may occur under certain failure conditions, and to prevent

Table 5.2
Format of the Synchronization Status Message (SSM)

QL1	0010	G.811 source
QL2	0000	Unknown quality level
QL3	0100	G.812 transit source
QL4	1000	G.812 local source
QL5	1011	G.813 (SDH equipment clock)
QL6	1111	Not to be used for synchronization

dependent subnetworks from taking their synchronization from the NE clock (G.813) when a higher quality source may be available.

5.2.4.12   Remote Defect Indication (RDI)

The RDI byte M1 is also a recent innovation. It applies only to the UNI at STM-1. The leading STM in the interleave was therefore not used to avoid confusion at the NNI for first-generation equipment, which would not, in general, support it. It provides a minimal level of single-ended remote management of customer premises equipment without the expense and security exposure of running an OSI communications channel.

5.2.4.13   Section Trace Byte (J0)

The section trace byte has been introduced to check the validity of the transmission media connectivity. This is a very important feature as misconnection of the physical media by human error is one of the most common problems in real-world installations.

5.2.5   Path Layer Adaptation

The path layers are the main vehicle for transmission networking in the SDH. SDH paths may transport client information across a complex network with many flexible transit nodes and use the link service of a variety of server layer types. At the other extreme, an SDH path may consist of a single link connection between two path termination points, with no intervening nodes.

5.2.5.1   Adaptation Between Section Layer and HO Path Layers

The higher order virtual container (HVC) is the structure used for transporting information in an HO path. The HVCs will, in general, have been assembled at points remote from either end of a particular multiplex section and will therefore have frame phases completely uncorrelated with the multiplex sections on which they are transported. Imperfections in reference sources and temperature-induced variations in total transmission delay are among the causes of frame phase variations.

The adaptation function between the MS layer and the HO path layer is required to combine a number of HVCs with different frame phases and locate them together in the MS payload with an encoded representation of the frame phase of each. Then, both data and frame phase can be recovered at remote HOP terminating nodes and transferred at HOP transit nodes despite the independent phase variations and offsets in the supporting sections.

5.2.5.2   Administrative Units and Administrative Unit Groups

When an HVC is located in an STM, the offset in bytes between its frame reference and the frame reference of the supporting MS is effectively measured and quantized to an integer number of bytes. This integer number corresponding to the frame offset value is located in row #4 of the SOH. The combination of an HVC and its encoded frame offset is called an administrative unit (AU) and the encoded value of the frame offset itself is termed the AU pointer. The administrative units AU-3 and AU-4 transport the higher order virtual containers VC-3 and VC-4, respectively, together with their respective frame offsets coded in the AU pointer. A group of three AU-3s when byte interleaved together occupies the same payload capacity as a single AU-4. The organization of an STM payload into either three AU-3s or one AU-4 forms an administrative unit group (AUG). The STM-*N* payload is notionally divided into *N* subdivisions corresponding to the STMs from which it was assembled. Each of these subdivisions marks the location of one or the other type of AUG. The two types of AUGs are illustrated in Figure 5.6 and Figure 5.7 and their assembly into STM-N is illustrated in Figure 5.4.

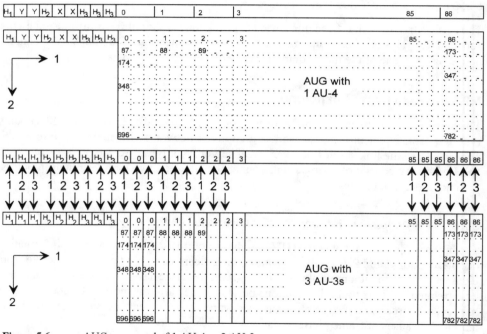

Figure 5.6   AUG composed of 1 AU-4 or 3 AU-3s

5.2.5.3   *Synchronous Payload Envelopes and Administrative Units*

A major difference between synchronous optical network (SONET) and European Telecommunications Standards Institute (ETSI) SDH is in the AU level at which these networks will be administered and in the consequent structure of the AUGs. In North America, AU-3s are used for normal (previously PDH) telecommunication traffic and AU-4 will be used for certain broadband payloads. In Europe, AU-4s are used for both these classes of traffic. The equivalent SONET terms are STS-1 synchronous payload envelope (SPE) for AU-3 and STS-3c SPE for AU-4.

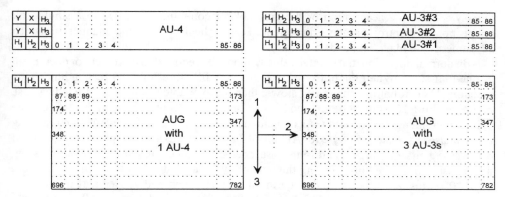

Figure 5.7   AUG with 1 AU-4 or 3 AU-3s from a SONET perspective.

This regional divergence owes more to the need to match well-established PDH administrative practice than to any attempt to optimize the SDH network itself. In the established PDH networks in North America, DS3 (44736 kbit/s) is the predominant networking level in the interoffice and long-haul networks while DS4 (139264 kbit/s) has achieved a similar status in Europe. These are well-matched to AU-3 and AU-4, respectively. The ITU-T has recommended that the AU-4 level be used for interworking on links between networks based on different AUs such as those across the Atlantic. There are now several installed examples of SONET SDH interworking testifying to the usefulness of this effort.

The two AUG structures and the AU pointer formats are illustrated in Figure 5.6 and Figure 5.7. The two bytes H1 and H2 allocated to the pointer can be viewed as one 16-bit word, as shown in Figure 5.11. The pointer value is carried in the last 10 bits (bits 7–16) of this word as a binary number with a range of 0–782. This represents the offset in frame phase between the start of the HVC frame and the H3 field of the AU, as indicated in Figure 5.8. The offset is measured in 3-byte increments for AU-4 and 1-byte increments for AU-3. This scaling in the frame phase offset quantization increment ensures compatibility between the American National Standards Institute (ANSI) and ETSI structures and hence relatively straightforward conversion.

5.2.5.4   AU Pointers

AU pointers provide a mechanism for dynamically accommodating the variations in frame phase between HVCs and the locally generated frame phase of the multiplex section into which they are to be multiplexed. The alignment process between the HO path layer and the MS layer requires that the HVCs to be multiplexed are buffered against the local reference with a hysteresis sufficient to suppress the effects of short-term phase variations. However just before the buffer overflows (or empties), the phase of the associated HVC must be incremented (or decremented) with respect to the MS reference and the pointer value adjusted accordingly. The dynamics of pointer processing and the accumulated network effects are described more fully in Chapter 11 and the detailed operation of a possible pointer processor implementation is described in Chapter 12. We will confine ourselves here to a description of the protocol for communication within a single link connection to control pointer adjustments across the NNI and the manner in which adjustments are made on the data stream.

In normal operation, the pointer value will only be required to increment or decrement in one-unit steps. When the buffer crosses its "low fill" threshold, corresponding to the incoming HVC running temporarily slower than the available channel rate, then the phase of the outgoing HVC must be slipped back in time by one unit relative to the MS and the pointer incremented by one unit. As explained previously, the unit by which the outgoing frame phase is adjusted is 1 byte for AU-3 and 3 bytes for AU-4. This operation is signaled by inverting bits 7, 9, 11, 13, and 15 that are labeled "I" and in the same frame-suppressing transmission from the buffer during the byte or bytes labeled "0" in the AUG. (There are three "0" bytes in AU-4 and only one in AU-3.) The next and subsequent frames will contain the new pointer value. This process of sending marked dummy information in a preassigned slot to equalize the offered rate to the available channel rate is called positive justification. Positive justification is illustrated in Figure 5.8 for the case of VC-4 in AU-4. This illustration uses the two-dimensional frame representation and illustrates the manner in which the VC-4 phase is adjusted with respect to the STM phase as a result of a change in pointer value.

Conversely when the buffer crosses its high fill threshold, corresponding to the HVC running temporarily faster than the available channel rate, then the phase of the HVC must be advanced in time by one unit relative to the MS and the pointer decremented accordingly. This operation is signaled by inverting bits 8, 10, 12, 14, and 16, which are labeled "D" and in the same frame transmitting the next byte (or three bytes) in the H3 byte positions. The next frame and subsequent frames will contain the new pointer value. This process of sending overflow information in a designated overflow channel is called negative justification. Negative and positive justification are illustrated in Figure 5.9 for the AU-3 and Figure 5.10 for the AU-4. In these diagrams, a single AU frame is illustrated as a stream of bytes in one dimension, with the pointers occupying the leading byte positions. The frame phase of the HVC in each frame described by the pointer value is then simply represented by the linear offset from the end of the H3 byte. Pointer activity is independent in each of the AUs within an STM-N except when the AUs are concatenated. Concatenation is described below.

Two other mechanisms are available for incrementing or decrementing pointers. The new data flag (NDF) occupies the first four bits of the pointer labeled N in Figure 5.11. When it is necessary to make an arbitrary, nonunit change in pointer value accompanying a reconfiguration, then the normal NDF value of 0110 is replaced by the value 1001. The new pointer value is contained in the same word as the NDF indication and takes effect immediately. If a new pointer value is detected unaccompanied by either of the mechanisms described previously, then it is ignored on the first occurrence; but if it is sustained for three consecutive frames then it is adopted as the new pointer value. Under all circumstances, consecutive pointer adjustments are constrained to be separated by three or more stable values.

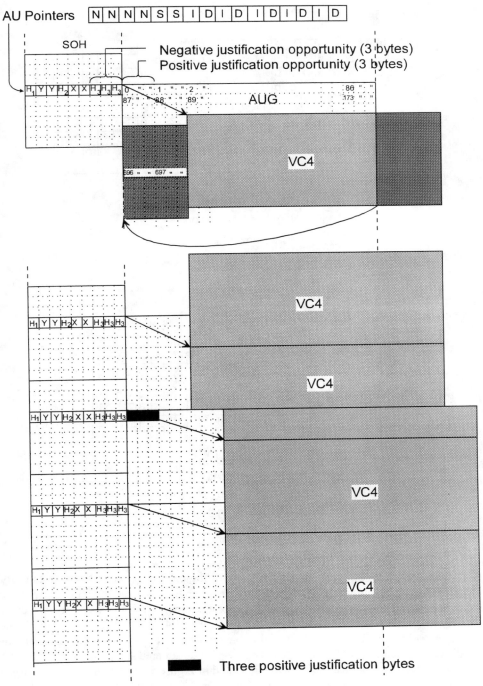

Figure 5.8   Positive justification in AU-4

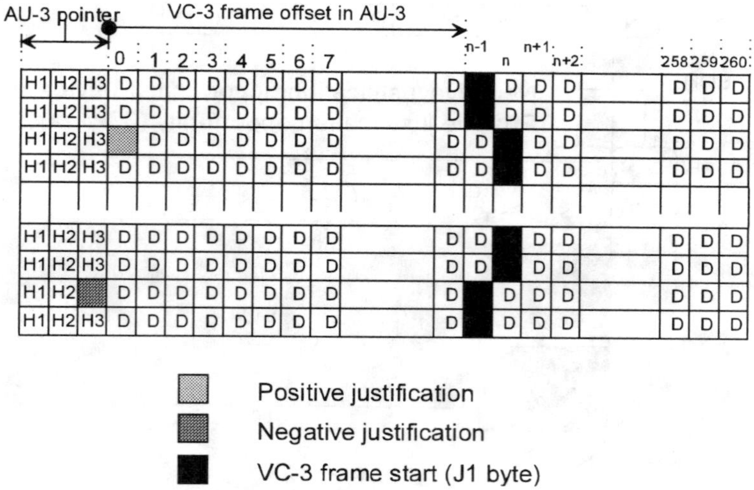

Figure 5.9   AU-3 pointer processing.

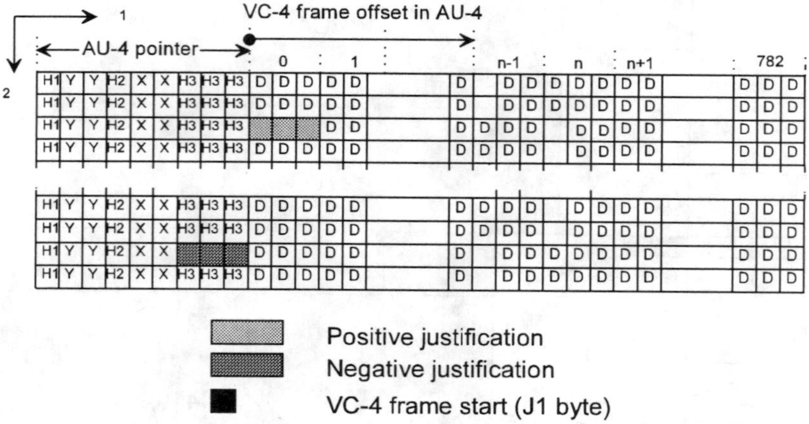

Figure 5.10   AU-4 pointer processing.

5.2.5.5   AU-4 Concatenation

Concatenation provides a mechanism for transporting payloads greater than the capacity of VC-4. Sets of X contiguous AU-4s may have their payloads locked by setting the pointer value in all but the leading AU-4 to a specific state, known as the concatenation indicator (CI). Pointer adjustments indicated for the leading AU-4 are then replicated in all the concatenated AU-4s in the set, maintaining bit sequence integrity over the whole broadband payload. Such a set of X concatenated AU-4s is designated AU-4-Xc. The CI pointer value held in all but the leading AU-4 is indicated in Figure 5.11. The frame phase offset quantization increment for a concatenated group is $3X$ bytes where X is the number of AU-4s in the concatenated group.

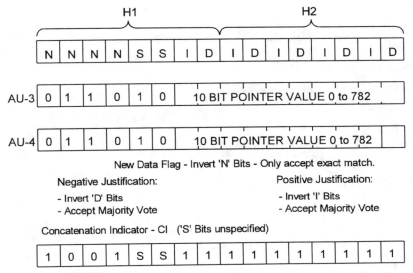

Figure 5.11   AU pointer format

The possible values of the concatenation number X is only constrained by the maximum payload capacity of the transporting structure. The value 4 is of particular interest as the AU-4-4c has been recommended for the transport of B-ISDN payloads.

5.2.5.6   AU-3 Concatenation

In SONET, the STS-1 synchronous payload envelope together with the STS-1 pointer is equivalent to the AU-3. These can be concatenated using exactly the same pointer mechanism as described above for AU-4. This was originally the only mechanism for transporting payloads greater than 50 Mbit/s in SONET. However, when ITU-T set the STM at three times the STS-1 capacity and made it the basic transport rate for B-ISDN, it became necessary to modify SONET to maintain compatibility. This involved reordering the byte-interleaving sequence used for multiplexing AUs into an STM-N. The SONET standard now describes a two-stage interleaving process whereby three AU-3s (equivalent to three STS-1 SPEs and their associated payload pointers) are first multiplexed using single byte interleaving to form an AUG (equivalent to STS-3 SPEs plus pointers). N AUGs (i.e., STS-3, SPEs, and their payload pointers) are then further multiplexed using single byte interleaving to form the payload of an STM-N (equivalent to an STS-3N SPE plus pointers). The two-stage interleaving process is illustrated in Figure 5.12 for the example of multiplexing three STS-3s into STS-12. The fields X and Y of the AU-4 pointer were then chosen to be the same as the AU-3 concatenation indicator. Thus the AU-4 is identical to the AU-3-3c, a feature that enables ease of interworking between AU-3 and AU-4 based networks and multivendor portability for broadband terminals.

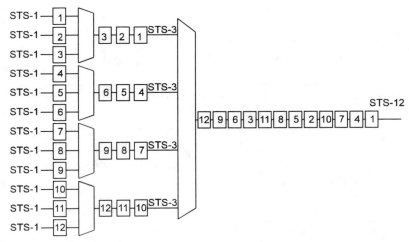

Figure 5.12   Byte interleaving STS-1s to form STS-12

5.2.6   The Higher Order Path Layers

The HOP Layers provide high-capacity transport across SDH networks. The HOP flexibility that can be expected in HOP transport node equipment will ensure that they become a major vehicle for SDH networking. Trail termination overheads are introduced at the source of a HOP trail and read at the sink for management purposes. The HOP layer in Europe will be based on the VC-4, while in North America, the VC-3 will fulfill this role. The overhead structure is the same for both and is described in the following.

5.2.6.1   Structure of VC-4 and VC-3

The VC-4 is a frame-synchronous structure of total data capacity (payload and overhead) equivalent to 150336 kbit/s. The structure suggests a presentation as a rectangular arrangement of bytes with 261 columns and 9 rows whose location within the AU-4 is indicated by the associated AU-4 pointer. Of the 261 columns, one is allocated to POH and 260 are allocated to client layer payload. It occupies the whole payload capacity of the AU-4.

The VC-3 is a frame-synchronous structure of total data capacity (payload and overhead) equivalent to 48960 kbit/s. It can be represented as a rectangular arrangement of bytes with 85 columns and 9 rows whose location within the AU-3 is indicated by the associated AU-3 pointer. Of the 85 columns, one is allocated to POH and 84 are available for client layer payload. The payload capacity of the AU-3 consists of 87 columns (equivalent to the STS-1 SPE in SONET). Columns 30 and 59 are allocated to fixed stuff (justification) to allow for this. The structure of the VC-4 and VC-3 and their disposition in the AU-4 and AU-3, respectively, are illustrated in Figure 5.13 and Figure 5.14.

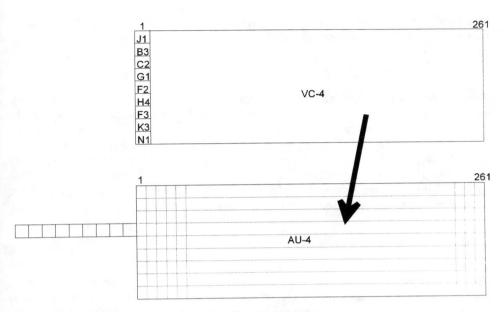

Figure 5.13 Structure of the VC-4 and its disposition in AU-4.

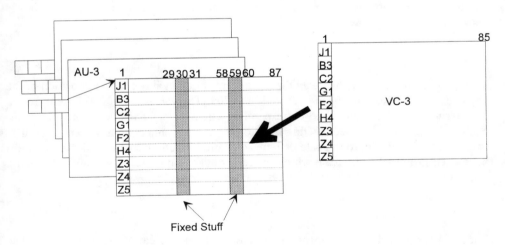

Figure 5.14 Structure of the VC-3 and its disposition in AU-3

5.2.6.2 Path Overhead (POH) in VC-3 and VC-4

VC-4 and VC-3 both share the same POH structure of one 9-row column. POH bytes fall in the same three categories as already discussed for SOH in Chapter 3. J1, B3, C2, and G1 are functions of the layer trail termination and are therefore client layer independent or payload independent. F2 provides a user channel and may be regarded as providing a link connection in some user layer. The H4 byte is used in different ways by different client layers and is client layer specific or payload specific. The bytes Z3, Z4, and Z5 are reserved for future use and may be allocated in either of the three categories.

5.2.6.3   Path Trace

The J1 byte provides a path validation and trace function. A unique identifier associated with the client layer access point at the path termination is inserted into the J1 byte at the path termination source function (a path termination is a specific example of the generic trail termination). This is recovered at the corresponding sink function at the other end of the path where it can be compared with the expected value, which may be locally stored or returned to the managing process for analysis. The specification provides for a 64-byte string, but the preferred format is the 15-byte ASCII option specified in E.164 preceded by a 1-byte frame start flag, thus forming a 16-byte pattern.

5.2.6.4   Error Monitoring

The B3 byte provides an error-monitoring function using bit-interleaved parity of depth 8 (BIP-8). A BIP-8 is calculated over the whole of the preceding VC frame and inserted in the B3 byte at the path termination source function. The remote receiver makes a similar computation and compares this with the B3 value received. Early specifications required that the total number of detected parity violations be reported. Current performance monitoring methodology favors a single error block report if one or more violations is detected in a single frame.

5.2.6.5   Path Label

The C2 byte provides a signal label that carries information about the composition of the payload. Two values have been specified in G.709. The value 0000 0000 is sent from a path terminating function if the corresponding client layer adaptation function is un-equipped and from a crossconnecting function if no crossconnection has been made. The value 0000 0001 is sent to indicate that the adaptation function is equipped. Other values have been proposed to indicate which particular adaptation option is equipped. Such a capability can be used to remotely set adaptation options.

5.2.6.6   Path Status

The G1 byte provides status information about the remote path termination sink. The format is illustrated in Figure 5.15. The four bits 1, 2, 3, and 4 are specified to return the remote BIP-8 violation count as a binary number. There are nine valid values correspond-

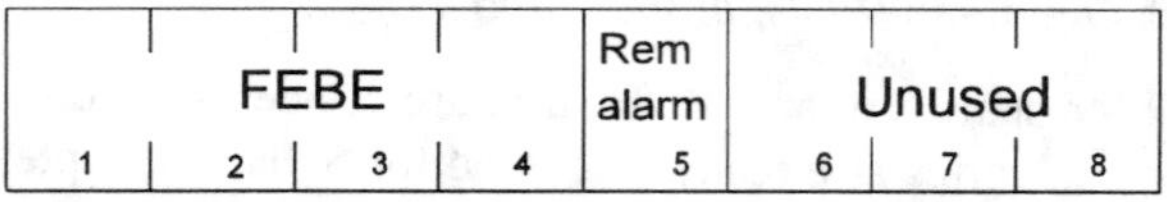

				Rem alarm		Unused	
	FEBE						
1	2	3	4	5	6	7	8

Figure 5.15   HCV path status byte format.

ing to 0 to 8 violations. One bit (via bit 5) denotes FERF and takes the value 1 when failure is detected and 0 otherwise. The criteria for sending FERF are the detection of AIS, path fail, or path trace mismatch. Bits 6, 7, and 8 are so far unallocated. The H4 byte is client layer specific or payload specific and will be described together with the associated client layer mappings.

5.2.6.7   New Tandem Connection Overheads

The K3 and N1 bytes, introduced in the latest specifications, may be terminated at tandem connection terminations. K3 provides a communication channel for coordinating protected path layer termination units and N1 provides the error monitoring and remote and forward error reporting for supervising tandem connections.

5.2.6.8   Adaptation to the Lower Order Path Layers

As indicated in the summary of SDH layer network client-server relationships illustrated in Figure 3.12, VC-3s can be used to transport VC-1s or VC-2s, and VC-4s can be used to transport VC-1s, VC-2s, or VC-3s. The VC-3 therefore has a dual role as a higher order path in a SONET network transported via AU-3 and as a lower order path in an ETSI SDH network transported in a tributary unit TU-3. Each of the LVCs is a frame-synchronous structure whose frame phase is uncorrelated with the supporting HVC. Dynamic LVC frame phase alignment information is transferred across the NNI using the same principle as for transferring HVC frame phase information explained previously.

The offset in bytes of the LVC frame reference from the supporting HVC frame reference is measured and encoded as a binary integer. The offset thus encoded is then transferred together with the associated LVC in the HVC payload. The combination of the LVC and the quantized frame offset is called a TU, and the frame offset indication itself is called the tributary unit pointer. The equivalent SONET terms are virtual tributary (VT) and VT pointer.

The TU structures and the LVCs from which they are derived have been chosen to have harmonic relationships with one another. This property allows flexible combinations of LVCs of different capacities to be transferred in the same HVC. The TUs are arranged in ordered groups, referred to as tributary unit groups (TUG), within the HVC payload area. The harmonic relationships between the TUs result in integer ratios between TUG capacities. This ordered arrangement of TUGs is fundamental to the implementation benefits associated with synchronous multiplexing.

5.2.6.9   Organization of HVC Payloads in TUG-2s and TUG-3s

The TUs allocated to carry the two primary groups at DS1 and E1 were designated TU-11 and TU-12, respectively, and allocated three and four columns, respectively, in the nine-row STM frame. This suggested the dimensions of the TUG-2 at 12 columns, which could be constructed from three TU-12s, four TU-11s, or a single TU-2. Seven TUG-2s may be packed in the VC-3 payload for the SONET structure, but a further level of grouping was needed for ETSI SDH because a VC-4 may need to carry one or two VC-3s as well as a combination of lower rate TUs. Therefore, the TUG-3 emerged constructed from seven TUG-2s or a single TU-3. This is the reason why the TU-3 has a structure that has more in common with an AU-3 than with the other lower order TUs.

The VC-4 payload is divided into three as illustrated in Figure 5.16. One column of POH and two columns of fixed stuff leaves 86 columns for each of three TUG-3s. This arrangement of TUG-3s can be considered to be a TUG-4, although this entity has not been explicitly described in G.709. Each TUG-3 can be used to transport one TU-3 or an assembly of smaller TUs.

The TU-3 structure is also illustrated in Figure 5.16. The three pointer bytes have exactly the same function and mode of operation as those in the AU-3 and the AU-4, even though they are disposed differently with respect to the payload. The range of the pointer value is smaller at 764 (rather than 782 for AU-3 and AU-4) because the TUG-4 is smaller than the AUG. As in the case of the AU-3, the pointer value indicates the byte position in the TU-3 at which the VC-3 frame start can be found. The VC-3 frame start coincides with the J1 byte of the POH.

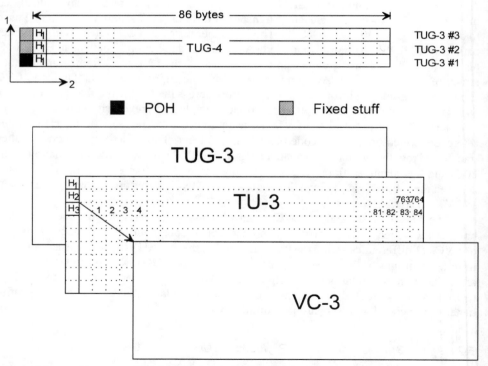

Figure 5.16   Transport of VC-3 in TU-3 and organization of VC-4 payload as TUG-3s.

When the TUG-3 is used to transport an assembly of smaller TUs, the TU-3 pointer is not used. In this case, it carries the null pointer indicator (NPI) value. This value is only valid for TU-3 pointers and not for AU pointers.

5.2.6.10   TUG-2s in TUG-3s and in VC-3s

The 86-column TUG-3 can alternatively be allocated to support seven TUG-2s. This is illustrated in Figure 5.17(a), where the corresponding TU-3 pointer carries the NPI value and one extra column of fixed stuff is introduced. The seven 12-column TUG-2s are byte interleaved in the remaining 84 columns.

In SONET, the 85-column VC-3 similarly supports seven TUG-2s, although the TUs are called VTs and the TUG is correspondingly referred to as a VT Group. This is illustrated in Figure 5.17(b), where the first column carries the POH as described previously and the seven 12-column TUG-2s are byte interleaved in the remaining 84 columns. The two diagrams of Figure 5.17 have been juxtaposed to emphasize the similarity of the TUG-2 structure in each case despite the differences in supporting structure.

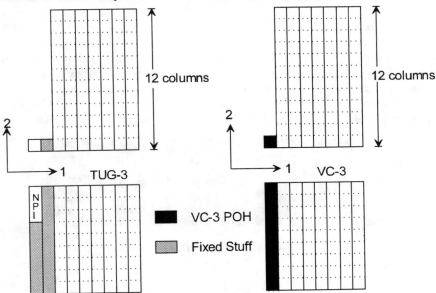

Figure 5.17   Equivalence of TUG-2s in TUG-3 or VC-3.

Although from a multiplexing viewpoint these two structures are identical, there is an important difference in the way they are administered. The VC-3 can be monitored as a single entity between its path (trail) terminations, while the TUG-3 can either be monitored as part of a VC-4 or as a collection of lower capacity monitorable entities, namely VC-1s and VC-2s. Nevertheless, the commonality of the multiplexing structure can be expected to simplify conversion at an administrative boundary.

5.2.6.11   Factors Influencing the Choice of TUG-2 and LVC Capacity

The choice of VC sizes for the lower capacities has been very much driven by the requirements of backwards compatibility with existing networks. Three major considerations served to focus attention on the primary rates of the existing PDHs:

- The 24- and 30-channel groups have become well-established as administrative units of capacity in North American and European networks, respectively.
- Most transition scenarios require the capability to transport the existing PDH structures, particularly the primary rates, across the islands of early SDH deployment.
- The primary rate channels have become well-established in Europe and North America for the delivery of leased line services to the business community.

5.2.6.12   Structure of the TUG-2

The TUG-2, with its 12 columns and 9 rows, was chosen because it can be arranged as four groups of 3 columns each (i.e., 27 bytes in 125 sec) or three groups of 4 columns each (i.e., 36 bytes in 125 sec). These capacities have become the basis of TU11 and TU12, respectively, that are used to transport frame-synchronous containers VC-11 and VC-12. G.709 describes how these containers may be used to transport the primary rates in a number of different ways, depending on application. It also describes how the 24- and 30-channel groups may be transported directly without the PDH overheads.

The 12 columns of the TUG-2 are allocated to four TU-11s, three TU-12s, or a single TU-2, as illustrated in Figure 5.18. The equivalent 4×27, 3×36, and 1×108

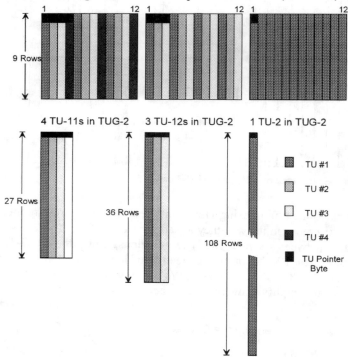

Figure 5.18   TU-11s, TU-12s and TU-2s organized as TUG-2.

representations are also illustrated to emphasize the numerical properties that make for ease of switching these assemblies. Any combination of these TUG-2 structures can be transported in the same VC-3, or as TUG-3 assemblies in the same VC-4.

5.2.6.13   Structure of TU-11, TU-12, and TU-2

The frame phase information required to define an LVC in its supporting TU is similar to that required to define an HVC in its supporting AU. Two bytes per TU frame equivalent to H1 and H2 in an AU are required to carry frame start information in the form of a TU pointer, together with a third byte per TU frame equivalent to H3 to act as a negative justification opportunity. This was considered extravagant in relation to the frame size for these smaller structures and was difficult to reconcile with the numerical compatibility just described. The TUs are therefore based on a four frame multiframe (i.e., 0.5 msec), thus simultaneously improving the channel efficiency by a factor of 4 and reducing the capacity at which a simple harmonic relationship could be achieved. The multiframe is identified by a 2-bit pattern in the two least significant bit positions of the H4 byte in the VC-3 or VC-4 path overhead.

One byte per TU in each HVC frame is used for the TU pointer. This is carried in the leading bytes of the corresponding TUGs, as illustrated in Figure 5.18. The pointer byte

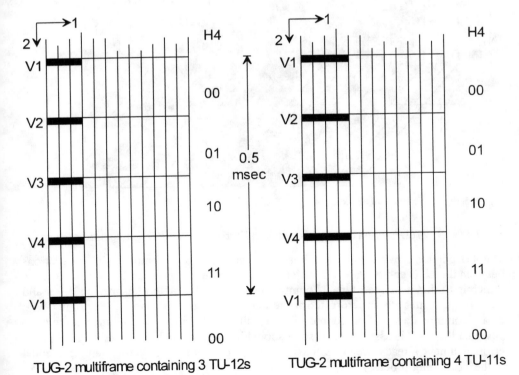

TUG-2 multiframe containing 3 TU-12s   TUG-2 multiframe containing 4 TU-11s

Figure 5.19   The TU multiframe.

is designated V1, V2, V3, and V4 in the first, second, third, and fourth frames, respectively, of the TU multiframe as shown in Figure 5.19. The S bits in the TU pointer format indicate the TU type. An LO-to-HO adaptation sink function can ascertain which TUG-2 arrangement is being received independently of local provisioning by examining the S bits of the pointer in the leading TU position of a TUG-2. The TU pointer processing mechanism is the same as the AU pointer processing mechanism already described, and the V1, V2 and V3 bytes play the same role in relation to the TU pointer as that performed by H1, H2, and H3 in relation to the AU pointer. The V4 byte is presently reserved by ITU-T for possible future use. Figure 5.20 illustrates an example of a mixed TU structured payload in VC-4, showing the fixed positions of the different pointers with respect to the supporting frames.

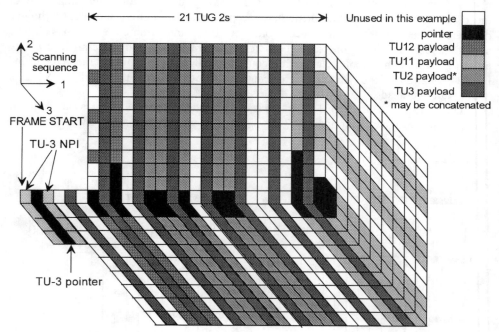

Figure 5.20   Example of a mixed payload in VC-4

5.2.6.14   *TU Pointers and the Adaptation Process Between the Logical Path Layers*

The TU pointer provides the mechanism to accommodate differential phase variations between the LVC and the HVC in which it is transported. This mechanism is identical in principle to that described for AU pointers previously. Variable transmission delay and clock irregularities are, as before, the prime source of these variations, but the pointers will also accommodate a plesiochronous offset resulting from a loss of reference at the node at which the LVC is assembled. Also, it should be remembered that these original phase errors will have already been quantized by the AU pointer process before being presented to the TU pointer process.

LVC data is buffered at the boundary with the HVC layer. If the average rate at which LVC data enters the buffer temporarily exceeds the rate at which the buffer is emptied by the HVC-derived channel clock, then the buffer will eventually cross its nearly full threshold. When this happens, an extra byte is read out and placed in the V3 location, and the pointer value decremented by one unit. This process is called negative justification, and the V3 byte is said to provide a negative justification opportunity.

Conversely, if the rate at which LVCh data enters the buffer temporarily falls below the rate at which it is passed to the HVC, then it will eventually cross the nearly empty threshold. When this happens, the clock that is reading the buffer to the HVC layer is suspended or gapped for 1 byte coincident with the byte immediately following the V3 byte in the TU, and the pointer value incremented by one unit. This process is called positive justification, and the byte immediately following the V3 byte is known as a positive justification opportunity. The data actually transmitted during the positive justification opportunity is not specified. The inverse process takes place when the LVC is recovered from the HVC.

The mechanism is illustrated in Figure 5.21 for TU-12. In this representation, a single TU-12 is observed over 21 multiframes—a period of time sufficient to demonstrate one

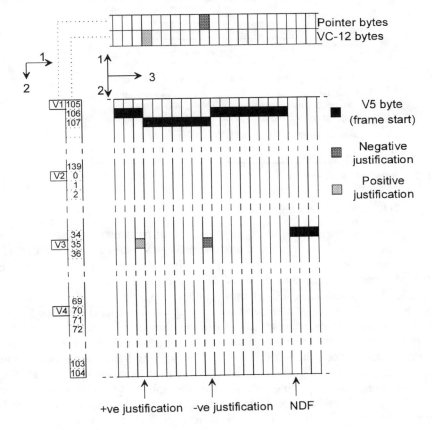

Figure 5.21   Operation of the pointer in TU-12.

positive justification, one negative justification, and one new data flag (NDF) assertion. The pointer bytes have been drawn in a separate column to more clearly show the TU capacity and its byte numbering. The order of transmission starts with the V1 byte and is followed immediately by the bytes at locations 105, 106, and 107, up to 139. That is followed by V2 and so forth to the byte in location 104, followed by 105 in the next multiframe.

The positive justification is illustrated accompanied by a TU pointer increment from 106 to 107. This is followed seven multiframes later by a negative justification with a corresponding TU pointer decrement from 107 back to 106. The TU pointer format is illustrated in Figure 5.22. Pointer increments and decrements are signaled by inverting the "I" bits or the "D" bits, respectively, in the same way as for AU pointers.

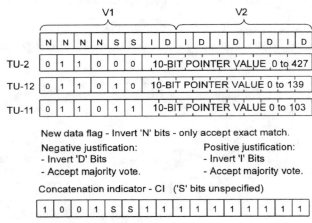

Figure 5.22   Format of the TU pointer.

5.2.6.15   TU-2 Concatenation

One of the applications foreseen for the next generation of transmission systems is the transport of new services at bit rates other than those well-established as part of the PDH. High-speed data and intermediate video rates are likely contenders. Such new service channels are, of course, ideal candidates for ATM, and the more interesting application for concatenation is in partitioning VC-4 bandwidth between a residual STM client base and the introduction of new ATM clients in parallel. Concatenation has been specified in G.709 for TU-2. Using this mechanism, TU-2s can be associated in such a way that bit sequence integrity is maintained across the multiple containers that constitute the payload.

In normal circumstances, the SDH accommodates varying transmission delays and other phase variations by allowing LVCs to "float" within the supporting HVC and HVCs, in turn, to "float" within the supporting STM-*N* under the control of the respective pointer mechanisms. However, to maintain bit sequence integrity across the transported payload, the resulting differential phase between the components of the concatenated group must either be suppressed or corrected at the receiver. In fact, both these methods are specified in G.709.

5.2.6.16   Contiguous Concatenation of TU-2s in a VC-3

This mechanism is the same as that specified for AU concatenation. The TU-2s that are contiguous in time within a supporting VC-3 may be concatenated by placing the concatenation indicator (CI) in the pointer field (see Figure 5.22) of all but the leading TU-2. All TU-2s in the concatenated group increment or decrement their pointers together under the control of the pointer value in the leading TU-2, thus suppressing the differential phase shifts that would otherwise result from uncoordinated pointer activity.

A difficulty arises if this concatenation mechanism is extended to TU-2s in VC-4. When TU-2s contiguously multiplexed in VC-3 are remultiplexed in VC-4 using the normal interconnection rules, they are no longer contiguous in VC-4. Therefore, if the same contiguous concatenation mechanism were used in VC-4 as in VC-3, it would not be straightforward to pass concatenated groups between an AU-3 network and an AU-4 network.

5.2.6.17   Virtual Concatenation

The mechanism selected for concatenation of lower order paths (LOPs) in VC-4 has been termed virtual concatenation because no concatenation indicator is sent and no action is taken at intermediate nodes to suppress the differential phase shifts resulting from normal pointer activity. Instead, responsibility for maintaining bit sequence integrity in the payload is left with the path terminating equipment. A concatenated group must be launched with all its pointers set to the same value. Then, provided they are always assigned to the same VC-4 (as one another) at LOP transit nodes, the receiving equipment can buffer the component payloads until the pointer values are again equal, thus ensuring payload integrity. With virtual concatenation, contiguity is not required, but the time sequence of TU-2s must be maintained.

Virtual concatenation has been adopted by ETSI for TU-2s and TU-12s in VC-4. One of the problems foreseen with contiguous concatenation was the increased blocking probability to concatenated groups in a situation where HVC bandwidth has become fragmented. This is not expected to be a problem with virtual concatenation. Some further work must be done to ascertain the maximum differential delay that can accumulate between the individual LOPs in a concatenated group before practical designs can be implemented. In access applications, this can be quite low. The practical advantages of this mechanism are significant and it seems likely that it will be widely adopted as ATM is introduced.

5.2.6.18   Structure of the VC-11, VC-12, and VC-2

These VCs are very similar, and so will be described together. They each have frame-synchronous structures consisting of 4 bytes of POH and, respectively, 103 bytes (VC-11), 139 bytes (VC-12), and 427 bytes (VC-2) of payload capacity in a 500-sec multiframe, consistent with the TU capacities discussed above. The V5 byte is defined to be the first byte of the VC frame, the byte to which the TU pointer points.

The V5 byte provides a trail termination function as illustrated in Figure 5.23. It provides two bits (bit 1 and bit 2) for a BIP 2 block error-monitoring function. This ensures

BIP-2		FEBE	Path trace	Signal label			Remote alarm
1	2	3	4	5	6	7	8

Figure 5.23   LO POH in byte V5 of VC-11, VC-12 and VC-2.

even parity separately in all the odd-numbered bits and all the even-numbered bits in the VC frame. One bit (bit 3) is provided for a far end bit error indication and bit 4 may be used as a remote defect indication (RDI). Three bits (bits 5, 6, and 7) are provided as a signal label and one bit (bit 8) is provided for a far end receiver fail (FERF) indication. The signal label has five defined values referring to the client layer adaptation function that is being used. These are "unequipped" when no client layer function is present, equipped (nonspecific), asynchronous, bit synchronous, and byte synchronous.

The LO path trace ID in the J2 byte takes the form of a 15-byte string of ASCII characters in the E164 format preceded by a 1-byte start flag as already described for the HO Path. The K4 byte is defined to provide a communication channel between path layer protection units and the N2 byte is reserved as a tandem connection monitor for performance supervision of tandem connections.

5.3   CLIENTS OF THE SDH PATH LAYERS

Any telecommunication signal to be transported in the SDH network must first be mapped into one or other of the synchronous containers described previously in this chapter. We will not attempt to describe comprehensively all of these in detail. Instead, we will discuss the main principles used in mapping client information into SDH VCs and consider some representative examples in more detail. SDH mappings fall into three main categories.

5.3.1   Mapping of Synchronous or Pseudosynchronous Signals into SDH VCs

This class of signals will have been derived from high-stability synchronized reference clocks. They are the most important and the most numerous because of the widespread influence of digital switching and the ISDN concept in shaping the modern telecommunications network. This is also the class of traffic for which SDH has been optimized. Circuit layer signals that are frame-synchronous (or nominally so) are mapped into fixed byte positions in an SDH VC whose frame phase is obtained from that of the client layer signal to be carried and whose offset from the local SDH system frame phase is used to set the pointer value at the source. Subsequent phase variations in the circuit layer signal with respect to the SDH reference at the VC originating node are accommodated by the pointer mechanism. The complete range of 64 kbit/s and $n \times 64$ kbit/s service rates can be considered in this class.

5.3.1.1 *Mapping of n × 64 kbit/s Services*

According to well-established current practice, isochronous 64 kbit/s channels are typically assembled into groups of 24 channels in North America, or 30/31 channels in Europe, for transmission over digital facilities at the primary rates of the PDH. These assemblies may originate within basic rate switching machines such as PSTN switches or basic rate digital crossconnect systems (DCS). Alternatively, they may originate within multiplex equipment that is part of a digital loop carrier (DLC) that has been loop-timed from the network or within a synchronized special services channel bank. The choice of the smaller SDH VCs has been strongly influenced by the administrative need to provide equivalent capacities to match these assemblies to ease the transition from PDH to SDH.

The so-called byte-synchronous mappings of G.709 have been developed primarily to meet this requirement. These are illustrated in Figure 5.24 for the case of the VC-11 defined for 24-channel groups, and VC-12 defined for 30/31 channel groups. The VC-11 has 7 reserved bytes per multiframe while the VC-12 has 15. These are normally regarded as fixed justification, but uses have been defined for specific applications. The bytes labeled R_0 for instance, can be used to transport the TS0 information from a 2048 kbit/s G.704 frame. This will be useful when it is necessary to transport such signals intact across an island of the SDH. The R_p byte is defined to provide an alternative mechanism for "out-of-slot" channel-associated signaling.

Number of frames in 4-frame multiframe

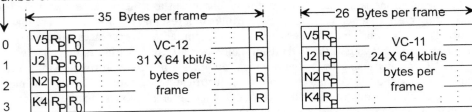

R_p	May be used as alternative outslot Channel associated signaling
R_0	May be used for TS0 of 2Mbit/s G.704 frame
R	Reserved byte
V5	POH

Figure 5.24   Byte-synchronous mapping of synchronous primary multiplex groups into VC-11 and VC-12.

The locked-mode mappings, in which pointers are frozen, and bit-synchronous mappings for synchronized but unframed signals, have also been specified but there has been little interest in implementing these in real networks.

5.3.1.2 *Higher Rate Isochronous signals*

There has been a tendency in some areas (notably broadcasting) to choose digital standards that are as compatible as possible with telecommunication standards. Ease of multiplexing high-quality audio with data and video and common transport and crossconnecting solutions are among the benefits expected. The digital video coding standard agreed upon by ETSI and the European Broadcasting Union (EBU) is a good example. In this standard, the digital video is multiplexed together with sound, teletext, forward error correction, and the many overheads required by the broadcast distribution network into a byte-organized 125-µsec composite video frame of 530 bytes. This is mapped directly into a VC-2-5c in a manner compatible to the synchronous payload structure defined in G.832 for transport over existing PDH paths.

5.3.2   Mapping of Unsynchronized Signals

These are signals whose rates are fixed but are not controlled closely enough to be adequately tracked by SDH pointers. All PDH signals are in this category, with the exception of the synchronized G.704 frames mentioned above. Generally, the PDH signals are themselves part of a transport network, and it is often more beneficial to terminate the PDH path or paths at a boundary with the SDH, recover the payloads, and transport these directly in the SDH VCs using one of the mappings described above. However, it is also useful to be able to transport such signals intact across SDH islands in transitional situations. More importantly, the PDH signals are frequently used to provide clear channel leased lines for private networks, many of which use proprietary frame structures internally. Contractual commitments will require that these continue to be transported by the SDH network in this transparent way.

   Plesiochronous signals that are not synchronized to a network reference, but are providing clear channel leased capacity or links in a residual PDH network, are mapped into SDH VCs by using a bit justification process similar to the processes already used for interlayer multiplexing in the PDH itself which was discussed earlier in this chapter. The SDH VC, created using the network clock, provides a justification frame with bit-level justification opportunities and error-protected justification signaling channels. The plesiochronous signal is justified into this frame using a similar adaptive synchronizing process to that described for plesiochronous multiplexing in Section 5.1.2.

5.3.2.1   *Mapping of Unsynchronized and Unstructured Low-Rate PDH Signals into VC-1/2*

At the primary rates, the justification frame is formed by using the reserved bytes to create one positive bit justification opportunity and one negative bit justification opportunity per VC frame of 500-µsec. A 3-bit justification signaling channel is provided for each and "two out of three" majority voting is used at the desynchronizer to provide some resilience to transmission errors. This process is termed positive-negative justification, and in this respect differs from the earlier plesiochronous multiplexers that used simply negative justification. The mapping jitter in the recovered signal produced by a positive-negative justification strategy is greater than that produced by a well-designed negative justification strategy. This requires a relatively narrowband desynchronizer to suppress the mapping jitter in the recovered signal to a level compatible with existing plesiochronous specifications. However, the desynchronizer performance required to attenuate pointer-induced jitter to similar levels is such that the mapping jitter is easily absorbed. Desynchronization and residual phase error impairments are discussed more fully in and possible implementations are considered, in Chapter 12. The VC locations used for tributary data are the same as those in the bit-synchronous mapping.

The two primary rate mappings are illustrated in Figure 5.25. These mappings are used when the input signal is asynchronous to the SDH clock, whether structured or not, and whenever the signal is unstructured, whether synchronized to a network clock or not. The North American DS2 rate (6312 kbit/s) is mapped into VC-2 using a mapping that is a simple extension of that used at VC-11, scaled by a factor of 4. No similarly simple mapping exists for the second level of the European PDH at 8448 kbit/s. This was a casualty of the compromise that produced the present standard, with its remarkably high level of global compatibility. This is not seen as a significant limitation as there is no significant role for either the 6312 kbit/s or the 8448 kbit/s rates in planned transition scenarios. The subjects of deployment and transition are discussed more fully in Chapter 13.

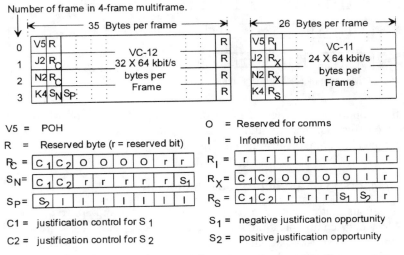

Figure 5.25   Mapping of plesiochronous primary rates in VC-11 and VC-12.

5.3.2.2   *Mapping High-Rate Plesiochronous Signals in VC-3/4*

The 34368 kbit/s, 44736 kbit/s, and 139264 kbit/s signals of the existing PDHs are the most obvious examples in this class, and all have defined mappings into SDH VCs. A less obvious example is provided by the fiber distributed data interface (FDDI) that has been standardized by the data-processing community as a high-speed LAN for distributed computing applications.

Mapping of plesiochronous 34368 kbit/s signals in VC-3 retains positive-negative justification as for the lower rates. Three justification subframes are created in rows 1–3, 4–6, and 7–9 of the 9-row structure. Majority voting on five justification control bits per justification subframe is used.

In the case of 44736 kbit/s and 139264 kbit/s signals in VC-3 and VC-4, respectively, negative justification only is used. There is a justification subframe with one negative

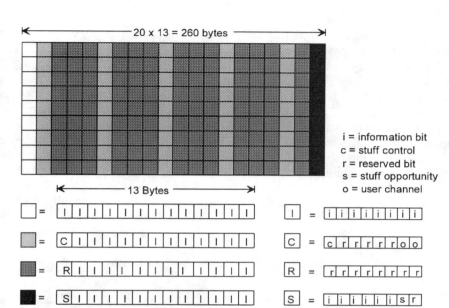

Figure 5.26   Mapping of 139264 kbit/s signals in VC-4.

justification opportunity in each of the 9 rows of the VC structure. Majority voting is performed on five justification signaling bits, C, in each subframe. The 139264 kbit/s mapping, typical of both, is illustrated in Figure 5.26.

5.3.3   Mapping of Asynchronous Signals in SDH

These are signals that do not have fixed information rates and are frequently bursty in nature. Most data communication signals (X.25, frame relay, etc.) are in this class. Asynchronous octet streams, which contain their own data block delineation information, are allocated the full payload capacity of the SDH VC. The adaptation process must implement flow control to ensure that the rate of the synchronous VC is not exceeded and justification with empty cells or packets to maintain approximately constant information flow when the source rate is low. In the case of ATM in SDH, additional cell delineation information was specified for STM-1 as part of the payload-specific overheads in the SDH VC, but this has been abandoned for all other synchronous containers in favor of self-delineation. This will be discussed in more detail after the principles of asynchronous multiplexing have been introduced.

5.4   PRINCIPLES OF ASYNCHRONOUS MULTIPLEXING

Asynchronous transport was first developed in the world of data communications where data sources are generally anything but periodic. The sporadic output from a keyboard or the stop-go mode associated with computer disk access exhibit no discernible periodicity. A typical data link protocol has its data organized in blocks of arbitrary length. Each block is preceded by a label containing a flag designed to uniquely identify the label in an otherwise random data stream. The flag identifies the start of the frame and implicitly the octet boundaries if, as is often the case, the stream is octet-structured. The flag used in the label must therefore be unique, or at least statistically rare. The block of data is termed a frame, and the label typically indicates the length of the frame and other information associated with operating and maintaining the link.

Refinements may include a terminating field containing a checksum or other validity monitoring mechanism. Addition of a channel identifier field to the frame start label provides the multichannel capability required for multiplexing different sources on one medium. The mechanism has also been called "label multiplexing" because channels are distinguished by the identifier in the label rather than the timeslot occupied in a periodic frame structure as in SDH and PDH. The channels are virtual in the sense that they exist by virtue of being allocated an identifier from a set regardless of whether data is transported or capacity used on the link.

A large number of variants of this basic procedure have been developed over the years. The data link layer (layer 2) of the ISO seven-layer model is typical of such data frames and LAPD, as specified for ISDN access, is a typical extension incorporating multiplexing features by virtue of the data link channel identifier (DLCI).

Large data transfers are organized as packets and allocated as payload to frames of a single virtual channel. The packet is formed in a similar way to the basic frame by adding a header label (and sometimes a trailing label) before allocating the packet to the virtual channel of the frame-mode data stream. Quite elaborate error-recovery protocols may be included. Such packets can be switched through a large data network on the basis of the identifier in the packet header. The X.25 packet-switching standard uses this principle and has become widely deployed in Europe, but less so in the United States. More recently, frame relay standards have been developed to support switching of individual frames. These may be seen either as a simplification of X.25 suitable for use in less error-prone environments or alternatively as an enhancement of the basic link layer protocol by adding some network level functionality.

5.4.1   Asynchronous Transfer Mode (ATM)

The asynchronous transfer mode (ATM), as opposed to the synchronous transfer mode (STM), can be seen as the ultimate refinement of this approach, reducing the structure to the bare minimum required for transporting data. In ATM, the data stream is octet-structured and the basic frames, called cells, are short and have fixed length. The short, fixed-length cell is well adapted for fast cell switching using hardware rather than software technology. Only a leading label is used in each cell, and this is called the cell header. The header carries the minimum of overhead to support multiplexing and switching only.

A lot of the existing traffic on the integrated digital network originating from analog sources, such as voice, broadcast video or audio, has been perceived as periodic and essentially synchronous because it is derived using the simple fixed sample rate, pulse code modulation schemes standardized in the early years. However this perception is misleading because the apparent periodicity is a property of the channel coding process, not of the information sources themselves. In fact, more recent developments in efficient coding and compression of voice and video material only serves to emphasize the essential nonperiodicity or burstiness of the information sources themselves. Speech is full of pauses and gaps at syllabic and conversational rates, while video material exhibits a much wider range of nonstationary statistical behavior. More powerful source coding mechanisms are now available, capable of exploiting the ability of ATM to absorb the essential burstiness that characterize these analog sources. ATM has the essential capability to multiplex and switch data from many sources with widely varying rates and information statistics, and is therefore the truly universal transfer mode for multimedia, information, voice, or video services whether originating in the broadband integrated services digital network (B-ISDN) or the fast-growing Internet and intranet segments.

5.4.2   Structure of ATM Cells

The basic tradeoff in dimensioning ATM cells is between multiplexing efficiency and delay performance for a given target channel multiplexing ratio. The longer the cell, the higher the ratio of client data to header data and hence the more efficient it is. The shorter the cell, the shorter the cell assembly delay and the lower the phase error effects associated with processing and routing. Delay and the inevitable variations in delay resulting from queuing in a busy network are more critical for real-time signals. Two-wire voice telephony circuits, in particular, are subject to echo impairments if the delay exceeds certain limits.

This basic ATM dimension was the subject of intense argument right at the start of the standardization process. Europeans, in general, supported a cell payload of 32 octets to limit the need for extra echo control for the voice component in an ISDN. North America favored a 64-byte payload to maximize efficiency, and took a more relaxed view of the echo problem. The Japanese, as in the earlier confrontation over SDH, took on the role of mediators. The convictions in both camps reached a quite religious intensity. A compromise proposal with 40 bytes per cell, with some interesting numerological properties in the SDH context, was narrowly defeated, and the vote at the end of the 1988 plenary was a cliffhanger. It ended in a typical ITU-T compromise of 48 bytes, combining either the best or the worst of both positions according to religious convictions. The agreed ATM cell size of 53 bytes made up of 5 header bytes and 48 information bytes is illustrated in Figure 5.27.

5.4.2.1   The Cell Header

The main functions of the header are the label itself, whose size determines the available multiplexing ratio, and a mechanism to guarantee resilience of the multiplexing process. An n bit label will support 2^n separate channels in the aggregate cell stream. Resilience in the presence of errors is achieved by means of the header error check (HEC) mechanism.

Within the cell label, 28 bits are allocated to the routing field. A three bit payload type identifier (PTI) distinguishes between particular classes of information flow and 1 bit is used to signal cell loss priority (CLP) in case it is necessary to discard cells when congestion occurs. A further octet is used for header error control (HEC).

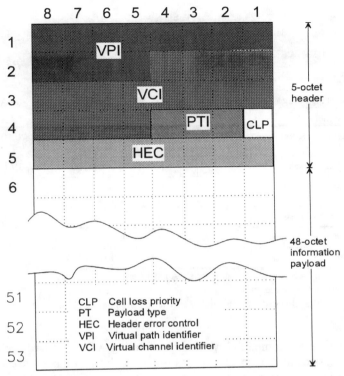

Figure 5.27 ATM cell structure.

5.4.2.2 Routing Field (VPI/VCI)

The ATM layer is organized as two sublayers. The virtual channel layer is a client of the virtual path (VP) layer as explained in Chapter 3. ATM virtual channels are normally designated as VCs within the ATM community; we shall use the form VCh to avoid confusion with SDH virtual containers. The 12-bit VP indicator field (VPI) and the 16-bit VCh indicator field (VCI) theoretically allow for 4,096 virtual paths and 65,536 virtual channels, although some are reserved as messaging channels for intralayer processes as indicated in Table 5.3.

5.4.2.3 Generic Flow Control (GFC)

Generic flow control is only available at the UNI and uses the first four bits of the VPI field, thus reducing the number of VPIs supportable at the UNI to 256. Connections admitted at a UNI may be of a controlled transmission type or an uncontrolled transmission type. Uncontrolled or controlled in this context refers to whether or not the network is directly exercising flow control of data from the source. Historically, LANs and other shared media access systems controlled media access by various mechanisms (token passing, collision detection, etc.). Connectionless WAN systems being developed concurrently, such as switched multimegabit data service (SMDS), adopted the 53-byte structure

Table 5.3

Predetermined VPI/VCI Allocations (and Valid PTI/CLP)

Intralayer process usage	*VPI*	*VCI*	*PTI*	*CLP*
Metasignaling	note 1	00000000 00000001	0A0	C
General broadcast signaling	note 1	00000000 00000010	0AA	C
Point-to-point signaling	note 1	00000000 00000101	0AA	C
VP layer tandem connection OAM process		00000000 00000011	0A0	A
VP layer trail OAM process		00000000 00000100	0A0	A
VP layer tandem connection OAM process		applicable to all VCI values except '0'	100	A
VP layer trail OAM process		ditto	101	A
Resource management		ditto	110	A
Unassigned cell		00000000 00000000	-	0

A — may be '1' or '0' according to the appropriate ATM layer function

C — CLP set to '0' at the source but may be modified to 1 by the network

note 1: For VPI = 0, the allocated VCI channel is only used for signaling between user and local switch

in an effort to maintain compatibility with the emerging ATM standard and also required such media access control. Uncontrolled transmission, however, is more common in real ATM networks and in this case, all four bits are set to zero. Controlled transmission may be used by the network to control aspects of the user network, particularly (as in SMDS) to arbitrate access among several terminals using a shared access medium.

5.4.2.4   Payload Type

The payload type indication (PTI) is a 3-bit field that additionally distinguishes between different types of cells associated with a particular VCh. The PTI is used as follows:

- Bit 4 = 0 indicates that the cell contains client data. In this case:—Bit 3 is used between client access groups passing transparently through the ATM layer;—Bit 2 is used as a forward congestion indicator such that it is set to 0 at the connection source and is changed to 1 when the cell transits a node experiencing congestion. With these two mechanisms, it is possible for an ATM client to exercise some end-to-end flow control with a view to relieving the congestion.
- Bit 4 = 1 indicates that the cell contains control or management data associated with the VCh indicated. In this case:—Bit 3 is set to 0 to indicate that the cell is used to transfer OAM information within the VCh. With bit 2 set to 1, the cell is used to connect source and sink trail termination functions in the VCh trail. When set to 0, it provides a similar function for VCh tandem connections, which are trails in a tandem monitoring sublayer between administrative boundaries.—Bit 3 set to

1 and bit 2 set to 0 indicates that the cell is used to connect agents of a resource management process.

5.4.2.5   The Cell Loss Priority Field (CLP)

ATM networks are normally designed so that the probability of cell loss is negligibly small. To achieve this, traffic offered to each connection must conform to a set of traffic descriptors that determine the statistical distribution of cell intervals in the admitted cell streams. The operator may, at his discretion, also accept cells that momentarily exceed the negotiated parameters. If congestion then occurs the operator will be allowed to discard those cells in excess of contracted commitments that were nevertheless accepted initially.

The CLP is also used to enable controlled cell discard in the presence of congestion and hence a more graceful degradation. When the onset of congestion is detected, cells whose CLPs are set to 1 are considered lower priority and may therefore be discarded before cells whose CLPs are set to 0, which are higher priority.

The mechanism may also be exploited in certain classes of coded analog information, such as video coding, where the source coder generates two classes of information, one of which is more essential than the other. In this case, information can be segregated into two classes of differing priority and allocated to VCIs with CLPs set accordingly.

5.4.2.6   Header Error Control (HEC)

The asynchronous multiplexing procedure is totally dependent for its integrity on its ability to reproduce the cell labels accurately and to discard as invalid any known to be in error. When transmission is degraded, and hence causing errors, it is acceptable to discard errored cells but not acceptable to interpret cells wrongly as, for example, belonging to a different VCh. One octet in the header is allocated for header error control. The polynomial formed by the first four octets of the header is divided by the polynomial $x^8 + x^2 + x + 1$. The 8-bit remainder is placed in the HEC field. At the receiver, the process is repeated and the locally calculated remainder compared with the received HEC.

This particular error-detecting code has the property of being able to detect and correct single errors and being able to detect (but not correct) multiple errors. To exploit this, the receiver is operated in two modes: correction mode and detection mode. In the absence of errors, the receiver is in the correction mode. If a single error is detected, a correction is made and the receiver passes to the detection mode. Subsequent error-free headers cause the receiver to return to correction mode. If multiple errors are detected, the cell is declared invalid and discarded, and the receiver passes to the detection mode.

Single errors (in practice, this means large mean time between errors) are rare, but can occur in optical transmission systems under marginal line conditions. Experience has shown that burst error events are far more common in practice. Therefore, it is probable that the correction mode will seldom be used in real life. Nevertheless, it is considered to provide a useful extension of cell transmission performance at no extra cost. The error detection is very powerful and the probability of invalid cells being passed is very low.

5.4.2.7   *Cell Delineation*

The recovery of cells from a multiplexed stream requires first that the octet and cell boundaries be identified before the client information can be recovered. If an SDH server is used, octet and cell start information can be carried by the server as part of the underlying transport service, as described above for VC-4. However, in the general case, from PDH servers or unstructured bit transmission systems, this service is not available. It is therefore necessary to provide an independent cell delineation mechanism.

Cell delineation is achieved easily on an error-free stream by using the correlation between the (32) header bits to be protected and associated (8) control bits introduced in the header by the HEC function. That is to say, the receiver inspects candidate 5-octet sequences from the incoming stream until it finds the sequence for which the HEC computes to error-free, discarding all others. Once aligned, a flywheel mechanism can be used to increase resilience of the cell delineation mechanism. That is to say, more than one errored HEC is required to initiate a new search. As with all such processes, the parameters of the search algorithm are not standardized, but can be chosen to achieve a performance level appropriate to the service being transported by making a tradeoff between time to recover alignment and resilience to transmission errors in a given environment.

5.4.3   The Asynchronous Multiplexing Process

The PDH and SDH multiplexing processes both involve the deterministic placing of input channel data units in specifically reserved slots in the aggregate stream. Small, deterministically dimensioned buffers are needed to hold input data while inserting overheads and the justification data needed to compensate frequency difference (in PDH) and phase difference (in SDH) of the input streams. In the generalized broadband context, the input streams are characterized as stochastic processes whose rates are not predetermined, and input channels compete for slots in the aggregate stream according to need to achieve the connection QoS parameters negotiated with the network.

An ATM network is an example of a queuing network and it is the behavior of the queues under varying traffic load conditions that is the main determinant of connection behavior. Input cells are placed in queues and allocated to cell slots in selected output ports according to availability and a policy expressed as an algorithm, for sharing available bandwidth when there is contention. Contention for resources causes some cells to be held longer than others, until the scarce resource becomes free. Because it is of finite length, a queue may overflow if traffic is too heavy. This causes loss of cells, and this loss is not necessarily confined to the connection causing the problem. Queues must be dimensioned to make the probability of cell discard acceptably low and to keep cell delay variation within acceptable bounds. A cell loss ratio of 10^{-10} is the normal target, but higher values may be considered for certain classes of service, while allowable cell delay variation is very much service-dependent.

The main data transfer service categories were introduced in Chapter 2, and the traffic characterization for the purposes of connection admission control is analyzed extensively in Chapter 9. The declared intention to offer a data stream conforming to a particular set of traffic descriptors is not sufficient to adequately protect the network so that it can guarantee adequate service to all users. The actual characteristics of the stream must be monitored throughout the network to ensure that the contracted conditions are met. This activity is known as "policing." Usage parameter control (UPC) is applied at the edge of the MS network to monitor and control the admitted data streams so that the network is protected from overload, whether malicious or accidental. Network parameter control (NPC) is applied within the network, particularly at administrative boundaries, for the same purpose.

5.4.3.1 ATM Trail Termination Processes

The ATM trail termination processes provide for monitoring the integrity and quality of individual connections by communicating status information from source to sink in the designated OAM overhead channel routed over the same server route as the client information. The so-called F5 information flows provide trail supervision in the VCh layer and the F4 information flows provide the same function in the VP layer. VCh cells with the value 100 in the PTI field are used to provide a communication channel between trail termination source and sink to achieve this. The F5 flow shares the same VPI/VCI channel identifier as the virtual circuit it is supporting, and therefore shares the same route as all other cells in the connection. The PTI value 101 indicates that the cell is used to communicate between points delimiting a tandem connection or connection segment of a VCh connection. Such monitored tandem connections are effectively trails in a tandem connection-monitoring sublayer.

The equivalent trail and tandem connection supervision functions in the VP layer use preassigned VCs within each VP. VCI = 4 provides communication end to end between

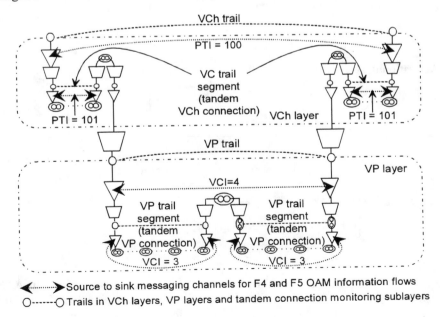

◀┈┈┈▶ Source to sink messaging channels for F4 and F5 OAM information flows
O┈┈┈O Trails in VCh layers, VP layers and tandem connection monitoring sublayers

Figure 5.28   OAM flows in ATM VPs and VCs.

trail termination source and sink, while VCI = 3 is used over the tandem connection. The OAM flows, their relationships, and the communication channels they use are illustrated in Figure 5.28.

5.4.3.2   Information Formats Within ATM Trail Processes

The various ATM layer trail termination processes indicated in Figure 5.28 communicate bidirectionally within the trails using a common cell payload format. This is illustrated in Figure 5.29. The OAM cell type field indicates the type of management function as listed in Table 5.4, namely fault management, performance management, and activation/deactivation. The function type field indicates the specific function being performed and the EDC field carries a CRC-10 error-detecting code based on the generator polynomial $x^{10} + x^9 + x^5 + x^4 + x + 1$.

Table 5.4

OAM Cell Type and Function Field Codes

OAM Cell Type	Value	Function Type	Value
Fault Management	0001	AIS	0000
		RDI	0001
		Continuity check	0100
		Cell loopback	1000
Performance management	0010	Forward monitoring	0000
		Backward monitoring	0001
		Monitoring & reporting	0010
Activation/deactivation	1000	Performance monitoring	0000
		Continuity check	0001

Header	OAM type	Function type	Function-specific field	Reserved for future use	EDC (CRC10)
5 octets	4 bits	4 bits	45 octets	6 bits	10 bits

Figure 5.29   Standard payload format for OAM cells.

5.4.3.3   Fault Management

The basic network requirements for in-service fault management are met in ATM by application of the established mechanisms of alarm indication signal (AIS) and remote defect indication (RDI). On detection of a service affecting failure, an AIS is sent downstream and an RDI is sent upstream in a designated OAM channel, thus informing all nodes up to the connection termination that it is unavailable. In addition, a 1-octet fault type field and a 15-octet fault location field are defined within the function-specific field to provide fault diagnostic information almost instantaneously wherever it is needed.

In ATM, a connection can exist even when no resources are allocated and there is no information flow. To check the integrity of such a connection, a continuity check function is provided. The loopback function is useful for single-ended integrity checking during connection setup, for fault location, and also as a means of estimating round-trip delay before accepting a connection. The loopback cell function-specific field includes the following:

- Loopback indication (1 octet) to signify whether loopback has occurred;
- Correlation tag (4 octets) to identify the cell for later correlation;
- Loopback location (15 octets) to indicate the locations to be looped;
- Source identifier (15 bytes) to allow the originator to recognize its own cells.

5.4.3.4   Performance Management

Performance-monitoring (PM) cells are introduced by the trail source into a virtual connection of path at intervals of N user cells. The block size N may be 128, 256, 512, or 1,024. The PM cell is inserted in the next unassigned cell slot. If no unassigned slot is available after $N/2$ cells have been sent, the monitoring cell may be forced by taking higher priority than other contenders. PM cells are examined at the downstream sink and the results reported to management processes. A backward PM cell is sent from the downstream sink back to the upstream source containing information about the downstream connection. The PM cell format is illustrated in Figure 5.30.

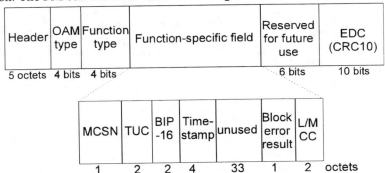

Figure 5.30   Format of the OAM performance management cell.

The fields are defined as follows:

- The monitoring cell sequence number (MCSN) is a 1-byte (modulo-256) indication of cell order to confirm that cells have been received in the correct order without any missing (or indeed misinserted).
- The total user cell count at 2 bytes indicates the block size used.

- The bit-interleaved parity method of error detection has already been described in Section 5.2.4.4. Here, BIP-16 is computed over the N user cells and the result transmitted in the BIP-16 field.
- The timestamp indicates the time of origination of the cell.
- The block error result is a 1-byte summary for backward reporting the number of BIP-16 errors detected at the sink.
- The lost or misinserted cell count (L/M CC) is the number of lost or misinserted cells as monitored at the sink and reported back to the source.

Performance monitoring may be initiated or terminated according to the needs of the process that is supported. Proper functioning requires that cell counters and sequence counters be synchronized within the trail process sources and the sinks aligned according to the parameters used. This is a real-time process that cannot be left to messaging within a TMN environment. The activation/deactivation cell, illustrated in Figure 5.31, is designed to perform the initiation and deactivation from one end, denoted the activator (or deactivator). The message identifier (MID) indicates the message function, as shown in Table 5.4:

- Direction of actions (DOA) indicates forward or backward monitoring.
- Correlation tag (CT) uniquely identifies the cell.
- The PM block sizes A-B indicate the block size chosen in the direction away from the activator/deactivator source.
- The PM block sizes A-B indicate the block size chosen from the remote sink toward the activator/deactivator.

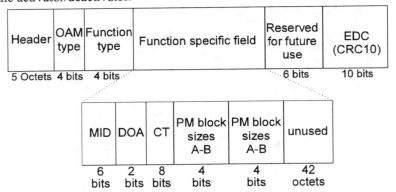

Figure 5.31   Format of the OAM activation/deactivation cell.

The trail termination processes in ATM comply with the same architectural model discussed in Chapter 3. The flexibility of ATM has enabled the provision of rather more comprehensive capabilities than will be found in SDH. Tandem connection monitoring, for example, may be introduced as required without revisiting basic standards. There is a danger in allowing gratuitous flexibility into the operational environment of a creeping complexity, which will generate its own problems for operational cost and service quality, but the ability to provide enhanced operational facilities such as continuity and loopback tests will significantly extend the diagnostic power available to the operator.

5.4.3.5 Adaptation of ATM to an SDH Server Layer

The ATM cell stream is presented to the SDH layer as an octet stream by an adaptation function that inserts idle cells if the offered rate is not sufficient to fully load the SDH channel and restrains the source, discarding cells if necessary, when the offered rate becomes too high. A special label value is assigned to identify idle cells. This has the value VPI = 0, VCI = 0, TPI = 0, and CLP = 1. In the reverse process, only nonidle cells are returned from the server to the ATM layer and idle cells are discarded. Therefore, the actual transmitted cell stream has a rate that is synchronous with the SDH container in which it is transported. The actual information rate is defined by the source, but limited by the maximum capacity of the SDH container.

It is possible for a malicious user to compromise the network from her terminal by inserting particular sequences in the information field. It is possible, for instance, to complement the SDH frame-synchronous transmission scrambler and thereby induce transmission corruptions. It is therefore necessary to scramble the cell information field before mapping. This is achieved with a self-synchronizing scrambler with generator polynomial $x^{43} + 1$. The scrambler operates on the information field only, being suspended during the cell header and resumed in the next information field. Long, single-tap

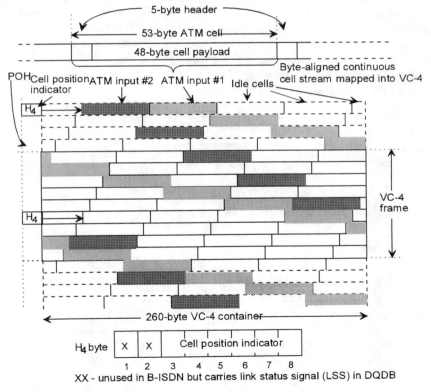

Figure 5.32   Mapping an ATM cell stream into VC-4.

scramblers of this type are not effective as transmission scramblers as they do not have a sufficient randomizing effect. They are, however, sufficiently secure to prevent malicious simulation. The length of 43 has been chosen as the lowest prime number greater than 40 to prevent error multiplication within the cell header.

5.4.3.6   *Mappings in SDH Containers*

The ATM mappings have been defined into VC-4 and VC-4-4c. Neither of these payloads carries an integer number of ATM cells, and therefore cell boundaries will not be in the same relative position in successive SDH frames. This does not matter, as cell streams carry their own cell delineation mechanism in the form of the header error check (HEC) already described. Nevertheless, the VC-4 mapping provides a cell position indicator encoded in the H4 byte of the path overhead that may be used in the cell delineation process. This takes the form of a binary number value that equals the number of bytes from H4 to the first byte of the first cell after H4. The maximum value is 52. This cell position indicator can be used to assist in cell delineation at the corresponding adaptation sink function.

In the VC-4.4c mapping, the payload available for cell transport is exactly four times the VC-4 payload. Three columns of fixed stuff are included in the positions where the POH of the interleaved VC-4s would have been. This is to ensure that if, in the future, virtual concatenation is introduced at this level, then the payload will not have to be reduced to maintain compatibility between explicit contiguous concatenation and virtual concatenation. Figure 5.32 illustrates two ATM channels in a single VC-4 and Figure 5-33 illustrates a single ATM channel mapped into a VC-4-4c container, as will commonly be found at an STM-4 trunk port of an ATM switch.

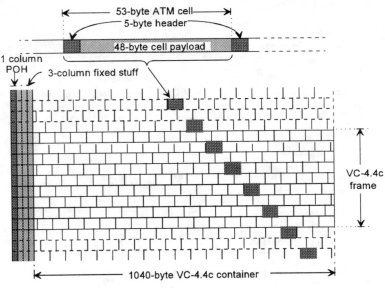

Figure 5-33   Mapping an ATM cell stream into VC-4-4c

Metropolitan area networks (MANs) based on the distributed queue dual bus (DQDB) protocol have the same cell structure as ATM, but incorporate extra data to control the operation of the bus. They therefore use the same cell-mapping procedure, except that bytes F2 and Z3 of the POH are used to carry the DQDB layer management (M1 and M2) bytes and bits 1 and 2 of the cell position indicator (H4) byte are used for the link status signal (LSS).

5.4.4   Use of PDH for the Transport of ATM and SDH

SDH and ATM have the potential between them of obsoleting first PDH as a major transport technology, and then isochronous switching as the main service-switching technology. Nevertheless, whenever new technology is introduced in the network, strict investment criteria will ensure that existing investment is protected and maximally exploited.

In many cases, the benefits of the move to SDH will be sufficiently great that existing PDH facilities will be replaced. The planning and management benefits are not fully realized if there is a substantial quantity of nonconforming elements remaining in the network. On the other hand, some PDH transmission facilities are very costly to replace and have many years of useful life ahead of them. Major long-haul submarine systems and high-capacity radio systems are often found to be in this category. In such cases, the precious facilities may be successfully integrated in a managed SDH network by allowing them to carry SDH containers.

ATM, on the other hand, is more service- and network-oriented. To achieve the service benefits of an ATM network, it seems to be important to achieve early wide area penetration. This cannot be achieved if it were necessary to await the arrival of SDH infrastructure everywhere. ATM cell streams mapped directly in PDH transmission systems allow wide area connectivity from the start. Thus ATM and SDH deployment plans may be effectively decoupled. This service-oriented motivation also applies to SDH to a lesser extent in that SDH continues to provide wide area managed connectivity for leased line services. Therefore exploitation of the embedded PDH base where feasible and economic is an important feature of most deployment plans.

5.4.4.1   *SDH Modems*

These pieces of equipment allow SDH path layer formats to be carried in PDH paths in the same way as the original high-speed modems of 20 years ago allowed digital primary groups to be carried in analog FDM groups. Mappings are defined in G.732 for 139264 kbit/s, 34368 kbit/s and 97728 kbit/s PDH paths. Each mapping consists of two parts: a pseudo SDH path layer at the PDH rate used that mimics the functionality of the SDH HOP layer and a payload mapping that uses an assembly of the standard TU and AU mappings as defined in Figure 5.6 and Figure 5.18. The pseudo path layer formats are illustrated in Figure 5.34 for 34368 kbit/s and 139264 kbit/s. The 97728 kbit/s mapping follows the same principles.

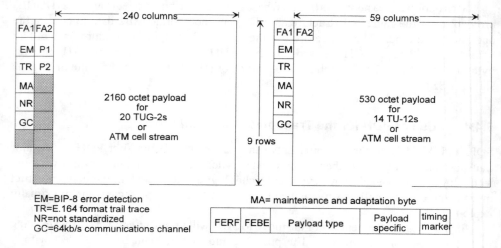

Figure 5.34   Pseudopath layer used for mapping SDH TUs or ATM cells in 139264 kbit/s and 34368 kbit/s.

5.4.4.2   Mapping ATM Cell Streams in PDH Paths

All the mappings of ATM cell streams into plesiochronous transport paths are documented in G.804. There are three distinct classes of solutions represented there: the higher rate PDH levels that use the G.832 pseudopaths described above, the primary (and secondary) PDH rates that retain the original G.704 synchronous frame and payload definition, and an alternative DS3 mapping that predated these but is retained for backwards compatibility with some earlier designs.

5.4.4.3   Mapping in DS3

The first mapping to be developed commercially, and perhaps the most bizarre, was that for DS3 in the United States. This defines a 125-μsec frame, termed the physical layer convergence protocol (PLCP), aligned with nibble1 boundaries derived from the DS3 frame itself and containing its own alignment information. Each 125-μsec PLCP frame contains exactly 12 ATM cells together with the alignment information and some path overheads. This was developed before the native ATM cell delineation mechanism based on HEC (described above) was known. Subsequently, a second alternative DS3 mapping has been agreed on (also documented in G.804) in which ATM cell octets are nibble-aligned in the total DS3 payload without a PLCP frame. In this version, cell delineation is achieved by HEC correlation as described above and defined in I.432.

---

[1] A nibble is 4 bits or half an octet.

5.4.4.4   *G.832-Compliant Mappings*

ATM cells are carried in the 139264 kbit/s, 34368 kbit/s, and 97728 kbit/s PDH paths using the same payload area as for the SDH mappings in the PDH defined in G.832. The pseudopath structures for 139264 kbit/s and 34368 kbit/s have already been described and illustrated in Figure 5.34. In all cases, the ATM cells are octet-aligned with the designated payload area and scrambled using the same self-synchronizing scrambler as described in Section 5.4.3.5. Cell delineation is achieved using the HEC correlation method defined in I.432 and described in Section 5.4.2.7.

The available payload used by the ATM cells represents exactly the same cell rate as an equivalent SDH payload. This facilitates bridging ATM carrying SDH and PDH links, which can thereby be performed without any cell rate adaptation.

5.4.5   ATM in Low-Rate PDH Paths

The DS1 (1544 kbit/s), the E1 (2048 kbit/s) and the DS2 (6312 kbit/s), with their standard synchronous frames, already provide octet-structured payload areas previously used for synchronized, 64 kbit/s information. These are used to accommodate the ATM streams with octet boundaries derived from the synchronous container and cell delineation achieved by the same HEC correlation method defined in I.432 and described in Section 5.4.2.7. Scrambling to combat malicious data injection is achieved with the same self-synchronizing scrambler described in Section 5.4.3.5.

The F3 OAM flows used to support the termination processes in the PDH server layer are naturally derived from those already in use in the PDH when other payloads are being carried. The two PDH primary rate mappings are illustrated in Figure 5.35.

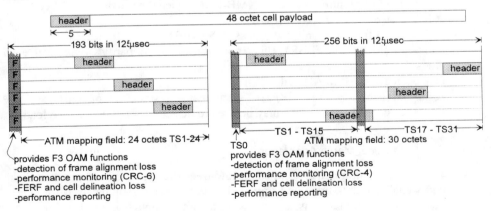

Figure 5.35   ATM cell mapping within the primary PDH rates.

5.4.6   The ATM Cell-Based Section Layer

To complete the array of transport possibilities available to carry ATM cell streams, the ITU-T has also defined a cell-based section layer. There is no intention at present to generate a whole new range of NNIs comparable to those in SDH. There is, however, considerable interest in providing a pure, cell-based physical connection at the user interface, where there is expected to be some scope for cost savings in a dedicated BB installation and where there is no significant embedded base of SDH or PDH transmission facilities to complicate the legacy situation. The requirements for such a layer are determined, as in SDH section layers, by the limitations of the transmission media with respect to timing transparency and spectral power density, and the well-established operations and maintenance requirements of physical installations.

5.4.6.1   Section Layer and Transmission Path Layer Cell Structure

Provision has been made for additional cell-based layers to perform regenerator section and the transmission path roles. These support the F1 and F3 OAM flows, respectively. For completeness, an F2 flow is mentioned equivalent to the multiplex section, but this is not considered necessary at the UNI and is not being pursued at this time. The capacity available for the ATM layers (VP and VCh) is limited to 149760 kbit/s and 599040 kbit/s in the STM-1 and STM-4 equivalents as for SDH, so physical layer cells (whether OAM or idle) must be inserted to ensure the correct ratio of physical layer cells to ATM layer cells.

The OAM cells must be transmitted at a rate that is high enough to provide adequate error monitoring and speed of response for failures. They contain the following fields:

- An 8-bit (PSN) sequence number operated modulo-256;
- An NIC field holding the number of cells included since the previous OAM cell;
- A monitoring block size (MBS); maximum 64 cells;
- The number of monitored blocks (NMB-EDC) since the previous OAM cell;
- A BIP-8 error-detecting code (EDC);
- A number of far-end monitored blocks (NMB-EB);
- A far-end block error (FEBE). This reports the number of parity violations in each block;
- One octet allocated to AIS (all "1"s);
- A far-end receive failure (FERF)—1 bit allocated to signal remote failures;
- Cell error control—a CRC-10 may be used to detect errors in the cell payload.

5.4.6.2   Physical Media Adaptation

To achieve high transition density and small low-frequency spectral components in the transmitted signal, the data must be scrambled. The $x^{43} + 1$ scrambler used for access security is not good enough for this, and, anyway, cannot be applied to the HEC. To obtain adequate transmission properties, the cell stream is scrambled by the modulo-2 addition of a pseudorandom sequence (PRS) generator defined by the polynomial $x^{31} + x^{28} + 1$. The original, unmodified cell stream can be recovered at the receiver by subtracting (equivalent to readdition modulo-2) the output of an identical, synchronized PRS generator.

It is well-known that a scrambler of order r may be synchronized by sending r linearly independent samples of the source PRS generator without error over the transmission channel. There are two basic requisites for this:

- A simple deterministic method of selecting r independent samples of the source PRS generator,
- A means of transferring these samples over the transmission channel without altering the rate or structure of the cell stream.

Statistical independence is simply achieved by selecting bit samples from the source PRS, which are spaced at 212 bits, equivalent to half a cell. To transfer these 2 bits per cell to the receiver they are used to modify, by modulo-2 addition, the first 2 bits of the HEC field. Once the descrambler is synchronized, the original HEC bits can be recovered by a complementary modulo-2 addition and the cell stream returned to the ATM layers corrected to normal. However, at the ATM transmission media (physical) layer, it is necessary first to recover cell delineation and then descrambler synchronization.

Cell delineation is easily achieved on the basis of the remaining 6 bits of the scrambled cell stream. These are still available to compute the header error check. Remember that because the modified HEC is itself calculated on the scrambled version of the header, the cell delineation is not dependent on the synchronization state of the scrambler. Once cell delineation is achieved, the descrambler enters the acquisition phase, in which the transmitted source PRS samples are recovered by modulo-2 addition with the correctly computed values for these 2 bits.

The receiver is naturally capable of calculating and regenerating the correct original HEC, enabling it to be used to recover the correct transmitted sample values. The recovered values are then compared with the values predicted by the local PRS generator at the receiver. If they agree, then no action is taken; if they disagree, then the appropriate values are reset in the local PRS. If the HEC indicates an errored cell header, then the recovered values are not used. At this rate, scrambler synchronization can be achieved in the absence of errors, within an elapsed time equivalent to 16 cells. Scrambler acquisition can commence before cell delineation is achieved, but it will not start to converge until the cell stream is itself correctly synchronized.

Once the descrambler is synchronized, it enters a verification state. This is necessary because undetected errors may have occurred during the acquisition phase. To achieve a confidence level such that the probability of false synchronization is 10^{-6} in the presence of an error rate of 10^{-3} requires 16 verifications.

After verification, the scrambler enters the normal state. In this state, a correctly descrambled cell stream is returned to the ATM layer; the properties of the HEC are completely unaffected. The descrambler continues to monitor the transmitted offset samples of the source PRS to detect subsequent loss of synchronization. Loss of synchronization will cause the descrambler to return to the acquisition state, restarting the cycle again.

As usual with such mechanisms, a balance must be achieved between robustness and speed of response in the presence of errors. A confidence counter, C, is used to trigger transitions between states.

- In the acquisition state, 16 consecutive error-free cells must be counted before passing to the verification state. A single errored cell returns C to zero.
- In the verification state, C is incremented for each error-free cell in which both predicted samples are correct and decremented for those in which one or both are incorrect. Errored cells are disregarded. When C reaches 24, the descrambler passes to the normal state. If C falls below 8, it reverts to the acquisition state.
- In the normal state, incorrect predictions cause decrements and correct predictions cause increments (up to a maximum of 24). When C drops below 16, the descrambler returns to the acquisition state.

The distributed sample scrambler/descrambler is simple to implement and is capable of very fast acquisition, even in badly errored environments.

5.5   ADAPTATION INTO THE VIRTUAL CHANNEL LAYER

The B-ISDN recommendations describe client adaptation as occurring in the "ATM adaptation layer" (AAL). We prefer the view of a layer network given in G.805 and described in Chapter 3, in which adaptation is the function performed between layers. Client layer characteristic information is adapted into a form suitable for transport in a server layer.

The AAL functions are essentially client-oriented in that they exist to safeguard the needs and guaranteed service quality of the client. This is in contrast with the traffic shaping and suppression associated with CAC and UPC/NPC policing, which exist primarily to safeguard the network against misoperation, accidentally or maliciously, by a user. These latter are considered intralayer functions normally located at administrative boundaries of a partitioned network.

The adaptation function is composed of a segmentation and reassembly (SAR) function and a convergence sublayer (CS) function. SAR is concerned with mapping client data in cells and, conversely, recovering cell payload as client data. This may additionally involve one or more of the following:

- Sequence numbering (used to detect missing or misordered cells);
- Indicating CS structure;
- Error monitoring and indication;
- Indication of multiplexed clients on a single ATM connection.

The so-called convergence sublayer is concerned with quality and integrity of the client information and may involve one or more of the following:

- Sending and recovering synchronization or framing information if required to a standard demanded by the client;
- Dealing with errored, out of sequence, or misinserted cells;
- Reordering cells, if required (e.g., in connectionless operation).

5.5.1   Classification of ATM Clients

The ATM client layers have been classified in I.363 according to their timing and statistical properties. Five classes of AAL were initially defined in I.363 for types of data streams whose traffic characteristics could be classified according to meaningful statistical properties. During the development of the standard, AAL3 and AAL4 have become merged in a single adaptation function, referred to as AAL3/4.

AAL1 is used for constant bit rate signals. This includes all the PDH rates and any proprietary signals when they are operated in clear channel mode. Mechanisms are also provided to transfer accurate timing or frame information, if necessary. AAL2, initially envisaged for new videocoding algorithms, has recently been redefined for composite cells carrying multiple, low-speed, delay-sensitive channels such as 64 kbit/s voice circuits, which benefit from the lower delay thereby achieved. AAL 3/4 was originally designed for CL data types as used in the PC LAN world, and its dominant use today is in shared media MANs such as SMDS/CBDS. AAL5 has emerged instead as the main procedure for carrying data traffic in ATM, and most streamed traffic such as video and audio is now directed towards AAL5.

5.5.2   Adaptation According to AAL1

AAL1 was developed initially to transfer clear channel signals of fixed rate, but with no restraint on exactly what rate. It was then extended to cover synchronous rates (i.e., derived from a network reference) with a structured bit stream, in particular, the 64 kbit/s octet structure and all its $n \times 64$ kbit/s derivatives (384 kbit/s, 768 kbit/s, 1920 kbit/s). Any of the SDH containers may be transferred by this AAL. Segmentation and reassembly is the same for all clients in this class, but CS functions are also specialized to deal with specific service features.

5.5.2.1   *Segmentation and Reassembly (SAR)*

The SAR protocol uses a single octet header in the leading position in each ATM cell. The structure is illustrated in Figure 5.36.

The sequence number (SN) field comprises one convergence sublayer indicator (CSI) bit, which indicates the existence or not of a CS function, and a 3-bit sequence count provided by the CS function. The sequence number protection field (SNP) provides a degree of resilience to errors in the SAR-PDU header. The protection mechanism is a two-stage process. First, a 3-bit CRC code is generated from the 4-bit SN, and this becomes the first three bits of SNP. The last SBP bit is set to create even parity over all 7 bits. Most signals in this class need the CS function to recover timing and/or structure.

5.5.2.2   *Convergence Sublayer for Circuit Emulation*

Digital circuits are characterized by their constant bit rate. They may be unstructured (e.g., primary rate clear channel) or they may be structured in frames of octets like the ISDN hierarchy of synchronized signals. In the unstructured case, the PDU is 1 bit mapped in

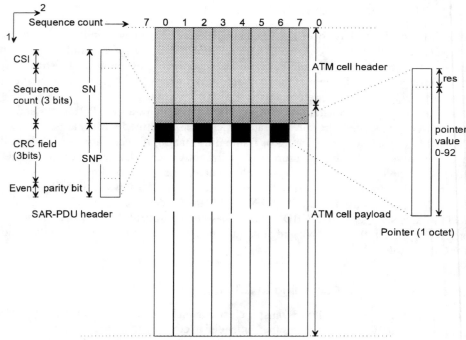

Figure 5.36   Adaptation of CBR signals to ATM according to AAL 1.

the complete 47-octet SAR-PDU payload. That is to say, data is transferred one bit at a time. In the octet/frame structured case, a 7-bit pointer is used in alternate cells to indicate the frame reference in the client payload. In this case, the structured data transfer method described below is used.

Cell delay variations at the input are buffered with headroom enough to allow for waiting time effects. The sequence count is monitored at the receiver to detect lost or misinserted cells. In these cases, clock transition integrity is maintained by passing dummy data to the client with the correct number of bits, albeit errored. Finally, the source clock is recovered and dejitterized to a level demanded by existing specifications.

5.5.2.3   Synchronous Residual Time Stamp (SRTS)

Source clock recovery is based on the synchronous residual timestamp (SRTS) method. In a fixed time span determined by N service clock cycles, the number of derived network clock cycles M_q is counted at the source. If the value M_q is transmitted to the sink, then the original service frequency can be reconstructed exactly since the derived reference frequency is also available there. However, M_q is actually made up of a nominal part and a residual part. The nominal part M_{nom} corresponds to the nominal number of cycles of the derived clock in T seconds and is fixed for the service. The residual part conveys the frequency difference information as well as the effects of quantization, and hence will vary. Since the nominal part is a constant, depending on the nominal service bit rate, it can be provisioned at the receiver. Only the residual part, the residual timestamp (RTS), is transmitted. A 4-bit RTS is sufficient to convey accurate synchronization information for all the G.702 rates. This is transmitted in the channel created by the CSI bit in the odd numbered cells (1, 3, 5, and 7) of the 8-cell sequence. The other four, which are not used for SRTS, are set to 0.

In a BB network, the 155520 kHz network clock will be available, and this can be used to derive references at the following frequencies:

$$155520 \times 2^{-k} \text{ kHz, where } k = 0,1,2,...11$$

For example, the 64 kbit/s service rate requires a derived clock of 155520×2^{-11} kHz or 75.9375 kHz. An analysis of the timing performance of the SRTS is contained in Chapter 11.

5.5.2.4   Structured Data Transfer (SDT)

The SDT procedure is required for transfer of octet or frame start information, particularly for (but not restricted to) $n \times 64$ kbit/s signals and SDH containers. Like SRTS, it exploits the 8-cell structure defined by the 3-bit sequence counter. In even number frames, an 8-bit pointer field is added after the SAR-PDU header. This field actually carries a 7-bit pointer that points to the start of the frame if it occurs in the 92-octet payload available between pointers in the 8-cell sequence after making allowance for the SAR-PDU header octet in every cell and the pointer octet in even numbered cells. Where the referenced payload does not contain a frame start, then the 7-bit pointer is set to the dummy value of all 1s. The 8th remaining bit of the pointer field is not used, but is nevertheless reserved for some future use. The SDT mechanism and the SRTS mechanism can be used together in combination or independently.

5.5.2.5   Convergence Sublayer for Video Circuit Emulation

Although new videocoding procedures are under study that are better adapted to the ATM transfer characteristics, there are a large number of standard and proprietary codecs designed for operation on the prevalent constant bit rate channels. This class of codecs is well-adapted to the AAL1 type of procedure. Many video coding schemes are based on a sampling frequency locked to the video subcarrier or frame rate. These frequency standards are available on client premises (video studios), and in these cases, the video frequency standard may be used as the derived reference for the SRTS procedure.

Compressed video is very vulnerable to transmission channel errors, and it is normal practice to include relatively heavy error correction. This is often included in the codec package. Where this is not the case, a powerful forward error-correction procedure is available in AAL1 based on the Reed-Solomon (128, 124). A forward error-correcting overhead of 4 bits is added for each 124-bit block of video data capable of correcting two errored octets or four erasures per block. An erasure is an errored octet whose location is known. The extra overhead of this method is 3.1% and the processing delay is 128 cells.

5.5.3   Adaptation According to AAL2

AAL2 was originally intended for digital video. It was initially thought that a new generation of bursty codecs could better exploit the nature of the ATM channel and that these would need a special adaptation layer. This approach has been dropped and the present AAL1 proposal, the most recent addition to the ATM range, provides for an additional cell-based multiplexing level to carry multiple low-rate channels within a single VCh. The cells are variable length and no switching is intended at this level. The motivation derived from mobile telephony is essentially to allow compatible transport of digital services including data, mobile telephony, videotelephony, multimedia conference, etc. efficiently but with minimum delay. The new minicells are placed contiguously in the bearer cell stream and effectively form a new ATM layer, termed here the common part sublayer (CPS).

The structure is illustrated in Figure 5.37. A single byte is placed in the cell payload in the first position after the cell header. This is the start field (STF). It contains a 6-bit offset indicator in the offset field (OSF), which is a number between 0 and 47 indicating the offset in bytes until the start of the first minicell. There is a 1-bit sequence number (SN) that counts cells in the same stream modulo 2, and, finally, a single parity bit (P) is used as an error check on the STF itself.

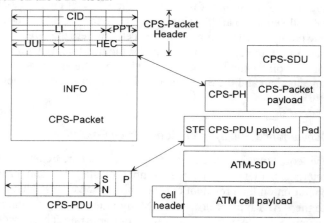

Figure 5.37   ATM Adaptation Layer 2 (AAL2).

The minicells that form the common part sublayer are formed in CPS packets, each with a 3-byte header and a variable payload. The 8-bit CID identifies the channel within the new cell stream and the 6-bit length indicator (LI) indicates the length of the CPS packet payload. Subsequent packets from other channels follow contiguously. If there is no information to send, then padding (all zeros) is introduced into the unused payload.

The CPS packet header also contains a 2-bit payload type field that serves to distinguish regular data from channel-specific maintenance information. The header is completed by a 3-bit end-to-end, user-to-user indication (UUI) field and a 5-bit header error check (HEC), which minimizes the possibility of misinterpretation due to error.

5.5.4   Adaptation According to AAL3/4

AAL3 and AAL4 were initially addressing connection-oriented and connectionless aspects of data transmission, but were eventually consolidated in the so-called AAL3/4 adaptation function to provide a comprehensive data transmission capability, capable of replacing many existing protocols. It was developed concurrently with similar work on broadband wide area networks in IEEE and addresses interworking issues with IEEE 802.6, which also formed the basis of SMDS (switched multimegabit data service), promoted by Bellcore in the United States and CBDS (connectionless broadband data service) in Europe:

- It may be used in streaming mode for continuous data commonly found in synchronously operated, connection-oriented packet networks or message mode where packets or datagrams are transmitted independently, in the manner used in connectionless networks.
- In message mode (MM), a blocking mechanism is provided for multiple, small, fixed-length messages in a single cell, but also a segmentation and reassembly mechanism is provided for large messages running across multiple cells.
- A multiplexing channel identifier is used to identify multiple channels, corresponding to end users in a terminating local area network.
- Per cell error checking is provided to support error recovery at the ATM layer and cell sequence numbering enables detection of missing cells or reordering where cells are received out of sequence.

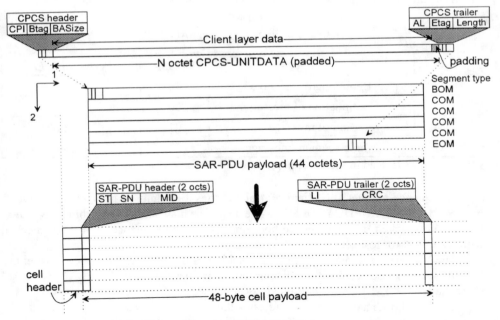

Figure 5.38   Adaptation of packet data in ATM by AAL3/4.

The structure of AAL3/4 is illustrated in Figure 5.38 . The client layer data is first padded out to be an exact multiple of 4 octets. Headers and trailers of 4 octets each are appended to form the so-called common part convergence sublayer protocol data unit (CPCS-PDU). The common part indicator field (CPI) is a single octet that conveys information needed for interpretation of BA size and message length. Only one value is currently agreed upon, namely the all-zero value that indicates that the counting unit for BA size and message length is octets. The Btag and Etag are 1 octet each and carry the same value, which must be changed for each successive CPCS-PDU. The receiver matches the Etag and Btag for a single PDU, but does not check sequence in successive PDUs.

The BA size indicates to the receiver the buffering requirements for the subsequent message. In message mode, the buffer allocation will always be equal to the message length, but in streaming mode it may be longer. The alignment field (AL) is merely included as padding to make the trailer up to 4 octets.

The main function of AAL3/4 is the segmentation and reassembly (SAR) of long user packets for transportation in ATM cells. An SAR-PDU is 44 octets long and is preceded by a 2-octet header and followed by a 2-octet trailer. The first 2 bits of the header are used to designate the segment type. There are four types of segment: one used for the segment containing the beginning of a message (BOM); continuation of message (COM) segments, containing the body of the message; and an end of message (EOM) segment containing the last part of the message. The remainder of the EOM cell after the end of message is empty; a new message starts at the beginning of a new BOM cell. For short messages that fit completely in a single cell, the single segment message (SSM) segment type is used. The ST coding and permissible segment length is illustrated in Table 5.5.

Table 5.5

Coding of Segment Types in AAL3/4

Type	Usage	Coding	Permissible LI
BOM	Beginning of message	10	44
COM	Continuation of message	00	44
EOM	End of message	01	4.....44, 63*
SSM	Single segment message	11	8.....44

*the value 63 is used in the abort procedure

A sequence number (SN) field of 4 bits carries a modulo 16 number that is incremented for each SAR-PDU in an SDU. The sequence may start at any arbitrary value as no account is taken of the sequence between SDUs. The sequence number is checked within an SDU for lost cells.

The MID field acts as a multiplex channel indicator. This was introduced to identify the data streams associated with particular terminals on a shared medium LAN, which all share the same ATM VCh. If multiplexing is not used, MID is set to all zero.

The length indicator (LI) field indicates the length of the SAR-PDU in octets. The permissible values are indicated in Table 5.5. Finally, the 10 bit cyclic redundancy check (CRC) is calculated over the whole of the SAR-PDU up to the CRC field.

AAL3/4 is now regarded as excessive for most applications, where much of the service specific functionality is already provided in the legacy data client.

5.5.5   Adaptation According to AAL5

AAL5 was developed in response to a widespread demand for something simpler and more efficient than AAL3/4. TCP/IP and frame relay formats were already in widespread use for data networking, and these place quite modest demands on the underlying transport layer.

The AAL5 protocol structure is illustrated in Figure 5.39. The most obvious difference compared with AAL3/4 is the complete absence of a per cell overhead in the SAR function. The signaling capability in the payload type identifier (PTI) field of the ATM cell header (see Figure 5.27) is exploited to provide SDU delineation. The cells containing beginning of message (BOM) or continuation of message (COM) are distinguished from the cell containing the end of message (EOM) by means of the user-user (PTI-UU) signaling capability in the ATM cell header. The end of message cell contains a PTI UU value = 1, while cells containing beginning or continuation of message have a PTI UU value = 0.

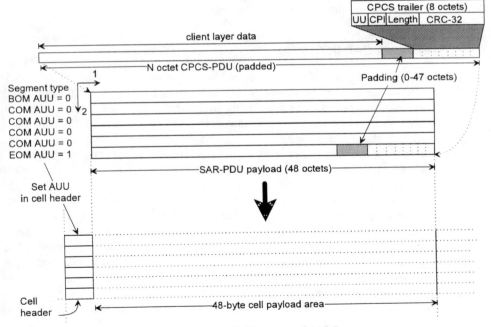

Figure 5.39   Adaptation of packet data in ATM by means of AAL5

AAL5 clients may also signal congestion information via the ATM layer using the congestion information bit of the PTI field in the ATM cell header. This may also be overwritten by congested ATM nodes, as explained previously, so that the information received at the AAL5 source reflects congestion detected at one or more nodes in the ATM layer or the AAL5 data client layer. This may be interpreted by the AAL5 client source as a request to reduce the rate of the data submitted for transmission. The CLP bit (i.e., cell loss priority bit) in the ATM header is similarly treated. That is, it may be set by the AAL5 client to reflect high or low priority as judged by the source, but may subsequently be marked down by the ATM layer from high to low priority if the call admission contract is violated at an ATM transit node.

Above the SAR level, the convergence sublayer is similarly lightweight. First, the SDU is justified to a multiple of 48 so that it may be segmented into an integer number of SAR PDUs with a padding field which will accordingly be 0–47 octets in length. There is no header, but the PDU starts with the first octet of the fist cell after an EOM cell. The trailer comprises a 1-octet CPCS-UU field available for user-to-user signaling; a 1-octet common part indicator (CPI) that currently acts merely as padding to bring the trailer up to 8 octets and is coded to zero; a 2-octet length field that enumerates the length of the PDU payload measured in octets; and the CRC-32 field (4 octets), which is computed over the whole of the CPCS-PDU including the payload and first 4 octets of the trailer.

Where multiplexing is required, this is defined in a service-specific manner, depending on the data client. For example, a frame relay client may be multiplexed by virtue of the DLCI in the FR layer. IP clients may be mapped directly into ATM VCs or may use a frame-based encapsulation derived from the frame relay format.

5.6   SDH PHYSICAL INTERFACES

The pressures behind the SDH development and evolution were discussed in Chapters 1 and 2. The need to standardize high-capacity physical interfaces to cope with the increased level of functional integration in network node equipment was in many ways the most pressing. This was necessary to allow the introduction of cost-effective, high-capacity equipment without the risk of locking operators into proprietary interface solutions. The medium of choice for interconnecting network node equipment, whether collocated or not, is increasingly optical, due to the ever-reducing cost and seemingly limitless bandwidth available. Hence, the emphasis on midfiber meet. Indeed, SONET meant "standard optical network" before it came to mean "synchronous optical network," and this is reflected in the details of the signal format. Extension to other media followed subsequently.

Electrical interface standards are required mainly for intrastation interconnections because initial deployments have to fit into a digital distribution frame regime dominated by coaxial plesiochronous interconnections at 44736 kbit/s and 139264 kbit/s, as defined in G.703. Although radio interface standards are desirable to meet similar transverse compatibility goals as for optical line systems, this need is not so pressing as there is no comparable pressure to integrate radio interfaces into terminal or nodal equipment.

5.6.1   Optical Interfaces and the Midfiber Meet

The physical characteristics of optical interfaces for the SDH are defined in G.957. They have taken on a somewhat wider scope than their electrical equivalents as specified for the PDH NNI in G.703. This is mainly because of the greater capability of optical technology in the distances that can be covered without regeneration. G.703 interfaces were confined to interconnecting equipment in a single station due to the physical limitations of low-cost, unrepeatered, coaxial transmission systems at 139264 kbit/s. The optical equivalents, however, can provide connections over quite large interstation distances just as easily as the few hundred meters across the station.

A further major difference is the requirement to provide transverse compatibility for regenerated line systems in a multivendor environment. This had not been attempted on any wide scale for the PDH. Transverse compatibility implies the ability to mix terminals and regenerators from different suppliers within one line system. Longitudinal compatibility, on the other hand, implies the ability to mix line systems from different suppliers within one cable. G.957 is also compatible with G.955 and G.956, which provide for longitudinal compatibility between optical sections at comparable rates. This is an important requirement that allows SDH to be introduced into the network alongside PDH.

Transverse compatibility, or midfiber meet, has been achieved (not surprisingly) at the cost of performance in terms of reach. This sacrifice is not significant, however, as the intrastation and short-haul interstation versions are easily achievable and cover the vast majority of applications. The recommendation gives some guidance on making extensions by mutual agreement to obtain greater performance, but there is no intention to restrain suppliers or operators from exploiting the technology to the full in achieving this. It can be expected that long-haul line systems will continue to press performance to the limit to minimize the number of repeaters and reduce cost.

The following discussion of SDH optical interfaces is confined to the general principles used in the recommendation to achieve an open and vendor-independent specification. Readers who wish to pursue the subject of optical transmission in more depth are referred to one of the many excellent books dealing with the subject of optical transmission.

5.6.1.1   Classification of Optical Interfaces by Application

The recommendation is framed in terms of the requirements for the optical section termination source (the transmitter), the optical section termination sink (the receiver), and the optical path between the optical section TCPs that are called the S and the R reference points. This is illustrated in Figure 5.40. The transmission distances that it is possible to achieve will depend on the fiber characteristics and the operating policy with regard to splices, connectors, and fiber maintenance margins. The following broad classifications based on distance were used as a guide in developing the recommendation:

- *Intraoffice (I)*—This corresponds to interconnection distances up to 2 km. Only 1,310-nm sources using fiber to recommendation G.652 are specified.
- *Short-haul interoffice (S)*—This corresponds to interconnection distances of approximately 15 km. Only 1,310-nm sources on fiber to G.652 and 1,550-nm sources for use on fiber to G.652, G.653, and G.654 are specified.
- *Long-haul interoffice (L)*—This corresponds to interconnect distances of approximately 40 km in the 1,310-nm window and approximately 60 km in the 1,550-nm window using fiber as specified in G.652, G.653, and G.654.

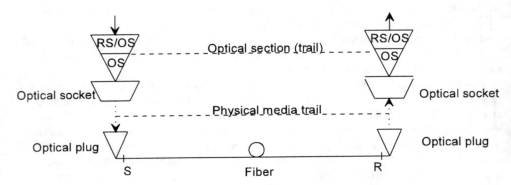

Figure 5.40   Reference points in the SDH optical section at which the physical interface is defined.

The parameter values specified depend on application, bit rate, and fiber type. The interface specifications are therefore designated by a three-part application code of the form A-N.x where A is the application reference I, S, or L as defined above; N is the STM level 1, 4, or 16; and x is the fiber-optical source type as follows:

1 (or blank) designates a nominal 1,310-nm wavelength source on G.652 fiber;

2 designates a nominal 1,550-nm wavelength source on G.652 for short-haul applications and either G.652 or G.654 for long-haul applications;

3 designates a nominal 1,550-nm wavelength source on G.653 fiber.

These classifications and the corresponding interface application designations are summarized in Table 5.6, reproduced from G.957.

Table 5.6
Classification of Optical Interfaces by Applications.

Application		*Intra-office*	*Interoffice*				
			Short haul		*Long haul*		
Nominal source wavelength (nm)		1,310	1,310	1,550	1,310	1,550	
Type of fiber		G.652	G.652	G.652	G.652	G.652/4	G.653
*Distance (km)		<2	~15	~15	~40	~60	~60
STM Level	STM-1	I-1	S-1.1	S-1.2	L-1.1	L-1.2	L-1.3
	STM-4	I-4	S-4.1	S-4.2	L-4.1	L-4.2	L-4.3
	STM-16	I-16	S-16.1	S-16.2	L16.1	L-16.2	L-16.3

* These distances are used for classification and not for specification.

5.6.1.2   *Optical Section Termination Source*

The recommendation is consistent with the use of light-emitting diodes (LED), multi longitudinal mode (MLM) lasers, and single longitudinal mode (SLM) lasers where appropriate. As these three classes of transmitting devices are characterized by decreasing spectral widths, it follows that the narrower spectral width devices can substitute for the wider ones, but not vice versa. The dispersion penalty associated with the anomalous, dynamic behavior of lasers, which has been called "chirp," has not been specified. However, this is only a problem for high bit rate, long-haul systems, and can be resolved in specific ways in each design by agreement between the parties involved.

The mean launch power is defined for a pseudorandom data sequence of full-width (i.e., nonreturn to zero or NRZ) transmitter pulses. A maximum and a minimum have been specified to allow for a range of implementations and some flexibility in optimizing the various degrading effects. Many operators require that an automatic laser shutdown capability be used for safety reasons.

By convention, a logical "1" is equivalent to light being emitted and logical "0" is equivalent to no light emission. The extinction ratio defines the ratio between the "On" power and the "Off" power. A minimum value is specified to ensure sufficient excursion of the optical signal level. The transmit pulse shape is defined by a conventional "eye diagram" (see Figure 5.41) that specifies the limits on allowable imperfections such as rise time, fall time, overshoot, undershoot, and ringing. A test configuration for the measurement of the transmit eye is given in an annex to the recommendation.

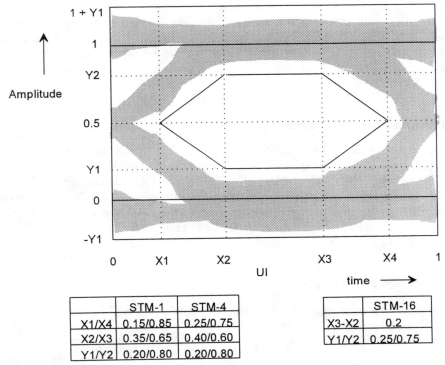

	STM-1	STM-4
X1/X4	0.15/0.85	0.25/0.75
X2/X3	0.35/0.65	0.40/0.60
Y1/Y2	0.20/0.80	0.20/0.80

	STM-16
X3-X2	0.2
Y1/Y2	0.25/0.75

Figure 5.41   Specification of the transmit optical pulse shape.

5.6.1.3   The Optical Medium

The fiber between the reference points is specified in terms of a permissible range of attenuation and a maximum dispersion where the application is considered to be dispersion-limited. The system planner must provide for fiber performance, splices, and connectors within these limits. Other factors that must be taken into account include aging effects and allowances for extra splices as a result of maintenance activities, and environmental degradations of passive devices such as connectors.

Any fiber discontinuity such as a splice, a connector, or any other passive component will produce reflections, and these can be a serious source of degraded performance. Reflections are controlled, where appropriate, by defining a minimum optical return loss and a maximum discrete reflectance between the reference points.

5.6.1.4   Optical Section Termination Sink

The receiver is specified primarily in terms of the minimum acceptable sensitivity and the overload level. Both are defined in terms of the ability to achieve a 1×10^{-3} binary error rate (BER) under thermal noise-limited conditions. The values were derived assuming worst case transmitter pulse shape, return loss, and extinction ratios. The reflectance of the receiver alone is specified independently of the reflectance of the medium, and an optical path penalty is specified to allow for the combined effects of dispersion resulting from intersymbol interference, mode partition noise, and laser "chirp." Methods for measuring reflections in the fiber and at the receiver are described in appendices to G.957. The relationship between the three elements of the specification are illustrated in Figure 5.41.

5.6.2   Electrical Interfaces

Electrical interfaces on coaxial cable are required primarily for backwards compatibility with the digital distribution frames that were installed for 139264 kbit/s and 44736 kbit/s intrastation connections. The STM-1 electrical section interface has been defined to be exactly equivalent to the 139264 kbit/s interface already defined in G.703. This makes for a particularly attractive transition scenario where the SDH and PDH equivalents are implemented as provisional options on the same multiplexer tributary, thus allowing a smooth transition from a PDH application to an integrated SDH network.

5.6.3   Radio Interfaces

Radio system recommendations have been the province of the CMTT, but there has been some considerable liaison between the ITU-T and CMTT on the subject of SDH. This has centered mostly around the architectural issues of path transparency and section monitoring as applied to radio systems. From the networking viewpoint, it is essential that the path layer is specified compatibly so that end-to-end paths can be supported over sections based on different transmission technologies. The argument for using the same section layer format for radio as for optical fiber revolves mainly around the economies available from reusing functions that may have already been developed for use in optical systems. Accordingly, CMTT has reserved some of the unused bytes in the section overhead for "media-specific use." However, the information capacity required for forward error correction (FEC) is so great that it is necessary to encapsulate the whole STM-N frame within a special radio frame, thus creating a special radio section layer. In this situation, no benefit is derived from using the SDH section overhead to define the radio section.

5.6.4   Submarine Systems

Submarine systems capable of transporting SDH formats will be an important feature of the global network of the future, particularly for broadband services. Coaxial submarine systems relied on closely spaced, carefully engineered, high-reliability, submerged repeaters. With the advent of optical transmission, a large number of the shorter routes no longer need submerged equipment. It is generally worth going to considerable lengths in terminal design to achieve an unregenerated span. Coherent optical transmission is but one of the techniques that will minimize the need for regenerators in the future. Such optical developments can easily be applied to conventional SDH transport equipment to provide the level of SDH networking required.

The rate and format dependency of digital regenerators is a barrier to future upgrades, whether to higher bit rate systems; WDM (wavelength division multiplexing); or to new, as yet unforeseen, transport technologies. For this reason, many planned long-haul submarine systems will use optical amplifiers to reduce or eliminate the regenerators. Thus, from an SDH architectural viewpoint, the amplified optical section provides a single unregenerated multiplex section for the transport of SDH paths. Such systems are upgradable to higher bit rates or WDM without retrieving the submerged plant.

Another popular approach to SDH-compatible transport on submarine systems is to encode the entire STM-N signal using a transmission code that has been optimized for simple, low-power remote maintenance of submerged regenerators. A 24B+P code as used on TAT-8 is an example. This was the technique used for the first generation of PDH-based, submerged optical systems. When used below the SDH section layer, it is equivalent to introducing an alternative RS layer into the architectural model of Chapter 3. It has the considerable advantage of allowing the reuse of the existing low-power, high-reliability regenerator designs by simply increasing the bit rate to allow for the extra layer of line coding.

References

The following standards are the main source documents for the material in this chapter. These deal with formats and signal structures:

ITU-T Recommendation G.707, *Network Node Interface for the synchronous digital hierarchy.*
ITU-T Recommendation I.361, *B-ISDN ATM Layer Specification.*
ITU-T Recommendation I.363, *B-ISDN ATM Adaptation Layer (AAL) Specification.*
ITU-T Recommendation I.413, *B-ISDN User-Network Interface Specification.*
ITU-T Recommendation I.610, *B-ISDN Operation and Maintenance Principles and Functions.*

The following recommendations deal with mappings between the various layer networks:

ITU-T Recommendation G.804, *ATM Cell Mapping into Plesiochronous Digital Hierarchy (PDH)*.

ITU-T Recommendation G.832, *Transport of SDH Elements on PDH Networks: Frame and Multiplexing Structures*.

ITU-T Recommendation G.802, *Interworking Between Networks Based on Different Digital Hierarchies and Speech Coding Laws*.

The following recommendations deal with issues relating to media layers and physical interfaces:

ITU-T Recommendation G.957, *Optical Interfaces for Equipments and Systems Relating to the Synchronous Digital Hierarchy*.

ITU-T Recommendation G.958, *Digital line systems based on the synchronous digital hierarchy for use on optical fibre cables*.

Connection Control Interfaces 6

The control of connectivity is central to the operation of a telecommunications network. In this chapter, we consider some of the protocols that have been developed, or are still under development, that can be used for the control of connectivity. Chapter 4 presented the set of control functions required to successfully set up and release connections across a network, and described them along with the associated logical interfaces in such a way that did not imply any particular protocol. In principle, many protocols can successfully implement the logical interfaces between these functions.

Originally, connectivity was controlled by human operators who took requests from customers in the form of a destination town and number and, by talking and cooperating with other operators in other exchanges, would find an available route and connect the customer with their desired distant end. Many of the features of modern automated protocols are still based on those used by human operators in the days of manual exchanges. This demonstrates that the control of connectivity can be effected in many different ways, differing only in the message syntax used. The essential semantics of connection control show striking similarity in their many manifestations, and there have been many over the years. The following is a brief list of the more important categories:

- Human dialogue between operators which, while retaining some of the informality and inventiveness of human interaction, was largely formalized, even down to a stylized pronunciation of the decimal digits of numbers;
- "Loop disconnect" where a direct current (dc) loop in a copper pair is broken and closed to signal connection control information (the method originally developed by Strowger 100 years ago, and still in widespread use for control between a residential customer and a local exchange);

- "Alternating current" (ac) signaling systems where the makes and breaks of loop disconnect signaling are used to modulate an analog carrier, which is normally within, or just outside, the speech band of an audio channel;
- Channel-associated signaling (CAS) systems in primary rate digital signals where, again, the makes and breaks of loop disconnect signaling are directly converted into the binary value of a bit in the primary rate frame structure directly associated with each narrowband channel;
- Message-based common channel signaling (CCS) systems, which form the backbone of signaling in most modern public switched telephone network/integrated services digital network (PSTN/ISDN) networks, and are normally based on International Telecommunications Union (ITU-T) Signaling System No. 7;
- Manual planning and operations processes working on manual distribution frames in the plesiochronous digital hierarchy (PDH) transmission network;
- Network management protocols based on the open systems interconnection (OSI) network management protocol, common management information service element (CMISE), used in synchronous digital hierarchy (SDH) and other networks to control the provision of private circuits;
- There has been much discussion recently in ITU-T Study Group 15 about the possible use of the protocol format developed by the Object Management Group (OMG) for distributed object systems called the common object request broker architecture interface definition language (CORBA-IDL) instead of CMISE;
- Finally, the Internet Engineering Task Force (IETF) has developed a set of protocols for connection control for the Internet that could also be used for the control of asynchronous transfer mode (ATM) networks.

ATM has its history in the switching side of telecommunications, and the signaling protocols used to control connectivity originate from the development of automatic switching, which replaced manual operators in exchanges many years ago. The signaling protocols followed essentially the same process that the operators used, reflecting the way operators would dialogue and then pass the call from one to the next. Even the modern international standard signaling system, C7, is not so far removed from this, and when following the message diagrams in this chapter, it is still possible to imagine the messages being passed by human operators.

Network management is a more recent development and has stemmed from the desire of telecommunications network operators to bring operations and maintenance staff out of exchange buildings and into central operations and maintenance centers. The technology to achieve this requires that all the information needed for operations and maintenance be relayed from the exchanges to the central site. Network management has therefore been largely a problem of collecting, relaying, sorting, and displaying information, with the protocols for network management largely reflecting this. In the past, transmission equipment was "dumb" in that it could not be remotely controlled in any meaningful way, and the connectivity of the transmission network was established and controlled on manual distribution frames (not dissimilar in principle to those of the old manual exchanges). The main status information available was an alarm to indicate that the signal it was receiving was lost or indicating a fault further up the transmission system (alarm indication signal, AIS). This information was the single most important piece of information to relay to the network operations center. As a result, transmission equipment, including SDH, has evolved in a managed network environment and has assumed the use of network management protocols.

However, the advent of SDH has brought about a major change in transmission network operating practice with the introduction of digital crossconnects. We now see, from our study of layering and partitioning, that control of connectivity is generically applicable in all layers and that SDH connectivity control is essentially the same as the control of connectivity in any other layer. The network management protocols that were designed to relay information to a central site for maintenance purposes are, with SDH, acting as connection control protocols.

The Internet is currently a connectionless network and so does not use connection control protocols. However, the desire to support real-time audio and video signals on the Internet has led the IETF to develop a set of protocols that work within the connectionless network, allowing it to offer connection-oriented services. At the time of writing, these protocols are still at the stage of "Internet draft," and are being used on a trial basis on many IP-based networks throughout the world. Ordinarily, as IP networks are clients of ATM or SDH networks, this would not be of direct relevance to networks based on ATM and SDH; however, there has been much discussion in the IETF on the use of these protocols to control ATM networks and hence we discuss them briefly here.

The focus of this chapter is primarily the signaling protocols designed for the control of connectivity in the ATM VCh layer network. However, we also present the latest development of the network level management protocols applicable to SDH virtual containers (VCs) and ATM virtual paths (VPs), which are still being developed in ITU-T Study Group 15 and show the semantic similarities and the important differences in emphasis. In addition, we briefly present the Internet connection control protocol, RSVP, and show where it too has similarities and differences to the signaling protocols.

6.1   TRANSPORTING CONNECTION CONTROL PROTOCOLS

Since a connection control protocol is controlling a telecommunications network, which is by definition geographically distributed, the connection control protocol must work across distributed parts of the network. The different ways of distributing the connection control functionality were discussed in Chapter 4, however, in all cases, connection control messages require transport across the network. These messages are exposed to corruption or even loss in the transport network, and so the protocol must be robust for these circumstances. In addition, messages may not arrive as expected and spurious messages can appear out of normal sequence, which, without careful design, can "lock up" the connection control functionality.

To manage this adverse environment for message transport, a *protocol controller* function is needed to sit at each end of the exposed message path. These can track each message and check that the expected responses are received within a reasonable time, and if they're not received, report an exception to the function initiating the message. This is illustrated in Figure 6.1.

Under normal circumstances, the protocol controller is semantically transparent as the message primitives coming from the messaging function result in external protocol

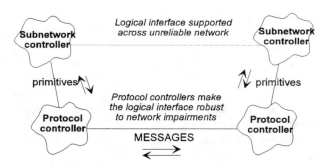

Figure 6.1   Use of a protocol controller.

messages, and vice versa. The protocol controller handles all forms of exceptions, including unexpected or inappropriate messages and a lack of response from the far end. In addition, the protocol controller will generate any ancillary messages required to manage the protocol itself, including the activation of the messaging link if it has not been provisioned or has been suspended. In general, it provides all the housekeeping needed to ensure that the connection controller sees only a simple, well-behaved messaging channel to a connection controller in another subnetwork controller. While there needs to be a protocol controller for every protocol that transits an unreliable transport network, there are a number of more generic standards for protocol controllers, including the following:

- The OSI application environment, including the transport, session, and presentation layer protocols as well as the association control service element (ACSE), the remote operations service element (ROSE), and the common management information service element (CMISE);
- The object request broker of CORBA is a protocol controller giving protection of generically developed applications from a distributed environment;
- The "stubs" required by the generic remote procedure call (RPC), defined by the Open Group, are also protocol controllers;

Each of these has come in for discussion in ITU-T Study Group 15 for the SDH network-level connection control protocol.

The signaling standards of ITU-T only specify the protocol controller and do not specify any of the basic functions of connection control described in Chapter 4. Indeed, to build a switch, the nature of the functions must be inferred from the specification of the protocol controllers and some additional information in "interworking" between signaling protocols. With signaling, there is a protocol controller specification for every signaling protocol, and only a small portion of any specification is common to another protocol. The normal vocabulary used is that signaling *messages* pass between the protocol controllers and signaling *primitives* pass between the connection controller and the protocol controller.

6.2   SIGNALING PROTOCOLS FOR THE CONTROL OF CONNECTIVITY

Connection control in modern telecommunications networks has developed through the use of transactional, message-based signaling systems. The signaling protocol is made up of a set of well-defined messages that work within an intralayer process characterized by connection control activities each comprising a particular set of well-defined message sequences. These message sequences establish connections and release them when no longer required.

As signaling systems have become more sophisticated over the years, they have also become more flexible. Flexibility was the primary focus of the ITU's signaling system, C7, which now accounts for the greater part of the Q series Recommendations. The messages of the early signaling systems carried very little information, for example, whether a handset was on the cradle or off the cradle, and this could be simply represented by the presence or absence of a voltage or the presence or absence of a tone. However, the messages of C7, while still retaining much of the simple semantics of the earlier systems, contain extra parameters and messages that allow the signaling system to support a wide variety of additional services beyond the simple setting up and releasing of connections.

Here, we concentrate on the messages associated with the simple setting up and releasing of connections. When it comes to the handling of "supplementary" services, there are currently conflicts between implementations that use the basic flexibility of C7 and implementations that use the "intelligent network" (IN). The IN uses an object-like approach in which reusable service-independent building blocks are defined, from which a wide variety of services can be constructed quickly and easily and without altering the basic IN protocol message set. Since the aim and scope of the book is oriented towards building broadband transport networks offering basic connectivity, neither IN nor B-ISDN are discussed directly. However, we do question whether—in a world of broadband, where all terminals will have reasonably powerful processing capabilities—these extra supplementary and/or IN service features are really useful to users who can potentially support many of these features more effectively within the applications running on their own processing platform.

6.2.1   Relationship Between Signaling Protocols

Signaling protocols for broadband networks are still evolving, although most of the basic functionality is stable. They have been developed both in the ATM Forum and also in ITU-T Study Group 11, and while there are some differences between the specifications of each body, they are largely compatible. The ATM Forum has tended to emphasize private network signaling, while the ITU-T has tended to emphasize public network signaling.

The ATM Forum has currently defined two signaling protocols that are very closely related. Each is associated with one of the primary interfaces identified by the ATM Forum:

- *User-network interface* (UNI 3.1 and UNI 4.0) covers the signaling between a terminal equipment and a private network, between a terminal equipment and a public network, and also between a private network and a public network;

- *Private network node interface* (P-NNI) covers the signaling between private networks and between the switches within a private network.

While the ATM Forum has looked closely at the broadband intercarrier interface (B-ICI), which covers the signaling between public networks and switches within a public network, at the time of writing there has been no clear agreement on the signaling protocol for this interface. Some people argue that the ATM Forum should simply adopt ITU-T signaling protocol B-ISUP; however, others favor a development of P-NNI for use in public networks.

Figure 6.2 illustrates the interfaces in ATM Forum and notes both the ATM Forum and the ITU-T signaling specifications associated with each.

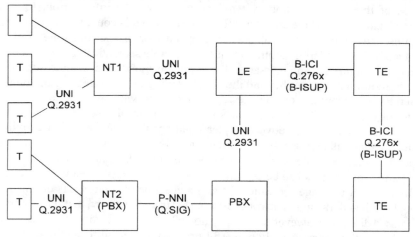

Figure 6.2   ATM Forum interfaces and associated signaling protocols.

The ATM Forum interface specifications cover all aspects of the interface, not just the signaling. They include the transport, management, synchronization, and performance aspects as well. In addition, since the Forum regards its activity as complementary to and assisting the work of the ITU-T, their specifications make reference to ITU-T Recommendations, where they exist.

In the ITU-T, the Recommendations are complete in their own right and generally do not make reference to anything other than other ITU-T Recommendations (and ISO standards). As they concentrate on public network signaling, there is no equivalent to P-NNI in ITU-T as yet, although it has been discussed. The main signaling aspects of the UNI, called digital subscriber signaling system No. 2 (DSS2) is covered in ITU-T Recommendation Q.2931, while those of the intercarrier interface are in the B-ISUP Recommendations Q.2761, Q.2762, Q.2763, and Q.2764. Both of these draw on other Q.2000 series Recommendations for the specification of the adaptation layer aspects of signaling. The full range of the currently agreed upon ITU-T Recommendations on broadband signaling are summarized in Table 6.1, Table 6.2., Table 6.10, and Table 6.14.

6.2.2   Lower Layers of the Broadband Signaling Protocols

The B-ISDN signaling protocols are built up in layers. This is illustrated in Figure 6.3. At the bottom is the ATM virtual channel, which carries the signaling messages. (The default signaling channel has been allocated a VCI = 5.) Above this, there are the common aspects of adaptation for any signaling protocol, the segmentation and reassembly (SAR), and the common part convergence sublayer (CPCS). These simply use AAL5, and so are described in the ITU-T Recommendation on the ATM adaptation layer (I.363) and only referenced in the Q series. These can support a number of service-specific convergence sublayers (SSCS). It should be remembered that a service in this context is a signaling protocol. Within the currently defined SSCSs, there is one for UNI signaling and one for NNI signaling, and within each there are two distinct functions:

- The service-specific connection-oriented protocol (SSCOP) ensures the integrity of the signaling link.
- The service-specific coordination function (SSCF) provides the layer interface to the specific signaling protocol.

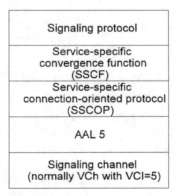

| Signaling protocol |
| Service-specific convergence function (SSCF) |
| Service-specific connection-oriented protocol (SSCOP) |
| AAL 5 |
| Signaling channel (normally VCh with VCI=5) |

Figure 6.3   Signaling protocol stack.

Above an SSCS is one of the signaling protocols itself. In addition, there is a low-level protocol called meta signaling that can establish a signaling channel under the condition where there are several pieces of terminal equipment all connected to the same interface on a switch using a multiplexer. Under these conditions, only one piece of terminal equipment can use VCI = 5 and the meta signaling protocol is needed to assign a VCI for the signaling channel with each terminal equipment. (Note the current releases of UNI signaling, both in the ITU-T and the ATM Forum, do not support meta signaling and assume that the signaling channel will be on a VCC with VCI = 5).

The signaling protocol layers are only approximately similar to the layers of the OSI stack. The meta signaling is approximately equivalent to OSI layer 3, while the SSCOP provides the OSI layer 4 function and the SSCF the OSI layer 6 function. These lower layers are reflected in the organization of the B-ISDN Q series Recommendations, as shown in Table 6.1.

Table 6.1

Signaling ATM Adaptation Layer (SAAL)

	Title	*Content*
Q.2100	B-ISDN Signaling ATM Adaptation Layer (SAAL) Overview Description	Gives a short overview of the lower layer aspects of B-ISDN signaling and references I.363 for details of how the signaling messages are carried in ATM cells
Q.2110	B-ISDN ATM Adaptation Layer - Service Specific Connection Oriented Protocol (SSCOP)	Describes the protocol that maintains and assures the integrity of a signaling link
Q.2120	B-ISDN Meta Signaling Protocol	Describes the protocol that assigns a VCI to a terminal equipment when there is more than one terminal equipment on the interface
Q.2130	B-ISDN Signaling ATM Adaptation Layer - Service Specific Coordination Function at the UNI	Describes how the signaling protocol at the UNI (Q.2931) interfaces to the signaling adaptation layer
Q.2140	BISDN Signaling ATM Adaptation Layer - Service Specific Coordination Function at the NNI	Describes how the signaling protocol at the NNI (B-ISUP) interfaces to the signaling adaptation layer

6.2.3   User-Network Interface Signaling (UNI 3.1, UNI 4.0, Q.2931, and Q.2971)

UNI signaling is oriented towards the transfer of information and the transactional status required for the control of connectivity. The emphasis is on the functionality and security of the protocol as it is used across commercial boundaries, and so needs to be able to fully negotiate the service required by the user. UNI signaling can explicitly request a wide variety of services, which the local exchange and/or private network must interpret and initiate appropriate actions to meet. It is also designed to avoid any user inadvertently, or even deliberately, affecting the operation of the network for other users. As a result, the emphasis is less on the speed and throughput performance. UNI signaling does not support the same level of maintenance features that are found in B-ISUP, largely because these could potentially be a security risk to a public network if a malicious user exercised them in inappropriate ways. The ATM Forum Specifications and ITU-T Recommendations for UNI signaling are listed in Table 6.2.

An important feature of UNI signaling is that it is asymmetric. It has a user side and a network side, and the behavior of the protocol in the two directions is slightly different. As a result, UNI signaling talks about the user-to-network direction or the network-to-user direction.

The messages of UNI signaling are divided into a number of categories. The first category is used for normal connection setup and release, the second of legacy working with the manual dialing of numbers, the third supports point-to-multipoint connections, and the final group is a set of maintenance and ancillary messages.

Table 6.2

Specifications of UNI Signaling

	Title	*Content*
UNI 3.1	ATM User Network Interface Specification Version 3.1	Complete specification of the full UNI - including transport aspects, management, addressing, and signaling
UNI 4.0	ATM User Network Interface Specification Version 4.0	Enhancements to UNI 3.1 primarily to the signaling specification to include leaf-initiated join point-to-multipoint
Q.2931	B-ISDN - DSS2 - UNI Layer 3 Specification for Basic Call/Connection Control	More than 250 pages giving the functional description, specification of the message formats and parameter formats, and specification of the behavior of the protocol controller in response to messages across the interface and primitives from the call controller
Q.2971	B-ISDN - DSS2 - UNI Layer 3 Specification for Point to Multipoint Call/Connection Control	Specifies the additional messages, parameters, and procedures needed to support point-to-multipoint services

6.2.3.1 *Basic Connection Setup and Release Messages*

These messages are listed in Table 6.3. Each message is set in response to a primitive and is sent if the state of the protocol controller determines it is appropriate and safe to send the message.

An example with two users connected directly through a single switch is shown in Figure 6.4. This shows the response to the initial "setup request" primitive from the originating user's connection controller and all the subsequent actions that ensue under normal conditions. The local exchange must look up its route table to find the link that it can use for the destination address and then decide if the resources on the link to destination user are adequate to support the desired connection. If all of this is confirmed, the "setup request" is forwarded to the destination user and an acknowledgment message "CALL PROCEEDING" passed back to the originating user in the interim.

The full set of signaling primitives and messages for a simple connection setup is shown in Figure 6.5, while those of the release (assuming it is in the forward direction) are shown in Figure 6.6. (The release sequence is identical, but reversed, if the release is initiated by the destination user.)

The ALERTING message is something of a holdover from telephony, as it is sent if the destination terminal equipment needs human intervention to complete the connection (e.g., lifting a handset). The ATM Forum did not include the ALERTING message in UNI 3.1 because under most broadband examples, an intelligent terminal will receive the connection setup indication and can automatically answer. However, it is now included in UNI 4.0, as it could be useful for examples such as videotelephony and videoconferencing.

The attributes of these messages in UNI 4.0 and Q.2931 (release 2/95) support a limited set of supplementary services, referred to as Capability Set 1. These are:

Table 6.3
UNI Signaling — Basic Connection Control Messages

Message	Purpose and Meaning
SETUP	This message initiates a call request across the interface and can be in either direction. Under normal broadband calls, it would be expected to contain all address information needed to complete the call.
CALL PROCEEDING	Sent in acknowledgment of a SETUP message to indicate that all the address information needed to complete the call has been received and the call request has been passed on.
PROGRESS	This message is used when a private network is interworking signaling systems and indicates the progress of the call.
ALERTING	Sent when a terminal equipment does not have automatic answering capabilities and must be manually answered. This message indicates the terminal equipment is awaiting the manual answer.
CONNECT	Sent when the terminal equipment answers the call request. In the network-to-user direction, this can be either directly in response to a SETUP message or following an ALERTING message, when the terminal equipment is manually answered.
CONNECT ACKNOWLEDGE	Sent in response to a CONNECT message.
RELEASE	This message can be sent in either direction in order to initiate the release of a call.
RELEASE COMPLETE	Sent to indicate the release of the call is complete and all the resources used by the call are available for future calls.

Table 6.4
Some of the Information Fields in the SETUP Message

Information Element	Type	Information Element	Type
Protocol discriminator (in case non-UNI messages exist)	M	Call reference	M
Message type (SETUP in this case)	M	Message length	M
ATM traffic descriptor	M	Broadband bearer capability	M
QoS parameter	M	AAL parameters	O
Called party number	O	Called party subaddress	O
Calling party number	O	Calling party subaddress	O
Connection identifier	O	End to end transit delay	O

M = mandatory   O = optional

- Calling line identification/restriction (CLIP/CLIR);
- Direct dialing in (DDI);
- Connected line identification/restriction (COLP/COLR);

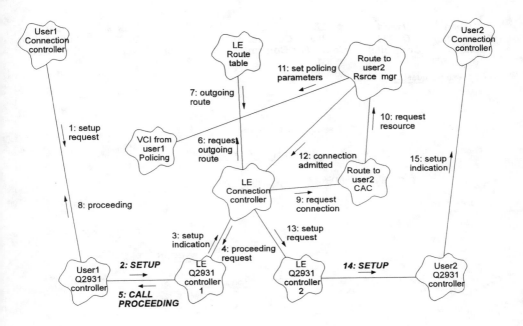

Figure 6.4   Response to a "setup request" primitive.

- Subaddressing (SUB);
- Multiple subscriber numbering (MSN);
- User-to-user signaling (UUS).

Some of the attributes that can be carried in a SETUP message are listed in Table 6.4. Some of these are mandatory, and the SETUP message is invalid without them. Most of the attributes, however, are optional, and are included if the requested service needs to use the attributes to fully define the service.

6.2.3.2   Messages to Support Manual Dialing of Addresses

Signaling grew up with telephony, where the destination was normally dialed by a human using a dial or keypad. Narrowband signaling allows progress to start on a connection setup before all the digits of the destination address have been sent by the user. This has the advantage that if the route is blocked or the address unknown, an indication can be returned from the point at which routing stopped, thus giving the human user an indication where the difficulty occurred.

The ATM Forum does not support this feature in either UNI 3.1 or UNI 4.0 as it adds complexity, and their assumption is that an intelligent terminal will store addresses and so can forward the full address without human intervention. These messages are listed in Table 6.5 and are only found in Q.2931.

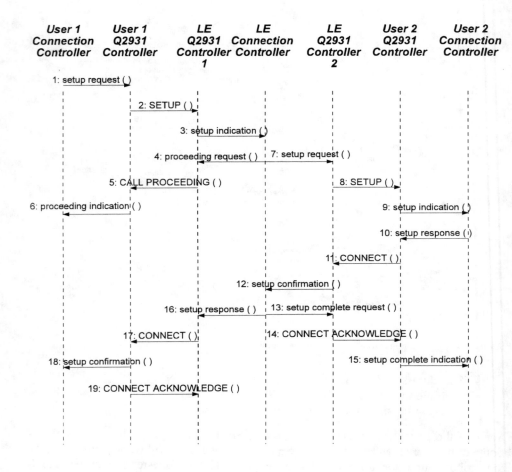

Figure 6.5   Primitives and messages for a normal connection setup.

Table 6.5

UNI Signaling Messages Supporting Manual Dialing

Message	Purpose and Meaning
SETUP ACKNOWLEDGE	Sent in response to a SETUP message if it does not contain sufficient address information to complete the call.
INFORMATION	Used to send additional information about a call. In particular it is used to forward additional address information if the SETUP message did not contain complete information.

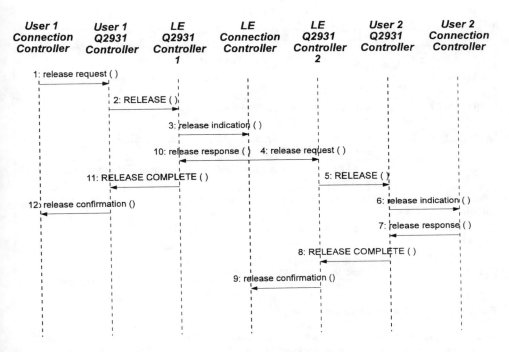

Figure 6.6　Primitives and messages for a normal connection release.

6.2.3.3　*Point-to-Multipoint Connection Setup and Release Messages*

As a point-to-multipoint connection is asymmetric, there are two forms of point-to-multipoint service, depending on which end initiates the request. In the first category, the source end (or root end) of the point-to-multipoint connection initiates the adding of sinks to the connection and dropping of sinks from the connection. In the second category, each sink end (or leaf end) can initiate the request to join an existing point-to-multipoint connection and will then attach to the point-to-multipoint connection at a point close to the leaf. However, even in this case, the root end must first of all have established an initial connection to a sink using a SETUP message so that there is an existing connection to which a leaf can attach.

Within the leaf-initiated join, there are also two possibilities. With root leaf-initiated join, the root is informed and has control over whether a leaf can attach to the connection, while in a network leaf-initiated join, the root is not informed, and so the root is not aware of the leaves that are connected. The messages associated with point-to-multipoint connections are listed in Table 6.6. The ADD and DROP messages are in both UNI 4.0 and Q.2971, while the LEAF messages are currently only in UNI 4.0.

Table 6.6

UNI Signaling — Additional Messages to Support Point to Multipoint Connections

Message	*Purpose and Meaning*
ADD PARTY	Sent by a root to request that a new leaf be added to a point to multipoint connection.
ADD PARTY ACKNOWLEDGE	Sent in response to an ADD PARTY message to indicate that the new leaf has been successfully added to the connection.
ADD PARTY REJECT	Sent in response to an ADD PARTY message to indicate that the new leaf cannot be added to the requested connection.
DROP PARTY	Sent by a root to request that a leaf be dropped from a point to multipoint connection.
DROP PARTY ACKNOWLEDGE	Sent in response to a DROP PARTY message to indicate that the leaf has been successfully dropped.
LEAF SETUP REQUEST	Sent by a leaf to request attachment to a point-to-multipoint connection.
LEAF SETUP FAILURE	Sent in response to a LEAF SETUP REQUEST to indicate that the leaf cannot be attached to the requested connection.

6.2.3.4   Maintenance Messages

These messages can send and request information about the state of a protocol controller if it has reported in a message some kind of error or exception. They can also force a reset if the protocol controller at one end detects that the protocol controller at the other end has "locked up" and is not responding in a normal way to messages. These messages are listed in Table 6.7 and are particularly important to maintaining connection control on a 24 hours a day, 7 days a week basis.

Table 6.7

UNI Signaling — Other Messages

Message	*Purpose and Meaning*
NOTIFY	This is a general message that can be used in either direction to volunteer information about a call.
STATUS ENQUIRY	This message is used to request the report of error condition on the signaling interface.
STATUS	Sent is response to a STATUS ENQUIRY message to indicate the presence of any defined error conditions.
RESTART	This message will request the full reinitialization and return to the idle condition of either a specified VCC, all VCCs within a VPC, or all the VCC controlled by the signaling interface.
RESTART ACKNOWLEDGE	Sent in response to a RESTART message when the request restart has taken place.

6.2.4   Private Network Node Interface Signaling (P-NNI)

Signaling between private networks is currently only being progressed in the ATM Forum with the P-NNI specification, and there is, as yet, no parallel work in the ITU-T. The basic signaling protocol is essentially the same as the UNI signaling protocol, with a few omissions and additions. P-NNI signaling does differ from UNI signaling in that it is symmetric (i.e., peer to peer) and so there is no directionality based on user to network or vice versa. The current omissions are the messages and features that support supplementary services and leaf-initiated point to multipoint. Parameters to support source routing and "crankback" alternate rerouting have been added.

Private networks often have a complex but sparse topology, reflecting the fact that the links are usually purchased as private circuits from a public network operator and are relatively expensive. As a result, routing can involve many transits through intermediate switching nodes in between the routing source and destination. For this, and many other reasons, source routing can be advantageous. The main addition is a new parameter in the SETUP message, which is a designated transit list (DTL) that allows the source route information to be carried forward from the source subnetwork to the transit subnetworks designated by the source route. With source routing, any alternate rerouting, if the primary routing encounters congestion, should take place from the source subnetwork. This means that the routing must crank back to the source subnetwork from the subnetwork in which the congestion was encountered. Crankback is support by adding a new parameter to the RELEASE message.

P-NNI is based on a rigorous and open application of partitioning. The routing is properly hierarchical in that a DTL only lists subnetworks that are at a peer level of partitioning. When a subnetwork in a DTL is partitioned, a new DTL is formed that contains the routing through the subnetwork.

Another feature added in P-NNI is the ability to request a connection setup from within the private network across a management interface as well as across a UNI interface. This feature is added to support "soft" private virtual channel connections (S-PVCs), which have all the properties of PVCs but can be set up using the P-NNI signaling protocol. This takes advantage of the routing exchange protocol and crankback to improve the resilience of the connection, as well as obviating the need to develop a parallel network management-based connection control and restoration process.

6.2.5   P-NNI Routing Exchange Protocol

The primary addition in P-NNI is that of a routing exchange protocol with which nodes can automatically construct route tables with the source's routes, which are carried in the DTL. Its operation is illustrated in Figure 6.7. The routing exchange protocol, unlike most that are in use today in the Internet and OSI, is inherently scalable as it is based on the partitioning principle described in Chapter 3. The use of partitioning means that the routing exchange protocol is limited in scope to peer subnetworks within a partition and neighboring subnetworks to the partition. This unlocks a considerable restriction on previous routing exchange protocols, which were normally restricted to a maximum of around 50 nodes.

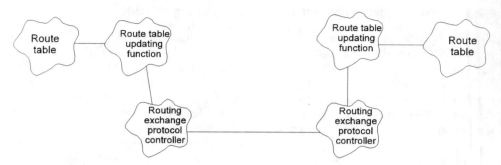

Figure 6.7   Operation of a routing exchange protocol.

The P-NNI routing exchange protocol is based around three principle messages: the HELLO message, the PTSP (P-NNI topology state packets) message, and the PTSE (P-NNI topology state element) acknowledge message. There are two other messages that are used in a startup phase when a node is first commissioned in the network or when some exception occurs in a transfer. These P-NNI routing messages are listed in Table 6.8.

Table 6.8
P-NNI Routing Exchange Protocol Messages

Message	*Purpose and Meaning*
Hello	Regularly broadcast by a node to its neighbors in order to maintain the status of the link between nodes and discover new nodes.
PTSP	Information message to carry many PTSEs.
PTSE acknowledgement	Used to acknowledge the receipt of PTSEs with one message carrying many receipts.
Database summary	Used during an initial route table synchronization when a new node is first discovered.
PTSE request	Used to request the retransmission of a PTSE.

The information carried in the HELLO is principally status information relating to the node and the links surrounding the node. It is used to ensure that two nodes are in full and correct communication across a link. The PTSP message carries the main route table information, and this is packaged into PTSEs, which are treated by the exchange protocol as single entities. In this way, reasonably large amounts of data can be passed and verified in one go. The PTSE is made up of information groups (IGs), and these are listed in Table 6.9.

P-NNI is a very powerful and very flexible protocol for the control of connectivity. Its use of partitioning renders it scalable; a certain lack of scalability has always been a drawback of previous private network signaling systems. It seems not unlikely that P-NNI may be used by public network operators instead of B-ISUP, at least for business-related applications. The P-NNI signaling protocol, with its routing exchange protocol using the rigorous partitioning of the VCh layer network, is probably the first specification to allow a network to fully exploit the potential power and flexibility inherent in the transport

Table 6.9

Information Carried in P-NNI Topology State Elements (PTSEs)

Information Group	*Purpose and Meaning*
Nodal state parameters	Used to advertise identity of a transit group (a Port ID).
Nodal	Used to advertise the node's identity and the status of the node with respect to its peers in the subnetwork.
Internal reachable ATM addresses	A list of ATM end-system addresses that are immediately reachable by the node.
Exterior reachable ATM addresses	A list of ATM end-system addresses that can be reached by the node through a non-P-NNI network.
Horizontal links	Use to advertise links to other peer nodes in a subnetwork.
Uplinks	Use to advertise links to nodes at the next level up of partitioning.

network architecture. At the time of writing, the P-NNI specification is receiving its final approval and it is hoped that the ITU-T will give serious consideration to the potential of this signaling and routing exchange protocol.

6.2.6   Inter and Intra Public Network Signaling (B-ISUP)

The broadband ISDN user part (B-ISUP) is based on the equivalent signaling protocol for narrowband networks and, like P-NNI, is symmetric. It is oriented towards large switches processing large numbers of calls in order to support fast call setup times and high call throughput. The extent to which this is appropriate to broadband services is not immediately clear. Current call tariff structures are not well-suited to many broadband applications, as they are based around the normal times for a telephone conversation. A tariff structure that charges calls linearly with time is likely to favor very short holding times (for example a few seconds) for connections in transactional applications. However, if the tariff structure is modified, say by charging significantly for a call setup and less for the time element of the call duration, then these transactional applications would be more likely to favor much longer holding times, say of several hours. Such a choice in the tariff structure could therefore profoundly affect the level of connectivity control translations that the network must process. Under one set of tariff conditions, the number of transactions could be very much higher than the current level associated with telephony, while under other conditions, the number of transactions could be significantly lower.

This is likely to affect the way B-ISUP develops and, indeed, whether P-NNI may become more significant. There is no clear consensus on this, with the traditional private networking community favoring P-NNI and many in the existing public network operators favoring B-ISUP (and the ATM Forum equivalent, B-ICI). The standards that make up B-ISUP are listed in Table 6.10.

B-ISUP is the set of messages that is used for the control of connectivity between two switches in a public network. The specification in the ITU-T Recommendations covers the messages and their parameters, their formats, and the behavior of the protocol controller.

Table 6.10

Signaling Protocol at the NNI (i.e., Broadband ISDN User Part (B-ISUP))

	Title	*Content*
Q.2761	B-ISDN - Functional Description of the B-ISUP of Signaling System C7	Describes the main part of the protocol; divided into separate parts for different aspects of the protocol
Q.2762	B-ISDN - General Function of Messages and Signals of the B-ISUP of Signaling System C7	Lists and describes the messages of B-ISUP as well as the parameters of the messages
Q.2763	B-ISDN - B-ISUP - Formats and Codes	Precise specification of the syntax of each B-ISUP message and possible parameters (also given in ASN.1 notation)
Q.2764	B-ISDN - B-ISUP - Basic Call Procedures	Defines the behavior of the protocol controller in response to messages across the interface and primitives from the call controller (given in both text and SDL diagrams) Signaling Protocol at the UNI (i.e., Digital Subscriber Signaling System No2 (DSS2))

As with other signaling protocols, it is the protocol controller that is specified. The functionality and behavior for the connection controller and subnetwork controller can be inferred from this.

The protocol controller is divided into a number of different aspects, and the behavior of each is specified independently. This clarifies the description and makes it simpler to add new aspects to the protocol or select only some aspects under special circumstances. Indeed, the version current at the time of writing of B-ISUP covers an initial set of capabilities, capability set 1, which includes messaging for call control, an initial set of supplementary services, and various housekeeping activities. The intention is to extend these capabilities by adding new functions to the B-ISUP specification, and the structure of the specification means this can be done without creating a large number of interactions with the existing capabilities. In many ways, this has many of the features of a fully object-oriented approach. The main parts are illustrated in Figure 6.8, which is based on Figure 1 of ITU-T Recommendation Q.2761.

6.2.6.1   Basic Connection Setup and Release Messages

As with UNI signaling, there is a set of messages that support normal connection setup and release. These are listed in Table 6.11. These, like Q.2931, include the additional set of messages that support the interworking with manually dialed addresses.

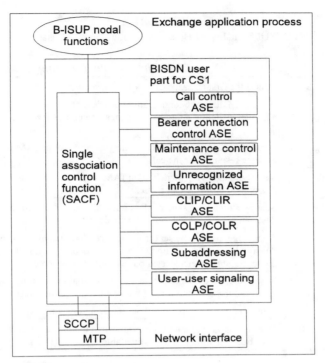

Figure 6.8   Elements of the B-ISUP protocol.

6.2.6.2   *Other Messages*

A variety of messages support ancillary services, from user-to-user signaling to transferring a connection setup attempt to a manual operator. These messages are listed in Table 6.12.

6.2.6.3   *Maintenance Messages*

This is a fairly large group of messages that are used for detailed maintenance of the link. Their existence in B-ISUP creates one of its difficulties in an increasingly deregulated environment, as their misuse can cause significant damage to the operation of a public network. As a result, some network operators have been reluctant to use narrowband ISUP, which has the same set of messages, outside their own network or to interconnect with other large and experienced operators. It remains to be seen whether this model of large public network operators is maintained, and this may well influence the extent to which B-ISUP is used. The B-ISUP maintenance messages are given in Table 6.13.

Table 6.11

B-ISUP Signaling — Basic Connection Control Messages

Message	Purpose and Meaning
Initial Address (IAM)	This is the first message of a call setup sequence.
Subsequent Address (SAM)	If the IAM does not contain all the address of the far end, for example, when a human is dialing a number, this message carries the extra parts of the address.
IAM Acknowledgment (IAA)	Sent by a receiving switch in acknowledgment of an IAM message if it can allocate resource for the call.
IAM Reject (IAR)	Sent if the receiving switch of an IAM message cannot allocate sufficient resource to meet the call request.
Address Complete (ACM)	Sent by a receiving switch when it has all the address information to reach the destination address (but not any subaddress).
Call Progress (PRG)	A message that can be sent in either direction to convey extra information during call setup or during the call. In particular, it conveys "alerting" when the terminal has been alerted but has not yet answered.
Answer (ANM)	Sent in the backward direction to indicate that the terminal equipment has answered the call attempt and has opened the connection.
Release (REL)	Sent in either direction to initiate releasing the connection.
Release Complete (RLC)	Sent in response to a REL message when the resources of the connection have been released and made available for another connection.

Table 6.12

B-ISUP Signaling — Other Control Messages

Message	Purpose and Meaning
Forward Transfer (FOT)	A message sent in the forward direction when operator assistance is required in an international call in order to bring in the operator.
User to User Information (USR)	Used for transferring user-to-user information associated with the user to user signaling supplementary service.
Segmentation (SGM)	Used, as a national option, if the lower signaling layer cannot support the length of a message. Subsequent parts of the message are carried in a segmentation message.
Network Resource Management (NRM)	A message sent during a connection in order to change the resource allocation (see resource management protocol below).
Confusion (CFN)	Sent if a message, or a parameter within a message, cannot be understood by the receiver.

6.2.7   Interworking Between UNI and B-ISUP Signaling

The closest that signaling standards get to specifying the connection controller is in the interworking specifications. These are listed in Table 6.14. A connection controller with UNI signaling on one side and B-ISUP on the other is considered to be interworking

Table 6.13

B-ISUP Signaling — Maintenance Messages

Message	*Purpose and Meaning*
Suspend (SUS)	A message sent in either direction to indicate a temporary disconnection.
Resume (RES)	A message sent following a suspension to indicate reconnection.
Blocking (BLO)	Sent to indicate that the receiving exchange should not allocate the indicated resource to new connections.
Blocking Acknowledgment (BLA)	Sent in response to a blocking message to indicate that the resource has been blocked.
Unblocking (UBL)	Sent to indicate that a blocked resource should be freed again for allocation to new connections.
Unblocking Acknowledgment (UBA)	Sent in response to a blocking acknowledgment message to indicate that the blocked resource has been freed.
Consistency Check Request (CSR)	Sent to request the activation of a consistency check function on a virtual path to ensure the correct allocation of VPC identifiers.
Consistency Check Request Acknowledgment (CSR)	Sent to indicate the consistency check function has been activated.
Consistency Check End (CCE)	Sent to terminate a consistency check session.
Consistency Check End Acknowledgment (CCE)	Sent to indicate that the consistency check function has been deactivated.
Reset (RSM)	Sent if one end of the signaling link gets out of phase with the other in order to release a resource.
Reset Acknowledgment (RAM)	Sent to confirm that a resource has been properly released.
User Part Test (UPT)	Sent to check whether a user part (e.g., B-ISUP) is available and working.
User Part Available (UPA)	Sent in response to a user part test message to indicate that the user part is working and available.

Table 6.14

Interworking Between Signaling Protocols

	Title	*Content*
Q.2650	BISDN, Interworking Between Signaling System C7, B-ISUP and Digital Subscriber Signaling System No2 (DSS2)	Describes the general semantics of the interworking in local switch with Q.2931 on customer side and B-ISUP on the network side.

between signaling systems, and is therefore covered by these recommendations. An example of a connection setup initiated within UNI signaling and forwarded using B-ISUP is illustrated in Figure 6.9. As can be seen, the basic semantics of the protocols is essentially the same for normal call setup and release.

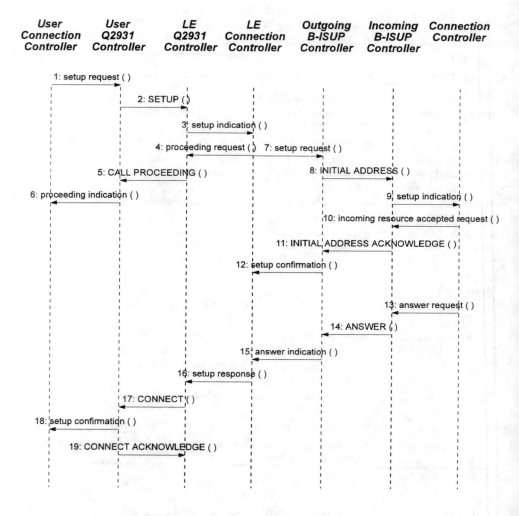

Figure 6.9   Primitives and messages to establish a connection using UNI signaling and B-ISUP.

6.2.8   Resource Management Protocol

The ATM ABR service uses a closed loop control mechanism to vary the cell rate during the connection to make maximum use of the capacity available within the network. This is achieved by using messages defined in a resource management protocol that the ATM Forum and the ITU-T have agreed on and is defined in the ATM Forum Traffic Management Specification and ITU-T Recommendation I.371.

Resource requests and allocations are exchanged, on a regular basis, for each connection on a link between the resources managers on either end of the link. From the point of view of each connection, a message is relayed through each resource manager through which the connection passes. As with the other protocols, this requires a protocol controller to manage the passing of the messages between the resource managers. The basic arrangement is shown in Figure 6.10. The background and context of this resource management protocol is described in more detail in Chapter 9.

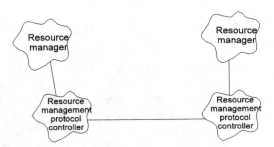

Figure 6.10   Operation of the resource management protocol.

This protocol is low level, and it does have extensive adaptation layers. The messages are carried directly in ATM cells. For VCHs, the cells have the same VCI as the VCh and are differentiated from the user's cells by a specific value of payload type indicator (PTI = 110). For VPs, a specific VCh is reserved from resource management cells (VCI = 6). There are a number of parameters that can be carried in the cells, and these are listed in Table 6.15.

Table 6.15

Parameters that can be Carried in Resource Management Message Cells

Field	Purpose and Meaning
Explicit cell rate (ECR)	The desired cell rate, which is reduced if required by each resource manager.
Current cell rate (CCR)	The cell rate at which the source is currently sending.
Minimum cell rate (MCR)	The minimum cell rate negotiated at connection setup.
Direction (DIR)	Indicates whether the cell is traveling in the same direction as the information to which it refers or in the return direction.
Backward notification (BN)	Indicates if the resource management cell has been generated at a downstream node and sent immediately in the backward direction to indicate congestion
Congestion indication (CI)	A flag that is set if a resource manager is indicating congestion and instructs a source to reduce its rate.
No increase (NI)	A flag to explicitly indicate to a source that it cannot increase from its current cell rate.

As a cell is passed to a resource manager, it examines the requested cell rate (i.e., the explicit cell rate, ECR) and decides whether that cell rate can be fairly allocated out of the capacity available on the link. If not, it reduces the value of the ECR field to the maximum it can allow to that connection.

There are also parameters set at connection setup, some of which control the time constant by which the cell rate can change, both up and down. These parameters are especially important if the ABR service is carrying traffic that also has a dynamic rate control loop, such as Internet traffic controlled by the windowing of TCP.

6.3   NETWORK MANAGEMENT PROTOCOLS FOR THE CONTROL OF CONNECTIVITY

ITU-T Study Group 15 has already generated NE management standards based on CMIP for both SDH and ATM as discussed in Chapter 7. The ATM forum has adopted these for ATM equipment management. The current focus in ITU-T SG15 is on network level standards, initially to control the connectivity of path layer networks based on SDH VCs and ATM VPs. This uses a methodology developed for open distributed processing which aims to define the important semantics of the interface without constraining the physical distribution of functionality in an implementation and remaining, as far as possible, independent of the actual protocols used at real interfaces.

6.3.1   The Open Distributed Processing Reference Model (ODP-RM) applied to transport connection management

The ODP reference model attempts to separate concerns within an information processing system by separately defining the system from five different viewpoints as illustrated in Figure 6.11. SG15 has developed formal notations based on simple adaptations of existing specification techniques, to express the ODP-RM viewpoints in recommendations. A standardized interface may refer to some or all of these viewpoint specifications to achieve the level of specification required.

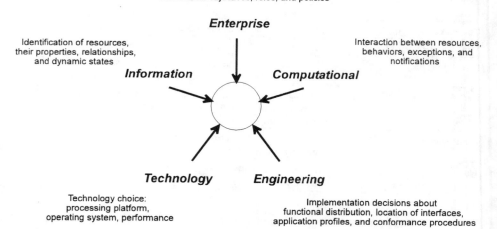

Figure 6.11   The five viewpoints of the ODP reference model.

At the time of writing, mature drafts exist for the enterprise, information, and computational viewpoints for the simple control of connectivity. This present scope is insufficient for even the most basic applications, and the all-important engineering viewpoint, which determines the protocols and syntax of a compliant implementation, is far from agreement. The proposed recommendation structure is indicated in Table 6.16

Table 6.16

ITU-T Recommendations for SDH Network Level Management

Recommendation		Content
G.851 series	Use of the ODP Framework	
G.851-01	Management of the transport network - Application of the RM ODP Framework	This recommendation describes the way that the ODP reference model has been applied to the management of transport networks.
G.852 series	Enterprise Viewpoint	
G.852-01	Management of a Transport Network - Enterprise Viewpoint for Simple Subnetwork Connection Management	This recommendation describes a number of "communities", which are each defined by their purpose, a number of roles, a set of policies, a set of actions, activity aspects, and contract aspects. The communities currently identified are - Simple subnetwork connection configuration - Simple failure monitoring - Simple monitored subnetwork connection
G.853 series	Common Information Viewpoint	
G.853-01	Common Elements of the Information Viewpoint for the Management of a Transport Network	This recommendation describes a number of information object classes (note these are not the same as the GDMO objects described in the next chapter, which are implementational viewpoint objects, although there can be a close relationship). The objects common to aspects of the management of transport networks, though currently with a bias to the control of connectivity, are specified using informal English language, semiformally using ASN.1, and formally using Z notation.
G.853-02	Subnetwork Connection Management Information Viewpoint	This recommendation defines the specific information objects required for subnetwork connection management and inherits from objects in the common elements.
G.854 series	Computational Viewpoint	
G.854-01	Management of a Transport Network - Computational Interfaces for Basic Transport Network Model	This recommendation is concerned with the specification of the operations that can be carried out in the activity of managing transport networks. The recommendation currently covers subnetwork connection configuration and monitored subnetwork connection configuration. The way the operation is defined is independent of the protocol to implement the operations.

6.3.2   Enterprise Viewpoint

The enterprise viewpoint focuses on purposes, scope, and policies and in this way catures the requirements of the application. Formally it is defined in terms of communities of enterprise objects which take particular roles in cooperating to provide specific management objectives. The following communities are currently identified:

- Simple subnetwork connection configuration;
- Simple failure monitoring;
- Simple monitored subnetwork connection.

The enterprise viewpoint defines the purpose of each community, the roles played by the actors within the community, the policies applied within the community, actions associated with roles which may be performed in support of the community purpose, activities consisting of combinations of actions, and service contracts constraining actor roles within the community.

For example, five roles have been identified within the simple subnetwork connection community defined in G.852-1. They are caller, provider, port, subnetwork, and subnetwork connection. The first two identify the principal client and principal server roles, respectively, within the community, while the latter three refer to objects defined in the transport architecture.

Two actions are currently defined in G.852-1 within the simple subnetwork connection configuration community:

- Setup point to point subnetwork connection;
- Release point to point subnetwork connection.

6.3.3   Information Viewpoint

The information viewpoint focuses on the semantics of the information; the information object classes, their relationships, their attributes, their states, and the valid state transitions. Information objects are defined using a GDMO-like object template (GDMO object templates are described in Chapter 7). A reduced ASN.1 attribute syntax is used that avoids complex data types (set of sequence, etc.) in favor of more (but simpler) objects, and the general relationship model (GRM) is used to express relationships. An information object may play one or more of the roles identified in the enterprise viewpoint. States and the valid state transitions are defined in terms of static and dynamic schema which include pre- and post-conditions applicable to each state value.

The description of the viewpoint is divided into generic objects, which are described in G.853-1, while objects that are specific to simple subnetwork connection configuration community are in G.853-2. Transport network resource objects are drawn from G.805. Objects that are specific to other communities will be added in future recommendations.

An outline definition of the information viewpoint subnetwork object is given in Figure 6.12.

The subnetwork information object reflects only the information aspects relevant to the control of connectivity in a subnetwork (i.e., the information aspects of the subnetwork controller).

subnetwork

Informal description

DEFINITION

"A subnetwork information object represents a G.805:1995 sub-network (see G.805:1995 definition)."

ATTRIBUTE

signalIdentification
"A sub-network carries a specific format. The specific formats will be defined in the technology specific extensions."

Semi-formal description

subnetwork INFORMATION OBJECT CLASS

DERIVED FROM networkInformationTop;

CHARACTERIZED BY

subnetworkPackage PACKAGE
BEHAVIOUR
subnetworkPackageBehaviour BEHAVIOUR
DEFINED AS
"<DEFINITION>";;
ATTRIBUTES
signalIdentification;;;

Formal description

_____ *subnetwork_Static* _____
subnetwork : $\mathbb{F}$ *OBJECT*
networkInformationTop_Static
signalIdentification_Static

---

subnetwork $\subseteq$ *networkInformationTop*
subnetwork $\subseteq$ *dom signalIdentification*

_____ *subnetwork_Dynamic* _____
Δ *subnetwork_Static*
networkInformationTop_Dynamic
signalIdentification_Dynamic

Potential relationships

<linkBinds>

<linkConnectionIsTerminatedBySubnetworks>

<sNIsPartitionedBySn>

<subnetworkHasSubnetworkConnections>

<subnetworkIsDelimitedBy>

Figure 6.12   Information viewpoint of the "subnetwork" object.

6.3.4   Computational Viewpoint

The computational viewpoint constrains the functional distribution possible within an ODP system. Computational objects are defined by their computational interfaces which are equivalent to the logical interfaces discussed in Chapter 2. They are defined in terms of operations designated by operational signatures. Invoking an operation on a computational object across a computational interface results in a defined behavioral response which may include invoking an operation on another computational object. Operations imply message passing between computational objects, however, only the operation on the receiver is defined. Message sources must naturally comply by generating the appropriate signature as part of its behavioral response. Operations may be derived from the actions defined in the enterprise viewpoint but may also be derived from other interobject collaboration necessary to apply the policies of the enterprise viewpoint or maintain the integrity of the information viewpoint such as status queries or initial parameter configuration. G.854-1 defines, for both "setup point-to-point connection" and "release point-to-point connection," the following:

- Parameters of the operation;
- Parameters of an operation response message;
- Exceptions that can be raised in response to an operation invocation;
- Computational object behavior in response to an operation.

The behavior can be dependent on preconditions existing in the subnetwork and is defined by the preconditions and the postconditions existing after the operation has been invoked. Any exceptions which must be raised if the preconditions are not met or if some subsequent activity is not completed are also defined for each operation.

The computational viewpoint is the key new feature of ODP which allows functional distributions other than the simple manager-agent distribution. Thus we can imagine that, in the context of connection control, fully distributed processes like C7 signaling and centralized configuration management systems as envisaged in TMN could be computational variants sharing the same enterprise and information viewpoint recommendations.

6.3.5   Development of an Engineering Viewpoint

In the engineering viewpoint actual decisions are made about the communications environment, performance and storage objectives and the actual physical distribution considered necessary to meet these objectives. There is little doubt that the engineering viewpoint recommendation will include a CMIP variant (see Chapter 7 for a fuller consideration of the common management interface protocol — CMIP). In this case the operations are constrained by the ability to map them onto standard CMIP operations. CMIP is now widely deployed and there is a strong industry-wide capability with tool chain support including reusable, "off the shelf" components.

There is a growing interest in using the protocols developed by the OMG (Object Management Group) based on the common object request broker architecture (CORBA) and its interface definition language (IDL), and there is no doubt that this has some advantages over the use of CMISE. However, many manufacturers and operators have a large investment in CMISE, which now has a good level of development support allowing enhancements to standards to be rapidly compiled into operational code. They argue that the advantages of using CORBA are insufficient to obsolete all these development tools and expertise.

Whatever the outcome of the discussion in Study Group 15, it is the view of both authors that any of the signaling systems described in this chapter, especially P-NNI with its soft permanent virtual channel, are perfectly valid engineering viewpoints for the enterprise, information, and computational viewpoints already defined. Whether this connection is taken up is still a matter for speculation; however, as ATM products start to find their way into the marketplace, it is likely that they will have signaling control before they have standardized network level management capabilities. Table 6.17 illustrates some of the important differences in approaches between signaling protocols and CMISE.

Table 6.17
Comparison of Signaling and Use of OSI for Connection Control

	Signaling Protocol Controllers	*OSI Protocol Controller with CMISE, ACSE, and ROSE*
Scope	Dedicated	Wide variety of applications
Orientation	Transactional	Information polling and event reporting
Primitives to application	Large number and specific to control of connectivity	Small number and very general
Speed	Fast	Slow
Throughput	High	Low
Security from malicious access	High — can only perform the specific actions implied in messages	Very low — security has to be an external feature to the protocol controller
Reliability of message transfer	Medium — relies on application as well as the protocol controller	High — built in as a basic service of the protocol controller

6.4 INTERNET CONNECTION CONTROL PROTOCOLS

As discussed in the introduction to this chapter, there has been discussion, especially in the IETF, to use the emerging Internet connection control protocols to control ATM connections. There are some significant differences in the way in which these protocols work, as they retain some of the features of the Internet that make them attractive in any equipment that combines IP and ATM functionality.

At the time of writing, there is much heated debate on the subject, and some of the discussion seems to be driven by both a certain competitive element between the ATM Forum and the IETF. In addition, there is a danger of assuming only one universal monolithic network is possible and that Internet is the only choice. This was the original vision of B-ISDN, and as operators have looked at the practicalities of deployment and evolution, it has become clear that there will never be a single universal network for all services. This is equally true for the Internet, and there will be many networks and many services that do not use Internet protocols. However, it does seem likely that these Internet connection control protocols will be used to control some ATM connections. The extent to which they are used in comparison with signaling protocols is very hard to determine at this point in time.

6.4.1   Resource Reservation Protocol (RSVP)

While the end objective of RSVP is basically the same as that of the other connection control protocols described in this chapter, it has some notable differences. These differences reflect both the connectionless nature of the IP layer network and also the "engineering" rather than "commercial" approach often associated with the IETF. The resource reservation protocol (RSVP) is currently at the stage of an Internet draft. The primary features are reasonably stable and operators are testing implementations.

In the context of the IP layer network and in the strict terms we defined in Chapter 3, we might expect RSVP to establish a *flow* of independent connectionless packets than a connection. However, the routing of the IP packets in the flow are fully constrained to follow the route established by the RSVP setup request, so that RSVP does truly setup up connections and this makes is easily translatable to connection-oriented networks. By fixing a routing for the packets and reserving capacity on the links over which they will travel, RSVP is the mechanism by which connection-oriented service can be offered on the Internet.

The introduction of RSVP is closely coupled by the introduction of differentiated quality of service across the Internet and by offering truly connection-oriented service, RSVP can give a level of guarantee that the specified packets of a connection will successfully transit the network. There is still much discussion on the nature and levels of quality of service, and the service classes currently under discussion do not match those developed by the ATM-Forum and ITU-T for ATM connections. This discussion will continue, and only time will tell the final set of service parameters which will achieve common currency.

The following are the main features which distinguish RSVP from signaling and network management protocols:

- RSVP is a unidirectional protocol. If a bidirectional connection is required, then two independent connections must be requested, one in each direction. (Signaling protocols assume a point-to-point bidirectional connection unless a point-to-multipoint connection is explicitly requested.)
- RSVP routes in the reverse direction to the direction of transmission; it routes from sink to source.
- RSVP makes use of the ability of IP to create flows, as described in Chapter 3, and the request from a sink can be for a number of connections which will arrive at the sink in the form of a single flow. The traffic management and capacity reservation is therefore made in terms of the flow, not in terms of the individual connections.
- The connection states in RSVP are "soft" as, unless the capacity reservation for the connection is continuously requested, the resource manager will drop the

reservation after a short period of time. This makes the protocol potentially robust in terms of avoiding "lock up."

- RSVP is "lightweight" in terms of the software development required to implement it when compared to signaling protocols. This builds on the general principles of the Internet for simple robust protocols.

RSVP is inherently a point-to-multipoint protocol with leaf-initiated join, however, the addition of the flow management feature means that it is not just inherently point-to-multipoint but also multipoint-to-multipoint. This makes it well suited for audio and video conferencing, one of the applications for which it was orginally envisaged. (Each terminal is acting as its own conference bridge as it receives the signals from all parties in the conference.)

The soft state mechanism is a good engineering solution to the problem of protocol controller "lock-up" which requires the RESTART message in ATM UNI signaling and the RESET message in B-ISUP. Each protocol controller effectively will restart if a regular message is not received from the far end. However, this mechanism also makes it much harder to track connections for the purposes of usage monitoring and billing. This has normally been a prime requirement for any connection control mechanism within the old telecommunications community. The Internet has challenged this assumption with its current mechanism of subscription based tariffing although subscription tariffing with the differentiated quality of service inherent in RSVP may be more difficult to sustain. This is discussed in more detail in Chapter 13.

RSVP is based on the messages that are listed in Table 6.18.

Table 6.18

The RSVP Message Set

Message	*Purpose and Meaning*
Path	The message sent in the same direction as the information indicating the flow parameters of the source.
Resv	The message sent from the sink requesting attachment to a flow from a source.
PathErr	This message results if a path message starts traveling back towards its originatating source and thus prevents routing loops.
ResvErr	This message results if a resv message starts traveling back towards its originating sink and thus prevents routing loops.
PathTear	This message initiates the release of a connection instigated by the source.
ResvTear	This message initiates the release of the leaf of a connection used by a sink to attach to the flow from the source.
ResvConf	Sent back to a sink when a resv request has been successfully completed.

6.5 DEVELOPMENT AND EVOLUTION OF CONNECTION CONTROL PROTOCOLS

In this chapter, we have presented three basic classes of connection control protocol; signaling protocols, network management protocols applied to connection control, and

Internet's emerging RSVP. We have concentrated on the signaling protocols, particularly UNI signaling and P-NNI, which seem likely to be the basis for most early ATM switched services.

The way in which these different classes develop and their relative importance into the future is a matter of much speculation. There are other classes beyond these which may become a major part of connection control, particularly those based on the emerging techniques of distributed computing based on the common object request broker architecture (CORBA) that has been recently developed within the Object Management Group (OMG).

It seems likely that, at least for the medium term, that all these protocols will be used in somewhere in broadband networks. As a result, connection controllers will need to work with this wide variety of different protocol controllers. In this case, it should become increasingly important to first develop the intralayer connection control process at a semantic level as described in Chapter 4, and independent of any protocol implementation, before committing to detailed implementation of one particular protocol. This gives the maximum opportunity of supporting broadband transport services across a wide variety of platforms. It remains to be seen the extent to which interworking between connection control protocols will be an obstacle to the development of broadband networks.

References

The following standards are the main source documents for the material in this chapter. These are general signaling standards and signaling adaptation layer (SAAL) standards:

ITU-T Recommendation Q.2010, *Broadband Integrated Services Digital Network Overview — Signalling Capability Set, Release 1.*

ITU-T Recommendation Q.2100, *B-ISDN Signalling ATM Adaptation Layer (SAAL) Overview Description.*

ITU-T Recommendation Q.2110, *B-ISDN ATM Adaptation Layer — Service Specified Connection Oriented Protocol (SSCOP).*

ITU-T Recommendation Q.2120, *B-ISDN Meta-Signalling Protocol.*

ITU-T Recommendation Q.2130, *B-ISDN Signalling ATM Adaptation Layer — Service Specific Coordination Function for Support of Signalling at the User Network Interface (SSFC at UNI).*

ITU-T Recommendation Q.2140, *B-ISDN ATM Adaptation Layer — Service Specific Coordination Function for Signalling at the Network Node Interface (SSCF at NNI).*

ITU-T Recommendation Q.2144, *B-ISDN Signalling ATM Adaptation Layer (SAAL) — Layer Management for the SAAL at the Network Node Interface (NNI).*

ITU-T Recommendation Q.2210, *Message Transfer Part Level 3 Functions and Messages Using the Services of ITU-T Recommendation Q.2140.*

The following are user network interface (UNI) signaling standards:

ATM Forum, *User Network Interface Specification Version 3.1*, 1994.

ATM Forum, *User Network Interface Specification Version 4.0*, 1996.

ITU-T Recommendation Q.2931, *Broadband Integrated Services Digital Network (B-ISDN) — Digital Subscriber Signalling System No. 2 (DSS 2) — User-Network Interface (UNI) — Layer 3 Specification for Basic Call/Connection Control.*

ITU-T Recommendation Q.2932.1, *Digital Subscriber Signalling System No. 2 — Generic functional protocol: Core functions.*

ITU-T Recommendation Q.2951, *Stage 3 Description for Number Identification Supplementary Services Using B-ISDN Digital Subscriber Signalling System No. 2 (DSS 2) — Basic Call.*

ITU-T Recommendation Q.2957, *Stage 3 Description for Additional Information Transfer Supplementary Services Using B-ISDN Digital Subscriber Signalling System No. 2 (DSS 2) — Basic Call; Clause 1 — User-to-User signalling (UUS).*

ITU-T Recommendation Q.2961, *Broadband Integrated Services Digital Network (B-ISDN) — Digital Subscriber Signalling System No.2 (DSS 2) — Additional Traffic Parameters.*

ITU-T Recommendation Q.2962, *Digital Subscriber Signalling System No. 2 — Connection Characteristics Negotiation During Call/Connection Establishment Phase.*

ITU-T Recommendation Q.2963.1, *Peak Cell Rate Modification by the Connection Owner.*

ITU-T Recommendation Q.2971, *B-ISDN — DSS 2 — User-Network Interface Layer 3 Specification for Point-to-Multipoint Call/Connection Control.*

The following are public network to public network signaling standards:

ITU-T Recommendation Q.2730, *Broadband Integrated Services Digital Network (B-ISDN) Signalling System No. 7 B-ISDN User Part (B-ISUP) — Supplementary Services.*

ITU-T Recommendation Q.2761, *Broadband Integrated Services Digital Network (B-ISDN) — Functional Description of the B-ISDN User Part (B-ISUP) of Signalling System No. 7.*

ITU-T Recommendation Q.2762, *Broadband Integrated Services Digital Network (B-ISDN) — General Functions of Messages and Signals of the B-ISDN User Part (B-ISUP) of Signalling System No. 7.*

ITU-T Recommendation Q.2763, *Broadband Integrated Services Digital Network (B-ISDN) — Signaling System No. 7 B-ISDN User Part (B-ISUP) — Formats and Codes.*

ITU-T Recommendation Q.2764, *Broadband Integrated Services Digital Network (B-ISDN) — Signalling System No. 7 B-ISDN User Part (B-ISUP) — Basic Call Procedures.*

The following are private network to private network signaling standards:

ATM Forum, *Private Network-Network Interface (P-NNI) Specification Version 1.0,* 1996

The following are interworking between signaling system standards:

ITU-T Recommendation Q.2610, *Broadband Integrated Services Digital Network (B-ISDN) — Usage of Cause and Location in B-ISDN User Part and DSS 2.*

ITU-T Recommendation Q.2650, *Broadband-ISDN, Interworking Between Signalling System No. 7 Broadband ISDN User Part (B-ISUP) and Digital Subscriber Signalling System No. 2 (DSS 2).*

ITU-T Recommendation Q.2660, *Broadband Integrated Services Digital Network (B-ISDN) — Interworking Between Signalling System No. 7 — Broadband ISDN User Part (B-ISUP) and Narrow-Band ISDN User Part (N-ISUP).*

The following are standards for connection control using network management:

ITU-T Draft Recommendation G.851-01, *Application of the RM-ODP Framework to the Management of the Transmission Network.*

ITU-T Draft Recommendation G.852-01, *Management of the Transport Network — Enterprise Viewpoint for Simple Subnetwork Connection Management.*

ITU-T Draft Recommendation G.853-01, *Common Elements of the Information Viewpoint for the Management of a Transport Network.*

ITU-T Draft Recommendation G.853-02, *Subnetwork Connection Management Information Viewpoint.*

ITU-T Draft Recommendation G.854-01, *Management of the Transport Network - Computational Interfaces for Basic Transport Network Model.*

The following are Internet connection control draft specifications

Braden, R., Zhang, L., Berson, S., Herzog, S., and Jamin, S., *Resource ReSerVation Protocol (RSVP) Version 1 Functional Specification*, Internet Draft, 1997

Management Interfaces **7**

The International Standards Organization (ISO) has presented a framework for developing network management standards as part of the initiative for open systems interconnection (OSI). The framework allows for monitoring and control of network resources, which are represented as "managed objects." It provides a model for understanding management concepts; a common structure of management information (SMI) for defining, identifying and registering managed objects and a set of services and related protocols for performing remote management operations.

7.1   ARCHITECTURAL OVERVIEW

In the OSI management framework, as illustrated in Figure 7.1, there exist application processes called "managers" which reside in managing systems and application processes called "agents" which reside in managed systems. Managers and agents co-operate via standard protocols within a shared schema of management knowledge to achieve the operational objectives of management. The managed objects are system and network resources that are subject to management. Management activities are effected through the interrogation and manipulation of the managed objects. Using the management services expressed as messages and the messaging protocols, a manager can direct an agent to perform an operation on a managed object for which it is responsible. Such operations may involve setting an attribute, retrieving an attribute value or performing some action and returning the result. In addition the agent may spontaneously generate notification messages to the manager indicating the occurrence of events in the managed objects under its control.

Although the protocols themselves make no assumptions about the asymmetry or otherwise of the relationships between the communicating application processes, the

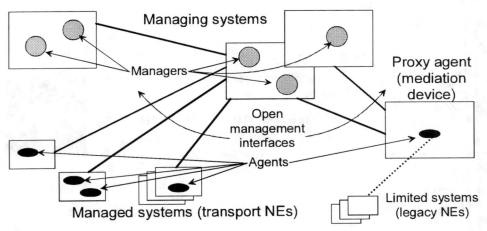

Figure 7.1   Architectural framework for the management of NEs.

manager is considered to have the superior role and the agent correspondingly the subordinate role. Many systems are restricted in their design to one role or the other but it is also possible for agent and manager roles to reside within a single system. A particular type of agent is that residing on a mediation device which acts as a proxy agent for resources which may not be able to provide the agent process themselves either for administrative reasons (e.g. insufficient access rights) or practical reasons (e.g. the resource may only provide access through a proprietary protocol).

7.1.1   Management Models

Management may be modeled in different ways to emphasize different viewpoints or perspectives. Three views are modeled in the OSI management framework:

- The organizational model introduces the concept of a management "domain" as an administrative partition of the network. Management domains are defined by the operator to meet his organizational needs. Management standards aim to minimize constraints on organizational aspects.
- The functional model describes the management functions and their relationships classified under the five specific management functional areas: fault, configuration, performance, accounting and security.
- The information model provides guidelines for the definition of managed objects (GDMO) and the associated management information. All information relevant to network management is considered to reside in a management information base (MIB) which is a conceptual repository of management information.

7.1.2   The Management Information Model

The transport object model derived from the functional architecture discussed in Chapter 3 describes the things involved in the transport processes to be managed. A computer scientist might call it a semantic description of the processes in the problem space. For management purposes it is necessary to identify the information associated with each object considered relevant for management and to express this in a syntactically rigorous manner as a formal specification. This is what we call the management information model.

ITU-T and ISO, as part of a joint standards activity, have produced three draft recommendations giving syntactic rules and guidelines for the definition of information models:

- X.720—Management Information Model,
- X.721—Structure of Management Information, and
- X.722—Guidelines for the Definition of Managed Objects.

These recommendations reflect the fact that information models require a formal grammar and standardized templates. This is necessary if the information is to be understandable to a remote process which may also be in another administrative domain or implemented by a different manufacturer.

7.1.3   The Generic Network Information Model

Using the ISO framework and methodology ITU-T Study Group 4 has developed M.3100 as a generic telecommunication network information model. From the beginning this enterprise was struggling to meet two competing objectives. On the one hand there was an attempt to study and classify current practice across the various telecommunication categories with the objective of abstracting a common or generic core of functionality. Many operators saw this as meeting their need for a standardized approach to management integration of existing systems - the manager of managers - in which the generic model was instantiated at a coordination level dealing with many different actual objects on the basis of their generic core of functionality. This is essentially a retrospective classification and taxonomy of a body of knowledge and the result can justifiably be termed "generic" in a true or natural sense.

On the other hand there was a recognition that much of the potential management capability which made SDH and ATM especially attractive was new. It did not in general exist in legacy systems and was largely undocumented. It was to fill this gap that the development of G.803, G.831 and subsequently G.805 was mainly directed. The quite different objective here was to generate a set of agreed superclasses with a common set of properties reflecting these new insights and to use these purely as a mechanism to direct subsequent development by using the principle of inheritance to define instantiable subclasses containing technology specific refinements but in addition to that functionality inherited from the "generic" superclass. In this approach the generic superclasses themselves are never intended to be instantiable. That is to say they are never used to represent a real instance of a managed object. This is not the natural meaning of genericity as discussed above. In this sense the so called generic model would be used as a sort of prototype to constrain unnecessary variety in the real technology specific models to be implemented. The tension between these conflicting objectives has never formally been resolved and as a result M.3100 contains quite a lot of extraneous information which is something of a burden and often leads to confusion.

All objects, including managed object class specifications, attribute specifications, etc. in the ISO framework are registered with an accredited administrative authority using a unique object identifier. Each accredited authority is allocated an identifier which it uses as the root of a class identification hierarchy. ISO and ITU-T each act as registration authorities and can allocate subordinate name space to other organizations. They also act jointly to maintain a register of objects of common interest. The result is a tree of object identifiers locating particular objects in the globally unique registry. ITU-T for example uses the recommendation number (e.g. M.3100) as the first level followed by the registered object type (e.g. Managed object, Attribute etc.) followed by an integer allocated serially to sequential registration requests.

The generic model then is intended primarily as a parent from which to derive by inheritance, technology specific instantiable models. It was envisaged that there would still be some scope for further vendor or operator specific extensions while still respecting the core functionality represented by the parent technology specific specification. A further source of confusion lies in that the technology specific models (or rather their developers) have not exploited the generic parent in a uniform manner.

In the following description we attempt to explain how, despite its shortcomings, this suite of recommendations is used to generate implementable specifications. In passing we will point out some of the anomalies resulting from the complex history alluded to above but in general the level of detail required to do this comprehensively is beyond our scope. Nevertheless, despite the muddle and thanks to some pioneering work by certain operators, consortia and industry interest groups, there are several good implementable, and indeed implemented, specifications in existence based on this methodology which have been generated from these recommendations. The level of specification thus achieved in the ITU-T model, while not alone sufficient to support real multivendor interworking provides nevertheless sufficient constraint to allow relatively easy accommodation amongst its descendants.

7.1.4   Managed Objects

Draft Recommendation X.720 gives a basic description of information modeling and the way information may be exchanged across an open management interface. This recommendation introduces the managed object, distinct from the plain objects described in the object model. A managed object can be observed and manipulated by a manager across an open interface and defines the information associated with the object in the agent. The managed objects of the information model are derived from the objects of the object model.

The simplest way of deriving an information model as defined in X.720 from the object model would be to take all the objects implemented on one managed system (NE processing host) and define a managed object for each. The open interface is then automatically created between agent and managers by declaring which objects in the managed system are visible through the agent. This approach has been followed to a large extent but there are several factors that restrict a direct mapping from objects in the object model to managed objects in the information model generally resulting from the restrictions placed on managed objects. The characteristic features of managed objects are restricted in order to simplify their definition and the protocol needed on the open interface. The features of a managed object are described in X.720 and include:

- a single unique name that cannot change during the life of the managed object;
- a predefined set of attributes according to its class with defined data type and allowed value range;
- a defined set of operations according to its class that may be activated by messages received across the open interface;
- a defined set of notifications which the object is required to send as messages across the open interface in response to the occurrence of certain specified events;
- only the managed aspects of a managed object may be expressed in its definition and object behaviors which express cooperation with the manager are avoided in the interests of preserving independence for the managing processes which were themselves out of scope.

The first and last features place significant restrictions on the information model.

7.1.4.1   Naming of Managed Objects

X.720 introduces a "contains" relationship to define naming. One managed object can contain many others, however, every managed object can be contained by only one managed object thus forming a one-to-many relationship tree between all managed objects. The "contains" relationship does not necessarily imply physical containment, as managed objects can be logically contained by another managed object. In some ways, the choice of the term "contains" has been unfortunate as it has often been wrongly taken to uniquely imply physical containment. Because its purpose is naming, a term such as "names" might have been more appropriate for this relationship.

The result of the "contains" relationship is a tree of managed objects that has a root at the top. Each managed object has one "parent" above it in the tree and can have many "children" below it. Each managed object is named according to its position in the tree. Each managed object has a *relative distinguished name* (RDN) that is different from the RDNs of all its siblings at the same level in the tree. The full name or *distinguished name* (DN) of an object is formed by generating a sequence of all the RDNs of the managed object and those of all its ancestors right back to the root as illustrated in Figure 7.2. This DN generated by a sequence of RDNs gives each managed object a unique name by which it is referenced by all management systems.

7.1.4.2   Characteristics of Managed Objects

Managed object attributes are in general visible across the management interface. They can only be accessed via the defined operations. They may be read or changed depending on the operations defined for the object. The attributes can either take values reflecting the parameters of the resource being modeled, values of calculated parameters useful for management, or the values representing the state of the managed object. The attribute data type is defined as, for example, a boolean, an integer, a printable ASCII string, a set of

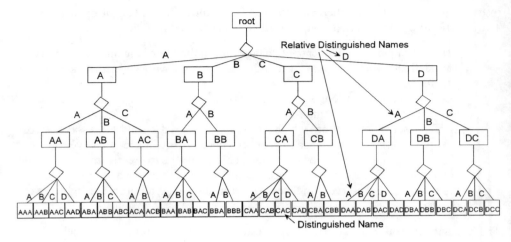

Figure 7.2   Naming of managed objects.

integers, a sequence of printable ASCII strings, etc. and the syntax for defining data types has been standardized in X.208 and is called the *abstract syntax notation No.* (ASN.1).

A managed object has a set of operations that reflect the interactions of the managed object with other parts of the management system. Draft Recommendation X.720 defines that a managed object can have the following operations expressed by messages from the managing side of the interface:

- create the managed object;
- delete the managed object;
- get an attribute value;
- set an attribute value;
- action on the managed object.

Managed objects may also be defined so that they respond to the occurrence of specific events in their scope by sending an event notification message towards the open interface.

- event notification

These are defined in the *common management information protocol* (CMIP) which defines the protocol aspects of the messaging.

Finally every managed object has behavior that describes the process encapsulated in the managed object. For example, if a change in a state attribute initiates a message to a manager only under certain circumstances, then this is stated in the behavior definition of the managed object.

7.1.4.3   Managed Object Template

Draft Recommendation X.722 gives a template for specifying managed objects. This provides a syntactic rigor which allows objects to be compiled automatically in what amounts to a first cut of a software implementation. The template is defined as follows:

```
class-label  MANAGED OBJECT CLASS
   [DERIVED FROM <<class-label>> [,<<class-label>>]* ;]
   [ALLOMORPHIC SET <<class-label>> [,<<class-label>>]* ;]
   [CHARACTERIZED BY <<package-label>> [,<<package-label>>]* ;]
   [CONDITIONAL PACKAGES
      <<package-label>> PRESENT IF <<condition-definition>>
      [,<<package-label >> PRESENT IF <<condition-definition>>]*;]
REGISTERED AS <<object-identifier>>;
```

Note:- Items in square brackets [ ] are optional items and items marked with an asterisk * may be repeated.

The key word DERIVED FROM identifies the list of superclasses from which the object class is derived and forms part of the managed object class definition. Deriving an object from multiple superclasses is known as multiple inheritance and is a feature supported by many object oriented languages. However, use of multiple inheritance from superclasses which are themselves non instantiable as in the ITU-T model, is no more than an editorial convenience and does not imply that a multiple inheritance mechanism must be supported in an implementation.

The managed object "top" is a superclass of all managed object classes. It provides generic attributes (e.g. object identifier) and generic methods (e.g. create and delete) required by all object classes.

ALLOMORPHIC SET identifies the list of superclasses to which the object will respond as being a member. Allomorphism is a property which allows an object to appear to its manager as though it were an object of another class having all the attributes and behaviors of the other class contained within it. Allomorphism was originally seen as an important tool to allow the integration of varying functionalities from different parentage. In fact there are outstanding technical difficulties in supporting it at the implementation level and this feature is no longer used.

CHARACTERIZED BY provides a list of packages that define the characteristics of the managed object over and above those inherited from the superclasses. Each package fully describes a homogeneous set of characteristics. The need for using both the concept of aggregation of packages and multiple inheritance in defining a managed object has often been questioned by those using the template as, in the absence of implementational significance both concepts support the editorial modularity which is their main justification. Multiple inheritance has a somewhat stronger influence in maintaining adherence to the generic core of functionality while aggregation of packages allows more flexibility to depart from this generic core.

CONDITIONAL PACKAGES have been even more contentious. It has been frequently and strongly argued that any object class definition that has optional or conditional features is not a good object class definition and that the optional or conditional behavior should be added as a subclass for the objects that have those features. If the conditional package described features that a particular object may have at some times and not others, then the package is not optional, its parameters simply have a null value at that time. Conditional packages are mainly confined to the non instantiable superclasses. Considerable effort has been directed towards elimination of conditionality from the instantiable, technology specific information models. X.722 also provides a template for the package definition that is:

```
<package-label> PACKAGE
    [BEHAVIOR <behavior-definition-label>
        [,<behavior-definition-label>]* ;]
    [ATTRIBUTES <attribute-label> propertylist [<parameter-label>]*
        [,<attribute-label> propertylist [<parameter-label>]* ]* ;]
    [ATTRIBUTE GROUPS <group-label> [<attribute-label>]*
        [,<group-label> [<attribute-label>]* ]* ;]
    [ACTIONS <action-label> [<parameter-label>]*
        [,<action-label> [<parameter-label>]* ]* ;]
    [NOTIFICATIONS <notification label> [<parameter-label>]*
        [,<notification-label> [<parameter-label>]* ]*;]
[REGISTERED AS <object-identifier>];
```

behavior definitions describe the logical behavior of the package and can include statements of whether some attribute values may be restricted by the value of other attributes, the way in which attributes respond to incoming messages, and when outgoing messages are sent. Unfortunately, all behavior definitions are currently in informal English language and so lose logical rigor.

ATTRIBUTES represent the information that the object holds and will have a data-type such as real, integer, string, etc. for complex structures of these types. The property list of the attribute defines whether an incoming message can change the value of the attribute directly or not. Attributes make use of CMIP GET, SET, ADD, and REMOVE operations messages to operate on information from complex list attributes.

ATTRIBUTE GROUPS are simply convenient groupings of attributes that may be re-used in other packages and defining them as a group means the member attributes need be defined only once.

ACTIONS give the definition of the CMIP actions which the package can undertake. The parameters of the action, both the information received by the object and the reply sent back by the object are defined in the parameter definitions.

NOTIFICATIONS define the CMIP event report from the package.

There are templates for parameters, name bindings, attributes, attribute groups, behaviors, actions, and notifications in X.722 that are necessary for a complete information model. In the interest of clarity much of this detail is not reproduced. We have confined ourselves to a few selective illustrations sufficient to explain the general principles and the nature of the specialization necessary to generate an implementable specification. While every effort is made to maintain accuracy, the examples are primarily for illustration and for the sake of topicality many are from draft agreements. The reader should always refer to the recommendations themselves where accuracy is important.

7.2   THE EQUIPMENT FRAGMENT

The generic equipment fragment of M.3100 provides three managed object classes all derived from "top" and illustrated in Figure 7-3. This fragment carries perhaps the greatest baggage from the past. Each of the objects has a long list of conditional packages which are present if "an instance supports it". The generic superclass in this case has become a holdall for attributes or features which may be required in instantiable subclasses and it is the responsibility of the specifier of the instantiable subclass to state if the condition for inclusion has been met.

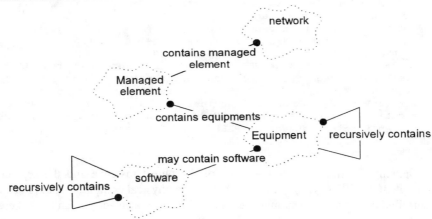

Figure 7-3   Equipment fragment from M.3100

The managed element is defined very loosely so as to include physically distributed entities as well as a traditional integrated NE product. Its main function in the information model is to give a focus to this NE entity and to provide a naming root for all the contained objects. It carries environmental, equipment, communication and processing error alarm notifications as well as administrative, operational and usage states from X.721:1992, although the semantics attached to each of them must still be defined at the level of the instantiable subclass in which they are used.

The equipment object class is a generic object class contained and named by the managed element which reflects the notion of physical network elements as consisting of assemblies of replaceable units regardless of their network function. The single mandatory package embodies this simple property. The M.3100 class definition is reproduced below.

```
equipment  MANAGED OBJECT CLASS
   DERIVED FROM top;
   CHARACTERIZED BY
      equipmentPackage PACKAGE
         BEHAVIOR equipmentbehaviour;
         ATTRIBUTES
            equipmentId                                    GET,
            replaceable                                    GET;;
   CONDITIONAL PACKAGES
      createDeleteNotificationPackage
```

```
        PRESENT IF "an instance supports it",
    attributeValueChangeNotificationPackage
        PRESENT IF "an instance supports it",
    stateChangeNotificationPackage
        PRESENT IF "an instance supports it",
    administrativeOperationalStatesPackage
        PRESENT IF "an instance supports it",
    affectedObjectsListPackage
        PRESENT IF "an instance supports it",
    equipmentsEquipmentAlarmPackage
        PRESENT IF "an instance supports it",
    environmentalAlarmPackage
        PRESENT IF "an instance supports it",
    tmnCommunicationsAlarmInformationPackage
        PRESENT IF "an instance supports it",
    processingAlarmPackage PRESENT IF "an instance supports it",
    userLabelPackage PRESENT IF "an instance supports it",
    vendorNamePackage PRESENT IF "an instance supports it",
    versionPackage PRESENT IF "an instance supports it",
    locationNamePackage PRESENT IF "an instance supports it",
    currentProblemListPackage PRESENT IF "an instance supports it";
REGISTERED AS {m3100ObjectClass 2};
```

The equipmentId attribute holds the RDN of the object instance and the replaceable attribute (of BOOLEAN type) indicates whether the physical equipment that is managed and controlled by the equipment managed object instance is detachable and can therefore be replaced or not.

The elements of the createDeleteNotificationPackage, attributeValueNotification-Package, and stateChangeNotificationPackage packages are defined in X.721 and cover the general syntax of notifications resulting from the creation and deletion of objects, and changes in attribute values. The basic CMISE M-CREATE and M-DELETE together with the createDeleteNotificationsPackage package may be used to support a wide range of instantiation behavior which is not captured in the model at all. This is not just an oversight but represents the wide range of behaviors which may be used in practice, each having benefit in different circumstances. Autoinstantiation by the agent followed by notification on association with or on request by the manager is one possibility. At another extreme objects can be created in the MIB even before the equipments required to support them have been plugged in. The set of behaviors associated with removing equipment instances for repair and plugging them back in, perhaps in the wrong location, are the real issues which determine the effectiveness of a manager agent relationship. All these interesting but inevitably contentious aspects are left for implementers and users to resolve to their mutual satisfaction. The various solutions which emerge in the marketplace are of course a barrier to multivendor working but so long as they are all simple derivatives of the same basic model adaptation should not be difficult. Survival of the fittest will no doubt rule in this aspect of standardization with operators and users insisting on their own preferred operational habits and vendors striving to limit variety while keeping all parties satisfied.

The two attributes administrativeState and operationalState of the administrativeOperationalStatesPackage package respectively are defined in X.721. The administrativeState is the means by which a managing system can lock or unlock the physical equipment. The operational state reflects the operability of the equipment itself as determined internally and revealed to the manager on request or via a state change notification. The behavioral semantics attached to these states in the equipment fragment is still part of the evolutionary process referred to in the last paragraph.

The affectedObjectsListPackage package has a single attribute affectedObjectsList that lists all the managed objects whose operation may be affected if the physical equipment were to be removed or otherwise disabled. In particular this is important to identify the transmission resources likely to be impacted if an equipment item is unplugged for maintenance purposes or conversely to locate a possibly unplugged equipment by correlation amongst transmission alarms.

The equipmentsEquipmentAlarmPackage has an attribute alarmStatus that simply reflects whether the physical equipment has anything currently wrong and a notification associated with this attribute. The environmentalAlarmPackage and processingErrorAlarmPackage packages have notifications associated with these specific types of alarm conditions. The current problem list package has an attribute currentProblemList that is a list of all the faults current on the physical equipment. There are options for the syntax of these problems in the attribute that makes the interpretation of the list attribute difficult. The tmnCommunicationsAlarmInformation package duplicates the alarmStatus attribute and the currentProblemList attribute of the above packages. More importantly it defines alarm notifications with a probable cause parameter in the form of an enumerated list. If all the packages are present in an object instance, then it has only one alarmStatus attribute and one currentProblemList attribute but all the notifications, as this is the logical OR of the packages.

The vendorNamePackage, versionPackage, locationNamePackage, and userLabelPackage packages each contain a single attribute that allows the object to record useful information about the physical equipment, that is respectively, the name of the manufacturer, the version of the equipment, the physical location of the equipment, and an arbitrary, user friendly name given by an individual operator for his own purposes.

The software managed object class represents executable code or data in the form of a file or module regardless of its function. It is treated in much the same way as equipment and is in fact contained and named by the instance of equipment on which it is stored. Much of the important behavior associated with software in a managed environment concerns the software maintenance processes and in particular remote download of software to a network element from a host or load server. These aspects are not included at the generic level.

7.3   TECHNOLOGY SPECIFIC MANAGED OBJECT CLASSES

Technology specific managed object classes are those that would normally be instantiated in real transport network elements. The SDH and PDH specific information model is defined in ITU-T Recommendation G.774 and the G774-xx series of supplements, the ATM specific information model in I.751. They are both based on SMI and GDMO as described above. They both use the generic model defined in M.3100 (10/92) to provide structure and a set of generic superclasses from which the technology specific models are derived.

The original SDH model in G.774 was developed in the same time frame as M.3100 and both benefited from a considerable level of mutual iteration in the later stages of the study period. It contained an equipment fragment and a termination point fragment which together with alarm management and resource management drawn from the generic G.700 series and crossconnection and equipment fragments from M.3100 covered a large part of the functionality required of managed SDH network elements. But substantial areas of prime importance for SDH deployment had to wait for the supplements in G.774-01, G774-02, G.774-03, G.774-04, G.774-05 and G.774-06 to provide a comprehensive specification of NE manageability including performance, supervision and protection.

A most controversial member of the G.774 family G.774-07 titled the Implementer's Guide contains a comprehensive list of changes made to the other recommendations in the family during the 92-96 study period ranging from editorial clarifications through bug fixes to functional modifications. This highlights one of the central problems of formal specification supported in this way through registered objects. In the past ITU-T recommendations evolved over the years by a series of small improvements, additions or corrections. Formal clarity was achieved for contractual purposes by citing the date of the referenced recommendation. Each managed object is in itself a fragment of specification and has a life outside of the recommendation which originally specified it. Therefore unless the change is accepted as being purely editorial a new object class must be registered no matter whether it is a bug fix or a genuine functional extension. In the examples below we have referred to the latest G.774-07 revisions where appropriate. A first revision is usually recognizable by the convention of including the string "R1" in the new registered classname.

7.4   THE TERMINATION POINT FRAGMENT

The termination reference points defined in the functional architecture described in Chapter 3 each justify the definition of a corresponding managed object class. The partitioning structure defines the generic classes independent of transport layer and the client-server relationship between layers provides the basis for the naming hierarchy. The class inheritance hierarchy of the termination point managed object classes is illustrated in Figure 7.4. At the top of the hierarchy is the terminationPoint (TP) managed object class. The only required attribute of this managed object class is a list of equipment and other managed objects that support the correct operation of the termination point. After this there is a large number of conditional packages most of which fall into the "present if an instance supports it" category. The presence of conditional packages, as with the equipment managed object class, mean that the standard is not as well defined as it should be, however, in this case the conditionality is generally suppressed at the technology specific level by redefining conditional packages as required or leaving them out. This means that the managed object classes that are actually implemented are rather less ambiguous. Object instances cannot be members only of termination point or even the trail and connection point managed object classes as these are superclasses that simply express the common features of the members of classes. The minimum level at which terminationPoint managed objects may be instantiated is at the technology level (SDH, PDH, ATM etc.).

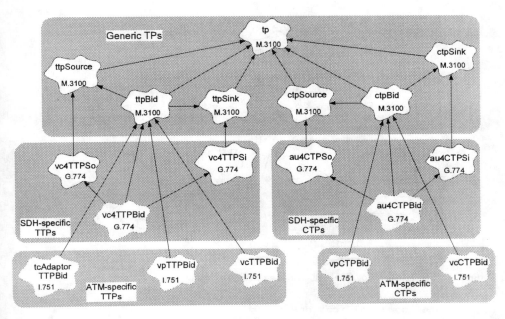

Figure 7.4   The termination point class hierarchy

One of the primary purposes of the terminationPoint managed object classes is to express connectivity. The basic connectivity relationships are expressed with pointer attributes as illustrated in Figure 7.5. The connectivity is managed and controlled by the fabric managed object and crossconnection managed object. These are both discussed in more detail below. Crossconnections are established and removed by actions on fabric objects while the general status of the crossconnection is managed and controlled by the crossconnection managed object. There are additional pointer attributes that describe the relationship between the terminationPoint instances and the crossconnection instance that controls the connection existing between them as shown in Figure 7.5. The various pointer attributes allow the manager to interrogate the connectivity of a network element from many different perspectives thus making the interface presented to the manager flexible and informative.

```
terminationPoint  MANAGED OBJECT CLASS
    DERIVED FROM "Recommendation X.721:1992":top;
    CHARACTERIZED BY
        terminationPointPackage PACKAGE
            BEHAVIOR terminationPointbehaviour;
            ATTRIBUTES
                supportedByObjectList                    GET;;
    CONDITIONAL PACKAGES
        createDeleteNotificationPackage
            PRESENT IF "an instance supports it",
        attributeValueChangeNotificationPackage
            PRESENT IF "an instance supports it",
```

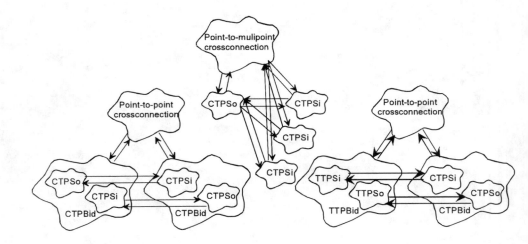

Figure 7.5   Examples of crossconnections between termination points.

```
stateChangeNotificationPackage
    PRESENT IF "an instance supports it",
operationalStatePackage
    PRESENT IF "the termination point is capable of determining
    the integrity of the signal passing through"
affectedObjectsListPackage
    PRESENT IF "an instance supports it"
crossConnectionPointerPackage
    PRESENT IF "an instance can be crossconnected",
characteristicInformationPackage
    PRESENT IF "an instance supports it",
networkLevelPackage PRESENT IF "an instance supports it",
tmnCommunicationsAlarmInformationPackage
    PRESENT IF "an instance supports it",
alarmSeverityAssignmentPointerPackage
    PRESENT IF "the tmnCommunicationsAlarmInformationPackage
    is present AND an instance supports alarm severity
    assignment";
REGISTERED AS {ccittObjectClass 8};
```

The createDeleteNotificationsPackage, attributeValueChangeNotificationPackage, stateChangeNotificationPackage, operationalStatePackage, affectedObjectsListPackage and tmnCommunicationsAlarmInformationPackage packages are the same as those in the equipment managed object class.

The characteristicInformationPackage package has a single attribute, characteristicInformation, that records the characteristic information terminating in a trail termination point or passing through a connection termination point. The crossconnectionPointerPackage package has an attribute crossConnectionObjectPointer that points to the crossconnection object that manages the actual crossconnection.

The instantiable termination points defined in the NE information models are all contained within a network element and are all part of the network element management level. Termination point objects will also exist at the network level. The network level package contains an attribute that can point to an object in the network level. The way in which the network level is modeled by managed objects is still very ambiguous.

A unidirectional trail is delimited at one end by its source termination and at the other end by its sink termination. A bidirectional trail is equivalent to a contradirectional pair of unidirectional trails. To model this situation trailTerminationPointSource and trailTermi-nationPointSink managed object classes are specified. The trailTerminationPointBid (bidirectional) managed object class is then derived by multiple inheritance from both. The trailTerminationPointSource managed object class has a list attribute to point to the downstream TTP sinks or CTP sources to which it is connected. The list data type is needed to allow for the multiple associations of a point-to-multipoint broadcast. The TTP sink object has a single valued attribute to point to the upstream TTP source or CTP sink to which it is connected.

```
trailTerminationPointSource  MANAGED OBJECT CLASS
   DERIVED FROM terminationPoint;
   CHARACTERIZED BY
      operationalStatePackage,
      trailTerminationPointSourcePackage PACKAGE
         BEHAVIOR trailTerminationPointSourcebehaviour;
         ATTRIBUTES
            downStreamConnectivityPointer                    GET;;
   CONDITIONAL PACKAGES
      administrativeStatePackage
         PRESENT IF "the termination point can be placed out
         of service using the management system",
      supportableClientListPackage
         PRESENT IF "the trail termination point can adapt to
         connection points from more than one client layer",
      ttpInstancePackage
         PRESENT IF "the name binding requires this attribute";
REGISTERED AS {ccittObjectClass 11};


trailTerminationPointSink  MANAGED OBJECT CLASS
   DERIVED FROM terminationPoint;
   CHARACTERIZED BY
      operationalStatePackage,
      trailTerminationPointSinkPackage PACKAGE
         BEHAVIOR trailTerminationPointSinkbehaviour;
         ATTRIBUTES
            upStreamConnectivityPointer                      GET;;
   CONDITIONAL PACKAGES
      administrativeStatePackage
         PRESENT IF "the termination point can be placed out
         of service using the management system",
      supportableClientListPackage
         PRESENT IF "the trail termination point can adapt to
         connection points from more than one client layer",
      ttpInstancePackage
```

```
                  PRESENT IF "the name binding requires this attribute";
REGISTERED AS {ccittObjectClass 10};


trailTerminationPointBidirectional MANAGED OBJECT CLASS
    DERIVED FROM trailTerminationPointSource,
        trailTerminationPointSink;
    CHARACTERIZED BY
        trailTerminationPointBidirectionalPackage PACKAGE
            BEHAVIOR trailTerminationPointBidirectionalbehaviour;;;
REGISTERED AS {ccittObjectClass 9};
```

The inclusion of the operational state package in the TTP classes means that the operationalState attribute must be present in every instance of a trail termination. This is an example where the conditional package in the TP superclass has been rendered unconditional in a subclass. The administrativeState attribute, however, is still optional. The supportable client list package has a list attribute that lists the CTP managed object classes that may be supported by the TTP instance. The TTP instance package is only included if the instance is not a member of any subclass of the TTP managed object classes and therefore is instantiated at this level in the inheritance hierarchy. The package has the attribute ttpId that contains the RDN of the instantiated managed object.

The same principle is used to define managed objects for connection termination points, namely; source, sink, and bi-directional with the bi-directional object class formed by the multiple inheritance from the source and the sink object classes. The connection-TerminationPointSource managed object has a single valued attribute to point to the up-stream TTP source or CTP sink to which its resource is connected. The CTP sink object class has a list attribute to point to the down-stream TTP sinks or CTP sources to which it is connected since this could be a point-to-multipoint broadcast.

```
connectionTerminationPointSource  MANAGED OBJECT CLASS
    DERIVED FROM terminationPoint;
    CHARACTERIZED BY
        connectionTerminationPointSourcePackage PACKAGE
            BEHAVIOR connectionTerminationPointSourcebehaviour;
            ATTRIBUTES
                upStreamConnectivityPointer                    GET;;
    CONDITIONAL PACKAGES
        ctpInstancePackage
            PRESENT IF "the name binding requires this attribute",
        channelNumberPackage PRESENT IF "an instance supports it";
REGISTERED AS {ccittObjectClass7};


connectionTerminationPointSink  MANAGED OBJECT CLASS
    DERIVED FROM terminationPoint;
    CHARACTERIZED BY
        connectionTerminationPointSinkPackage PACKAGE
            BEHAVIOR connectionTerminationPointSinkbehaviour;
            ATTRIBUTES
                downStreamConnectivityPointer                  GET;;
    CONDITIONAL PACKAGES
        ctpInstancePackage
```

```
         PRESENT IF "the name binding requires this attribute",
         channelNumberPackage PRESENT IF "an instance supports it";
REGISTERED AS {ccittObjectClass 6};


connectionTerminationPointBidirectional  MANAGED OBJECT CLASS
    DERIVED FROM
        connectionTerminationPointSource,
        connectionTerminationPointSink;
REGISTERED AS {ccittObjectClass 5};
```

The CTP instance package is only included when the CTP is instantiated at this generic level in the inheritance hierarchy (layer specific Ids are included in the technology specific subclasses). The package has a single attribute ctpId that carries the RDN of the instantiated object. The channel number package has a single attribute channelNumber, an integer that records the channel number of the connection in the server trail.

The semantics of operational state in TPs has been problematic throughout the standards development activity and it is noteworthy that operational state is excluded from the CTP subclass but included in the TTP subclass. There is a widespread understanding that the operational state of a TP somehow represents the "good health" of the terminated connection. The trail termination point had become associated with layer boundary supervision to monitor among other things connection integrity. Hence its operational state had an obvious significance and in practice the transition to disabled could be linked with receipt of signal fail or excessive errors for example. In general of course, a trail termination point may not have the capability to determine its own operational state in this way.

A connection termination point on the other hand is not normally supervised to this level and furthermore its operational state (if it had one) would be directly derived from that of its server TTP hence an operational state attribute in each CTP may be considered redundant and inefficient. In general of course instantiated CTPs may well have associated monitors. Therefore although we may think it illogical to introduce the operational state in the superclass it is not inconsistent with the normal usage in SDH and ATM.

7.4.1   SDH Specific Termination Point Classes

Although TPs in different SDH layers have much the same characteristics there are minor differences. More importantly it is probable that evolution over time will require extended features which may not be generic across all layers. It was mainly for such reasons of extensibility that SDH TP managed objects have been specialized in each layer. The VC4 TTP source, sink, and bi-directional are illustrated and described here as examples of SDH TTP managed object classes while the AU4 CTP source, sink, and bi-directional are illustrated and described as examples of SDH CTP managed object classes. These managed object classes provide Id attributes so the objects may be instantiated and override some of the optionality embodied in the conditional packages of the generic managed object classes from which they are derived. In addition they add the SDH specialized attributes necessary to control the path trace, signal label, and FERF functions in the TTPs and recognize the 'payload pointer' functionality in the CTPs (not to be confused with upstream and downstream connectivityPointer attributes).

```
vc4TTPSourceR1 MANAGED OBJECT CLASS
    DERIVED FROM
        "Recommendation M.3100:1992":trailTerminationPointSource;
    CHARACTERIZED BY
        "Recommendation X.721:1992":administrativeStatePackage,
        "Recommendation M.3100:1992":createDeleteNotificationsPackage,
        "Recommendation M.3100:1992":stateChangeNotificationPackage,
        vc3-4SourcePackageR1 PACKAGE
            BEHAVIOR
                vc3-4SourcePackageR1behaviour BEHAVIOR
                DEFINED AS *When 16 bytes are supported, the 16 bytes
                of the path trace shall be conveyed at the management
                interface.*;;
            ATTRIBUTES
                "Recommendation G.774-05:1994":
                    j1PathTraceSend                        GET-REPLACE,
                "Recommendation G.774:1992":c2SignalLabelSend   GET;;
        vc4TTPSourcePkgR1 PACKAGE
            BEHAVIOR
                vc3-4TTPSourcePkgR1behaviour BEHAVIOR
                DEFINED AS
                    *This object class originates a vc4 trail, i.e. the
                    point at which the SDH VC-4 is originated.*;;
            ATTRIBUTES
                "Recommendation G.774:1992": vc4TTPId        GET;;;
REGISTERED AS { g77407ObjectClass 24 };


vc4TTPSinkR1 MANAGED OBJECT CLASS
    DERIVED FROM "Recommendation M.3100:1992":trailTerminationPointSink;
    CHARACTERIZED BY
        "Recommendation X.721:1992":administrativeStatePackage,
        "Recommendation M.3100:1992":createDeleteNotificationsPackage,
        "Recommendation M.3100:1992":stateChangeNotificationPackage,
        "Recommendation M.3100:1992":
            tmnCommunicationsAlarmInformationPackage,
        vc3-4SinkPackageR1 PACKAGE
            BEHAVIOR vc3-4SinkPackageR1behaviour
            ATTRIBUTES
                "Recommendation G.774:1992":j1PathTraceExpected
                    DEFAULT VALUE SDHIMPASN1.Null
                                    GET-REPLACE REPLACE-WITH-DEFAULT,
                "Recommendation G.774-05:1994":
                    j1PathTraceReceive                      GET,
                "Recommendation G.774:1992":
                    c2SignalLabelExpectee                   GET,
                "Recommendation G.774:1992":
                    c2SignalLabelReceive                    GET;;;
        vc4TTPSinkPkgR1 PACKAGE
            BEHAVIOR
                vc4TTPSinkPkgR1behaviour BEHAVIOR
                DEFINED AS  *This object class terminates a vc4 trail,
                i.e. the point at which the SDH VC-4 is terminated.*;;
            ATTRIBUTES
```

```
              "Recommendation G.774:1992":vc4TTPId            GET;;;
REGISTERED AS { g77407ObjectClass 23 };


vc4TTPBidirectionalR1 MANAGED OBJECT CLASS
    DERIVED FROM
        "Recommendation M.3100:1992":
            trailTerminationPointBidirectional,
        vc4TTPSinkR1,
        vc4TTPSourceR1;
    CHARACTERIZED BY
    vc3-4BidirectionalPackageR1 PACKAGE
        BEHAVIOR
            vc3-4BidirectionalPackageR1behaviour BEHAVIOR
                DEFINED AS
                *A communicationsAlarm notification shall be issued if
                a far end receive failure (G1 byte) is detected. The
                probableCause parameter of the notification shall
                indicate FERF (Far End Receive Failure).*;;;;
REGISTERED AS { g774ObjectClass 22 };


au4CTPSource  MANAGED OBJECT CLASS
    DERIVED FROM connectionTerminationPointSource;
    CHARACTERIZED BY
        createDeleteNotificationsPackage,
        au4CTPSourcePackage PACKAGE
            BEHAVIOR au4CTPSourcePackagebehaviour;
            ATTRIBUTES
                au4CTPId                                      GET,
                pointerSourceType                            GET;;;
REGISTERED AS {g77xObjectClass x};


au4CTPSinkR1 MANAGED OBJECT CLASS
    DERIVED FROM
        "Recommendation M.3100:1992":connectionTerminationPointSink;
    CHARACTERIZED BY
        "Recommendation M.3100:1992":createDeleteNotificationsPackage,
        "Recommendation M.3100:1992":operationalStatePackage,
        "Recommendation M.3100:1992":stateChangeNotificationPackage,
        "Recommendation M.3100:1992":
            tmnCommunicationsAlarmInformationPackage,
        au4CTPSinkR1Pkg PACKAGE
            BEHAVIOR au4CTPSinkR1Pkgbehaviour behavior;
            ATTRIBUTES
                "Recommendation G.774:1992":au4CTPId          GET,
                "Recommendation G.774:1992":pointerSinkType   GET;;;
    REGISTERED AS { g77407ObjectClass 4 };
```

The behaviors associated with the packages define such characteristics as response to path trace mismatch, label mismatch, operational state change and the behaviors associated with AIS and FERF.

Note that the operationalStatePackage package which had been excluded in the superclass has been reintroduced in the au4CTP and indeed all the other SDH path layer

CTPs. This reflects the fact that the pointer function is monitorable and provides a form of supervision of the preceding link connection. The tmnCommunicationAlarmPackage package is also introduced. This provides among other things the capability to notify receipt of AIS at a CTP sink which is often used as a service alarm at administrative boundaries. It is one of the minor inconsistencies that the HOP CTPs reinforce this AIS behavior with an explicit statement while the LOP CTPs omit any reference in the behavior definition. The pointerSinkType and pointerSourceType attributes indicate whether the TP in question takes the form of a concatenated group or not.

7.4.2   ATM Specific Termination Point Classes

The VP and VCh layer TTPs and CTPs defined in I.751 follow a similar pattern to their SDH counterparts in the manner in which they inherit from the M.3100 TP super class hierarchy. There are however some interesting variations in style which probably owe more to the preferences of the modeling group rather than any real differences in principle.

In ATM the transmission convergence function represents the adaptation between the ATM VP layer and its server layer and is responsible for cell delineation and scrambling. It is capable of notifying the status of cell delineation within the ATM stream and enabling or disabling the scrambling. This has been modelled by the tcAdaptorTTPBidirectional managed object class reproduced below.

It has a one to one relationship with the server layer termination to which it is associated. As a subclass of TTP it is named from the managed element and indicates cell delineation via a communication alarm from the tmnCommunicationAlarmInformation-Package package. The vPCTPs are named from this object.

```
tcAdaptorTTPBidirectional MANAGED OBJECT CLASS
    DERIVED FROM
        "Recommendation M.3100:1992":trailTerminationPointBidirectional;
    CHARACTERIZED BY
        "Recommendation M.3100:1992"
            tmnCommunicationsAlarmInformationPackage,
        "Recommendation M.3100:1992":createDeleteNotificationsPackage,
        "Recommendation M.3100:1992":stateChangeNotificationPackage,
        tcAdaptorTTPBidirectionalPkg PACKAGE
            BEHAVIOR tcAdaptorTTPBidirectionalBeh;
            ATTRIBUTES
                tcTTPId                              GET;;;
    CONDITIONAL PACKAGES
        cellScramblingEnabledPkg
            PRESENT IF "cell scrambling may be activated and deactivated
            for the supporting ATM interface.";
REGISTERED AS { i751ObjectClass 16 };
```

A group of objects all derived from "top" are defined to be associated with tcAdaptorTTP instances to designate the network role played by an instance. The uni managed object class, the interNNI managed object class and the intraNNI managed object class indicate that the associated object tcAdaptorTTP is located at a user to network interface, a network node interface between two offices or a network node interface between two equipments in the same office. These are instantiated by the management system and associated with the TTP to which they refer by the underlyingTTPPointer GET attribute. In SDH such network roles are not explicitly expressed inside the NE.

Constraints on the interlayer relationships are expressed by the atmAccessProfile managed object class which is attached by naming to instances of tcAdaptorTTPBidirectional or vpTTPBidirectional managed objects. A tcAdaptorTTPBidirectional managed object instance must have an atmAccessProfile before it can support instances of vpCTPBidirectional. It has vpLevelProfilePackage, vcLevelProfilePackage and max-BandwidthPackage conditional packages. The vpLevelProfilePackage package is only present in instances attached to a tcAdaptorTTPBidirectional instance. It has read only attributes which define the maximum VPI field supportable by the NE and the maximum number to be supported in the application which may be set by the manager. Two more attributes record similar information pertaining to the maximum number of active connections which are potentially supported by the NE and the maximum number agreed to be supported within the application. The vcLevelProfilePackage records exactly the same attribute data for the VCh level. Presence of the maxBandwidthPackage is conditional on the vpLevelProfile being also present.

The model allows instances of atmAccessProfile also to be attached to vpTTPBidirectional. In this case there is no vpLevelProfilePackage nor maxBandwidthPackage and the values in the vcLevelProfilePackage take precedence over a competing instance attached to the tcAdapterTTPBidirectional instance. The model is illustrated in Figure 7.6.

The ATM TP managed object classes differ hardly at all in the VP and VCh layers. The TTP objects reveal their OAM significance in the presence of attributes and behaviors associated with their location at the layer boundary. The vpTTPBidirectional managed object class is reproduced below:

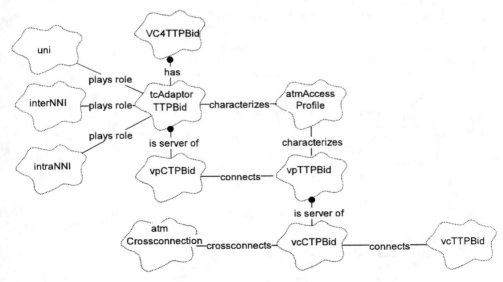

Figure 7.6   ATM termination point hierarchy.

```
vpTTPBidirectional MANAGED OBJECT CLASS
    DERIVED FROM "ITU-T M.3100":trailTerminationPointBidirectional;
    CHARACTERIZED BY
        "Recommendation X.721:1992":administrativeStatePackage,
        "Recommendation M.3100:1992":
            attributeValueChangeNotificationPackage,
        "Recommendation M.3100:1992":createDeleteNotificationsPackage,
        "Recommendation M.3100:1992":
            tmnCommunicationsAlarmInformationPackage,
        vpTTPBidirectionalPkg PACKAGE
            BEHAVIOR vpTTPBidirectionalBeh;
            ATTRIBUTES
                vpTTPId                                          GET;;;
    CONDITIONAL PACKAGES
        loopbackOAMCellPkg
            PRESENT IF "the VPC termination point supports OAM
                cell Loopbacks";
REGISTERED AS { i751ObjectClass 23 };
```

The loopbackOAMCellPkg package contains an action which requires the agent, in response to the manager, to insert a single loopback OAM cell and verify its return. The loopback cell can be injected into the connection at any point and identifies whether it is a segment or end-to-end flow and the precise location where it should be looped. To provide a check on a complete trail end-to-end in both directions the loopback cell is injected into the trail from the source at one boundary, transported along the trail, looped at the remote layer boundary and monitored at the associated sink located at the originating layer boundary.

The ATM CTP is distinguished by cell transfer QoS parameters and the need to set the traffic descriptors by which the UPC/NPC functions may be configured. The vpCTPBidirectional managed object class is reproduced below:

```
vpCTPBidirectional MANAGED OBJECT CLASS
    DERIVED FROM "Recommendation M.3100:1992":
        connectionTerminationPointBidirectional;
    CHARACTERIZED BY
        "Recommendation M.3100:1992":
            attributeValueChangeNotificationPackage,
        "Recommendation M.3100:1992":createDeleteNotificationsPackage,
        "Recommendation M.3100:1992":crossConnectionPointerPackage,
        vpCTPBidirectionalPkg PACKAGE
            BEHAVIOR vpCTPBidirectionalBeh;
            ATTRIBUTES
                vpCTPId                                          GET,
                segmentEndPoint
                    DEFAULT VALUE AtmMIBMod.booleanFalseDefault
                                                         GET-REPLACE;;;
    CONDITIONAL PACKAGES
        egressTrafficDescriptorPkg
            PRESENT IF "supplied by the managing system. This
            package must be present at points where egress UPC/NPC
            functions are performed.",
        ingressTrafficDescriptorPkg
```

```
        PRESENT IF "supplied by the managing system. This
        package must be present at points where ingress UPC/NPC
        functions are performed.",
    oamEgressTrafficDescriptorPkg
        PRESENT IF "supplied by the managing system. This
        package must be present at points where egress OAM
        UPC/NPC functions are performed.",
    oamIngressTrafficDescriptorPkg
        PRESENT IF "supplied by the managing system. This
         package must be present at points where ingress
        OAM UPC/NPC functions are performed.",
    "Recommendation X.721:1992":administrativeStatePackage
        PRESENT IF "supported by the Network Element",
    qosClassesPkg
        PRESENT IF "QOS Class information is supplied by
        the managing system",
    loopbackOAMCellPkg
        PRESENT IF "the VPL termination point supports
        OAM cell Loopbacks";
REGISTERED AS { i751ObjectClass 22 };
```

The segmentEndPoint attribute allows a CTP to be designated as a segment end point. This is where the F4 information flows terminate. The segment is equivalent to a monitored tandem connection in SDH.

The main distinguishing feature of ATM CTPs is their cell transfer capabilities under contracted traffic conditions. The parameters describing the contracted traffic conditions are held for each instance. They are set by the manger usually at connection setup and traffic may subsequently be compared against the measured characteristics of the offered traffic to ensure compliance with the contract. For VPs the manager sets the required parameters at connection set up. VChs will normally (but not necessarily) be set by the call control process but the same object structure is used. The egressTrafficDescriptorPkg package applies to the ctpSource part of the ctpBidirectional. It provides SETable attributes for peak cell rate and sustainable cell rate, the cell delay variation tolerances based on each of these and the maximum burst size. The ingressTrafficDescriptorPkg package provides the same capability at a sink. The OAM cell traffic descriptors packages provide for peak rate and CDV settings only. All the traffic descriptor packages follow the same pattern. The ingressTrafficDescriptorPkg package is reproduced below to illustrate this pattern.

```
ingressTrafficDescriptorPkg PACKAGE
    ATTRIBUTES
        ingressPeakCellRate                    GET-REPLACE,
        ingressCDVTolerancePCR                 GET-REPLACE,
        ingressCDVToleranceSCR                 GET-REPLACE,
        ingressSustainableCellRate             GET-REPLACE,
        ingressMaxBurstSize                    GET-REPLACE;
REGISTERED AS { i751Package 11 };
```

The qosClassesPkg package provides a QOS attribute with an enumerated list of five classes (0-4). The actual quality specification referenced is not specified but this attribute is used to determine UPC/NPC procedure. The relationship of QoS class to UPC/NPC procedure is considered an individual operator's decision.

ATM TTPs and CTPs may be created and deleted by the management system or by the agent itself as a side effect of a management action on the crossconnect fabric.

7.5   THE CROSSCONNECTION FRAGMENT

The crossconnection fragment deals with the management of connectivity within a network element. Initially the notion of a matrix in a network element being also the lowest level of subnetwork in a given layer was the focus for crossconnection. However crossconnect machines often support more than one layer and the notion of a crossconnect fabric containing resources from several layer networks emerged as a more general concept. The crossconnection fragment is illustrated in Figure 7.7.

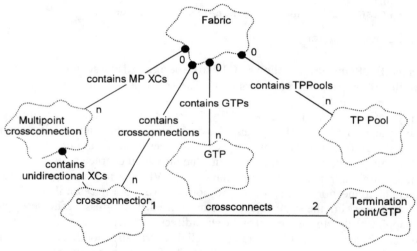

Figure 7.7   The generic crossconnection fragment

The generic fabric managed object class manages and controls the potential connectivity in the network element. It provides actions to create a crossconnection managed object instance between a designated set of TPs. The crossconnection instances may equally be deleted by an action on the fabric managed object.

```
fabric MANAGED OBJECT CLASS
    DERIVED FROM top;
    CHARACTERIZED BY
        fabricPackage PACKAGE
            BEHAVIOR fabricbehaviour;
            ATTRIBUTES
                fabricId                        GET,
                administrativeState             GET-REPLACE,
                operationalState                GET,
                availabilityStatus              GET,
```

```
                listOfCharacteristicInfoType                    GET,
                supportedByObjectList             GET-REPLACE ADD-REMOVE;
            ACTIONS
                addTpsToGTP,
                removeTpsFromGTP,
                addTpsToTpPool,
                removeTpsFromTpPool,
                connect,
                disconnect;;;
REGISTERED AS {ccittObjectClass16};


crossConnection MANAGED OBJECT CLASS
    DERIVED FROM top;
    CHARACTERIZED BY
        crossConnectionPackage PACKAGE
        BEHAVIOR crossConnectionbehaviour;
        ATTRIBUTES
            crossConnectionId,
            administrativeState                        GET-REPLACE,
            operationalState                           GET,
            signalType                                 GET,
            fromTermination                            GET
            toTermination                              GET,
            directionality                             GET;;;
REGISTERED AS {ccittObjectClass 15};
```

There was considered to be a generic requirement to manage bundles or groups of connections as single entities and this has led to the definition within the crossconnection fragment of the gTP managed object which represents a set of TPs which are treated for configuration purposes as a single TP. In SDH this construct is also used to manage concatenation. The tpPool managed object provides a mechanism to interact with a group of TPs which share some characteristic. TPs may be placed, by action on the fabric, in tpPools that are used to model access and transit groups within the network element. This is performed by a manager on a fabric are illustrated in Figure 7.8.

By setting the administrative state of the crossconnection managed object, it is possible to switch the crossconnection on and off without losing the information that the particular crossconnection has not been removed. This can be the case if a trail is set up ahead of its requirement by a client. When all connections in the trail are complete the crossconnections can be switched on.

The generic crossconnection fragment objects from M.3100 have been used directly in the SDH instantiable model. However in retrospect this is widely recognized as being overspecified for a generic model. In ATM the atmFabric managed object is rederived from "Recommendation X.721:1992":top. It supports the connect and disconnect actions which are redefined for the ATM environment.

```
atmFabric MANAGED OBJECT CLASS
    DERIVED FROM "Recommendation X.721:1992":top;
    CHARACTERIZED BY
        "Recommendation M.3100:1992":stateChangeNotificationPackage,
        atmFabricPackage PACKAGE
```

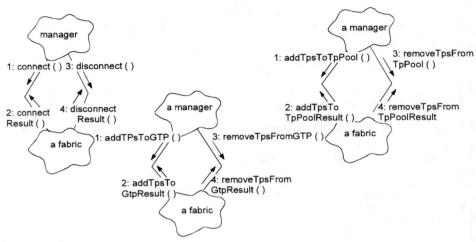

Figure 7.8   Actions and results on a crossconnect fabric.

```
BEHAVIOR atmFabricBeh;
ATTRIBUTES
    atmFabricId                                              GET,
    "Recommendaion X.721:1992":
        administrativeState                             GET-REPLACE,
    "Recommendation X.721:1992":operationalState    GET,
    "Recommendation X.721:1992":availabilityStatus  GET;
ACTIONS
    connect,
    disconnect;;;
REGISTERED AS { i751ObjectClass 4 };
```

The atmCrossconnection managed object is derived from the generic crossconnection of M.3100 but as only bi-directional crossconnections are considered in this model the directionality attribute is set to bi-directional.

The common assumption for routing in an SDH path layer was that the server layer topology has been created and the links are populated with link connections and hence tpPools can be loaded with CTPs etc. Crossconnection requests may then be formulated in terms of existing TPs to be connected or any equivalent TP in a designated pool. The common assumption for routing in the ATM layers is that the server layer topology has been created and the links have a certain available bandwidth resource but this is not allocated to specific TPs until the connection request which specifies the resource to be allocated in terms of the ATM traffic parameters. On receipt of a connect action the atmFabric may instantiate a TP using the VPI/VCI as its RDN. When CTPs are created the traffic descriptor attributes are set to reflect the resource allocation agreed. The atmFabric may reject a request if there is insufficient capacity to meet the cell transfer requirements or if the parameter combination is not valid. The ASN.1 syntax for the ConnectInformation parameter of the connect action is illustrated:

```
ConnectInformation ::= SEQUENCE OF SEQUENCE {
   fromTermination        [0] CtpOrDescriptor,
   toTermination          [1] CtpOrDescriptor,
   administrativeState    [2] AdministrativeState OPTIONAL}

   CtpOrDescriptor ::=CHOICE {
      ctp              [0] ObjectInstance,
      ctpDescriptor [1] Descriptor}

   Descriptor ::= SEQUENCE {
      interfaceId [0] ObjectInstance, -uni, intraNNI, or interNNI
      vpi [1] VpiValue OPTIONAL, —ass..d by m..d system if absent
      vci [2] VciValue OPTIONAL, — set to 0 for VP cross connect
      egressCDVTolerancePCR [3] CDVTolerance OPTIONAL,
      ingressCDVTolerancePCR [4] CDVTolerance OPTIONAL,
      egressCDVToleranceSCR [5] CDVTolerance OPTIONAL,
      ingressCDVToleranceSCR [6] CDVTolerance OPTIONAL,
      egressMaxBurstSize [7] MaxBurstSize OPTIONAL,
      ingressMaxBurstSize [8] MaxBurstSize OPTIONAL,
      egressPeakCellRate [9] PeakCellRate OPTIONAL,
      ingressPeakCellRate [10] PeakCellRate OPTIONAL,
      egressSustainableCellRate [11] SustainableCellRate OPTIONAL,
      ingressSustainableCellRate [12] SustainableCellRate OPTIONAL,
      egressQosClass [13] QosClass OPTIONAL,
      ingressQosClass [14] QosClass OPTIONAL,
      oamIngressPeakCellRate [15] PeakCellRate OPTIONAL,
      oamEgressPeakCellRate [16] PeakCellRate OPTIONAL,
      oamIngressCDVTolerance [17] CDVTolerance OPTIONAL,
      oamEgressCDVTolerance [18] CDVTolerance OPTIONAL,
      segmentEndPoint [19] Boolean}
```

7.6   THE PERFORMANCE FRAGMENT

The NE performance fragment deals with the recording by the network element of the incidence of error events and unavailable time. There are essentially two management clients for such information, the service management layer that has the responsibility to intercede on behalf of the end user and the network management layer which has the responsibility to locate and repair network faults, including the sort of faults which cause intermittent failure or degraded transmission. Both the SDH and ATM standards are quite rich in defect reporting capability; the problem is how much detailed information to retain and for how long to meet the requirements of both clients. The model as always must leave enough scope in regard to what and how data is processed for system designers and operators to implement performance management strategies to suit their needs. From a service viewpoint individual records are aggregated for accounting purposes and they must be detailed enough to resolve any conflict with the customer. From the operational viewpoint the historical archive must be long enough to resolve patterns occurring over days or weeks yet there must be sufficient resolution to allow diagnostic processes to distinguish between events occurring at about the same time.

   In Chapter 8 we discussed the underlying causes of performance degradation events in the network and the rationale behind the temporal filtering of collected data to identify significant events leading to the choice of the performance monitoring parameters errored seconds (ES), severely errored seconds (SES), background block error rate (BBER) and unavailable time (UT) as defined in G.826. It has been recognized that there is potentially useful diagnostic information within the substructure of individual error bursts but there is insufficient understanding at this stage to take any systematic account of it much less to standardize it. Operators are interested in distinguishing short interruption events (SIE) in the region of 3-10 seconds from a service viewpoint because many data transfer protocols are particularly sensitive to degradations of this order but an SIE parameter has not been standardized.

There has been no attempt to standardize diagnostic processes but we can assume that where in-service monitoring is widely deployed as in SDH/ATM networks it will be exploited by performance management systems which provide some level of automatic support to fault location procedures. The general principle is to count each of the standardized events over a fixed interval. If a predetermined threshold count is exceeded in this interval then a threshold crossing event is issued which may also be designated an alarm. This may also initiate a diagnostic procedure in the management system aimed at isolating the root cause. In this case, once a root cause has been established it is recorded and there is no further need to retain a record of the associated events. If no threshold is exceeded during the measurement interval the accumulated result at the end of the measurement interval is nevertheless retained in a time stamped record. This may be forwarded immediately or may be placed in a time stamped historical record. Historical records in the NE are necessary for two main reasons:

- the manager can schedule recovery of nonurgent data in an efficient manner, and
- it may not be possible to access the NE for an extended period due to failure or congestion of management system or DCN (data communication network).

The duration of the count intervals, the number of counters, the thresholds and size of the archive are parameters which are chosen to try and maximize the successful diagnosis of most problems without excessive demands on data storage and processing. The choice of 15 minute and a 24 hour recording intervals for unavailability and performance event data has been made on a rather arbitrary basis but represents nevertheless a widely supported compromise between the conflicting requirements of resolution and comprehensiveness. NEs should be capable of retaining 15 minute history records for up to about four hours; a period long enough to isolate and restore management system failures and build up missing archive. The 24-hour record should be retainable for one day.

An information model for generic (i.e., technology independent) performance management which also draws on object classes specified in the X.700 series has been recommended in G.822. This forms the basis for the technology specific models required for SDH and ATM which are defined in G.774-01 and I.751 respectively. The performance fragment is illustrated in Figure 7.9.

The Scanner defined in X.739, represents a generic class of objects designed for summarization of numerical attributes. The scanner GDMO syntax is summarized below.

```
scanner MANAGED OBJECT CLASS
    DERIVED FROM "Recommendation X.721:1992":top;
    CHARACTERIZED BY
```

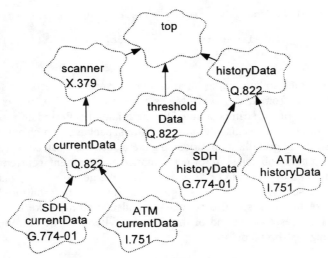

Figure 7.9 Performance management MO class library.

```
scannerPackage PACKAGE
BEHAVIOR
    scannerbehaviour;
ATTRIBUTES
    scannerId                                                    GET,
    administrativeState                                  GET-REPLACE,
    granularityPeriod                                    GET-REPLACE,
    operationalState                                           GET;;;
CONDITIONAL PACKAGES
    availabilityStatusPackage
        PRESENT IF  "the managed object can be scheduled",
    duration
        PRESENT IF   "the scanning function is to be enabled
        between specified start and stop times",
    dailyScheduling
        PRESENT IF   "daily scheduling is required and the
        weekly or external scheduling package is not present",
    weeklyScheduling
        PRESENT IF "weekly scheduling is required and the
        daily or external scheduling package is not present",
    externalScheduler
        PRESENT IF "reference to external scheduler is required
        and the daily or weekly scheduling pkg is not present",
    periodSynchronizationPackage
        PRESENT IF "configurable agent internal synchronization
        of repeating time periods is required",
    createDeleteNotificationsPackage
        PRESENT IF "notification of object creation and object
        deletion events is required",
    attributeValueChangeNotificationPkg
        PRESENT IF "notification of attribute value change
        events is required",
    stateChangeNotificationPackage
```

```
            PRESENT IF "notification of state change event is required"
REGISTERED AS { joint-iso-ccitt ms(9) function(2) part11(11)
   managedObjectClass(3) 7};
```

Scanners are designed to summarize metrics from managed objects. Observed attribute values are retrieved from designated objects during a "scan," which is initiated periodically. The period of the scan is denoted by the granularityPeriod attribute which is set at object creation. The scanner also defines conditional packages which manage scheduling and synchronisation of granularity periods. In general, scanners can collect and summarize information from many objects, but for performance management a currentData managed object class has been defined in Q.822 derived from the generic scanner above. This is a particular type of scanner which scans its own attributes. The monitored attributes are defined in the technology specific subclasses and commonly take the form of event counters which are reset at the end of the granularityPeriod. The currentData managed object class syntax is shown below.

```
currentData MANAGED OBJECT CLASS
   DERIVED FROM "Recommendation X.739:1993": scanner;
   CHARACTERIZED BY currentDataPkg PACKAGE
      BEHAVIOR currentDatabehaviour ;;
      ATTRIBUTES
         suspectIntervalFlag REPLACE-WITH-DEFAULT DEFAULT VALUE
         Q822-PM.defaultCurrentDataSuspectIntervalFlag      GET,
         elapsedTime                                        GET;;;
   CONDITIONAL PACKAGES
      filterSuppressionPkg
         PRESENT IF "an instance supports it and the
      zeroSuppressionPkg is not present",
      historyRetentionPkg
         PRESENT IF "historyData objects are to be created at the
         end of an interval.",
      maxSuppressedIntervalsPkg
         PRESENT IF "an instance supports it and at least one of
         zeroSuppressionPkg or filterSuppressionPkg is present.",
      measurementListPkg
         PRESENT IF "an instance supports it or the object class
         is currentData",
      numSuppressedIntervalsPkg
         PRESENT IF "suppression counts are required and the
         filterSuppressionPkg or zeroSuppressionPkg is present.",
      observedManagedObjectPkg
         PRESENT IF "an instance supports it.",
      scheduledPMReportPkg
         PRESENT IF "scheduled notifications are to be emitted.",
      thresholdPkg
         PRESENT IF "a Quality of Service Alarm Notification is
         to be emitted for threshold crossing.",
      zeroSuppressionPkg
         PRESENT IF "an instance supports it and the
         filterSuppressionPkg is not present.";
REGISTERED AS {q822ObjectClass 1};
```

A typical operational scenario is illustrated in Figure 7.10. The currentData is assumed to be contained in the monitored object and this is indicated in the model by naming the currentData instance from the monitored object instance containing the current data. The measurementListPkg package is only used if a technology specific subclass is not available. The scheduledPMReportPkg package is responsible for emitting the scanReport notification at the end of the measurement interval. It contains a list of performance measurements, each one represented by an attribute *identifier*, from amongst those supported by the monitored object and its current *value*. The identifiers of the attributes required to be measured are listed in the scanAttributeList attribute. The listed attribute must of course exist in the technology specific currentData object. Attributes of type counter have a count type syntax and may be resetable or wrap round.

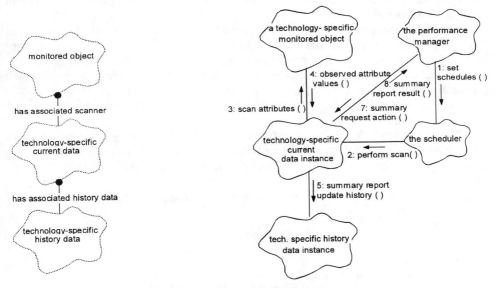

Figure 7.10   The monitoring model and interaction diagram.

The historyRetentionPkg package is responsible for the creation of a historyData object instance containing the contents of the technology specific currentData object at the end of the measuring interval as specified in the granularityPeriod attribute. It must be retained for a number of measuring intervals equal to or more than the integer value indicated in its historyRetention attribute.

The filterSuppressionPkg package contains a discriminatorConstruct attribute which can be used to suppress the creation of historyData objects under certain conditions. For example a record already correlated to a root cause may result in suppression of the historyData creation. The zeroSuppressionPkg package similarly suppresses the creation of empty performance history records. The maxSuppressedIntervalsPkg package has an attribute which allows the manager to limit the number of suppressed intervals so that the currentData object is forced to generate history objects at some minimum rate even when there is nothing to report. The number of suppressed intervals is recorded in the read only attribute numSuppressedIntervals in the numSuppressedIntervalsPkg package.

The thresholdPkg package is responsible for issuing a quality of Service Alarm notification when the value of a PM parameter violates its threshold setting. The threshold-DataInstance attribute is a pointer to a thresholdData object that contains the threshold data for the performance parameters. The thresholdData object contains a counterThresholdAttributeList of PM attribute identifiers with corresponding threshold values (integers) and optional alarm severity.

The historyData object holds an exact copy of the currentData at the end of the scan. A time stamp is added in the form of the periodEndTime attribute and it also carries an observedManagedObjectPointer attribute pointing to the monitored object. Instances of currentData and historyData objects are created for each of the recommended measurement periods, 15 minute and 24 hour.

Technology specific subclasses sdhCurrentData and an atmCurrentData are derived from the generic currentData managed object class. Both contain a mandatory zeroSuppressionPkg package. The sdhCurrentData contains a mandatory thresholdPkg while the atmCurrentData does not. This implies that all SDH specific currentData objects will support thresholding while ATM specific current data may require thresholding in specific subclasses. Thresholding is required in cellLevelProtocolCurrentData for example but not in vpCurrentData.

The SDH pathTerminationCurrentData managed object class is summarized below as an example of a technology specific subclass. An instance of this class is named from the monitored object which will be an SDH path layer TTPSink (HO or LO).

```
pathTerminationCurrentData  MANAGED OBJECT CLASS
    DERIVED FROM sdhCurrentData;
    CHARACTERIZED BY
        pathTerminationCurrentDataPackage  PACKAGE
        BEHAVIOR pathTerminationCurrentDatabehaviour;
        ATTRIBUTES
            "Recommendation X.739:1993": granularityPeriod
                REQUIRED VALUES SDHPMASN1.SDHGranularityPeriod  GET,
            bBE       REPLACE-WITH-DEFAULT                       GET,
            eS        REPLACE-WITH-DEFAULT                       GET,
            sES       REPLACE-WITH-DEFAULT                       GET;;;
    CONDITIONAL PACKAGES
        cSESCurrentDataPackage
            PRESENT IF "an instance supports it",
        farEndCSESCurrentDataPackage
            PRESENT IF "an instance supports it" ,
        uASCurrentDataPackage PRESENT IF "an instance supports it",
        farEndCurrentDataPackage
            PRESENT IF "if monitoring of the far end is supported and
            the monitored point is Bi-directional";
REGISTERED AS {g774-01MObjectClass 9};
```

Binary block errors (BBE), errored seconds and severely errored seconds as defined in G.826 are each counted over the measurement interval designated by the granularityPeriod attribute. The conditional packages cSES and uAS are supported if individually time stamped consecutive severely errored second and unavailable second events are recorded; much more useful for diagnostics and more relevant to service impact.

In SDH the presence of performance metrics are implicit in the transport specifications. The trail termination capability to monitor error events and detect various anomalous conditions is a mandatory aspect of the specification. In ATM, however, the OAM information flows may be configured or not. The end-to-end monitoring which uses the F5 flow will be associated by definition with the endpoint TTPs but the segment monitoring using the F4 flows may be associated with a limited number of specifically nominated CTPs. Therefore the bidirectionalPerformanceMonitor managed object classes is introduced to represent the performance monitoring process. It implements activation and deactivation mechanisms and carries specific information such as monitored block size. Defect monitoring which is implicit to the process, such as discarded cell counts, are modeled in the cellLevelProtocolCurrentData and tcAdaptorCurrentData managed object classes in the same way as their SDH counterparts.

The main technology specific distinguishing feature of ATM is the manner in which VChs and VPs are defined by their traffic descriptors as described in Chapter 5. The UPC and NPC functions are responsible to monitor traffic flows to ensure that the traffic parameters defined in the descriptors and contractually agreed at connection acceptance are not violated. If violations occur cells may be discarded or tagged as suitable to discard later if congestion is encountered. The service impact of cell discards, whether resulting from the user exceeding his contract or network congestion from unrelated causes, is much the same and performance records must be kept about this specific aspect of ATM network performance.

The upcNpcCurrentData object counts cells passed and cells discarded or tagged for violation of traffic descriptors and can only be associated with vcCTPs and vpCTPs as only CTPs carry the appropriate UPC and NPC monitoring attributes. The atmTrafficLoad-CurrentData object counts number of incoming and number of outgoing cells. This may be done at CTPs in general and also at the underlying TTPs (tcAdaptorTTP or vpTTP) at UNIs or NNIs. The upcNpcCurrentData object syntax is summarized below.

```
upcNpcCurrentData MANAGED OBJECT CLASS
    DERIVED FROM atmCurrentData;
    CHARACTERIZED BY
        upcNpcCurrentDataPkg PACKAGE
        BEHAVIOR upcNpcCurrentDataBeh;
        ATTRIBUTES
            discardedCells REPLACE-WITH-DEFAULT DEFAULT VALUE
                AtmMIBMod.integerZero                          GET,
            successfullyPassedCells REPLACE-WITH-DEFAULT DEFAULT VALUE
                AtmMIBMod.integerZero                          GET;;;
    CONDITIONAL PACKAGES
        discardedCLP0CellsPkg
            PRESENT IF "the managed system performs UPC/NPC functions
            separately for high Cell Loss Priority (CLP) cells (i.e..,
            cells with CLP=0)",
        oamDiscardedCellsPkg
            PRESENT IF "the managing system requests it for OAM cell
            flow and the managed system supports it",
        oamSuccessfullyPassedCellsPkg
```

```
           PRESENT IF "the managing system requests it for OAM
           cell flow and the managed system supports it",
       successfullyPassedCLP0CellsPkg
           PRESENT IF "if the managed system supports high
           priority only policing and has the ability to count
           cells that are successfully passed by the CLP=0 UPC/NPC
           policing function",
       taggedCLP0CellsPkg
           PRESENT IF "the managed system supports Cell Loss Priority
           (CLP) tagging";
REGISTERED AS { i751ObjectClass 18 };
```

7.7   COMMUNICATION BETWEEN MANAGERS AND AGENTS

Network element management is a subprocess within a larger management process and has its own subprocesses and communication needs. It contains at least a manager and one or more NE agents but may also contain mediation device agents and presentation agents. The information model above defines the managed objects seen across an open interface on the NE agent. There is no similar definition of the objects in the manager or any of the other agents involved. However they are all constrained by the capabilities of the NE which is defined as receiver of messages from the manager and a sender of messages (notifications) to a manager typically designated by a forwarding address in an event forwarding discriminator. The syntax of the information model was defined to be compatible with the messaging capabilities of the *common management application protocol* (CMIP) defined by the joint CCITT and ISO work activity. CMIP defines the syntax and some very elementary parts of the message transport process for telecommunications management. For the NE management and control messaging interface an OSI seven layer stack has been standardized in Recommendation G.784. This stack with all the CCITT and ISO references is shown in Table 7.1.

The ISO data communication model can be seen as a refinement within one layer network of the recursive transport model presented in previous chapters except for the different usage of the term layer. This convergence is clearest at the lower layers where it is most important for the design of high quality integrated networks. The relative positioning of the transport network functional model and the ISO seven layer protocol model is discussed in Chapter 3.

7.7.1   Application OSI Layer

The application OSI layer is made up of a number of *service elements* (SE). These are the messages which an agent will recognize and for which it can execute the appropriate operation. The primary SE for this interface is the SDH network element management SE that is defined in CCITT Recommendation G.784. A number of other ancillary SEs are also required that help in the administration of the open management interface:

- *Common management information service element* (CMISE) gives the general format of the messages allowed in management messaging and is defined in ISO Standard 9595.

Table 7.1

Communication protocol stack used for communication between managers and agents.

OSI Layer	Function	ITU-T Recommendation or ISO Standard		
7	Application	CMISE - ISO 9595 ACSE - X.217, X.227 ROSE - X.219, X.229		
6	Presentation	X.216, X226		
5	Session	X.215, X.225		
4	Transport	X.214, X.224 ISO 8073		
3	Network	SDH Embedded Control Network ISO 8473	Public Packet Network X.25	LAN ISO 8473
2	Link	Supplied by client layer network		

- *Association control service element* (ACSE) performs the initial negotiation across the open interface in order to determine that the two parties to the communication are talking about the same thing, for example, SDH network element management. This is described and specified in CCITT Recommendations X.217 and X.227.
- *Remote operations service element* (ROSE) supervises the messaging transactions across the interface as described and specified in CCITT Recommendations X.219 and X.229 respectively. These ancillary SEs require their own messages that are defined in the recommendations.

7.7.2   Presentation OSI Layer

The OSI presentation (layer 6) and session (layer 5) layers are less well justified than the others. Presentation in this context, is nothing at all to do with presentation at a human machine interface but rather how data is represented at an application interface. It was originally envisaged that applications would be specified using several different syntax and data formats. These may have been optimized for storage efficiency, minimum compression processing or some other aspect. The presentation layer converts the application syntax into a standardized transfer syntax using what we now call the binary encoding rules (BER); transport engineers frequently confuse this acronym with that for bit error rate. The main task of the presentation layer was to negotiate between the presentation entities to agree on an appropriate transfer syntax to use between two application agents which may be using different (but recognized) data structures. In other words when ASN.1 specifies a "sequence of integer" for example, the binary encoding rules will determine exactly what bit level encodings are used to represent it.

In the case of transport NE management the ITU-T has specified *abstract syntax notation No.1* (ASN.1) as the application syntax for all management applications and this is defined in Recommendations X.208 and X.209. And only one transfer syntax is provided in the form of the BER therefore the presentation layer functionality is minimal.

7.7.3   Session OSI Layer

The session layer controls the scheduling of the communication; the timing and sequencing of the different parts of the communication. The session OSI layer is analogous to call control and includes functional units for half duplex dialogue control (i.e. only one side talks at a time) and insertion of synchronization markers to support recovery of failed applications. However the only functional units required by the transport management standard are the full duplex dialogue and the kernel which provides for starting and finishing sessions including an orderly close of a session after certain classes of lower layer communication failure.

General requirements for the session OSI layer are described and specified in ITU-T Recommendation X.215 and X.225 respectively but Q.812 specifies the profiles to be used in Q3 management interface applications.

7.7.4   Transport OSI Layer

The transport OSI layer provides transparent data transport between the session OSI layer entities. It can in fact support multiple simultaneous sessions involving separate applications although this is still a rare occurrence in NE management applications. In addition, the transport layer can optimize for transfer delay, quality or cost by choosing appropriate routing options from the network layer. The transport OSI layer sets up OSI transport connections across a network, monitors the accuracy of the data transport, and depending on the transport class, corrects any errors normally by requesting the retransmission of any corrupted data. There is a close equivalence between a transport OSI layer connection and a trail in a layer network. This equivalence is important for the integration of management messaging into the overall network architecture and is essential for the design of reliable homogeneous transport networks.

Five transport classes are specified designated TP0 through TP4. TP0 is effectively "do nothing" because there is still a widespread belief that if the underlying network resource is very high quality then the transport layer is redundant. TP1 provides error recovery from notified error conditions (network disconnect etc.). TP2 provides the multiplexing features to support multiple sessions and TP3 is a combination of TP1 and TP2. TP4 is the most comprehensive class and as well as multiple sessions it provides for encapsulation in a monitored frame in which the user data is explicitly monitored for errors and retransmission is requested accordingly. In practical terms of course this implies a degree of buffering and flow control with a negotiable windowing mechanism to provide the dimensional limits to these processes. The standard requires the support of TP0, TP2 and TP4.

X.25 connections may form part of a management network. It is useful in some areas to interconnect isolated managed subnetworks to the main network using the public packet network. X.25 have quite heavy link by link error control and are deemed by some not to require a comprehensive transport layer and hence may work adequately with TP0. The ability to support multiple sessions is of course independent of the error control issue. TP4 is considered mandatory over a connectionless network service due to the significantly high probability of corruption due to congestion and this means that TP4 is used almost exclusively.

The transport layer is defined in ITU-T recommendations X.214 and X.224, however, these recommendations currently do not cover all the network OSI layer scenarios of interest in transport management. In particular the connectionless network layer for LANs is not defined. The appropriate transport layer standard is therefore the ISO equivalent standard ISO 8073 together with ISO 8073/AD2. These are called for by Recommendation G.784 for SDH management interfaces. Recommendation G.784 calls for the capability of transport class 4, the highest transport class, for all SDH management interfaces.

7.7.5   Network OSI Layer

The OSI network layer provides network connections between transport OSI layer entities. The network OSI layer, however, may be partitioned into subnetworks for the purposes of routing. In particular for SDHNE management this partitioning concept is used to establish *SDH management subnetworks* (SMS) within an overall telecommunications management network. These are subnetworks made up of interconnected NEs capable of supporting management communications in the data channel embedded in the section layer.

An NE management system may communicate with its NE agents using the SMS. SDH management messaging may be carried on the *SDH data communication network* (SDCN) which is normally carried directly on the *data communications channels* (DCC) in the SDH section overhead (D1-3 and D4-12) using a LAPD based frame mode adaptation as specified in Recommendation Q.931. However, there is no architectural reason why other channels could not be used, for example, a VC-1 or an ATM VCh providing an adequate level of operational security can be provided. The managing system is connected to the SMS via a gateway NE (GNE) and the standard places no topological constraints on the SMS. That is the NEs and GNEs may be interconnected in any valid way.

The SDCN is operated as a connectionless packet based network. That is to say data is transferred between NSAPs according to routing behavior of the subnetwork and does not setup permanent connections between NSAPs. The advantages of connectionless operation were already well established in commercial TCP/IP networks and this mode of operation was considered well suited to the predominant transaction type of dialogue common in network management. This has been primarily documented by ISO and the network layer specification in ISO 8473, is specified in Recommendation G.784.

Another justification for adopting the connectionless mode in SMS was the desire to use LAN technology within the station to interconnect management systems with each other and with GNEs and it is still considered somewhat problematic to transit between connectionless and connection oriented environments. Despite this early objective it was not clear how to specify the SMS in order to support a standardized network layer. The ISO standards defined protocols and address formats but what was to be done about routing? Some transport NE types, a terminal multiplexer perhaps, may have no routing functionality but even quite simple NEs such as line terminals or ADMs will be required to transfer messages destined to other NEs on a choice of routes, and there was no enthusiasm for the additional management chore of administration of routing tables.

Within the LAN environment a stable, and mature standard already existed for automatically updating routing information between the systems connected to it. The so-called ES-ES routing exchange protocol is an automatic process whereby end systems periodically inform each other of their presence. In ISO terminology an end system (ES) is where the application agent resides and an intermediate system (IS) is merely a message switching function. Thus all NEs have an ES-functionality while many NEs in the SMS will also have an IS-functionality. Within the typical SMS the ES-IS routing exchange protocol was required. While this is now a stable and implementable standard, at the time of the first standards issue this was not the case and G.784 annexes describe alternative mechanisms to work around the routing administration issue. In particular a mechanism to recreate a shared media LAN environment within the generalized mesh of an SMS is described whereby each transport NE is required to execute a frame relay and broadcast toward all links except the one on which the message was received. Address filtering to the network layer meant that apart from this modified link layer procedure all NEs only require ES functionality. We do not know to what extent this option has already been implemented in the installed base but the ES-IS routing exchange protocol is now the norm for such management subnetworks.

Other *data communications networks* (DCNs) can be used to transport SDH management messages. Two other significant types of DCN today are the X.25 based packet switched network and *local area networks* (LAN) based on *carrier sense multiple access/collision detect* (CSMA/CD) and in the future, it seems likely that wide area DCNs will be carried on the ATM network. These are not carried on the SDH DCC and every SDH network element has an interface to one of these DCNs. The interfaces for DCNs are described in CCITT Recommendation G.773.

The X.25-based packet network, which is ubiquitous in some areas is useful for communicating with remote control centers not directly connected to an SDCN, although latency is often a problem. X.25 is also sometimes specified as a backup. The introduction of frame relay and ATM switching will significantly enhance the performance available to this portion of the solution in the future.

In the meantime the interworking between a connectionless SMS and a connection oriented extension is still problematic. There is one clear statement that internetworking shall be on the basis of a network layer relay under a peer to peer transport layer connection. G784 references an ISO report into this problem (DTR 10172) which contains the surprising disclaimer that "...it is not a standard and never will be." The interworking functional unit in the ISO technical report is required to examine the destination address of network protocol data units and allocate them to an appropriate X.25 switched virtual circuit. Despite the disclaimer, this is the only method recommended in G784 for achieving interworking between an SMS and an X.25 service.

7.7.6   Link OSI Layer

The link OSI layer as defined within the SMS is based on the LAPD protocol defined originally for the N-ISDN UNI. G.784 specifies a set of primitives which are mapped on the LAPD primitives of Q921. An acknowledged (AITS) and an unacknowledged (UITS) service are defined and required to be supported by the SMS. G.784 specifies such things as which SAPI (service access point identifier) and which TEI (terminal endpoint identifier) to use, as these may be allocated to different management applications which may share the SDCN and several default settings for timers and for options which are not exploited in this application. The choice of LAPD in this application may seem surprising but there were those in the debate at the time who were seeking to avoid the connectionless network layer functionality preferring instead a frame relay approach. It was largely to preserve this option as a possibility which led to the early specification of LAPD.

7.7.7   Physical OSI layer

The physical OSI layer is effectively provided by the DCC channels provided in the SDH section overhead. Other transitional formats such as those which use PDH formats to carry SDH or ATM create an equivalent channel within the transport overhead. Precise usage of these channels in SDH is not prescribed but it is a requirement that the D1-D3 channel be available at every regenerator section, that is to say at every node, while the larger D4-D12 channel is carried transparently through regenerators and accessed at multiplex sections only. With these limitations the network designer is free to create any suitable topology for the SDCN. The DCC channels must be explicitly enabled on a per section basis which leads to another interesting chicken and egg situation requiring this setting to be made before the DCN is itself initialized.

Failures in the lower layers are recognized by the ES-IS routing exchange protocol at the network layer, which will find an alternative route across the section layer network if one exists. This of course leaves the burden of providing alternative node and/or link disjoint routings to guarantee this with the network designer.

References

The basic specifications for network management referred to in this chapter are contained in the X.700 series of recommendations from ITU developed in association with ISO. Particularly relevant are:

ITU-T Recommendation X.680, *Abstract Syntax Notation One (ASN.1) — Specification of Basic Notation.*
ITU-T Recommendation X711, *Common Management Information Protocol (CMIP).*
ITU-T Recommendation X.720, *Structure of Management Information (SMI): Management Information Model.*
ITU-T Recommendation X.722, *Structure of Management Information: Guidelines for the Definition of Managed Objects (GDMO).*
ITU-T Recommendation X.725, *Structure of Management Information: General Relationship Model.*

Managed objects for use in generic management applications are defined in later recommendations of the X.700 series. The following have been referenced in particular:

ITU-T Recommendation X.733, *Systems Management: Alarm Reporting Functions.*
ITU-T Recommendation X.734, *Systems Management: Event Report Management Functions.*
ITU-T Recommendation X.735, *Systems Management: Log control function.*

ITU-T Recommendation X.736, *Systems Management: Security Alarm Reporting Function*.

ITU-T Recommendation X.738, *Systems Management: Summarization Function*.

ITU-T Recommendation Q.822, *Performance Management*.

ITU-T Recommendation G.773, *Protocol Suites for Q-Interfaces for Management of Transmission Systems*.

Technology specific management specifications define information models for SDH and ATM. These are referenced in:

ITU-T Recommendation G.774, *SDH Management Information Model for the Network Element View*.

ITU-T Recommendation I.751, *Asynchronous Transfer Mode Management of the Network Element View*.

Transport Performance 8

One of the principle motivations for network operators introducing SDH was the extensive performance monitoring features embedded within it. The same capability has been provided in ATM where the OAM cells may be enabled if needed. The demands placed on transport networks have increased significantly in recent years with an explosion in the number of data oriented services that are sensitive to transmission errors. As a result, most users are looking for not just improvement in transmission performance, but also performance guarantees that can be verified.

This chapter examines the principles that lie behind performance monitoring by looking at the basic requirements for performance monitoring in digital networks, the sources and characteristics of performance degradation, and the way in which such degradations are monitored in the SDH and ATM. In addition, the way transport network architecture affects performance monitoring is examined and it is shown that this has a significant influence on the way a layer network is managed.

8.1 REQUIREMENTS FOR PERFORMANCE MONITORING

The SDH offered a new opportunity to include performance monitoring into its structure that was not the case with PDH. It is worth listing the basic requirements that were used for the design of the performance monitoring capability built into SDH and were essential to the international agreements. These are:

- the performance of trails should be accurately monitored while the trail is in service;
- the performance of link connections and subnetwork connections that form a part of the trail should be monitorable while the trail is in service and without affecting the integrity of the trail;

- the parameters used to monitor the performance of trails should faithfully capture the characteristics of any degradation so that effective maintenance action can be taken;
- the parameters used to monitor the performance of the trails should be meaningful to the clients of the trail.

On the basis of these requirements, a fundamental principle of SDH performance monitoring is that every trail in every layer has its own inservice performance monitoring overhead. Figure 8.1 illustrates the general performance monitoring of trails and connections.

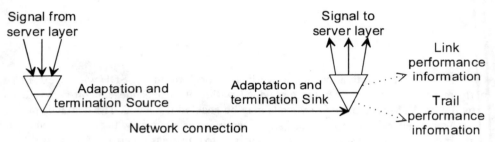

Figure 8.1   General performance monitoring of trails and connections.

8.2   PERFORMANCE PARAMETERS: ERROR PERFORMANCE AND AVAILABILITY

Error performance and availability are the two basic measures applied to digital networks for both maintenance purposes and performance contracts. Error performance is a measure of the number and statistics of errors in the digital signal while availability measures periods of time during which the digital signal is not present for the user. In the past they have been regarded as essentially separate subject areas; however, there are now several factors that make a unified treatment worthwhile. The requirement for a more unified treatment has been noted by ITU-T which has found that the definition of a formal distinction between error performance and availability is very difficult.

ITU-T agreed in 1984 that availability should be defined in terms of error performance. For a 64 kbit/s connection, any continuous period of time with severely errored performance of greater than 10 seconds would be classed as unavailable time. However, this definition has not satisfied many applications. For example, 50% of telephony calls are cleared down within five seconds of an interruption to the call and many data controllers will close down a data link within a few seconds of an interruption. In addition, for many applications it is useful to know how unavailable time is made up. For example, one long period of unavailability may be more tolerable than many shorter periods of unavailability. As a result, many network operators ignore the 10 second definition in favour of their own specialised metric sometimes using this to differentiate their service. This has stimulated a revision of this subject in ITU-T.

To understand the true difference between poor error performance and unavailability it is necessary to examine the common causes of both impairments and then present a means of modeling both in a common and consistent way.

8.3   DEVELOPMENT OF ERROR PERFORMANCE CONCEPTS

PCM and digital transmission were originally designed as a means of increasing the robustness of speech transmission. Most speech coding algorithms including A-law, μ-law, and *adaptive differential pulse code modulation* (ADPCM) are remarkably insensitive to transmission errors in the server digital connection. As a result, early digital telecommunications systems were not always designed or commissioned to give the best error performance the technology was capable of delivering. Data transmission was achieved by modulation to give an analog signal fitting in the speech band and again this signal was relatively insensitive to errors in the underlying digital transmission systems. The introduction of ISDN and 64 kbit/s data services carried directly on the digital transport network has significantly changed the design objectives of digital transmission systems and error performance has now become a major consideration. The PDH was designed to transport data bits efficiently with low cost equipment and did not have any real provision for error performance monitoring built into the frame structures. This has been a major difficulty for network operators in their efforts to improve the performance of transmission networks.

The study of error performance is as old as data communications. Historically, the subject has been closely tied to information theory and channel coding. It was studied under the branch of mathematics and engineering concerned with the theory of number groups and fields with great innovations coming from people like Shannon, Turing, and Hamming. Originally, the main parameter used to define error performance was *long term bit error rate* (LTBER) which is the number of bits received in error as a proportion of the total number of bits sent, calculated over a long period of time. In practice, this parameter has two major drawbacks. First, it contains no information on the distribution of the bit errors (it was assumed that thermal noise was the only source of errors and that the errors were therefore distributed accordingly). Second, the parameter is not capable of measurement by a network operator while the trail is in service as the significance of every bit has to be known. If this parameter is ever used now, it is normally for historical compatibility.

Despite this, a great deal of this early work is of very general application and is carried forward into SDH, but it may be expected that some extension to this pioneering work is necessary.

8.4   SOURCES OF ERROR IN HIGH CAPACITY TRANSPORT NETWORKS

Transmission equipment today typically performs to very high standards and with very high margins against errors. Errors will still occur however for various reasons beyond the designers direct control. The following is a brief survey of some of the more common mechanisms. The list is not exhaustive and unfortunately, a general feature of any study into error activity is that the more rare the mechanism, by definition, the more difficult it is to track down and characterize. Many network operators are very concerned to understand all the mechanisms that can cause errors in the transmission network in order to design them out of their network.

Each layer in the transport network has characteristic error producing mechanisms. Some of these are still not well understood and there is little published material from network operators with which to validate design targets. The following is a short survey of error producing mechanisms and their characteristics.

8.4.1   Noise in regenerated systems

Regenerator systems in the transmission media layers are limited primarily by noise processes. Average white gaussian noise is the type of noise that results from the random thermal motion of electrons in electronic materials and is often referred to as thermal noise. One of the fundamental principles of thermodynamics is that all matter that is above absolute zero temperature must have some form of energy normally in the form of kinetic energy of its particles including any free electrons. The thermal energy of the electrons means that the electrons will travel in a random manner and if a measuring set is used to measure the resulting electric current from the electron motion, it would show the following characteristics reflected in the description "average white gaussian noise":

- the long term level and characteristics of the noise do not change with time, thus, the term average;
- the power spectral density of the noise is constant up to very, very high frequencies, thus, the term white;
- the probability distribution of the noise current at any point in time is gaussian (or normal). This is a direct result of the fact the current is made up from the sum of the random motions of many electrons and that the sum of an arbitrary collection of distributions will always tend to a gaussian distribution.

It may be shown that:

$$Measured\ noise\ power\ =\ k\,T\,B$$

where T is the absolute temperature of the material in Kelvin, B is the bandwidth of the measuring set, and k is Boltzmann's constant.

The part of the transmission system most sensitive to thermal noise is the line system receiver where the signal level is at its lowest. This is true irrespective of the technology and so applies to optical receivers, radio relay receivers, and satellite receivers. Each technology has adopted different ways of reducing the effect of the noise. For example, optical systems have improved the optical signal level by reducing the loss in the fiber, while satellite receivers are often specially cooled with liquid nitrogen to reduce the thermal noise.

The characteristics of thermal noise are well understood and are relatively simple to predict. In the past, the study of thermal noise dominated the design of transmission systems for good error performance. As a result of all this work, most transmission systems are now designed to operate well within their thermal noise limits and so modern transmission systems, under normal operation, give effectively no errors as a result of thermal noise.

Some practical situations can occur where the power level of the incoming signal is reduced from its design level and thermal noise can result in transmission errors.

8.4.1.1   Degrading Lasers in Optical Systems

In optical systems, the laser used at the optical transmitter can degrade with time and the launched optical power can slowly reduce. Once this reduces below a certain level, the level of the thermal noise at the receiver can become significant relative to the received signal level leading to the generation of errors.

8.4.1.2   Rayleigh Fading in Radio Relay Systems

The atmospheric propagation in line-of-sight microwave radio systems is often more complex than the simple direct path between the transmitting dish and the receiving dish. Sometimes extra paths can be established by reflections from the ground or atmospheric "ducts" as illustrated in Figure 8.2. These additional atmospheric paths can vary considerably from occasions when they are not even detectable to some occasions when the signal strength on one of the additional paths is almost as strong as the direct path. This means that the signal arriving at the receiver may be a combination of the signal from the direct path and a number of signals from additional paths, each with a slightly different transmission path length.

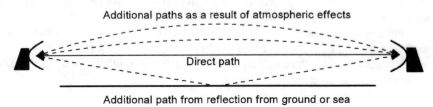

Additional paths as a result of atmospheric effects

Direct path

Additional path from reflection from ground or sea

Figure 8.2   Multipath Rayleigh fading.

Occasionally, the total signal from the additional paths can be of similar strength and opposite phase to that of the direct path, in which case the received signal will reduce significantly and this is referred to as multipath Rayleigh fading. One particular feature often seen during Rayleigh fading is that the signal fades and then recovers, then fades again and recovers, and so on. This occurs when the additional path varies in length with time, which is often the case with atmospheric "duct" paths. As the path varies in length the signal will cycle from in-phase with the direct path, to out-of-phase with the direct path. The scale of the length change need not be great as the wavelength used for radio-relay systems is usually between 10cm and 1cm. When it is out of phase, Rayleigh fading can occur.

When Rayleigh fading occurs, the level of the received signal relative to the thermal noise of the receiver reduces and transmission errors can occur.

8.4.1.3   *Rain Fading in Radio-Relay and Satellite Systems*

Some of the frequencies used for radio-relay and satellite systems are absorbed by water and moisture in the atmosphere. When there is heavy rain, systems using these frequencies can be affected. Much of the signal strength is absorbed by the rain and so the received signal appears to be attenuated compared to the normal clear sky conditions. In a similar way to Rayleigh fading, when rain fading occurs, the level of the received signal relative to the thermal noise of the receiver reduces and transmission errors can occur.

8.4.2   Electrostatic and Electromagnetic Interference

There are several sources of intense electromagnetic radiation that can interfere with normal transmission in several ways.

8.4.2.1   *Electrostatic Discharges*

Sometimes a significant build up of electrostatic current can discharge. This discharged current can sometimes momentarily raise the local earth potential sufficiently to disrupt the normal operation of the electronics and cause transmission errors. Closely coupled to this is the effect of the large burst of electromagnetic radiation that accompanies the electrostatic discharge and can induce currents in conductors at a level that can swamp the normal signal level and cause transmission errors.

The two most common sources of electrostatic discharges are lightning and people. Lightning comes from very large electrostatic build up between clouds discharging and can result in disruption to external electronics in line systems and even submarine systems when the lightning strikes the sea. People can cause electrostatic discharge when working in stations. The human body can build up electrostatic and when equipment is touched, unless precautions are taken, the electrostatic is discharged through the equipment disrupting the operation of the electronics in the equipment.

8.4.2.2   *Solar Radiation*

The sun is a source of both electromagnetic and electrostatic radiation that can cause interference. If the sun gets directly behind a radio transmitter, the electromagnetic radiation from the sun can swamp that of the transmitter. This periodically happens with satellites, particularly the geostationary satellites used for telecommunications. With geostationary satellites this is normally for a period of around five to ten minutes, twice a year, when the signal from the satellite is swamped by solar noise resulting in a period of heavy transmission errors. This is often referred to as a sun outage. Sun outages on satellite systems can be predicted from the satellites orbit and many satellite systems will transfer operation to another satellite for the duration of the sun outage.

While this is normally only a feature of satellite systems, radio-relay systems can occasionally suffer from sun outages at dawn or dusk. This will happen more often in the tropics where clear skies are more frequent and the sun rises and sets in the same place each day.

8.4.2.3   *Power Switching*

When high voltages or high currents are switched somewhere in a transmission station, there is often a short transient effect on the power supply voltage and earthing to transmission equipment. In addition there is often an associated burst of electromagnetic radiation. Both of these can affect transmission by momentarily swamping or distorting a signal in such a way that results in transmission errors.

8.4.3   High Resistance Electrical Connections

Mechanical connections between electrical conductors can become poor for several reasons including poor solder connections, metal aging and metal fatigue, accumulation of dust and grease, moisture from rain or flooding in an underground plant, and so forth. It is difficult to generalize what effects such mechanical connections have on the transmission system, however, such mechanical connections are normally sensitive to physical vibration.

A connection can appear to be good until it is vibrated, when it may change from a very low resistance to a high resistance temporarily. The high resistance may substantially reduce the signal level causing transmission errors. This period of high resistance may last only as long as the vibration and the connection may become good after the vibration disappears.

The opposite may also occur where a connection may be poor with a high resistance as a result of old age and/or dust and grease. The transmission system may be giving transmission errors as a result of the reduced signal level. Mechanically vibrating the connection may restore the connection to low resistance. This is behind the long established maintenance practice of "reseating the card" in equipment.

8.4.4   Regenerator Section Timing Impairment

Many of the timing impairments described in Chapter 11, if allowed to accumulate to excessive levels, can result in transmission errors. High frequency jitter has been the most sensitive parameter in the past and it may be expected that this will also be the case with SDH systems. The jitter causes the detector circuit at the end of the transmission system to misalign and effectively read the incoming signal at the wrong time. Thermal noise causes random variation in the y-axis of a receive "eye" diagram and high frequency jitter causes random variation in the x-axis of the receiver "eye" diagram as illustrated in Figure 8.3.

8.4.5   Marginal Design

Many of the mechanisms should not result in transmission errors if equipment and networks are designed correctly by leaving a sufficient margin in the important design parameters. This applies particularly to thermal noise and timing impairments.

There may be a few circumstances, however, where a system is operating nearer its limits, in which case it will be more prone to transmission errors; for example, an optical

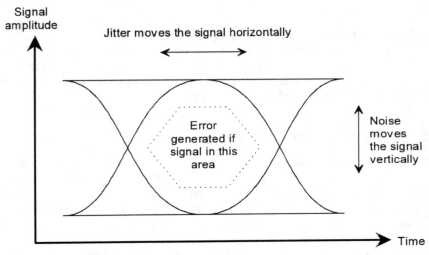

Figure 8.3   Jitter and noise as error-inducing mechanisms.

system working near the limits of its reach to span fixed locations without intermediate electronics, or an equipment on the margins of the network working close to or even above the normal jitter limits. Often these situations occur because a network operator may, for practical reasons, wish to stretch a system beyond its specification and accept some erosion of operating margin.

Equipment manufacturers may also erode design margins. For example, a component may be temporarily in short supply, and rather than stop production, a manufacturer may include an inferior component. Such practice again erodes the design margins in the equipment.

Similar effects can happen if equipment and networks are not comprehensively tested before they are used. Equipment may perform well within specification under normal test conditions, but under some circumstances, for example extreme cold, high humidity, or aging, the equipment fails to perform within specification and the design margins become eroded. Inevitably the occasion will occur where several marginal effects will coincide in time and place and transmission errors will result.

8.4.6   Path Layer Error Mechanisms and Error Propagation Between Layers

Synchronisation failure in the nominally synchronized 64 kbit/s transport layer leads to frame slips in a PDH network which will be perceived as error events by the client. Such events are constrained at extremely low levels by tight specification of standby clocks. Nevertheless double failures whereby a standby clock is out of limits when the synch failure occurs may still lead to unacceptable error activity. In SDH the pointer mechanism allows an accommodation of synchronization failure in the SDH layers. But again a coincidental failure of a desynchronizer, invisible under normal circumstances may give rise to error inducing phase excursions in the presence of pointers.

In a PDH network a quite short error burst in one layer may succed in corrupting the justification control channels in one or more client layers to the extent that a justification error is made. This will lead to a loss of frame alignment in the client layer followed by a frame search period. The resulting error event is thus significantly prolonged. A similar occurrence in an SDH network does not have this effect. The inherent clock stability ensures that the correct phase is maintained until error free transmission resumes after the burst. The correct pointer values should again be evident. In the unlikely event that a pointer change occurred during the error burst some error extension will result while the new pointer value is validated. No resetting of pointers is attempted until a new pointer value has become established.

In the ATM layers traffic parameters are specified according to the anticipated statistics of the signals and the resources available. If the offered traffic goes outside the agreed limit for whatever reason then congestion may occur and ATM nodes respond by discarding cells. Cell discard is perceived by the client as a burst of error activity indistinguishable from those produced by other mechanisms. Traffic related performance in asynchronous systems is a subject in itself which is discussed in Chapter 9.

Again in asynchronous systems a relatively small error burst in a lower layer can corrupt important cell or packet overhead information resulting in the loss of a complete cell or packet.

8.5   THE CONCEPT OF AVAILABILITY

In the classical analysis of availability, a long measurement time is divided into periods of time; the time when a system working is able to give service, called available time, and the time when the system is not working and unable to give service, called unavailable time, as is illustrated in Figure 8.4. The most common parameter is the proportion of the measurement time during which the system was working and is often calculated as a percentage.

$$Availability = \frac{total\ available\ time}{measurement\ time}$$

In the same way

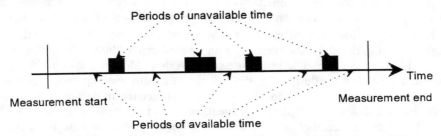

Figure 8.4   Divisions of measurement time into periods of available time and unavailable time.

$$Unavailability = \frac{total\ unavailable\ time}{measurement\ time}$$

$$= 1 - availability$$

This subject has always been closely tied to the study of component reliability and repair times. A period of unavailable time is caused by a failure event somewhere in the system and available time is re-entered when the system is repaired or possibly recovers by itself. The availability of the system is therefore determined by the reliability of the system and the time taken to restore service following a failure. It may be shown that:

$$Availability = \frac{MTBF}{MTBF + MTTR}$$

where

$MTBF$ = mean time between failures

$MTTR$ = mean time to restore

This is the classic equation of availability, but it is based on mean times and as systems have become more reliable this can prove misleading. For example, a piece of transmission equipment may have a *mean time between failures* (MTBF) of 30 years; however, its expected life may be only 10 years and it would be tempting to think that this means that the equipment will never fail. What the MTBF really means is that if there are 30 equipment items in a station, on average, one will fail per year. This suggests that a clearer description of reliability is in terms of the failure rate of the equipment.

8.5.1   Failure Rate

The failure rate is defined as the probability that a system will fail during a short period of time δt and is normally given the symbol λ. The failure rate is also a function of time and the probability that a system will fail can change during the life of the system and so is written $\lambda(t)$. Most systems exhibit a characteristic behavior where the failure rate is higher earlier in the life of the system during a period called "burn-in". The failure rate is normally lower in the middle of the life of the system during its main working period. Towards the end of the design life, the failure rate usually rises again and this period is normally called "burnout." When the failure rate is plotted for the design life of the system, it shows the characteristic "bathtub" shape illustrated in Figure 8.5.

The expected number of failures over a longer period of time is calculated by integrating the failure rate over the period of time:

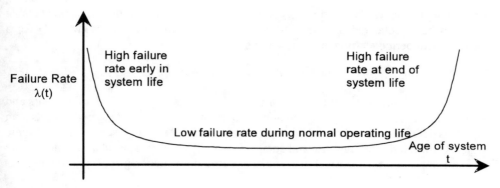

Figure 8.5   The "bathtub" curve of failure rate over the life of a system.

$$E\{failures:\ t_1 < t < t_2\} = \int_{t_1}^{t_2} \lambda(t)\delta t$$

if $\lambda(t) = \Lambda$ is constant over the period, which is not often the case;

$$= E\{failures:\ t_1 < t < t_2\} = \Lambda(t_2 - t_1)$$

8.5.2   Failure Rate and MTBF

Relating the failure rate to MTBF illustrates how using MTBF can give misleading descriptions of the system. There are three possible ways of defining MTBF. The most common is the inverse of the failure rate shown below. The second is the inverse of the mean failure rate over the design life of the equipment. The third is the inverse of the mean failure rate calculated over a particular period of interest during the design life. Each has its practical or mathematical difficulties. The first means that MTBF is a function of time and therefore it is misleading to call it a mean time. The second obscures the significance of the important bathtub characteristics of failure rate. The third ignores the burn-in and burnout failures of the overall bathtub characteristic, however, when only a specific part of the design life is of interest, this is the most useful definition.

$$1 \qquad MTBF = \frac{1}{\lambda(t)}$$

$$2 \qquad MTBF = \frac{t_d}{E\{failures:\ 0 < t < t_d\}}$$

$$3 \qquad MTBF = \frac{(t_2 - t_1)}{E\{failures:\ t_1 < t < t_2\}}$$

Assuming that the restoration times are largely independent of the failure rate, unavailable time can be calculated from the failure rate and *mean time to restore* (MTTR) for a measurement period starting at t_1 and ending at t_2:

$$Unavailable\ time\ =\ MTTR\ E\{failures:\ t_1 < t < t_2\}$$

In the earlier example, where the system was quoted as having an MTBF of 30 years and a design life of 10 years, the expected number of failures during the design life is less than one. This definition allows such a situation to be interpreted sensibly, suggesting that some systems will fail during their life and some will not. In addition, special assessments may be made during burn-in, normal working, and burnout by choosing appropriate measurement intervals.

8.6   DEFINITION OF TRANSPORT FAILURES AND RESTORATION TIME

Failures may be defined from two different viewpoints:

- a failure is recorded when a critical component of the system fails (a critical component is one on which the system relies for successful operation and cannot operate without it);
- a failure is recorded when the system is observed to cease working.

These two definitions are not the same and will not give the same results. The first is based on a synthesis of the system from its components, while the second is based on observing the system as a whole from outside without any reference to its constituent components.

If the components are considered, it is not always the case that a component failure will also mean that the system will be observed to stop working; still less that a particular connection will fail. For example, a component such as a laser may be designed with a tolerance margin and while its performance may have fallen below its design limit and therefore is deemed to have failed, it may still operate sufficiently for the system to work normally. Conversely, a component's performance may be within its design limits and either because of poor system design or some unexpected nonlinear effects, the overall system does not operate normally. Finally, the parameters and thresholds used to define the system failure and component failure may be different.

This distinction is especially important in transport networks. In this case, the network is a system, the service offered is the trail, and the components of the system are line systems, crossconnects, switches and multiplexers. The network operator must specify the components and operate the system within safe operating limits. The parameters and limits useful to achieve this may be different to those required by the client who only sees the performance of the end-to-end trail.

In the discussion on sources of error activity, it was seen that most mechanisms that produce degradation in error performance result in short periods of high error activity and long periods with no errors. As a result, it is possible to interpret error performance in terms of events that cause error activity. In a very similar way, unavailable time can be interpreted in terms of events that cause unavailability. Indeed, it is often difficult to distinguish between poor error performance and unavailability.

When the trail is measured, the restoration time is simply the time taken for good error performance to be restored and it is not normally possible to determine how this happened when viewed from the TTP. It could have been that the poor error performance was as a result of an error inducing mechanism and that the performance spontaneously recovered; it could have been that the poor error performance was as a result of a component failure and that the system automatically recovered the service using automatic protection switching; or it could have been that the poor error performance was as a result of a component failure and that the performance recovered when human action was taken to repair the system.

We can draw two very important conclusions from this. First, a single set of parameters is required for the specification of trails of any bit rate and for the specification of equipment. Second, when considering the performance of a trail, the distinction between poor error performance and unavailability is arbitrary. ITU-T Recommendation G.826 defines the new set of parameters that will be used on all SDH trails and also defines unavailable time as a threshold on poor error performance.

8.7   A MATHEMATICAL MODEL FOR ERROR PERFORMANCE AND UNAVAILABILITY

The purpose of a mathematical model is to help in the design of systems and to assist in the analysis of measured results. The mathematical model for error performance and unavailability described here is a significant refinement of those found in many textbooks on telecommunications and is based on the requirements of SDH transport networks and the physical environment in which they will be deployed. It has been used within ITU-T to assist the setting of maintenance and commissioning limits for both PDH and SDH trails. The model is based on the following assumptions partly illustrated in Figure 8.6:

- trails operate largely error free with the exception of short periods of error activity or unavailability;
- periods of error activity and unavailability are caused by discrete events;

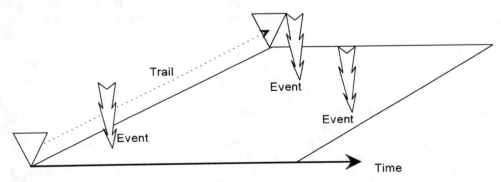

Figure 8.6   Error and unavailability events are distributed in time and place.

- such events are not independent and hence the occurrence of one event means that the imminent occurrence of another event is more likely.

This last assumption is based on recent discoveries in mathematics in the fields of nonlinear dynamics and chaotic systems, with applications in such diverse areas as macro-economics, subatomic physics, and weather forecasting. There are now several popular textbooks on this subject that clearly illustrate the areas in which this new mathematics has found useful application. Error performance was one of the first areas to receive attention. These early studies were carried out before the digitalization of transport networks, but may be modified to take account of the different error and unavailability inducing mechanisms that now affect digital networks.

The assumption asserts that the events, but not necessarily the individual bit errors, occur as a result of chaotic processes. New chaos theory provides a mathematical foundation for what has been anecdotally recognized for a very long time: it never rains, it pours; bad things come together; everything depends on everything; black Monday in the stock markets.

8.7.1   Independence of Events and the Distribution of Events

The idea that all events affecting a system are independent of one another and that the occurrence of one event has no effect on the occurrence of any other event is one of the most commonly used assumptions in simplifying the statistics of error performance and unavailability. However, most of the natural world does not behave like this. In statistical terms, there are many more clusters of events and much longer periods without any events than would be suggested if this assumption were true.

For example, consider the distribution of rainy days in a year. Supposing that there were 36 rainy days in a year, we can now look at when those days occur in relation to each other. If all days were independent, then each day would have a 1 in 10 chance of being rainy. The probability of a week of continuous days of rain would be 10^{-7} or about once every 30,000 years and the chance of a four month drought without rain 3×10^{-6} or about once every 1,000 years. In practice both would be much more common. The conclusion is that the occurrence of rainy days is not independent and that the fact that one day is rainy will make the fact the next day is rainy more likely. In other words, rainy days will tend to cluster together.

This example from the realm of meteorology is in fact directly relevant as some of the mechanisms that affect error performance are in fact weather dependent, especially those affecting radio and satellite systems. All electrical communication systems are sensitive to atmospheric electrical activity associated with storms and so we would expect to see this clustering of events reflected into the error activity.

This type of distribution of events can be seen in most of the mechanisms that produce error activity and unavailability. For example, a common source of error activity directly results from human activity in a station. A member of the maintenance staff working to repair a fault on a piece of equipment may by his very presence induce error activity, or even cause a fault on a neighboring piece of equipment. The occurrence of the new event is therefore dependent on the original fault thus leading to a clustering of the events. Indeed, a brief analysis of every mechanism shows that there is some level of dependence between and amongst mechanisms as well as the dependence of the multiple events caused as a result of a single event.

Chaos mathematics gives a means of describing the clustering of error and unavailability events in terms of the fractal, or Hausdorf dimension. If the events are all independent then the fractal dimension is zero, and if all the events are totally dependent on each other (that is, they all occur at the same time in a single cluster) then the fractal dimension is one. In practice, the fractal dimension is somewhere between these two extreme values.

8.7.2   Event Interval

The distribution of error inducing events is essential to the description of the clustering and a mathematical way of representing this distribution is required. This is achieved with the concept of event interval that is the length of time between events that cause error activity. To measure the event interval, short periods of time τ are used. The τ period could be the bit period or a number of bit periods. However, if there is one or more errored bits in the τ period, then the τ period is considered to be errored. The event interval is the total number of continuous τ periods that do not have an error following an errored period (including the errored τ period). An elementary property of event intervals is that the total of the event intervals during a measurement is simply the measurement time. In addition, the number of event intervals recorded is the number of errored τ periods. From this it is possible to say:

$$Proportion\ of\ errored\ periods\ = \frac{number\ of\ event\ intervals}{total\ of\ event\ intervals}$$

A continuous record of event intervals allows the position of every errored τ period to be marked in time and so the error activity is recorded without a reduction in the statistical information. The distribution of event intervals (that is, the number of short event intervals compared to the number of long event intervals) gives information on the clustering of events and hence, information on the fractal dimension of the events.

8.7.3   Distribution of Event Intervals

In general terms, the distribution of event intervals can be modeled as a stochastic process. This implies that the probability of occurrence of the next errored τ period is a function of the positions of all the previous errored τ periods. However, it is possible to construct a good distribution that fits well with observed results by considering only the position of the previous errored τ period. This type of stochastic process is called a renewal process.

The length of the τ period should be selected to be at least of the same order as the duration of the majority of error events in order to properly resolve the events. In practice, this is between 1 second and 100 seconds. The duration of a period of unavailability is largely determined by the restoration systems and should be excluded from the statistics, but the occurrence of the event itself should be included.

There are three principle probabilities in the consideration of the distribution of event intervals as a renewal process:

$$P(ei \geq n) = \text{the probability that the event interval is greater or equal to } n \text{ periods}$$

$$P(ei = n) = \text{the probability that the event interval is exactly } n \text{ periods}$$

$$= P(ei \geq n) - P(ei \geq (n+1))$$

$$P(ei = n \mid ei \geq n) = \text{conditional probability that the event interval is } n \text{ periods given it is known that it is not less than } n \text{ periods}$$

$$= \frac{P(ei = n)}{P(ei \geq n)}$$

The third probability can be related to the case of independent events as this probability excludes the correlation between the successive τ periods. In the case of independent events giving rise to the errored τ periods, this probability is constant.

In the case of independent events:

$$P(ei = n \mid ei \geq n) = \text{constant}$$

$$= \lambda\tau$$

and from this can be derived

$$P(ei \geq n) = (1 - \lambda\tau)^{(n-1)}$$

$$\approx e^{-\lambda\tau(n-1)} \qquad \text{when } \lambda\tau \text{ is small}$$

$$P(ei = n) = \lambda\tau(1 - \lambda\tau)^{(n-1)}$$

where λ is the normalized probability that the τ period is errored. When $P(ei \geq n)$ is plotted as a log/log graph, as illustrated in Figure 8.7, an exponential curve is produced. However, in the practical case where the fractal dimension is greater than zero, events are not independent and occur in clusters. This means that there must be a greater number of shorter event intervals and so $P(ei = n \mid ei \geq n)$ must decrease with increasing n under these circumstances. Thus in the case of clustered events:

$$P(ei = n \mid ei \geq n) = \text{decreasing function of } n$$

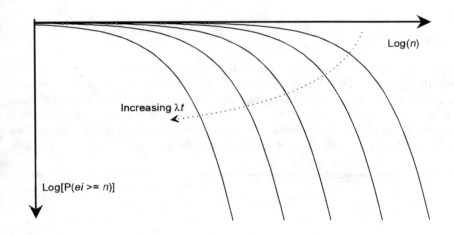

Figure 8.7   Poisson distribution of event intervals.

A direct application of chaos theory indicates that the probability is inversely proportional to n

$$P(ei = n \mid ei \geq n) = \alpha n^{-1}$$

where α, the constant of proportionality, is the fractal dimension. However, this function does not apply when n is small. The reason for this is a complex mathematical phenomenon governing the Hausdorf limiting process and is outside the scope of this book. A more general equation that is correct for all values of n is:

$$P(ei = n \mid ei \geq n) = \frac{-(n+1)^{(1-\alpha)} + 2n^{(1-\alpha)} - (n-1)^{(1-\alpha)}}{(n+1)^{(1-\alpha)} - n^{(1-\alpha)}}$$

$$P(ei \geq n) = n^{(1-\alpha)} - (n-1)^{(1-\alpha)}$$

$$P(ei = n) = -(n+1)^{(1-\alpha)} + 2n^{(1-\alpha)} - (n-1)^{(1-\alpha)}$$

it can be easily shown that, when n is large

$$P(ei = n \mid ei \geq n) \approx \alpha n^{-1}$$

$$P(ei \geq n) \approx (1-\alpha)n^{-\alpha}$$

$$P(ei = n) \approx \alpha(1 - \alpha)n^{(-1-\alpha)}$$

When $P(ei \geq n)$ is plotted as a log/log graph, it gives a straight line with gradient $-\alpha$ as illustrated in Figure 8.8. This is very similar to the Pareto distribution and was originally proposed by Mandelbrot for describing transmission error statistics. This distribution, however, has both practical and mathematical limitations. The mathematical difficulty is that the mean of the distribution is infinite. The practical difficulty is in the interpretation of the extremely long error intervals that it predicts; simply that they are not seen in practice.

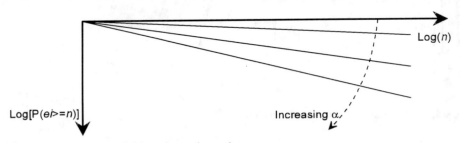

Figure 8.8   Pareto distribution of event intervals.

A distribution suggested recently by Bond overcomes these difficulties. The simple chaotic model suggests that clusters of clusters of clusters and so on will occur no matter how large a timescale is used. Bond suggests that there is a limit to the clustering of clusters at that point the very large timescale clusters become independent of one another. This leads to the following formula.

The Bond model requires two processes, one representing the clustering of events at shorter timescales, and a second representing independence of clusters at longer timescales:

$$P(ei = n \mid ei \geq n) = \text{decreasingfunctionof}n + \text{constant}$$

and the model is based on

$$P(ei = n \mid ei \geq n) = \alpha n^{-1} + \lambda\tau$$

however, this equation again does not apply when n is small. The precise formulae for the Bond distribution are as follows:

$$P(ei = n \mid ei \geq n) = \frac{-(n+1)^{(1-\alpha)} + 2n^{(1-\alpha)} - (n-1)^{(1-\alpha)}}{(n+1)^{(1-\alpha)} - n^{(1-\alpha)}}(1 - \lambda\tau)$$

$$P(ei \geq n) = \left[n^{(1-\alpha)} - (n-1)^{(1-\alpha)}\right](1 - \lambda\tau)^{(n-1)}$$

$$P(ei = n) = \begin{array}{l} \left[n^{(1-\alpha)} - (n-1)^{(1-\alpha)} \right] (1 - \lambda\tau)^{(n-1)} \\ - \left[(n+1)^{(1-\alpha)} - n^{(1-\alpha)} \right] (1 - \lambda\tau)^{n} \end{array}$$

when n is large these formulae can be approximated:

$$P(ei = n \mid ei \geq n) \approx \alpha n^{-1} + \lambda\tau$$

$$P(ei \geq n) \approx (1 - \alpha)n^{-\alpha}e^{-n\lambda\tau}$$

$$P(ei = n) \approx (\alpha n^{-1} + \lambda\tau)(1 - \alpha)n^{-\alpha}e^{-n\lambda\tau}$$

The mean of this distribution is well defined and is approximately:

$$E\{ei\} \approx \frac{(\lambda\tau)^{(1-\alpha)}}{\Gamma(2-\alpha)}$$

Figure 8.9 shows the effect on the distribution of event intervals of differing fractal dimensions while keeping the mean constant.

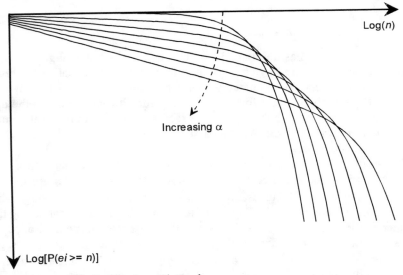

Figure 8.9   Bond distribution of event intervals

Much of the mathematics is relatively advanced and this presentation is intended merely to introduce the important concepts. These concepts are also relatively new and there is still much research work needed to fully examine their implications and the ways in which they may be fully exploited. For the purposes of this book they provide a means by which clustering of events that are very well recognized anecdotally, may be described and quantified. Before these concepts can be applied to real error distributions seen in transmission networks, it is necessary to look at the means available for detecting impairments in real networks.

8.8   THE TRAIL TERMINATION PROCESS FOR ERROR MONITORING

One of the basic requirements for performance monitoring is the ability to detect error activity and failures while a trail is in service. While this feature was not included in the PDH, it is an essential feature of SDH which has inservice monitoring built into the trail termination process at every network layer. The embedded trail termination process was introduced in Chapter 3. The termination illustrated in Figure 3.17 shows four termination processes two of which, the frame phase alignment process and block error monitor process are used in performance monitoring. If the frame phase information is lost, then it may be assumed that the data is not being passed correctly to the trail client. Loss of frame phase event notifications and error event notifications are available at each network layer in SDH.

8.8.1   Loss of Frame Phase

The loss of frame phase is normally simple to detect and, in SDH, it is available for the monitoring of connections as parts of trails in a layer.

8.8.1.1   Loss of Frame Alignment

The regenerator section has the frame phase encoded in the frame alignment signal and the frame phase is recovered by tracking the frame alignment signal. If the regenerator section no longer detects the frame alignment signal it enters a loss of frame alignment state and is deemed to have lost frame phase and hence it cannot pass on the data to its client, the multiplex section. If the regenerator section is in a loss of frame alignment state, then an *alarm indication signal* (AIS) is passed onto the multiplexer section - all the data bits are set to binary one.

The strategy that the regenerator section uses to detect and monitor the frame alignment word is not specified by ITU-T, however, some aspects are implied. The regenerator section must find the frame alignment within 250μs of receiving an error free signal (assuming there are no simulations of the frame alignment signal during that time) and under test Poisson errors at 10^{-3} must not lose alignment more than once every six minutes. If the signal turns to random data with no frame alignment signal, the regenerator section must detect the loss of frame alignment within 625μs. This means that the precise conditions under which the regenerator section passes into and out of the loss of frame alignment is not defined. In practice, however, the loss of frame alignment is a very good detector of hard failure, even short transient hard failures.

The frame phase of the multiplex section is derived directly from the regenerator section and there is no additional information carried about the multiplex section frame phase. However, the regenerator section passes AIS to the multiplexer section when it is in the loss of frame alignment state, the multiplex section is deemed to have lost frame phase when it contains AIS. When AIS is received by the multiplexer section, it passes it on to the HVC it is supporting.

8.8.1.2   Loss of Pointer

The frame phase of the HVC is not simply derived from the multiplex section frame phase as the AU pointer gives a dynamic offset in the HVC frame phase. The frame phase of the HVC can only be derived therefore, if there is both a good multiplex section frame phase and a good AU pointer. The AU pointer is detected in much the same way as the frame alignment signal and when the pointer is not correctly recovered the HVC enters a loss of pointer state.

The algorithm by which the HVC enters and leaves the loss of pointer state is more closely defined than the loss of frame alignment state. In general, a loss of pointer condition is entered when three AIS pointer values are received or between eight and ten invalid pointer values are received. The loss of pointer condition is left when three consecutive identical pointer values are received.

The loss of pointer state, like the loss of frame alignment state is good at detecting hard failure, and it should be anticipated that AIS in the pointer will be the only major source of the loss of pointer state. It should also be noted that the multiplex section will maintain the AIS in the AU pointer even when the HVC enters a new multiplex section. This is a service that the multiplex section performs for the HVC that allows a management system to use the detection of AIS in the AU pointer as a means of detecting AIS in the HVC.

If the HVC enters the loss of pointer state, then AIS is passed onto the client link connections including any LVCs. The LVC can then enter a loss of pointer state, in this case TU pointer, in the same way as the HVC.

The loss of frame phase indicators are given in Table 8.1 for all the different types of virtual container.

Table 8.1
Loss of Frame Phase Indicators in Each Layer

Virtual Container	*Loss of Frame Phase Indicator*
RS(STM-N)	Loss of frame alignment (LOF)
MS(STM-N)	MS-AIS
VC-4-xc	AU-4-xc loss of pointer (LOP)
VC-4	AU-4 loss of pointer (LOP)
VC-3	AU-3 or TU-3 loss of Pointer (LOP)
VC-2-vxc	TU-2 LOP (any one of x)
VC-2xc	TU-2-xc loss of pointer (LOP)
VC-2	TU-2 loss of pointer (LOP)
VC-12	TU-12 loss of pointer
VC-11	TU-11 loss of pointer

8.8.2   Block Error Monitor

Block error monitors are rather more complex than detectors for loss of frame phase and, in SDH, block error monitors are normally only used for the monitoring of trails and not for the monitoring of connections.

The SDH frame is 125µs long and is independent of the bit rate. SDH makes use of the frame for detecting transmission errors by forming the data into blocks based on the 125µs frame and dynamically checking the integrity of each block of data when its trail is terminated. All the section layer trails are formed into 125µs blocks along with the VC-4, VC-3, and VC-4-xc. The VC-2, VC-12, VC-11, and VC-2-xc are formed into 500µs blocks.

The content of each block of data can be arbitrary and in order to tell whether there is any error in the block, additional information must be added to the users signal that can be verified. The most common way in which additional information is added is in the form of a *cyclic redundancy check* (CRC) and SDH uses a special case of this code with several other useful properties.

A CRC is most easily thought of as the remainder of a division sum. The block of data is regarded as a very large binary number and this is divided by a specified, small number. The result of the division is a quotient and a remainder. The quotient is ignored (it is not normally even calculated) and the remainder is kept. In order to illustrate this, consider the large decimal number that is the data block that is ready for transport:

45674859

if we divide this number or any other number by 97, it has a remainder that is an integer between 0 and 96 that can always be represented as two decimal digits. To make room for these digits, we first multiply the data by 100 to give:

4567485900

this is now divided by 97 to give a remainder of 49. Having calculated the remainder it is now possible to adjust this new number so that it is divisible by 97 by subtracting the 49 in a way that keeps the data visible. This is done by adding the 97 complement of 49, that is 48. The number that would be sent is:

4567485948

The data is still clearly visible, but if we now divide the number by 97, the remainder is 0. If the number is corrupted during transport, it is unlikely that the corrupted number would still be divisible by 97. The integrity of the transported data can therefore be checked by dividing the number by 97 after it has been received and checking that the remainder is 0.

Practical CRCs use exactly the same technique, but they make use of special binary arithmetic that can be very easily implemented by shift registers and exclusive OR gates. The rules of the arithmetic are not the same as normal binary arithmetic and are derived from the special properties of number fields where the range of allowed numbers is fixed, which is the case where the number of binary digits in the arithmetic is fixed. For the CRC operation, the block is regarded as a fixed length binary number; however, it is normally represented as a polynomial. The degree of the polynomial is one less than the total number of binary digits while the coefficients of the polynomial are equal to the value of binary digits. For example, suppose the data block was six bits long and was

100110

the polynomial to represent this data is

$$x^5+x^2+x^1$$

The divisor is a binary number also written as a polynomial, for example 1011 which is written as $x^3+x^1+x^0$. The derivation of the arithmetic rules are complex, but the rules themselves are very simple: addition is performed by the exclusive OR of the coefficients of corresponding terms in the polynomials. Addition and subtraction are therefore the same operation while multiplication and division are defined as successions of additions/subtractions in the normal way. In this example the remainder of the division of the dividend by the divisor will require three binary digits and so the original polynomial must be multiplied by x^3 in order to make room for the remainder.

The division using the example would be:

$$
\begin{array}{l}
x^5+\ \ +x^3+\ \ +\ \ +x^0 \\
\overline{} \\
x^3+\ \ +x^1+x^0\,|\,x^8+\ \ +x^5+x^4+\ \ +\ \ +\ \\
x^8+\ \ +x^6+x^5 \\
\overline{} \\
x^6+\ \ +x^4+ \\
x^6+\ \ +x^4+\ x^3 \\
\overline{} \\
x^3+\ \ +\ \ + \\
x^3+\ \ +x^1+x^0 \\
\overline{} \\
x^1+x^0
\end{array}
$$

The remainder digits are therefore 011. These are subtracted (which is the same as adding) to the shifted data and

100110011

is transmitted. It is left for an exercise for the reader to show that this polynomial, when divided by the divisor gives a remainder of zero.

The way the remainder is added in SDH is shown in Figure 8.10. A part of the Trail overhead is used to carry the remainder calculated for the previous frame. The calculation of the remainder for the next frame then includes the remainder from the previous frame.

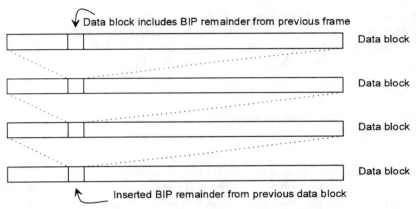

Figure 8.10   Insertion of BIP remainder bits in overhead of VC data block.

This overlapping of the CRC blocks does not affect the operation of the CRC and makes the implementation a little simpler. The overhead location used for carrying the remainder for each type of virtual container is described in Chapter 5.

The type of CRC used in SDH is called *bit interleaved parity* (BIP) and is a special type of CRC with a polynomial of

$$x^n + x^0$$

called a BIP-n. The values for *n* for the different types of virtual container are given in Table 8.2.

Table 8.2
Block Error Monitors for SDH Paths

Network layer	Block length	Order of BIP
RS(STM-N)	125 ms = 19,440N data bits + 8 remainder bits	BIP-8
MS(STM-N)	125 ms = 1,9224N data bits + 24 remainder bits	BIP24N
VC-4-xc	125 ms = 18,792 data bits + 8 remainder bits	BIP-8
VC-4	125 ms = 18,792 data bits + 8 remainder bits	BIP-8
VC-3	125 ms = 6,120 data bits + 8 remainder bits	BIP-8
VC-2-vxc	125 ms = 3,424x data bits + 2x remainder bits	BIP-2
VC-2-xc	125 ms = 3,424 data bits + 2 remainder bits	BIP-2
VC-2	125 ms = 3,424 data bits + 2 remainder bits	BIP-2
VC-12	125 ms = 1,120 data bits + 2 remainder bits	BIP-2
VC-11	125 ms = 832 data bits + 2 remainder bits	BIP-2

A BIP-n has an *n* bit remainder with the special property that each bit of the remainder is a single bit parity for a subblock formed by the every xth bit of the block, hence the term bit interleaved parity. Interleaving is illustrated in Figure 8.11. This polynomial is useful

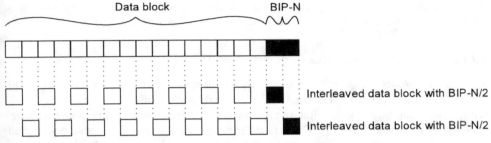

Figure 8.11   The interleaving of data blocks and BIP remainders.

for virtual concatenation payloads where each separate VC-2 has a BIP-2. The simple concatenation of the BIP-2 remainders gives a single BIP-2x remainder.

8.9   ERROR PERFORMANCE AND UNAVAILABILITY PARAMETERS

The mechanisms within SDH described above are used to derive the error performance parameters now given in draft Recommendation G.826 and unavailability parameters. They are all based on errored blocks and the BIP-n is used to assess whether the 125 μs or 500 μs blocks described above contains errors or not.

8.9.1   Error Performance

The parameters, described in draft Recommendation G.826, are only accumulated when the trail is in available time.

8.9.1.1   *Errored seconds (ES) and errored second ratio (ESR)*

An *errored second* is a one second interval in which there has occurred one or more errored blocks. A *severely errored second* contains 30% or more errored blocks or at least one severely disturbed period (SDP). For out-of-service measurements an SDP occurs when, over a period of time equivalent to four contiguous blocks or 1 msec, whichever is the larger, either all the contiguous blocks are affected by a high bit error density of 10^{-2} or worse, or a loss of signal is observed. For inservice purposes, an SDP is estimated by the occurrence of a network defect. In SDH detection of path AIS, loss of pointer or path trace mismatch are considered network defects.

SDPs lasting for several seconds may be precursors to periods of unavailability, especially when no automatic protection or restoration procedure is applied. Alternatively they may terminate spontaneously or as a result of protection or automatic restoration. Such short interruptions in the range 2 to 10 seconds are particularly troublesome, resulting in lost calls and severe disruption of data services, and they are also not uncommon. G.826 only constrains short interruptions indirectly via the SESR.

The *Errored Second Ratio (ESR)* is the ratio of *errored seconds* to total seconds during a measurement time. The *severely errored second ratio (SESR)* is similarly the ratio of SESs to total seconds during a measurement time. In both cases unavailable seconds are excluded both from the impaired seconds count and the total seconds count.

8.9.1.2   *Background Block Error Ratio (BBER)*

A *background block error (BBE)* is an errored block not occurring as part of an SES and the BBER is the ratio of total errored blocks to total blocks during a measurement time, neither total including blocks belonging to a severely errored second, or unavailable time. A period of *unavailable time* (UT) starts at the onset of ten consecutive severely errored seconds. These 10 SESs are considered part of the unavailable time. A new period of available time starts at the onset of ten consecutive non SESs. These 10 non SESs are considered part of available time. A bidirectional path is unavailable if one or both directions are unavailable.

8.10   ERROR PERFORMANCE AND UNAVAILABILITY OBJECTIVES

The definition of trail and connection performance objectives has long been a difficult subject for ITU-T as operators and manufacturers worldwide are reluctant to commit to limits that cannot be met by the worst performing trails in a network. This has led to the establishment of a set of objectives that are the lowest common denominator across all network operators. The limits given in G.826 should therefore be treated very much as a worst case and it may be expected that network operators will compete on the performance of their trails and this will drive up the actual quality.

The limits are quoted for an international hypothetical reference path (HRP) of 27,500 km. Allocation of this maximum limit is given to trails or connections of shorter length. The limits are given in Table 8.3 and the HRP illustrated in Figure 8.12. The evaluation period for each parameter is one month.

Table 8.3
ITU-T Performance Objectives for SDH Trails

	VC-11/VC-12	VC-2	VC-3	VC-4	VC-4-4c
ESR	0.04	0.05	0.075	0.16	for study
SESR	0.002	0.002	0.002	0.002	0.002
BBER	2×10^{-4}	2×10^{-4}	2×10^{-4}	2×10^{-4}	4×10^{-4}

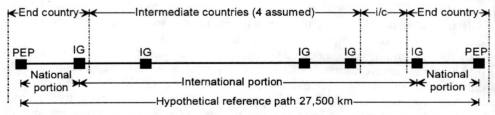

Figure 8.12   Hypothetical reference path for allocation of error performance objectives.

The boundary between a national and an international portion is defined to be at an international gateway (IG) which will normally correspond to a crossconnect, a multiplexing equipment, or a switch. The location of intermediate IGs is not taken into account. A block allowance of 17.5% is made in respect of all of the G.826 parameters to each of the national portions. A further distance-based allowance is made to take account of the relative size of the national portion calculated at 1% per 500 km of the estimated length. A satellite hop in the national portion is allocated 35%.

The international portion is allocated a block allowance of 2% per intermediate country plus 1% for each terminating country. A further distance allowance of 1% per 500 km is also allowed. A satellite hop is allowed 35% as in the national portion.

8.11   EXAMPLES OF OBSERVED RESULTS

Many network operators have undertaken extensive monitoring of their networks in order to characterize the type of error events that occur and their distribution, as the first step to improving the overall performance of the digital network. Figure 8.13 illustrates the distribution of events, based on published results from BT, of sample 139264 kbit/s trails in the United Kingdom which suggests a value of α of about 0.3. Published results from extensive monitoring by AT&T in the United States suggest that a value of α of about 0.2 gives a good fit to observed results.

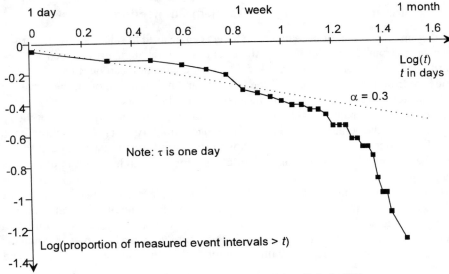

Figure 8.13   Distribution of error events on sample 140Mb/s trails in the UK.

8.12   TANDEM CONNECTION MONITORING

In the foregoing discussion we have been concerned mainly with monitoring integrity and quality of trails or elementary connections, but there are many circumstances where it is important to be able to monitor performance metrics for a segment of a trail corresponding to a tandem connection as defined in Chapter 3. These include localization of failures within the network, and retrospective apportionment of responsibility for outages or poor performance in a multi-operator contractual situation. Cost-effective strategies for dealing with these requirements must take into account the nature and statistics of the events being monitored in a high quality optical network as well as the contractual relationships with other operators and the regulatory provisions governing those relationships. This section considers the requirements, the solutions available, their advantages and weaknesses, and their architectural implications.

8.12.1   Requirements for the Monitoring of Tandem Connections

Any one of the tandem connections that make up a trail may need to be monitored for validity, integrity, or quality at some time or another, however, the tandem connections between operator boundaries are of particular interest because of the special contractual or regulatory conditions that are likely to apply. A number of methods have been considered to provide these features. Some are available continuously in service while some can be made available on demand. Given the rare and unpredictable nature of some of the phenomena to be monitored this latter type of provision offers a flexible and cost effective solution in many cases.

8.12.2   Inherent Monitoring of a Tandem Connection

The inherent monitoring of the supporting layers can be used to calculate the performance of a tandem connection. Because of the client server relationship between layers the integrity and performance of a client layer tandem connection can be inferred with high confidence from the logical union of the integrity and performance of all the server layer trails on which it is supported, and all the client layer subnetworks that it transits. This can be augmented in the path layers by information describing the integrity (but not the error performance) of the connection itself, available in the tributary unit (TU) or administrative unit (AU) pointer. This must be compared with similar information available at the node where the information entered the tandem connection. The architecture is illustrated in Figure 8.14 .

Correlation of all the information relating to a single tandem connection is a network level process requiring the collection of information from many different nodes and interpreting this with a knowledge of the network connectivity. This is one of the capabilities expected to be available in the TMN. Correlation can be achieved down to the resolution of the time stamp held with the individual record but this will be adequate in many cases.

8.12.3   Nonintrusive Monitoring

Nonintrusive monitoring relies on the possibility of applying 1 to 2 broadcast, Tee-connections at the input and the output of the tandem connection to compatible trail terminations capable of monitoring the transmitting signal and overhead. This system needs no additional standards support but transit node equipment must be specified to

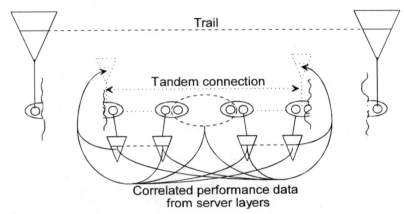

Figure 8.14   Inherent monitoring of a tandem connection.

provide the feature. Correlation is still required but now it is sufficient only to correlate the two ends of the tandem connection; no knowledge is required of the intervening server trails or their performance. The test access mechanism and the per-channel monitoring termination are considered expensive to provide comprehensively, therefore, the capability is often only made available on a statistical basis to be used on demand by a small proportion of the tandem connections terminating on a node.

The ability to apply nonintrusive monitoring to selected tandem connections provides a useful complement to inherent monitoring. Marginal error conditions due to problems near the layer boundary at the transit node that may evade detection by the inherent monitoring may be detectable in this way. Also, the calculation of error performance by comparing the error performance at either end of the tandem connection will provide a more accurate estimation of the error performance than that available by extrapolating from the server layer.

The path termination identifier and path label that will have been inserted at the source may be directly monitored on entering and leaving the tandem connection as a check on connectivity. This may then be used as part of the set up procedure to validate each tandem connection in turn until the whole path has been set up and may also be used to check continuing correct connectivity by applying the nonintrusive monitor periodically on a shared basis. The monitor itself may be more comprehensive and capable of monitoring clients of the layer being accessed. The functional architecture of nonintrusive monitoring is illustrated in Figure 8.15.

8.12.4   Intrusive Monitoring

Intrusive monitoring implies a full test break access enabling a valid test condition to be inserted at the input to the tandem connection and a non intrusive monitor to be attached at its output. This offers the possibility of more comprehensive tests but can only be used when the tandem connection is out of service. Like the nonintrusive monitor, it is likely that the test-break facility, the test generator, and trail termination source function will generally be shared and therefore only available on demand. To enable remote diagnostic

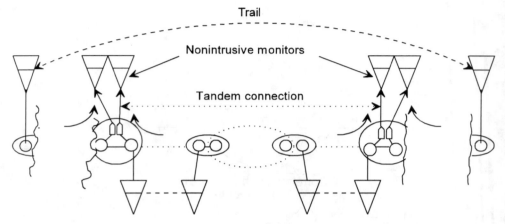

Figure 8.15   Nonintrusive monitoring of a tandem connection.

and test in a managed network, such facilities will be essential. This feature requires no additional standards support as the entity under test is in fact, a trail, as defined in the recommendations (albeit artificially created for the duration of the test). The information available is as complete as that at the end of any trail. This mechanism may be incorporated in the path setup process allowing a tandem connection from the middle of the final path to be set up and validated before connection of the terminating links.Figure 8.16 illustrates the functional architecture required to implement intrusive monitoring.

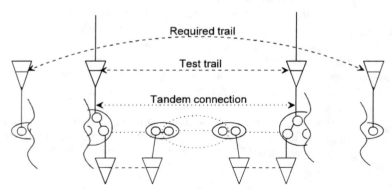

Figure 8.16   Intrusive monitoring of a tandem connection.

8.12.5   Sublayer Monitoring

Sublayer monitoring uses the procedure illustrated in Figure 3-14 (b) to decompose the existing trail termination to generate an extra sublayer for monitoring purposes. The resulting functional architecture is illustrated in Figure 8.17. The new sublayer trail is placed below the original layer because of its lower entropy. At a practical level this can

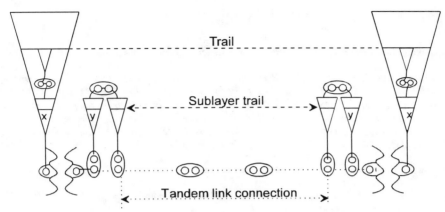

Figure 8.17   Sublayer monitoring of a tandem connection.

be achieved by overwriting part of the original overhead (one of the unused or reserved bytes) with a new sublayer trail overhead while at the same time correcting the parity of the original overhead to allow for the new overhead that has been added. In this way, any parity errors that had accumulated upstream of the tandem connection are preserved in the encoding of the original overhead. The original content of the unused or reserved bytes has not been specified. It has only been required that the receiver ignore them. This must be taken into account when recalculating the original parity at the sublayer boundary. The sublayer termination that is collocated with the original layer termination (marked x in Figure 8.17) is implied because the original content of the byte will have been constrained by G.709 to maintain even parity. Similarly, the sublayer termination at the other end of the terminating tandem connection (marked y) is also implied, as up to this point parity block errors in the original trail and in the sublayer trail are by definition the same.

There is considerable interest in providing this function in special equipments to be placed at interoperator boundaries and at operator-customer boundaries where the customer may operate a significantly large premises network such that the layer termination may not be part of the network terminating function. There is no doubt that sublayer monitoring offers better accuracy, resolution and fault coverage than the mechanisms that rely on correlation in the TMN. However, unless it is provided ubiquitously it will be less flexible than the TMN based mechanisms since it cannot directly support the monitoring of overlapping or nested tandem connections.

Time and experience will decide whether the benefits of sublayer monitoring justify its cost. If the commercial and regulatory environments engender a litigious approach to transport provision, then such capabilities will certainly be necessary to protect against misrepresentation of quality achieved against contract. It seems unlikely however that operators would willingly get drawn into exchanging penalty charges for every single error event. In this case statistically aggregated performance statistics computed to high confidence levels by the TMN would appear to be a more appropriate basis for such interoperator, quality-derived, accounting adjustments.

References

The following standards are the main source documents for the material in this chapter:

ITU-T Recommendation G.821, *Error Performance of an International Digital Connection Operating at a Bit Rate Below the Primary Rate and Forming Part of an Integrated Services Digital Network.*

ITU-T Recommendation G.826, *Error Performance Parameters and Objectives for International, Constant Bit Rate Digital Paths at or Above the Primary Rate.*

ITU-T Recommendation G.827, *Availability Parameters and Objectives for Path Elements of International Constant Bit Rate Digital Paths at or Above the Primary Rate.*

The following gives a presentation of the Bond distribution:

Bond, D. J., "A Theoretical Study of Burst Noise", *British Telecom Technology Journal*, Vol. 5 No. 4, October 1987, pp 51-60

Traffic Performance

Telecommunications networks aim to provide information transfer capability between the network's access points, on demand; that is, as soon as the user requests it. The immediacy of the service is one of the most noticeable aspects of the perceived quality. There are two properties of networks that make this a matter of probability and not of certainty:

- The ability to respond to service requests as they occur implies that the network operator must provide resources in the network in advance of the demand.
- The ability to share resources among many users and different uses by an individual user is an important factor in reducing overall costs.

When, as is generally the case, these two factors are both, the network operator cannot be certain that when a user makes a request for a connection, there are sufficient resources in the network to satisfy the request. The statistics of allocating capacity for users' demands determines the traffic performance of the network, and maintaining adequate traffic performance levels becomes a major preoccupation for the network operator.

In the narrowband network, the study of traffic performance has matured over many years following the seminal work of Erlang in 1909. However, the traffic management procedures that have evolved have relied heavily on the fact that only one type of connection is offered, a circuit-switched connection, and the order of magnitude of the call holding time is relatively stable and well-known, namely the few minutes of a telephone conversation. The diversity of broadband service connections and the variety of holding times leads us to expect that all this received wisdom from the traffic management of the narrowband world cannot be directly applied to broadband networks. Figure 9.1 and Figure 9.2 show some applications and the variation in holding time and burstiness that may be expected for each one.

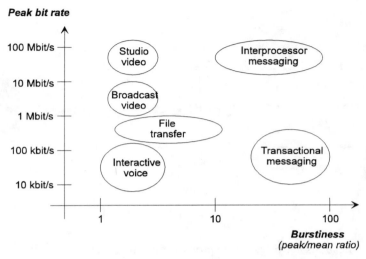

Figure 9.1   Position of applications with respect to bit rate and burstiness.

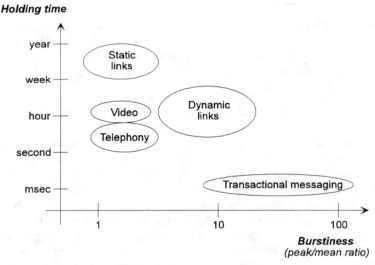

Figure 9.2   Applications with respect to holding time and burstiness.

To manage the traffic implications of all these types of connections, we must return to the basic principles of traffic statistics. This has been extensively studied in recent years, and there is a large volume of published work on the subject. One of the features of broadband traffic statistics is that even very simple scenarios become mathematically very complex, and it is easy to become preoccupied with detail while failing to understand the wider picture. In this chapter, we present some of the more important principles and results, with enough detail to indicate the assumptions and the methods behind derived results.

For readers interested in pursuing the subject in more depth, the European Commission's Cooperation on Science and Technology Report 242 gives a very good account of the main stochastic models. Much of the information presented here is drawn from this source.

9.1   TRAFFIC STATISTICS AND PARAMETERS

Before considering the sort of parameters useful in describing traffic, it is helpful to develop some understanding of the activities surrounding the design and operation of broadband networks likely to depend on traffic performance. We see the need for traffic objectives and specifications falling into three broad areas:

- In defining and forecasting services;
- In controlling the real-time operations and performance of the network;
- In design and planning networks and their equipment.

Each of these areas demands a clear understanding of the broadband traffic implications; however, each has rather different requirements. This is illustrated in Figure 9.3.

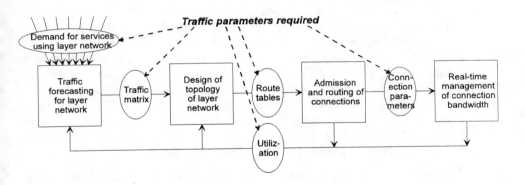

Figure 9.3   Use of traffic parameters.

9.1.1   Service Definition, Tariffing, and Forecasting

9.1.1.1   Requirements for Traffic Parameters

In moving to broadband, the range of transport services that the network can offer increases greatly, so the way in which we define these services must expand to keep pace. In Chapter 4, we noted that a transport service could be defined in five phases (i.e., attachment, connection setup, transfer, connection takedown, and detachment)—excluding a billing phase—and for each phase, we need to have a set of traffic descriptors. A number of requirements for such traffic descriptors can be identified. They should be or do the following:

- *Objectively measurable*—Traffic descriptors should be measurable in such a way that anybody carrying out the measurement will arrive at the same result. This allows both the client and the server to have a non-contentious description of the service that has been delivered.
- *Understandable by the user*—While some clients, especially internal and business clients, may be able to understand complex statistical parameters, many private customers will not. Clients must have confidence that they are getting what they want and are expected to pay for.
- *Repeatable and reliable*—If the client makes the same use of the network over several occasions, then the traffic descriptor should reflect this. Since traffic descriptors are usually statistical, a high level of averaging will generally be required to produce repeatable and reliable measurements. This may be problematic for low volume users.
- *Oriented towards the user and not the network*—Traffic descriptors defining services should be oriented towards the users' requirements and not to the way the network responds when supplying the service. As well as promoting understanding of the parameter by the user, this also ensures that the service definition will outlive network technology evolution.
- *Useable in the network design and planning process*—This is to ensure that efficient networks are built that will reliably meet future service demand.
- *Enable price comparisons between competitive services*—The traffic descriptors should allow a direct comparison to be made between services when they are both capable of supporting a client. This may involve the comparison between, for example, a circuit-switched service and an unspecified bit rate (UBR) service.
- *Enable a tariff structure that adequately reflects network costs*—This is necessary to effectively operate in a competitive marketplace where competition will drive prices down towards costs. An operator must, therefore, be able to assess the impact that meeting demand will have on the costs of the network.

These requirements are not simple to meet, especially when put together. This should not, however, deter us from trying to meet them. Any selection of traffic parameters should go some way to meeting all these requirements.

9.1.1.2   Forecasting

The traffic forecast for a service is the starting point for the network design and planning process. The way this forecast is expressed can have a profound impact on the design, and critical to this is the level of uncertainty in the forecast. A network designed to a very specific demand forecast may look quite different from a network designed to the same mean forecast but with a considerable uncertainty about the mean. Traffic forecasts are therefore most effectively expressed as a mean and variance of the particular traffic parameters at some given time in the future.

9.1.1.3   Separation of Traffic Descriptors From Tariff Structure

In the days of electromechanical technology, the only realistic method of billing for usage of the public switched telephone network (PSTN) was to directly count traffic units and then bill directly for these. This meant the tariff structure was embedded in the network and hence very difficult to change. For example, differential tariffing to preferred destinations, volume discounts, and option-based tariffs balancing subscription and usage components were all very difficult, and normally impractical, to offer.

Separating the recording of usage from the design of the tariff structure goes a long way to meeting many of the requirements described above. The recording of usage is only accurately recording usage of the network. The way in which the user is billed may be decided independently. This can allow for a wide variety of packaging options to be constructed by the network operator, which can go a long way to reflect the very large economies of scale a network operator enjoys when serving one large client as opposed to many small clients. This is illustrated in Figure 9.4.

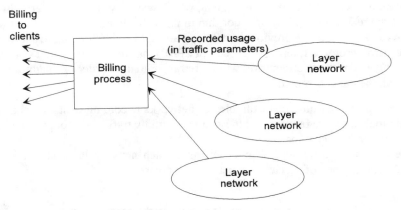

Figure 9.4   Separation of billing from recording of usage.

9.1.1.4   Real-Time Operations

The real-time control of traffic within a network is one of the most important and also one of the most difficult tasks for a network operator. The operator is trying to achieve a number of conflicting requirements:

- Satisfy all profitable requests for service.
- Maximize the utilization of the network.
- Meet the performance commitments given to service requests.
- Maintain stable operation of the network in the event of traffic overload and/or service requests.

It is the last of these that gives the most challenging problem, and this has also been the case for narrowband networks. There are some well-known traffic phenomena that give rise to instability, but techniques have been devised over the years to control the traffic and maintain stability, at least in today's PSTN/ISDN and data networks. As the number of connection types increases, so will the number of mechanisms that produce instability. Most of the mechanisms in the narrowband network were discovered and characterized largely by observation in real networks, and there is every reason to expect the same will happen with broadband networks.

In general, traffic instability results when the network responds to congestion by producing even more traffic. Below, we list some of the better known mechanisms:

- When a large destination in the PSTN approaches congestion, users start to make repeated call attempts, greatly increasing the number of call attempts.
- Following the failure of a switch, traffic is often redirected to neighboring switches, which may themselves become congested and so further redirect traffic. As this happens, the routings become extended, so the network becomes more congested.
- If a packet is dropped as a result of congestion and the packet is part of a file, higher layer protocols will normally require retransmission with consequent increase in traffic levels. At one point, the Internet experienced large on/off oscillations in loading as a result of this mechanism, which was only fixed by a careful redesign of the real-time windowing algorithm in the TCP client.
- As a variation of the previous example, if an ATM cell belonging to a transmission control protocol/Internet Protocol (TCP/IP) packet is dropped as a result of congestion, the whole packet will be retransmitted an order of magnitude greater than the data corrupted.

In addition, there are numerous examples of networks becoming unstable as a result of common mode software failures, but these are not strictly traffic control problems and are discussed in the next chapter.

The real-time traffic control mechanisms in each network layer aim to produce a "graceful degradation" of service as congestion increases.

9.1.2   Design and Planning of Network and Equipment

Certain characteristics of the telephony environment contributed to the generation of a relatively simple statistical model of traffic for design and planning. Many readers will be familiar with the use of Erlang statistics for planning and dimensioning a telephony network. The technique has been in use for over 85 years and has evolved to the extent that many telecommunications professionals use the results without being fully aware of where they come from.

One of the major differences with a broadband network is that the nature of the traffic is sufficiently different from telephony that the wealth of experience built up using Erlang statistics is no longer directly applicable. There are several new features of broadband networking that must be added to the simple telephony picture before we can successfully design and plan networks. These include the following:

- The multilayer nature of broadband networks requires that design and planning using traffic parameters be done in all layers.

- Holding times can vary from milliseconds for a connectionless packet to years for some links in client networks (e.g., private circuits) with examples of most holding times in between.
- The wide variety of applications makes forecasts more volatile;
- For many connections, their bit rate will fluctuate during the call.

Our traffic parameters for design and planning must work within this expanded environment, but there are certain properties of the simple Erlang model that contribute significantly to its ease of use and reliability of results, and we should seek to maintain these in the expanded broadband traffic model. Two particular properties we should seek to emulate are:

- A reliable probability distribution relating the offered traffic, the available network capacity, and the probability of successfully carrying the traffic on the available capacity (i.e., grade of service);
- An additive property enabling the simple aggregation of traffic from different sources and/or locations such that the aggregate value accurately reflects the effect of the combined traffic.

In the case of the Erlang traffic model, the first property is satisfied by the Erlang distribution while the second is satisfied because Erlangs have a linear addition property. Any new traffic descriptors for which similar properties cannot be derived will not be as useful in the network design process.

9.2   BASIC BROADBAND TRAFFIC MODELS

While there has been much discussion on broadband traffic parameters, consensus has been harder to achieve. This is partly because most operators expect that a level of practical experience is required before too many firm decisions can be taken. In this section, we look at broadband traffic as it exists on different timescales and look at some of the stochastic processes that can be used to describe traffic at these timescales. This method, first suggested by Hui, should give a complete picture of broadband traffic in more general terms. In the next section, we will look at some of the parameters that have been already agreed upon in the international standards fora.

9.2.1   Scales in Time

We are interested in timescales in which some form of stochastic process is taking place. In the case of the Erlang model, there is only one timescale that is formally acknowledged; this is the connection scale with two interrelated stochastic processes, one modeling the arrival of new connections and the other modeling the termination of connections. Informally, other timescales are also considered, for example, the daily "busy hour" and other seasonal fluctuations.

Broadband traffic has significant components at shorter timescales and hence we must consider more than the connection timescale of the Erlang model and try to develop reasonable stochastic models across a range of significant timescales. These stochastic models will to some extent be interdependent, and this presents an extra difficulty.

The timescales at which we can find stochastic processes that are important to our understanding of traffic are listed in Table 9.1. A complete picture of timescales, strictly, could also include a "byte" scale, which would consider the multiplexing of bytes in narrowband networks and in SDH. However, this multiplexing is fully deterministic and so there is no stochastic dimension. It is therefore not considered as an interesting traffic problem.

Table 9.1

Scales in Time

	Stochastic phenomenon	*Stochastic models*	*Traffic impairments*
Cell scale	Statistical multiplexing of cells from different sources	Various queue models	Cell loss
Burst scale	Statistical multiplexing of packets or groups of packets	Various queue and population models	Cell loss
Connection scale	Admission of connections	Various population models (e.g., Erlang)	Refused connection setup attempt
Human activity scale	Correlated human activity giving daily, weekly, and yearly cycles	Statistical inference of cycles and correlations	Refused connection setup attempt
Planning scale	Forecasting of traffic demand after planning lead time	Numerous forecasting techniques	Refused connection setup attempt

9.2.2   Queuing and Population Models

9.2.2.1   Description of Queues and Populations

Many of the stochastic processes in the study of broadband traffic can be modeled using queues and populations. Classic queuing and population problems can be found in many textbooks on stochastic modeling and there is a good body of knowledge already built up. Stochastic queues and populations are normally described in the form $A/S/n$ where

- A—Describes the process modeling the arrival of items into the queue;
- S—Describes the process modeling the servicing time of each item;
- n—Is the number of servicing stations.

The convention for describing different arrival and servicing processes is as follows:

- M—*Memoryless,* where the process generating the occurrence of an event is fully independent of any other event. This is a Poisson process where the interarrival distribution is an exponential distribution;
- G—*General,* to signify any distribution;
- D—*Deterministic,* where the arrival time or service time is fixed.

These are the commonly used terms, however, a number of extensions have been added to consider the cell queuing problem. These include:

- Geo—*Geometric,* which is similar to the memoryless with the exception that events can only happen at discrete intervals, according to a Bernoulli process rather than a Poisson process, so that the interarrival distribution is a geometric distribution rather than an exponential distribution;
- N×D—To denote the multiplexing of *N deterministic* event processes and where each event process has the same characteristics (i.e., N channels of the same cell rate);
- ΣD_i—To denote the multiplexing of *N deterministic* event processes where each event process has different characteristics (i.e., N channels of different cell rates).

9.2.2.2   *Mathematical Modeling of Queues and Populations*

The mathematical modeling of queues and populations is well-developed. They are generally modeled as stochastic processes. Indeed, most textbooks on stochastic modeling include many queuing and population examples. A queue or a population can be modeled using the simplest of stochastic processes, a Markov chain.

In the case of a queue, every possible length of the queue is given a state, and if the queue has a given length, it is said to be in the state associated with that length. The Markov chain is then defined as the possible transitions between states in a fixed time interval and is often written in the form of a square matrix. The rows and the columns are the states, and the entries in the matrix are the probabilities of moving from one state to another in the time interval, Δt. For example, in a queue that is modeling the buffering of ATM cells, a suitable time interval would be the time taken for a cell to pass out into the link. This is illustrated in Figure 9.5. The case of a population is equivalent, except that the state of the Markov chain is now defined to be the number of servers in use. (The arrival and serving processes are normally called the birth and death processes.) The specific cases corresponding to queues and populations are shown in Figure 9.6.

The transition probabilities can be calculated from the arrival and servicing processes. The Markov chain can then be formed by starting with a column vector, $p(t)$, where the entries are the probabilities of each state. The entries in the column vector will therefore total to unity. The probability of the states after time Δt can be calculated by multiplying the column vector by the matrix, $\mathbf{M}$, to arrive at a new column vector, $p(t + \Delta t)$.

$$p(t + \Delta t) = \mathbf{M}\, p(t)$$

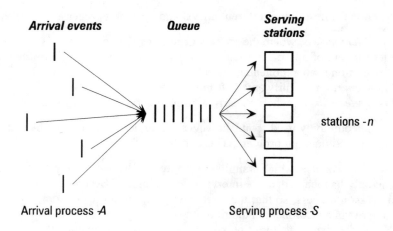

Figure 9.5   Basic modeling of queues and populations.

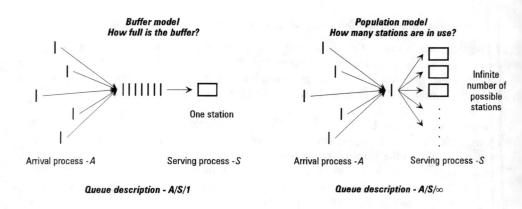

Figure 9.6   Modeling a queue, and modeling and population.

We are particularly interested in the case where the probability vector reaches a long-term steady state such that $p(t + \Delta t) = p(t)$ and this is determined solely by the state transition matrix, M. Assuming that it exists, we are interested in finding $p(\infty)$ such that

$$p(\infty) = M\,p(\infty)$$

and this is the probability distribution for the queue. The condition for $p(\infty)$ to exist normally depends on the arrival and service processes. For example, if the long-term mean arrival rate is higher than the long-term mean service rate, and the queue contantly grows in length, then $\mathbf{p}(\infty)$ will not exist. If the queue is modeling a buffer and the buffer can hold a maximum of B cells, then the probability of cell loss, $Q(B)$, can be determined by totaling the probabilities of all the fill states that are greater than B; that is,

$$Q(B) \; = \; \int_{k=B}^{\infty} \mathrm{p_k}(\infty)$$

9.2.3   Cell Scale

This timescale considers the multiplexing of cells using a first-in-first-out (FIFO) queuing buffer, which is at the heart of every ATM concentrator or switch. The process is illustrated in Figure 9.7. Traffic congestion at the cell scale results in the number of cells in the buffer exceeding the size of the buffer, and some cells are therefore lost.

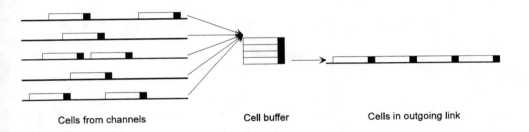

Cells from channels   Cell buffer   Cells in outgoing link

Figure 9.7   Modeling of cell scale buffering.

A number of queuing models have been studied and between them give an understanding of the relationships, at the cell scale, between offered traffic, buffer size, and cell rate on the outgoing link. In the following, we consider the M/D/1, Geo/D/1, N×D/D/1, and the $\Sigma D_i/D/1$ queues.

9.2.3.1   Modeling Using An M/D/1 Queue

The M/D/1 queue might reasonably describe the situation when all the cell streams that are being multiplexed have already passed through many previous multiplexing queues and the interarrival time between cells has lost any deterministic nature it might have had initially.

An advantage of the M/D/1 queue is that no special consideration needs to be given to the cell rates of the individual cell streams. Since all the cell arrivals are independent, the cell rates from all the cell streams can simply be added to give an aggregate cell rate.

If we take Δt to be the service time for a cell, τ_{cell}, and is therefore the reciprocal of the link cell rate in cells/sec, then, if there is at least one cell in the buffer, exactly one cell will be serviced in Δt. The number of cells that could arrive in Δt will be according to a Poisson distribution with a parameter $\lambda_{cell}\tau_{cell}$ where λ_{cell} is the long-term rate of cell arrivals. The state transition matrix for the M/D/1 queue is

$$\mathbf{M} = \begin{bmatrix} P(0) & P(0) & 0 & 0 & 0 & \cdot \\ P(1) & P(1) & P(0) & 0 & 0 & \cdot \\ P(2) & P(2) & P(1) & P(0) & 0 & \cdot \\ P(3) & P(3) & P(2) & P(1) & P(0) & \cdot \\ P(4) & P(4) & P(3) & P(2) & P(1) & \cdot \\ P(5) & P(5) & P(4) & P(3) & P(2) & \cdot \\ \cdot & \cdot & \cdot & \cdot & \cdot & \cdot \end{bmatrix}$$

where $P(n)$ is the Poisson probability of n events with the parameter for the Poisson distribution of ρ_{cell}, and where

$$\rho_{cell} = \lambda_{cell}\tau_{cell}$$

From this state transition matrix, it can be shown that for an M/D/1 queue,

$$Q_{cell}(B) = (1 - \rho_{cell}) \sum_{n=B}^{\infty} \frac{e^{-\rho_{cell}(n-B)} \left\{ \rho_{cell}(n-B) \right\}^n}{n!}$$

where $Q_{cell}(B)$ is the loss probability at the cell scale. Since τ_{cell} is the reciprocal of the link rate, this means that ρ_{cell} is the long-term loading on the link. If this is unity, then the link is fully loaded.

An approximate formula can be used when B is large and when ρ_{cell} is close to unity, which gives a better insight into the cell loss probability:

$$Q_{cell}(B) \approx e^{-\frac{2(1-\rho_{cell})}{\rho_{cell}B}}$$

from which we can see that the extreme tail of the distribution in which we are interested has an exponential shape. This means that the probability of cell loss will decline exponentially with increasing buffer size. Curves for cell loss probability are given in Figure 9-8.

9.2.3.2   Modeling Using an N×D/D/1 Queue

This queue is useful for examining the opposite extreme, where all the cell streams that are to be multiplexed are fresh and the cells from each stream arrive at regular intervals. The use of the N×D queue reflects the fact that the analysis is considerably simplified if the cell rate of each stream is the same. This is clearly not a realistic case; however, it gives a good comparison with the M/D/1 case. The main feature of this system is that while all the cell streams are deterministic and the same, the phase of each cell stream is random.

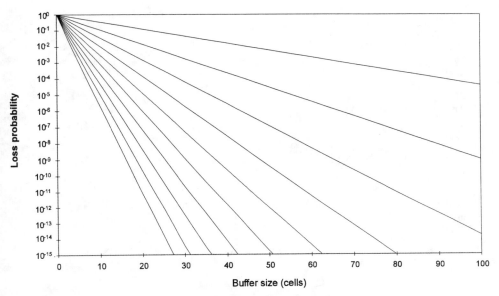

Figure 9.8   Cell loss probability with an M/D/1 cell-scale queue.

As with the M/D/1 queue, we take Δt to be the cell period at the link rate and, again, if there is at least one cell in the buffer, exactly one cell will be serviced in Δt. To determine the number of cells that could arrive in Δt, we consider that each channel will take one cell every R cells, where $1/R\tau_{cell}$ is the cell rate of the channel. In a period of R cells, each channel will contribute exactly one cell. The value of R must be greater than N to ensure that the channel loading is less than unity. Indeed, the loading on the link, ρ_{cell}, is N/R.

In this case, the queuing problem cannot be expressed in the form of a state transition matrix, as the arrival of cells during the period of R cells is not an independent process. For example, if all N cells arrive in one cell period, then they cannot arrive for the remaining $R - 1$ cell periods. However, it may be shown that

$$Q_{cell}(B) = \sum_{k=B+1}^{k=N} \binom{N}{k} \left\{ \frac{\rho_{cell}(k-B)}{N} \right\}^{k} \left\{ 1 - \frac{\rho_{cell}(k-B)}{N} \right\}^{N-k} \frac{N - \rho_{cell}(N-B)}{N - \rho_{cell}(k-B)}$$

Again, there is an exponential approximation; however, in this case it is less accurate, but is still very useful for indicating the general sensitivity of cell loss probability to buffer size and loading.

$$Q_{cell}(B) \approx e^{-2B \left\{ \frac{B}{N} + \frac{1 - \rho_{cell}}{\rho_{cell}} \right\}}$$

We can see the difference between the M/D/1 queue and the N×D/D/1 queue in Figure 9.9. We can note a number of points:

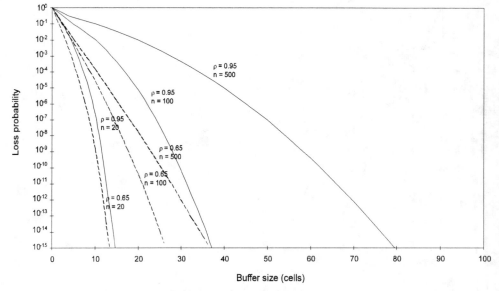

Figure 9.9   Cell loss probability with an N×D/D/1 cell-scale queue.

- Under all practical conditions, a buffer is required in the cell multiplexer;
- The size of the buffer depends on the loading of the multiplexed cell stream.
- A buffer sized on the M/D/1 queue could be significantly oversized if the individual cell streams are actually deterministic.
- At lower loading levels, the difference between the N×D/D/1 queue and the M/D/1 queue are much less than at higher loading levels.

We can see from the approximation equations that while the M/D/1 queue is a simple exponential distribution, the N×D/D/1 queue is of almost identical form except that it has an additional B^2 term in the exponent. This indicates that at light traffic loading, when this term is not dominant, the fill modeled by the two queues will be the same. However, at higher loading, this quadratic term will dominate, and, for the same cell loss probability, the N×D/D/1 queue will require a much smaller buffer.

9.2.4   Burst Scale

At the burst scale, we are interested in phenomena that can cause modulation of the cell rate over short time periods. These can include the following:

- Packetization in data protocols where a number of cells are required to support a data packet;
- Bursts of data as a result of interactive or real-time data protocols, for example, as a result of a person working at an on-line workstation;

- A change of scene in a VBR constant-quality video signal where an increase in the data rate is required to convey a completely new picture frame rather than the small changes from the previous picture frame;
- Periods of activity and inactivity in interactive voice (e.g., telephony) as a result of syllabic and conversational variations.

9.2.4.1   Different Approaches to Handling the Burst Scale

There are two different ways of managing bursts, and there has been much debate as to which is the most appropriate. These are

- Fully buffer all bursts.
- Use only the buffering required for cell scale multiplexing and load the system such that the probability cell rate from burst scale phenomenon is small (e.g., comparable to cell scale phenomena).

These two approaches reflect the histories of two different parts of the telecommunications market. The two approaches are illustrated in Figure 9.10.

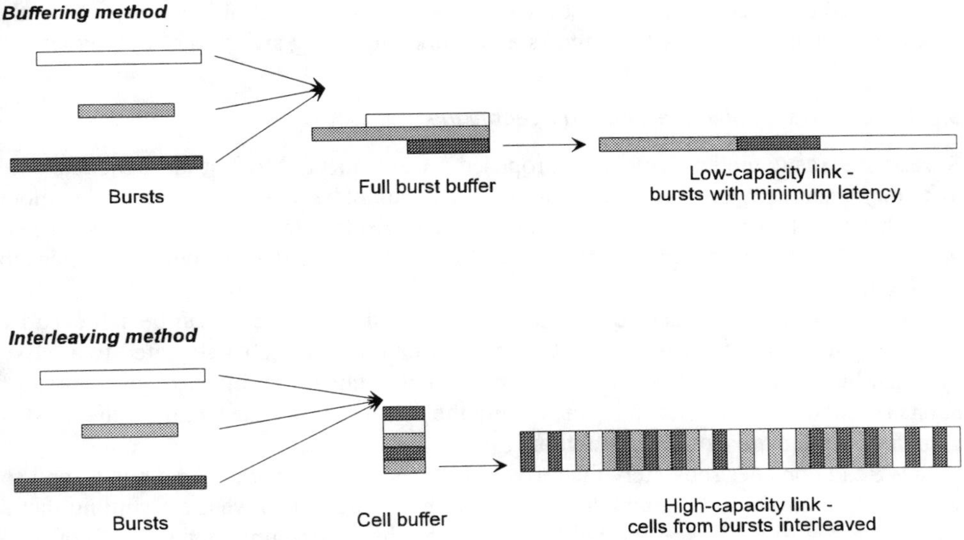

Buffering method

Bursts

Full burst buffer

Low-capacity link - bursts with minimum latency

Interleaving method

Bursts

Cell buffer

High-capacity link - cells from bursts interleaved

Figure 9.10   Different methods of managing the burst scale.

The buffering approach has its origins in private data networks where the networks are relatively small and link capacity is relatively expensive. In addition, a primary performance requirement in these networks has often been the "response time" seen by a user on a workstation working to a remote host system. This is governed by not just the time delay in the connection, but also the time it takes to convert the packet of information into a serial bit stream. This packetization delay can be very significant with low bit rate links, so private data networks have sought to reduce this to a minimum. This is achieved by allocating the full bandwidth of the link to the packet, and if any other packet requires transmission at the same time, it is buffered until the first packet has been fully transmitted. Since the links are both relatively expensive and of low bandwidth, the buffering costs are not the primary consideration. However, if the buffering becomes too great, then the buffering delay will reduce the response time. In general, this approach gives a system minimum possible response time when the network is lightly loaded and the response time will degrade as the loading on the system increases.

The second approach has much more in common with large, public circuit-switched networks. In these networks, the link capacities are large and the relative costs of nodes and links are governed only by the technology and not by commercial ownership as well. When the link capacity is large, it is possible to use the statistics of many connections to manage the burst scale without significantly increasing the latency of any one connection. In addition, in large networks with high link capacities, the cost of buffering becomes an issue, not just as an absolute cost but also because it can compromise the stability of the network.

The development of ATM has grown more complex with time, and this is one area where there is still no easy agreement emerging across the industry. Both sides still have strong lobbies, so it seems likely that both strategies will be deployed but oriented around the indicated markets. It remains to be seen where the service definitions are sufficiently robust to be handled successfully across a network employing switches of both types.

9.2.4.2   *Modeling the Buffering Technique*

Several stochastic models have been proposed for this queue. Most of these models tend to be mathematically complex, with solutions that cannot be easily stated or used without going beyond the scope of this book. (For the reader interested in pursuing this further, the COST 224 report is highly recommended reading and can also provide references to many other papers on the subject).

However, a good understanding of buffering at the burst scale can be achieved by examining the restricted case where the full link bandwidth is also allocated to a burst. This can be modeled by considering a simple M/M/1 queue. In this case, the arrival of bursts is modeled by a Poisson process while the distribution of the length of the burst is modeled by the exponential distribution.

We can consider separately the arrival of bursts and the length of the bursts, and so we initially count only the number of bursts in the queue. Since we are assuming that a burst is passed to the link at the full link rate, then the service time is the length of time for the top burst in the queue to play out at the link rate. The probability of the number of bursts in the queue therefore conforms to the standard state transition matrix for a M/M/1 queue,

$$
\mathbf{M} = \begin{bmatrix}
1 - \lambda\delta t & \dfrac{\delta t}{\tau} & 0 & 0 & \cdot \\[2ex]
\lambda\delta t & (1 - \lambda\delta t)(1 - \dfrac{\delta t}{\tau}) & \dfrac{\delta t}{\tau} & 0 & \cdot \\[2ex]
0 & \lambda\delta t & (1 - \lambda\delta t)(1 - \dfrac{\delta t}{\tau}) & \dfrac{\delta t}{\tau} & \cdot \\[2ex]
0 & 0 & \lambda\delta t & (1 - \lambda\delta t)(1 - \dfrac{\delta t}{\tau}) & \cdot \\[2ex]
0 & 0 & 0 & \lambda\delta t & \cdot \\[2ex]
0 & 0 & 0 & 0 & \cdot \\[2ex]
\cdot & \cdot & \cdot & \cdot & \cdot
\end{bmatrix}
$$

where λ is λ_{burst}, the arrival rate of bursts and τ is τ_{burst}, the mean size of the bursts, and δt is an arbitrarily short period of time. The steady-state solution for this state transition matrix is the geometric distribution:

$$
p(n) = (1 - \rho_{\text{burst}})\rho_{\text{burst}}^{n}
$$

where

$$
\rho_{\text{burst}} = \lambda_{\text{burst}}\tau_{\text{burst}}
$$

is the overall long-term traffic loading, and must be less than 1. The fill of the buffer depends on the number of bursts in the buffer and the size of those bursts. We know that the size of each burst is exponentially distributed, so the distribution size of n bursts will be the n fold convolution of this exponential distribution.

$$
s(t)\,|_{n=0} = \delta(t)
$$

$$
s(t)\,|_{n=k} = \frac{t^{k-1}e^{-t/\tau_{\text{burst}}}}{\tau_{\text{burst}}^{k}\,(k-1)!}\,u(t) \quad 1 \le k < \infty
$$

The probability that the buffer is filled to a time t is the sum across all values of n; that is,

$$
s(t) = \sum_{k=0}^{\infty} p(k)\, s(t)\,|_{n=k}
$$

$$s(t) = (1 - \rho_{\text{burst}}) \left[\delta(t) + \frac{\rho_{\text{burst}}}{\tau_{\text{burst}}} e^{-(1-\rho_{\text{burst}})t/\tau} u(t) \right]$$

From this we can determine the probability, $Q_{\text{burst}}(B)$, that a buffer of fixed size B will overflow and cause loss at the burst scale:

$$Q_{\text{burst}}(B) = \int_{\tau=B}^{\infty} s(t)\, dt$$

$$= \rho_{\text{burst}} e^{-(1-\rho_{\text{burst}})B/\tau}$$

We can see that this is already in the exponential form of the approximations for the models of cell scale buffering with the exception that now, for a given loss probability, the buffer size must be proportional to the mean burst size. In fact, to achieve the same cell loss probability as the M/D/1 model of cell scale buffering, we must increase the buffer by a factor of twice the mean burst size in cells. Since sizing based on the M/D/1 queue was very much worst case when compared with the N×D/D/1 queue, this indicates an even larger multiplication factor in the buffer size when compared to the nonbuffering burst scale technique.

This is illustrated in Figure 9.11 and means that the buffer will be many times the size required for just cell scale phenomenon, so the cell scale phenomenon becomes irrelevant in the analysis. The actual burst size is hard to predict since there are many mechanisms that can produce bursts and these may themselves change over time. This makes it difficult to produce any reliable design using this technique. However, in applications where very high loading factors are economically important, the inability to easily predict the cell loss has been treated as being of secondary importance, and as can be seen from the equation for $Q_{\text{burst}}(B)$, if the mean burst length does increase, cell loss probability can be restored by backing off the loading.

9.2.4.3   Modeling the Nonbuffering Technique

In this case, we are no longer concerned with modeling queuing in buffers. Instead, we model in the burst scale activity and then map its effects onto the cell scale, and from that infer the probability of cell loss.

We start by considering the instantaneous total cell rate at the burst scale, λ_{total} and its relation to the total link capacity, C. As λ_{total} is a stochastic process, we can calculate the overall probability of cell loss at the burst scale as the ratio of the mean rate of lost cells at the cell scale with given λ_{total} and C, to the mean cell rate, that is, $E(\lambda_{\text{total}})$. Therefore the loss probability at the burst scale, Q_{burst}, can be calculated as follows:

$$Q_{\text{burst}} = \frac{E(\textit{rate of cells lost with } \lambda_{\text{total}} \text{ and } C)}{E(\lambda_{\text{total}})}$$

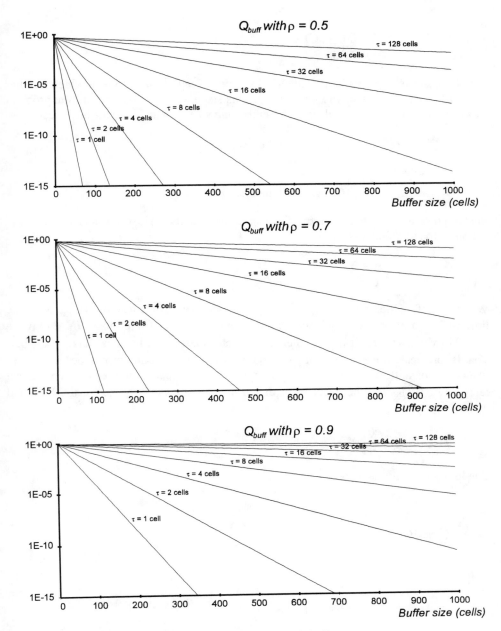

Figure 9.11   Cell loss probability with an M/M/1burst scale buffer

To map this onto the cell scale where the cell loss will occur, we define a cell scale stochastic process, $p_{cell}(\lambda_{total}, C)$ which is the instantaneous cell loss probability and so:

$$Q_{burst} = \frac{E(p_{cell}(\lambda_{total}, C)\ \lambda_{cell})}{E(\lambda_{total})}$$

The simplest case to consider is N identical sources of cell streams, each of which is either on or off as was illustrated in Figure 9.7. At this stage, we can also simplify the problem by assuming that the burst scale is sufficiently long when compared to the burst scale that:

$$p_{cell}(\lambda_{total}, C) = \begin{cases} \dfrac{\lambda_{total} - C}{\lambda_{total}} & \text{when } \lambda_{total} > C \\ \\ 0 & \text{when } \lambda_{total} < C \end{cases}$$

That is, all cells and only cells above the maximum capacity are lost. We also define ρ_{burst}, the mean loading factor, such that

$$\rho_{burst} = \frac{E(\lambda_{total})}{C}$$

We are therefore looking for the probability that there are sufficient sources on at the same time such that $\lambda_{total} > C$. Each source must have a mean cell rate $\rho_{burst}C/N$, and if the probability that a source is on is α, then the cell rate of a source when it is on is $\rho_{burst}C/\alpha N$. That is, the peak to mean ratio of each source is $1/\alpha$. The maximum number of sources that can be on at the same time is the maximum channel capacity divided by the peak cell rate (i.e., $\alpha N/\rho_{burst}$). The probability that n sources are on is given by the binomial distribution:

$$p_{burst}(n) = \binom{N}{n} \alpha^n (1 - \alpha)^{N-n}$$

The loss probability is therefore

$$Q_{burst}(\alpha, \rho_{burst}, N) = \sum_{n=\alpha N/\rho}^{N} p_{burst}(n)\ \frac{\dfrac{\rho_{burst}C}{\alpha N} - C}{\rho_{burst}C}$$

$$= \sum_{n=\alpha N/\rho}^{N} \binom{N}{n} \alpha^n (1 - \alpha)^{N-n} \left(\frac{n}{\alpha N} - \frac{1}{\rho_{burst}} \right)$$

In Figure 9.12, we can see the effect of varying each of the parameters and, in particular, we can see that for a given probability of cell loss, the number of channels that can be carried falls as their peak-to-mean ratio rises. We can see that by dividing the total link capacity by the number of channels that can be carried without exceeding a given loss

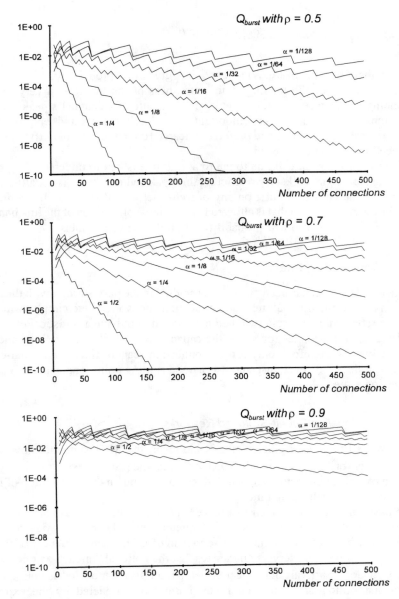

Figure 9.12   Cell loss probability with the nonbuffering method at the burst scale.

probability, each channel takes up an effective bandwidth on the link that is between the mean rate and the peak rate. This effective bandwidth is a very useful parameter and can be used by connection admission control in deciding whether to admit a bursty VBR connection onto a link. While neither the ITU-T nor the ATM Forum have formally defined effective bandwidth as yet, we can use the following working definition:

$$effective\ bandwidth\ =\ \frac{total\ link\ capacity}{N}$$

with N such that $Q_{burst}(\alpha, \rho_{burst}, N)$ is less than a given loss probability.

This sensitivity of the loading to the burstiness of the traffic is a very important understanding in the design of ATM networks. The ratio of a channel's peak cell rate to the maximum link cell rate is very important in loading a system and, empirically, this should be at least more than 10 and preferably nearer 100 if a reasonably efficient loading with low cell loss is to be achieved.

This ratio of the peak cell rate to the total link rate is key to understanding how the nonbuffering technique works. By restricting this peak cell rate, there is usually sufficient space between the cells of a burst on any one channel to still successfully perform cell scale buffering. Any channel that is allowed to take a large proportion of the link bandwidth will greatly increase the cell loss probability for all the active channels.

9.2.5   Connection Scale

This is the timescale that has been extensively studied in the past, starting with the seminal work of Erlang. At the time, all telecommunications networks were circuit-switched, so the analysis reflects this. Circuit-switched networks do not have a statistical cell scale or burst scale and, at the connection scale, the capacity of each channel is fixed and equal. Erlang's analysis is, therefore, only for the connection scale of fixed- and equal-capacity connections. It does, however, also serve as a good starting point for a general analysis of the connection scale.

9.2.5.1   Modeling Circuit-Switched Networks

This, following Erlang's original work, is based on the $M/G/\infty$ queue; however, the analysis is easier for the more specific $M/M/\infty$. In fact the results are the same; that is, the distribution of active connections on a link does not depend on the distribution of holding time, only on the mean holding time.

The basic model is illustrated in Figure 9.13. Each channel available to support a connection is analogous to a server in the queuing model, and we wish to find the probability distribution for the number of servers in use. Choosing a queue model with an infinite number of possible servers guarantees that the notional "queue" will never fill. In the $M/M/\infty$ queue, the arrivals of connection setup requests are modeled as Poisson events, while the distribution of connection holding times are modeled by the exponential distribution. The state transition matrix for the $M/M/\infty$ queue is

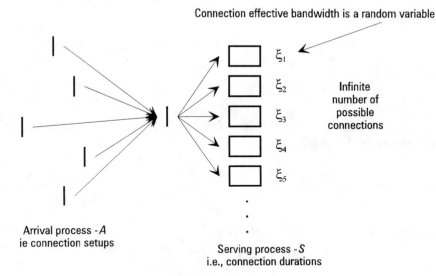

Connection effective bandwidth is a random variable

ξ_1

ξ_2

Infinite
number of
possible
connections

ξ_3

ξ_4

ξ_5

Arrival process - *A*
ie connection setups

Serving process - *S*
i.e., connection durations

Population description - A/S/∞

Figure 9.13   Modeling of the connection scale.

$$
\mathbf{M} = \begin{bmatrix}
1 - \lambda\delta t & \dfrac{\delta t}{\tau} & 0 & 0 & \cdot \\[2ex]
\lambda\delta t & (1 - \lambda\delta t)(1 - \dfrac{\delta t}{\tau}) & \dfrac{2\delta t}{\tau} & 0 & \cdot \\[2ex]
0 & \lambda\delta t & (1 - \lambda\delta t)(1 - \dfrac{2\delta t}{\tau}) & \dfrac{3\delta t}{\tau} & \cdot \\[2ex]
0 & 0 & \lambda\delta t & (1 - \lambda\delta t)(1 - \dfrac{3\delta t}{\tau}) & \cdot \\[2ex]
0 & 0 & 0 & \lambda\delta t & \cdot \\[2ex]
0 & 0 & 0 & 0 & \cdot \\[2ex]
\cdot & \cdot & \cdot & \cdot & \cdot
\end{bmatrix}
$$

where λ is λ_{conn} (i.e., the mean arrival rate of connection setup attempts), τ is τ_{conn} (i.e., the mean connection holding time), and δt is an arbitrary small time interval. In this case, the probability of a call terminating is proportional to the number of calls present. The steady-state solution to this state transition matrix is the Poisson distribution:

$$p_{conn}(n) = \frac{v^n}{n!} e^{-v}$$

with

$$v = \lambda_{conn} \tau_{conn}$$

therefore, if the link can support N channels, the probability of a new connection setup request being refused, $Q_{conn}(N)$, is

$$Q_{conn}(N) = \sum_{n=N}^{\infty} \frac{v^n}{n!} e^{-v}$$

We also know that for the Poisson distribution

$$\mu = v$$

$$\sigma^2 = v$$

The product v is a dimensionless parameter, usually called the Erlang, the normal measure of traffic used in narrowband networks. Since the Poisson distribution has the property that when two Poisson random variables are added, the sum also has a Poisson distribution, this is why the Erlangs from different connection traffic sources can simply be added.

This highly desirable feature depends on the original assumption of the M/G/∞ queue, and if this assumption is not valid, then we cannot expect to "add" connection traffic from different sources so easily. This may be the case when connection traffic of very different holding times are mixed together (for example, telephony connections and private circuit supporting links in client networks). In this case, it is the assumption of the steady-state distribution that is most suspect, as growth and other longer timescale factors are of the same time order as the holding times. The overlap in timescale between the connection scale, the human activity cycle scale, and the planning scale is a major complication in the design of broadband networks.

9.2.5.2   *Modeling the Connection Scale in More General Networks*

In this analysis, we can start with the analysis for the circuit-switched network as this gives a model for the number of connections present. We now need to consider the capacity of these connections, be they CBR, VBR, or ABR connections. In this analysis, we can treat ABR connections as CBR connections with the CBR rate set to the minimum bit rate. Any rate in excess of this is determined dynamically, after connection setup, by a closed loop control mechanism, which assures low cell loss. We assume that CBR connections are requested at a specified bit rate, which is also the CBR connection's effective bandwidth. Similarly, we assume VBR connections are specified with a peak bit rate and a mean bit rate from which an effective bandwidth can be calculated. From this, capacity on a link can be allocated using effective bandwidth.

We need the probability distribution of the connections' effective bandwidth, $s(\xi)$, and there is no obvious choice. This will depend on actual applications, of which there is only very early data. However, as we need to take the n-fold convolution of this distribution, any choice will tend, by the central limit theorem, towards a normal distribution following the convolution. The gamma distribution has many attractive properties in this case, particularly,

- It is a positive-only valued distribution.
- The mean and variance can be varied independently.
- The convolution of two like gamma-distributed random variables also has a gamma distribution.
- Several other positive-only valued distributions are special cases of the gamma distribution including the exponential and χ^2 distributions, and the Poisson distribution is also very closely related.

These properties make the gamma distribution very useful in many applications. Using the gamma distribution, we can write down the distribution for capacity required by n connections:

$$s(\xi)\mid_{n=0} = \delta(\xi)$$

$$s(\xi)\mid_{n=k} = \frac{b^{ak}\,\xi^{ak-1}\,e^{-b\xi}}{\Gamma(ak)}\,u(t)\quad 1\le k<\infty$$

where

$$a = \frac{\mu^2}{\sigma^2}$$

$$b = \frac{\mu}{\sigma^2}$$

Then,

$$s(\xi) = \sum_{k=0}^{\infty} p_{conn}(K)\,s(\xi)\mid_{n=k}$$

and the probability that the total effective bandwidth on the link is greater than the capacity of the link, C, is

$$Q_{conn}(C) = \int_C^\infty \sum_{k=0}^\infty p_{conn}(k) \, s(\xi) \, |_{n=k} \, d\xi$$

$$= \sum_{k=1}^\infty \frac{e^{-v} v^k \, \Gamma(ak, bC)}{k! \, \Gamma(ak)}$$

where $\Gamma(x, y)$ is the incomplete gamma function.

This can be expressed in terms of a loading factor ρ_{conn}, as for the burst and cell scales. We note again that v is the mean number of connections and define a parameter σ_μ, which is a normalized standard deviation of the effective bandwidth of the connections (i.e., $\sigma_\mu = \sigma/\mu$). In this case, we can set

$$a = \frac{1}{\sigma_\mu^2}$$

$$b = \frac{v}{\sigma_\mu^2 \, \rho_{conn} \, C}$$

Figure 9.14 shows the loss probabilities for different values of ρ_{conn}, v, and σ_μ. In the special case when "Symbol"$s_m = 1$, that is the effective bandwidths have an exponential distribution and v/ρ_{conn} is large, then $Q_{conn}(C)$ has an asymptotic approximation:

$$Q_{conn}(C) \approx \frac{\rho_{conn}^{3/4} \, e^{-v(1/\sqrt{\rho_{conn}}-1)^2}}{\sqrt{4\pi v}}$$

9.2.6   Human Activity Cycle Scale

As people, we are social beings, and we tend to do the same sort of thing at the same time every day. For example, we tend to work during the day on week days and enjoy ourselves in the evenings and at weekends. This tendency sets up strong correlations in the traffic flows in a network associated with these social habits. Many of these are well known from telephony, including morning, afternoon, and early evening busy hours, and a large increase in international traffic on Christmas day.

In statistical terms, these are predictable correlations in traffic patterns. We can select a suitable model for the predictable traffic flow, and by using statistical inference, fit it to historically observed traffic flows. The most common technique for this is least squares curve fitting.

In the method of least squares, we must start with an assumption that there is some underlying process that is generating the correlation in the traffic. From this, a general

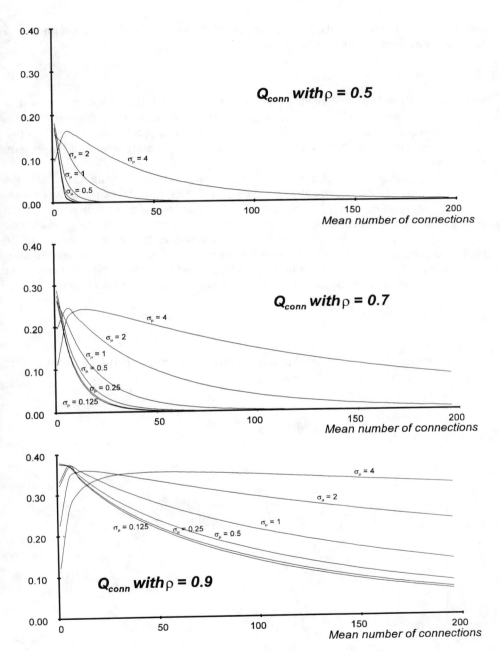

Figure 9.14   Probability of connection refusal through insufficient bandwidth.

function can be selected that describes the correlation mechanism and will be a function
with one or more parameters. For example, we could assume that the level of traffic, T, is

correlated with the time of day, *t*, and the day of the week, *d*, and so we assume that a function describing the correlation exists:

$$T(t, d) = f(t, d, a_1, a_2, a_3, a_4, a_5, ...)$$

where a_1, a_2, a_3 and so forth are the parameters of the function. It is important to start with a good understanding of what is causing the correlations and to use the least squares technique to quantify this understanding. This is because it is easy to find correlations that appear to be in observed data but that, in reality, are simply manifestations of longer term random effects.

Having started with such a function, we can now set about trying to evaluate the parameters based on the historically observed data. Figure 9.15 shows observed connections present on a link over a 24-hour period. Supposing we assume a simple correlation model where the connections present have a constant underlying mean during each hour of the day. We now have a function for our prediction of traffic as a function of 24 parameters and we solve for the 24 parameters using the standard process of least squares curve fitting. This finds the values of the parameters that gives the minimum variance of the observed points from the function; that is, we are looking for the minimum of:

$$\sum_{\text{all } i} (y_i - T(t_i, d_i))^2$$

where *i* is the count of all observed points. At this minimum, the partial derivatives with respect to each parameter will be zero; that is,

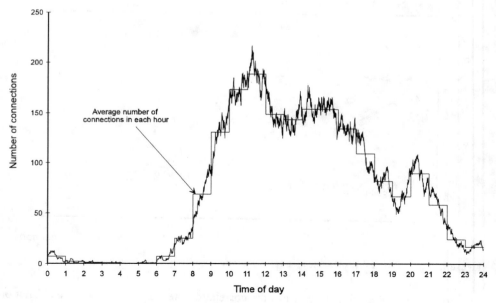

Figure 9.15   Example of daily traffic pattern.

$$\partial/\partial a_k \left(\sum_{\text{all } i} (y_i - T(t_i, d_i))^2 \right) = 0$$

for all a_k. For this particular example, the solution to this system of equations is trivial, however, the technique can be used with a wide variety of functions. In this case,

$$a_k = \frac{1}{n_k} \sum_{\text{all } i(k)} y_{i(k)}$$

where $i(k)$ are all the observed points that fall into the time interval associated with parameter a_k, and n_k is the number of such points. That is, a_k is simply the mean value of all the observed points falling into the time interval associated with a_k.

To use this, however, we must understand how good the correlation is between the assumed correlation function and the observed data. We can gauge this by looking at the variance around the correlating function; that is,

$$\sigma^2 = \frac{1}{n_k} \sum_{\text{all } i} (y_i - T(t_i, d_i))^2$$

where n_i is the total number of observed points. This variance is very important in the use of functions that describe correlation, as they allow confidence intervals to be defined. Several factors need to be taken into account, however, when using such a function and this variance:

The observed points falling within a single cycle and points falling at similar points in different cycles are treated equally in this analysis, and this may not reflect the true situation very accurately. For example, points on one day may be strongly correlated while there may be a larger variation between days; on the other hand, the correlation between points around 10 a.m. each day may be more strongly correlated than between those around 8 a.m. and those around 10 a.m. on the same day. If intercycle and intracycle correlations are different, then the calculated variance will depend on the number of points taken on each cycle. If this is the case, it is possible to carry a separate correlation exercise for each cycle and then to perform a correlation between the correlations. In this way, it is possible to separate intercycle variance from intracycle variance.

If the traffic present on a link is used as the basic measure, then this can only change at a rate determined by a time constant set by the holding time of the traffic. In the case of cells or bursts, this is very short, and often in the case of connections, it is still much shorter than changes as a result of the human activity cycles. However, many newer applications may have longer holding times, as noted at the beginning of the chapter, in which case, the integration effect of the holding time may mask the effect of the cycles. In this case, the arrival events and termination events should be observed separately.

9.2.7   Planning Scale

At this timescale, we are interested in the long-term growth or decline in demand for services. The planning must assess how much demand is likely to exist at the end of the normal planning lead time. If it takes one year to augment capacity in a network, then the planner needs to know what the nature of the demand is likely to be in one year's time. In the past, planners have plotted the total busy-hour Erlangs present in a network over time, assuming a general, slow, exponential growth over time. More sophisticated modeling would also take into account other parameters, such as predicted national economic growth, and modify the predicted growth in busy-hour Erlangs accordingly. However, an environment with many services and with competition makes such planning models increasingly unreliable and, as new services are introduced, account for less and less of the network.

When looking at the planning for new services, it is appropriate to look at the "S" curve used by marketers to describe the early growth and maturing phase of what they call the "product life cycle" (noting that, in marketers' terms, a telecommunication service is a product). A normal product life cycle is illustrated in Figure 9.16. At the launch of a new product, growth is slow but then may rapidly take off. There then follows a period of rapid growth, after which time the market saturates and the volume begins to level off. After a while, the product becomes outdated and the volume declines, often gradually.

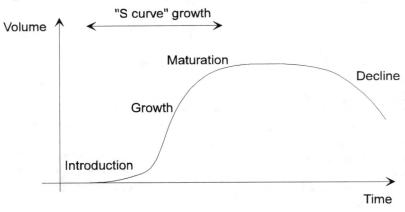

Figure 9.16   A classical product life cycle.

The difficulty raised by the "S" shape of the growth phase, is that the time at which the product "takes off" is uncertain. When the effect of this uncertainty in "takeoff" is added to the "S," as illustrated in Figure 9.17, the uncertainty in the forecast volume at the planning lead can be very large indeed. The expected range within which the forecast could lie may well range from zero volume to many times the mean, or quoted, forecast. And this would suggest that frequently in broadband networks, the uncertainty in any forecast will be more important than the mean. The planner must decide between the following options:

- Risk being unable to meet demand, possibly losing out to competitive offering;
- Risk having a heavily underutilized network, possibly for an extended period.

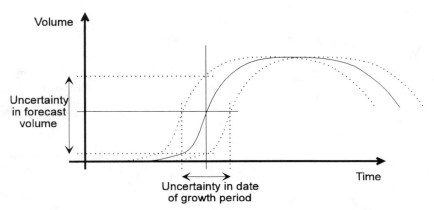

Figure 9.17   The effects of uncertainty in date of growth phase on volume forecasts.

- Work to shorten planning lead times to reduce uncertainty.

Clearly, the last of these is the most attractive; however, this is often difficult for planning departments built around the slow, exponential growth paradigm.

9.2.8   Assumption of Statistical Independence

In looking at models for traffic at different timescales, we are attempting to form a reasonably complete picture of traffic needed to meet all the requirements set out earlier in the chapter. One of the major issues with any of these models, is that, with very few exceptions, they assume that at the timescale longer than the one of concern there is a level in an underlying statistical steady state that allows the parameters of the probability distributions to remain fixed. For example, in the Erlang model, the Poisson distribution for the number of active connections assumes that the underlying rate of connection setup and the mean holding time are constant. We know this is not the case, as they can vary throughout the day.

The advantage of analyzing traffic over different timescales is that each timescale gives information on the validity of the assumptions used at a shorter timescale. To continue the example, the human cycle timescale gives an indication of the rate at which the rate of connection setups changes during the day. As we noted earlier, the integration time implicit in the $M/M/\infty$ queue is the mean connection holding time. We can therefore see that if the rate of connection setups changes faster than the mean connection holding time, then our assumption of an underlying steady state is no longer valid. We can generalize that when timescales start to overlap, the assumption of statistical independence in the shorter timescale becomes quite invalid. Equally, averaging at the shorter scale assumed by the longer timescale becomes invalid.

Trying to find traffic models that cover several timescales at once normally results in highly complex models that may not add greatly to our total understanding. However, a great deal can be learned by simply understanding the nature of interaction between timescales. Some of the more important interactions are summarized in Table 9.2.

Table 9.2
Interactions Between Timescales

Timescales	Interactions	Effect
Cell-burst	Very intense bursts saturating the link	The calculation of buffer size based on cell scale is no longer valid and must be based on burst scale.
Burst-connection	Single burst connections	As connection setup times reduce, it is possible, and economic under the tariff regime, for the user to setup a connection for each burst of traffic, thus greatly increasing the number of connection setups. This is already observed in single burst connections on the N-ISDN.
Connection-human cycle	Long holding time connections	These generally have the effect of flattening out the human cycles; however, they can have harmful effects if interleaving of demand is attempted based on complementary human cycles.
Connection-planning	Very long holding time connections	New connections in the network will not grow fast enough to take advantage of a new topology of the network.
Connection-planning	Very short planning lead times	New connections in the network will not grow fast enough to take advantage of a new topology of the network.
Human cycle-planning	Short planning lead times	Planning mechanisms can be used to follow the human cycles.

9.3   PARAMETERS FOR MEASURING BROADBAND TRAFFIC

In this section, we look at the formal parameters that can be used to measure traffic and traffic performance. Many of these stem from the traffic models described in the previous section, and, certainly, the parameters would be much less useful if they could not be predicted by the traffic models.

9.3.1   Generic Cell Rate Algorithm

The generic cell rate algorithm (GCRA) is an internationally agreed upon algorithm, defined by both the ITU-T and the ATM Forum, which measures cell rate at a specified timescale. It may be used for monitoring CBR services or any service where the cell rate is fixed over a given period of time (e.g., ABR service), or any service where a maximum cell rate is specified, including many VBR services.

The algorithm starts with the assumption that the cells will have a minimum time gap between them, set by the current peak cell rate at the time. The algorithm is called the "leaky bucket algorithm," a name that gives considerable insight into the way it works. It is illustrated in Figure 9.18.

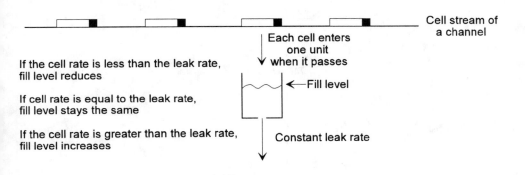

Figure 9.18   Leaky bucket algorithm to measure cell rate.

The bucket is filled with a cell's worth of water every time a cell arrives, while at the same time the bucket regularly leaks a cell's worth of water at the deterministic rate set by the peak cell rate. If the actual cell rate is below the peak cell rate, the buffer will never start filling beyond one cell's worth of water; however, if it starts rising above this rate, then the bucket will start to fill. The height to the top of the bucket is the second parameter. If this height is set high, then the bucket will absorb bursts and CDV without overflowing. However, if the bucket height is set much lower, then smaller bursts and even excessive CDV will cause the bucket to spill. Any cells that cause the bucket to spill can be marked for cell loss priority, as they have caused a violation in the GCRA.

The actual algorithm is defined by either of the flow charts in Figure 9.19 and is normally written as GCRA(T, τ), where T is 1/PCR and τ is the maximum acceptable excursion from the theoretical arrival time (i.e., the height of the bucket). The value τ can be much smaller than T if CDV is to be measured on a CBR service, or many times T if a mean cell rate is being measured.

9.3.2   Peak Cell Rate, Sustained Cell Rate, and Effective Bandwidth

At the time of writing, VBR services have been partially addressed by the ITU-T and the ATM Forum. This is partly because the early anticipated services for ATM are CBR and ABR. Currently, VBR services will be defined in terms of a peak cell rate (PCR) and a sustained cell rate (SCR), both of which can be monitored using the GCRA. The SCR is like a mean; however, it is measured over a defined period of time and so is a "moving mean." This peak to sustained ratio was central to the modeling of the burst scale nonbuffering technique. We could see that the bandwidth required by a channel—by looking at the number of channels with a given PCR and SCR that can be accommodated

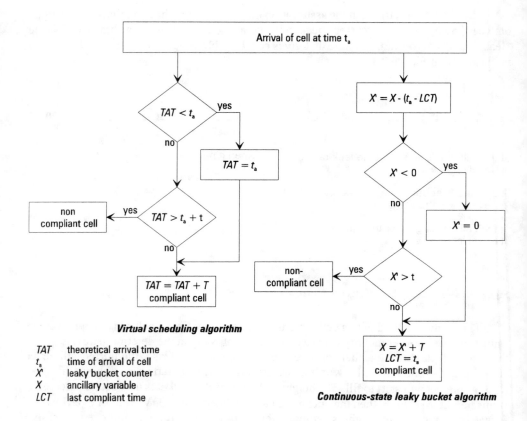

Figure 9.19   Formal definition of the generic cell rate algorithm.

on a link—was between the PCR and the SCR and also depends on the total link rate. The cell rate calculated in such a way is called the effective bandwidth of the VBR channel. This effective bandwidth can then be used by the connection admission control in deciding whether to admit a new VBR connection.

9.3.3   Cell Loss Ratio

This was the main parameter that the cell and burst scale models were attempting to predict. It is defined by both the ITU-T for ATM in Recommendation I.356 and the ATM Forum in the Traffic Management Specification. It is defined as follows:

"Cell loss ratio (CLR) is ratio of total lost cells to total transmitted cells in a population of interest. Lost cells and transmitted cells in cell blocks counted as severely errored blocks should be excluded from the population."

The exclusion of lost cells from severely errored blocks related to the transport performance is discussed in the next chapter. Here, it is emphasized that measured degradation of a parameter may have more than one cause. Even with the definition as above, the loss of a cell, while most likely to be caused by buffer overflow, could be caused by cell misrouting as a result of corruption to the header.

9.3.4   Grade of Service

This parameter has been in use for many years and is the connection scale equivalent of cell loss ratio. The grade of service is defined as the number of successful connection setup attempts as a ratio of the total number of connection setup attempts.

The ratio is normally quoted as a percentage. The grade of service is selected as a parameter when the Poisson distribution associated with Erlang's model is used by planners in deciding how much capacity to allocate to a given number of forecast Erlangs of traffic; that is,

$$Grade\ of\ service\ =\ 1 - Q_{conn}(N)$$

It is also directly measured and used as a very important quality measure for any network. We can extend the grade of service parameter to include CBR, VBR, and ABR connections by using the same definition. When planners wish to predict the grade of service using the effective bandwidth for such a connection, we can use

$$Grade\ of\ service\ =\ 1 - Q_{conn}(C)$$

as defined above.

9.3.5   Cell Transfer Delay

Buffering in ATM switches and concentrators doesn't merely cause the occasional loss of cells, it also delays the cells while they queue in the buffer. For many applications, this delay is not important; however, for some real-time applications, it is significant, as discussed in chapter 11 on timing and synchronization.

Cell transfer delay can be divided into two parameters, one relating to the absolute time taken by cells to cross a network from source to sink, while the second relates to the statistical variation in delay as a result of buffers filling and emptying over time.

9.3.5.1   Absolute Cell Transfer Delay

Again, the ITU-T and the ATM Forum are working on a set of parameters for cell transfer delay (CTD). Both define a mean absolute cell transfer delay as:

"Mean cell transfer delay is the arithmetic average of a specified number of cell transfer delays."

It is aimed at passive measurement during a connection. The ATM Forum, on the other hand, is also working on a maximum cell transfer delay (maxCTD) parameter that is aimed at use in negotiation during connection setup. This is defined currently as follows:

"The maximum cell transfer delay (maxCTD) specified for a connection is the (1-a) quantile of CTD. There may be a relationship between (a) and CLR. The relationship is for further study. a shall be smaller or equal to the negotiated CLR for the connection"

This relationship between the distribution of cell transfer delay and cell loss reflects the fact that both depend on the probability distribution of buffer fill.

9.3.5.2   *Cell Delay Variation*

The ITU-T and the ATM Forum have defined a set of parameters for cell delay variation (CDV) based on the requirement to measure CDV for different purposes and to specify CDV during the connection setup negotiation.

Two separate measurement scenarios are identified, one that measures the CDV with respect to a defined reference, and one that measures the relative CDV between two measurement points. These are one-point CDV and two-point CDV, respectively:

- The one-point CDV can be used for measuring the CDV of a CBR service where the cells should arrive at defined, equal time intervals, and so a time reference can be calculated for the negotiated CBR bit rate.
- The two-point CDV uses the cell arrival times at one point to give the reference times for the arrival of the cells at the second measurement point. This is illustrated in Figure 9.20.

A simple algorithm is now used to measure the difference in time of each cell for its reference time. This works by accumulating, over time, maxCTD and measuring how much each cell arrives in advance of its reference time. This is illustrated in Figure 9.20. If a maximum CDV has been specified, this algorithm will identify cells that are beyond this specified CDV.

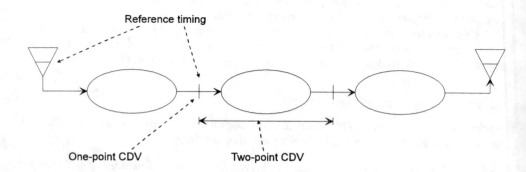

Figure 9.20   One-point and two-point cell delay variation.

9.3.5.3 Accumulation of Delay

In addition to CTD and CDV, the ATM Forum has also defined a set of parameters that define the way delay accumulates down a connection. These are, as defined by the ATM Forum:

- *Fixed Delay, F*—The fixed delay includes propagation through the physical media, delays induced by the transmission system, and fixed components switch-processing delay.
- *Upper bound, U*—The upper bound on the CTD.
- *Mean, M* — The mean represents the mean queuing delay.
- *Variance, V*—The variance of the queuing delays, which is the same as the variance of the CTD (adding in the fixed delay does not change the variance).
- *Discrepancy, D*—The discrepancy is the adjustment factor that is applied to a Gaussian approximation (based on the mean and variance) to better approximate the real delay distribution.

All of these parameters, with the exception of the discrepancy, can be added linearly for each component making up the overall connection, and so can be calculated for a connection. The discrepancy can be accumulated by keeping track of the maximum discrepancy. They can therefore be used in deciding whether to accept a connection setup request that specifies either maxCTD or CDV, or both.

The actual queuing delay distribution (i.e., the buffer fill distribution) depends on the design of the switch and the policy on admitting connections onto links. The queue models described earlier for cell scale buffering (or buffering at the burst scale if that is the policy in use) gives an indication of these distributions, which are clearly not Gaussian. Along with the mean and the variance, the discrepancy parameter is used to define the actual distribution by defining its difference from a Gaussian distribution with the same mean and variance. The actual distribution may well be arrived at empirically by direct measurement on the particular switch.

9.4 CONTROL OF TRAFFIC

Any network will deploy a number of mechanisms to control traffic. Many of these are statistical, as the resources must be provided before demands are made for them. We can describe these as "open loop" control mechanisms, as the timescales of these mechanisms do not allow feedback from congestion in the network to the sources of traffic. When it is possible to apply feedback we have a "closed loop" system. This is illustrated in Figure 9.21. The ABR service uses a closed loop feedback mechanism to vary the bit rate during a connection, and this is carefully designed to maintain overall stability in the network.

9.4.1 Statistical Control (Open Loop Control)

There are many forms of statistical control that can be used. Two of the most important, and most frequently discussed for broadband networks, are cell loss priority (CLP) and connection admission control (CAC).

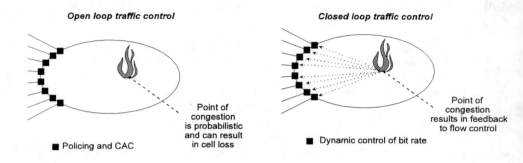

Figure 9.21   Open loop and closed loop control of traffic.

9.4.1.1   *Cell Loss Priority*

One of the bits on the ATM cell header is designated for marking cell loss priority (CLP). Most cells will have this set to 0; however, if it is set to 1, any buffer that is likely to overflow can first discard cells with the CLP bit set to 1 before dropping any cells with the CLP bit set to 0. This is a mechanism that operates at the burst scale.

When a connection is negotiated, GCRA parameters are set, which can then be policed by the GCRA. When the GCRA indicates a nonconforming cell, it can mark it by setting the CLP to 1. In this way, the cell need not be discarded straight away; however, if it encounters any congestion, it will be discarded.

9.4.1.2   *Connection Admission Control*

This is the principle control mechanism employed in a circuit-switched network. Connection admission control (CAC) is the mechanism by which each part of the network, and ultimately each link, has the opportunity to refuse a connection setup request. Its operation is illustrated in Figure 9.22. In a circuit-switched network, the general CAC mechanism was very simple—if a channel is available, the connection is admitted onto the link. If the connection can be admitted onto every link required to complete the connection, then the overall connection is admitted by the network. This is illustrated in Figure 9.22. This mechanism is nondiscriminatory in that all connection requests are treated equally. The chance of success for two connection requests made at the same time on the same route are equal. In broadband networks, there are two factors which distort this:

- The probability of admission will depend on the effective bandwidth requested.
- Priority may be given to some special connection requests, for example, restoration of connections that are being reestablished following a failure in the network.

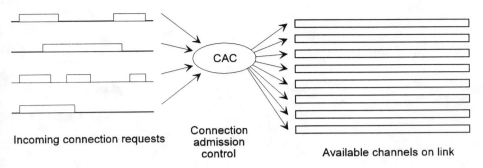

Figure 9.22   Connection admission control.

There is opportunity to deploy many different strategies in connection admission control. There is still much debate on the mechanisms to implement it, and since most are regarded as proprietary to an operator or manufacturer, there is little discussion in the open. One more general method that has been discussed in literature is the method of "implied cost" described by Kelly, where each connection setup request can be given a notional value and the resources it requires given a notional cost. And the notional cost of a link is set to be a function of the current utilization of the link. The cost is, in essence, formed by the probability that admitting the connection will cause congestion in the future.

9.4.2   Explicit Control (Closed Loop Control)

When the rate of change in traffic coming from sources of traffic is slower than the time needed to measure congestion and signal to the sources, stable closed loop control mechanisms are possible. A rather trivial example of this is when the lead time for a private circuit is longer than the planning lead time in the network. In this case, the capacity in the network can be explicitly regulated to meet demands.

9.4.2.1   *Dynamic Control by Indirect Rate Adaptation*

The Internet uses a mechanism that works as part of the transmission control protocol (TCP). TCP indirectly regulates the rate at which packets are transferred in the network via the windowing mechanism. This forms a closed loop feedback mechanism from points of congestion in the network to the sources of traffic, which motivates the sources to reduce their effective transmission rate in order to alleviate the congestion. The TCP source sets the packet size entering the network and transmits the window size towards the TCP sink. Each transmitted packet is numbered in sequence so the sink can check if a packet is missing from the sequence. The sink explicitly acknowledges all received packets towards the source. Lack of acknowledgment implies packet loss, and the source responds by reducing the packet size close to zero, from where it is increased slowly so long as receipt acknowledgments are received.

This control mechanism has evolved over the years and now allows the Internet to run at very high utilization levels while maintaining a reasonable level of stability. The basic operation of this protocol is illustrated in Figure 9.23. To avoid the capacity

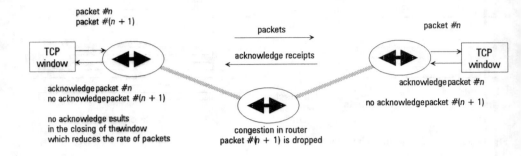

Figure 9.23   Dynamic rate control by TCP.

multiplication effect noted at the beginning of the chapter, the mechanism responds to dropped packets as the indicator of congestion. While the response is after the congestion has occurred, no new traffic is generated as a result of the congestion in the attempt to control it.

9.4.2.2   *Dynamic Control by Direct Rate Adaptation*

The ABR service defined by the ATM Forum uses a closed loop control mechanism to directly control the bit rate of the connection by means of the automatically operating resource management (RM) protocol. This is modeled on the TCP rate adaptation mechanism; however, since the ATM network allows a connection setup phase, this protocol is more complex, but may be expected to provide for smoother control.

The basic operation of the ATM RM protocol is illustrated in Figure 9.24. Information is passed in a loop from the traffic source to its sink, through the intervening switches, and then passed back to the source in the return channel, also through the intervening switches.

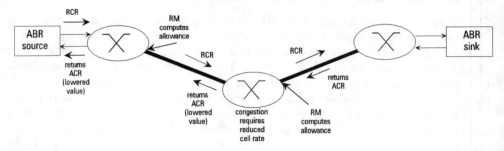

Figure 9.24   Dynamic rate control by ABR.

The protocol works in two phases. First, there is the connection setup phase, and second there is the transfer phase. At each phase, explicit negotiation for capacity takes place.

At connection setup, the user can specify two traffic parameters in the signaling protocol, which can then be used both during the connection setup phase and during the transfer phase. These are the minimum cell rate (MCR) and the peak cell rate (PCR).

The protocol will then negotiate an initial cell rate (ICR) from the MCR and PCR. The initial cell rate must be agreed to by the CACs for all the links on the route, and so will be effectively determined by the CAC that has least available capacity. If a CAC cannot allocate the requested MCR, it can offer an alternative lower value or the connection setup request can be refused. There are other parameters, which are set at connection setup, that control the operation of the transfer phase.

During the transfer phase, a real-time rate adaptation and flow control protocol works "in band" in that it is carried in cells belonging to the connection. The source is allowed to request an increase in cell rate. The request is assessed by the resource manager at each link in the route. Limitations at one link are sent to the next link. The responses are looped at the end of the connection and returned to the source, which may then be offered a new higher rate. This protocol can adjust the allowed cell rate (ACR), which is the maximum rate at which the source can send in response to growing congestion in the network, and, in principle, can respond before any cell loss is actually produced.

References

The following standards are the main source documents for the material in this chapter:

ATM Forum, *Traffic Management Specification Version 4.0*, 1996.
ITU-T Recommendation I.371, *Traffic Control and Congestion Control in B-ISDN*.

The following gives analysis of traffic aspects of ATM networks:

Roberts, J., Mocci, U., and Virtamo, J., (Eds), *Broadband Network Teletraffic, Final Report of COST 242*, Springer, 1996.

Network Resilience 10

Modern industrialized societies are becoming increasingly dependent on telecommunications. Major service outages for whatever cause, whether it be human error or component failure, can cause expensive disruption to large sections of the community. Maintaining service availability even under failure conditions has therefore become a prime objective. Survivability of the transport network in the presence of cable breaks and equipment failure is now a routine requirement, while graceful recovery from major disaster is a central operational and planning preoccupation for most network operators. Synchronous digital hierarchy (SDH) provides a number of standardized mechanisms to provide protection against most low-level failures while the managed flexibility inherent in both asynchronous transfer mode (ATM) and SDH provides the basic features to support effective wide area service restoration.

This chapter introduces the concept of availability, explores the factors that determine achievable availability, and reviews techniques and procedures that may be used to ensure complex layer networks can be operated with the expected availability.

10.1 SERVICE AVAILABILITY IN MULTILAYER NETWORKS

We can recognize two distinct responses to network failure. The first is service-oriented and is directed towards restoring interrupted services on failure by attempting to reestablish the failed service using spare resources, or resources released by dropping lower priority traffic. This has led to the development of processes that respond to loss of service by essentially renegotiating with the network to locate suitable alternative resources with which to restore an acceptable level of service. The term service "restoration" is often used to describe this class of activities.

The other is resource-oriented and is directed towards improving the basic availability of the network components so that failure does not result in loss of service. Naturally, design and process improvement plays a part in improving basic reliability, but the ultrahigh levels of component availability currently demanded has led to the provision of low-level autonomous mechanisms that exploit design redundancy. This ensures that failures are detected and repaired before a loss of service can be declared. These have been termed "protection" mechanisms.

We don't intend to make too much of the terminological distinction between restoration and protection because, as we shall see, it is not always easy to make such a distinction in practical systems. Nevertheless, it is interesting to note the difference between the top-down and bottom-up approaches to resilience that have developed in each of the network layers. The selection of appropriate strategies in each layer of a complex multilayer network is a major network design issue.

10.1.1   Service Impact of Network Failure

The service impact of the failure of a network component naturally depends on the scope of the failure but also on the tolerance of the supported services. Voice telephony, for example, can still operate satisfactorily in the presence of quite large transport failure because telephony networks are usually planned to provide physically diverse transport alternatives between public switched telephone network (PSTN) nodes. Existing calls using the failed facility are dropped and the correspondents must themselves reestablish communication on the remaining facilities. Providing this does not happen too often, the resulting inconvenience is usually considered minor and acceptable. If there is inadequate provision in terms of spare capacity such that congestion results from the call reattempts, the service impact may quickly become unacceptable. This is a very basic form of service restoration where the subscribers themselves are the agents initiating the activity, applying their own individual priority decisions about whether to redial or not and how long to persist, in the event of repeated failures, in attempting to reestablish connection. Alternate routing algorithms in the circuit layer may also be used to provide alternative routings to replace those lost with the failure. If the alternative resource is scarce, alternative routes are provided on a first come, first served basis. This is a powerful model that is also the basis of a class of path restoration systems finding application as an autonomous process in the transport network.

Data communication is more diverse in the rates and protocols used and also in the user services supported. Some low-rate transaction systems such as those used for electronic funds transfer at point of sale (EFTPOS) are in a similar class to voice telephony in levels of inconvenience and scope, but interruption of remote security alarm systems due to communications failure, even if the scope is not large, could be much more serious.

Connectionless data networks such as those that have developed to support the Internet employ automatic and adaptive procedures that make them resilient in a similar way to the telephone network, even when using very unreliable network resources, provided they have access to alternative transport routing possibilities that are unlikely to fail simultaneously.

Users of dedicated leased lines are generally the most vulnerable to failure. It is common in today's network for such leased line users to also lease spare capacity so they can implement their own recovery procedures. Guaranteeing diversity to such users is not a trivial problem for operators who for other reasons need to decouple physical network operations from services. This is made even more difficult in a multioperator situation where a user service may in fact be subcontracted in parts to several independent operators who have neither the capability nor the motivation to safeguard mutual diversity. Total loss of such service for rather long and uncertain periods can be catastrophic to many business users, whose businesses may have become dependent on them.

Where protection or restoration mechanisms are applied, the response time can be critical to the perceived severity of the impact from a service perspective. Again, telephony is not seriously impacted by interruptions short enough that no calls are lost. If, however, the interruption has wide scope and is long enough to drop a large number of calls in progress, the subsequent call reestablishment activity in the PSTN can cause transient signaling overload, with an impact comparable to that with no protection at all. Longer response times are, in general, tolerable for less frequent occurrences or for failures of smaller scope if this can be traded against other advantages such as cost efficiency.

Quality expectations are undeniably higher today, and the trend is towards high availability across a broad range of services. In parallel, the evolving market environment stimulated by deregulation and competition is demonstrating a considerable interest in a broader range of service availabilities. So, we can expect that ultrahigh availability will be obtainable at a price, but there will also be a place for "bargain price" services with "reasonable" but lower availability. Exactly where the balance will be struck in terms of absolute availability targets and range of tariff-related tradeoffs in differentiated services is, as usual, a complex interaction of technology, economics, and market dynamics, and only time will tell.

10.1.2 Availability in Layer Networks

As described in Chapter 8, availability A of a resource over a period of time is defined as the proportion of the time for which the resource is available for service. Measurement or computation over a very long period or a statistically significant number of instances, assuming statistically independent failure events, gives the asymptotic availability, which is used as a general measure for comparison. The unavailability **U** is the complement of A, thus $U = 1 - A$. If the mean time between failure ($= 1/\lambda$) and the mean time to repair ($= 1/\mu$) are both assumed constant with time and $\lambda << \mu$, then U is approximately given by the ratio λ/μ.

A transport layer path, or indeed a subnetwork connection, has an unavailability U, which is the sum of the unavailabilities of its component link connections U_l and subnetwork connections U_{sn}. The value U_l depends directly on the unavailability of the serving trail (HO path or multiplex section in SDH) and U_{sn} is the sum of the U_ls and U_{sn}s at the next level of the recursive decomposition in the same layer. Ultimately, U_{sn} depends directly on the transport equipment, which implements the basic matrix:

$$U = \Sigma U_l + \Sigma U_{sn}$$

Applying this expression recursively within and between layers the availability of a transport path can be calculated, given a knowledge of the availability of individual physical network components.

10.1.2.1   Self-Healing Systems

A self-healing system is one in which redundant resources are provided that can be substituted for failed resources on detection of failure. Provided that the failure rate associated with the detection and substitution mechanism, λ_s, is much better than that of the resource itself and the failure rate of the resource, λ, is low enough so that $\lambda^2 \ll \lambda_s$, then a significant improvement in availability can be achieved. This is illustrated in Figure 10.1. The principle is used in equipment protection where the resources are replaceable units within a piece of equipment and also in SDH network protection where the resources are the transport entities: connections, paths, or sections.

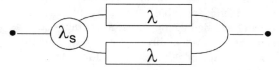

Figure 10.1   Simple protection model.

However, as described in Chapter 8, the simplified assumption of independent failure events is almost never true in practice. Failure events, like many other natural occurrences, are fractal in nature and tend to be strongly clustered in both time and place. A single cable break is likely to affect all the fibers in the cable, even though individual systems only see a single failure. It has also been observed that the mere presence of a human operator in an equipment station increases the incidence of multiple failures quite considerably.

Despite undoubted improvements in overall quality, high levels of functional integration together with the ever-increasing capacity of optical-fiber systems conspire to increase the degree of clustering. Put another way, the incidence of failure may be much improved but the consequence of such failures as do occur is correspondingly greater.

10.1.3   Protection of Network Components

The availability of individual network elements can be improved by applying local protection within the equipment. The detailed mechanisms need not be subject to standardization because the externally observable function of the equipment is unaffected by the protection mechanism. Cable breaks and all the other possible interruptions to the transmission media may be healed by network-level procedures applied to connections (link connections or subnetwork connections), selectively or in groups. The required cooperation between the network elements terminating protected connections justifies standardization of these procedures in a multivendor environment.

10.1.3.1   Equipment Protection

The study of reliability in the context of quality control is a mature discipline. In general, equipment is specified to meet certain quality targets that are measured in terms of mean time between failures. Computation of the reliability of a network element from the combined reliabilities of its many components, taking account of design-dependent interaction between components, is a well-established part of the design process. The failure rate of individual components is established by measurement and statistical sampling techniques. As a technology or component matures, its associated failure rate will normally stabilize at some low, irreducible value as design and process imperfections are corrected. This can be taken as the basis for equipment reliability computation. The mean time to repair a network element or replaceable module will be a characteristic of the operations and maintenance process. The reliability and repair time response metrics taken together define the availability of the component.

Good modular design can reduce the impact of failure, but even where reliability and mean time to repair are as good as can be reasonably expected, it is often not possible to meet the very high availability targets set for network elements without taking further steps. This implies replication of modules at various levels, continuously monitoring status in operation, and then switching over a faulty module for a working standby when necessary. Duplication of power supplies and common services such as synchronization or control units are typical features employed to enhance the resilience of network equipment. Where a design uses multiple modules of the same type, these can sometimes be configured in 1 for *n* protection arrangements to provide similar cover more efficiently.

10.1.3.2   Software Failures

Failure is not restricted to hardware components. Measurement and control of software quality, however, is a somewhat younger science. The software technologies still being currently applied are not as mature and reliable as their corresponding hardware analogs. Software can misoperate for a large variety of reasons. The ubiquitous software bug is a familiar aspect of latter-day technological demonology. The integration and validation phases of the design cycle should discover and eliminate most of these, but the "bug level" can seldom be assumed to be zero at first deployment. A large part of the design effort in software systems will typically be dedicated to design quality aspects; ensuring recovery from loops, lockups, and memory leaks, and otherwise reducing the bug level. In larger, more elaborate systems, microsynchronization of processes, rollback of transactions and subsequent reallocation of processes to alternative resources are among the techniques found to ensure resilience of software systems.

The advantage often claimed for software-based implementation over hardware implementation, namely flexibility and "ease of modification," is also its major weakness in the sense that quality, which is already difficult to achieve initially, is even more difficult to maintain in a context where a very large number of valid configuration options and periodic in-service field upgrades is the norm. One of the main claims for the next generation of software systems based on object-oriented design principles is a substantial improvement in attainable quality levels.

10.1.3.3   *Maintainability*

While redundancy is primarily justified in terms of improving system availability, there is usually an important secondary benefit in terms of in-service maintainability. Apart from autonomous recovery from failure, the existence of redundant resources allows hardware or software to be changed or upgraded in-service. It is this alternative application of redundancy that underlies the requirement for "hitless" changeover. While it is normally considered acceptable to take a short error burst while a detected failure is being filtered and validated, the injection of errors as a byproduct of routine maintenance intervention is not accepted so easily.

10.1.4   Network Protection

In the same way as a failed hardware component can be replaced in-service by a working unit installed as a spare in the network element for that purpose, so also a failed transport entity (a failed connection) may be replaced by a spare transport entity (a spare connection) reserved in the network for that purpose. The fact that the components of the protection process are distributed in the network between different network elements implies a degree of standardization if the processes are to be workable in a multioperator, multivendor environment. The main issues to consider are:

- Location and identification of components used by the protection process;
- Criteria for determining component failures;
- Behaviors of the protection process agents and components;
- Messages between process agents and their (message) semantics;
- Communication channels to carry messages between process agents;
- Message syntax/format.

A network protection process, then, preserves the availability of a protected connection by replacing previously allocated but failed resources by working spare resources configured for this purpose. The working and spare resources are interconnected by protection switches whose behavior is controlled by a corresponding set of communicating process agents located in the terminating elements, which, when triggered, cooperate to effect the replacement. We refer to the set of working and spare connections, the switches that change them over, and the agents that cooperate to effect the changeover as the components of the protection process. The protected connection is the abstraction seen by the client and is considered to remain unchanged after protection switching, even though its components may have been exchanged.

10.1.4.1   *Protected Components and Failure Detection Criteria*

There are protection processes directed towards the protection of link connections and others towards the protection of subnetwork connections. To protect either type of connection, it is necessary to identify the criteria from which its failure or unacceptable degradation may be inferred. In the absence of an integrated termination process, a connection failure may be ascertained from the individual failure states of its elementary components. The failure of an elementary link connection (the simple link connection of G.805) may be very reliably inferred from the state of its serving trail as revealed by the trail termination. In fact, the failure of the trail may be taken to reliably infer failure of all its client link connections. Protection of link connection sets associated with a single trail in this way is the basis for a large class of protection systems, usually identified by reference to the serving trail whose failure or degradation is used as the criterion to initiate protection switching. Since the trail is monitored between its source and sink, the trail status is discernible at the trail termination sink and no coordination is needed across the network with the source. The ability to quickly detect failure in this way by observing conditions at a single element is very useful, and enables implementation of simple fast-acting systems. Figure 10.2 gives a generic illustration of the components and their disposition. The general m:n case is seldom found today, but 1:n, 1:1, and 1 + 1 systems are common.

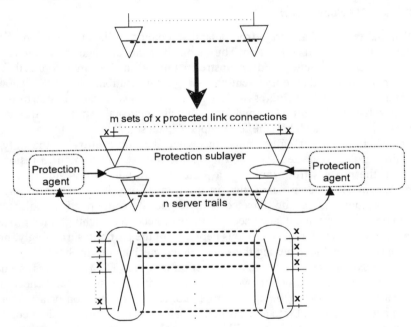

Figure 10.2   Protection of a set of elementary link connections associated with a single trail.

All the SDH multiplex section (MS) protection variants from the standard automatic protection system (APS) of G.783 through to the MS shared protection ring systems (discussed below) and systems that protect plesiochronous digital hierarchy (PDH) link connections by means of SDH path components are all in this basic class. It is possible to classify the link connections in the protected set into protected and unprotected subsets and only apply the protection switch selectively to the protected subset, although the only standardized option acts on the whole set. Protecting the whole set is usually the most attractive from an implementation viewpoint, as this may be done in a single cost-efficient switching action at the rate of the server trail client payload.

Tandem connections, in general, whether they be subnetwork connections or tandem link connections, have no such close relationship to a single monitorable entity that might form the basis for similar, simple switch criteria. The status of the tandem connection may be determined by independently reporting the status of all the component connections and computing from this the status of the tandem connection. There are several examples of centralized OS-driven systems based on this approach. The tandem connection status may also be determined by examining the signal at ingress using a nonintrusive monitor and comparing it with that at egress. If the signal is good at ingress and bad at egress, then the tandem connection itself is inferred to be bad and may be replaced. However, this still needs correlation between the status indications from the two remotely located NEs in which the connection is terminated.

10.1.4.2   AIS Substitution

The AIS signal measured at egress from a tandem connection is an indication of failure somewhere in the upstream connection. Thus, alone it cannot distinguish between a failure within the tandem connection and an upstream failure on the same connection. This situation can be improved by implementing a simple substitution. Thus AIS, denoting an upstream failure, is detected at ingress and substituted for onward transmission by a special dummy signal having the correct normal format but distinguishable from the normal signal in some way with a distinct label (the "unequipped" path signal in SDH, for example). The tandem connection sink at egress must recognize the substitution as the only valid interpretation in the connected state and make the reverse substitution to restore AIS, refraining in the process from initiating a protection switch. In this situation, an AIS generated within the tandem connection in the normal way by a failure in one of its elementary components can be detected, unambiguously interpreted as failure of the tandem connection component, and used to initiate protection switching. Such procedures have been implemented in plesiochronous networks in the past but, surprisingly, no such procedure has been included among the standardized mechanisms in SDH.

The recently standardized tandem connection monitor (TCM) achieves the desired effect by introducing a monitorable overhead at the connection source and checking it at the sink. This can, of course, be applied to a subnetwork connection or a tandem link connection and incorporated into a protection system. The monitored tandem connection is in effect equivalent to a trail in a TCM sublayer; the protected entity is then the link connection client of this new trail and behaves similarly within the protection architecture to the generic protected elementary link connection as described above.

In the special case of 1 + 1 subnetwork protection (described below) where the same signal is transmitted on both the working and the standby connection, upstream failures generating AIS outside the protected subnetwork can be distinguished by the fact that the resulting AIS will be detected on both working and standby SNCs at egress, assuming a validation delay sufficient to allow for differential transport delays on the two routes. In this case, detection of a single AIS on the working or the protection SNC alone can be reliably interpreted and the appropriate protection activity initiated. This is the protection switch trigger mechanism that is defined in the standardized subnetwork protection (SNC-P) architectures used in both HO and LO path layers.

10.1.5   Linear Protection Architectures

The term "linear protection" is applied in point-to-point topologies such as line transmission systems with protection agents at each end communicating over the link between them.

10.1.5.1   1 + 1 Linear Protection

Linear protection architectures provide protection for elementary links (groups of link connections) by means of replacing the supporting trail when it has failed. The simplest architecture in this class is the so-called 1 + 1 automatic protection switch (APS) architecture illustrated in Figure 10.3.

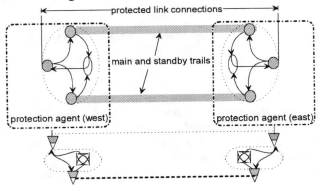

Figure 10.3   1+1 automatic protection switching.

In the 1 + 1 architecture, the protected link connections are bridged onto both trail termination sources while the counterdirectional connection is switched from the sinks under the control of the protection agent. The protection switching is achieved in a unidirectional manner, although the failure condition that triggers the switching in the case of a unidirectional failure will also, in general, send a remote defect indication (RDI) to the far end serving trail termination in the appropriate overhead channel and initiate a complementary switching response at the far end. The protection response requires no further communication between the protection agents themselves.

Figure 10.4 illustrates the behavior of the protection agent in the 1 + 1 architecture. In the initial state, both working and protection trails are enabled and the signal is bridged to both (that is to say, it is transmitted on both). In the other direction, the signal received from the working trail is switched to the client connection, while the spare or protection trail is not connected.

The simple state machine illustrates the behavior when either the working or protection channel fails, or both. Simple revertive behavior is illustrated in Figure 10.4. That is to say, when the initially working trail recovers it is returned to service. This has the side effect of causing an errored second while switching back. For this reason, nonrevertive behavior is often specified. In this case, the process can be left in the protected state indefinitely or may revert to normal when prompted. Naturally, this may be done "out of

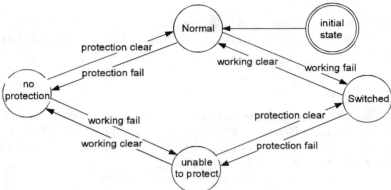

Actions:
-When protection switch state changes from normal to switched OR from switched to
 normal a protection switch event notification is issued.
-On entering the "unable to protect" state, the protected resource agent is notified as being disabled.

Figure 10.4   Protection agent behavior in 1 + 1 protection process.

hours" or during a predesignated maintenance period when minimum service impact can
be ensured. The nonrevertive behavior is illustrated in Figure 10.5. Reversion may be
triggered externally from TMN after the failed connection has been repaired. Alternatively,
because of the inherent symmetry, the connection may be left in the switched state, which
is henceforward regarded as the normal state; subsequent failure will then trigger the
reverse switch.

Actions:
-When protection switch state changes from normal to switched OR from switched to normal a
 protection switch event notification is issued.

-On entering the "unable to protect" state the protected resource agent is notified as being disabled.

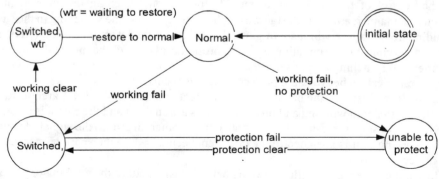

Figure 10.5   1 + 1 protection agent behavior with externally controlled reversion to normal.

The protection fail transition is omitted for simplicity. Such layer process specifications will normally contain considerably more detail concerned with exception handling
and maintenance. Exercising the process or locking it out during test procedures are

examples of the sort of activities frequently required to be supported. Our purpose here is to illustrate the distinguishing feature of the main protection architectures, therefore we confine our illustration to the essential principles sufficient to understand the core function.

10.1.5.2   1:N Shared Protection

The 1 + 1 architecture is effectively 100% redundant. For long, repeatered systems the 1:N shared protection architectures illustrated in Figure 10.6 have often been used for greater cost efficiency. In this class, N working trails share one protection trail. A failure or degraded condition on a working trail (#n) is detected by the trail termination sink. The end that detects the failure and initiates the protection procedure is called the tail end. The tail-end protection agent initiates the procedure by signaling to the head end.

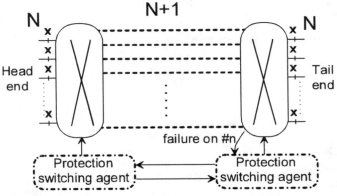

Figure 10.6   1:N protection architecture.

In the normal state, the working trails and the protection trail are all available. The process is symmetrical, but the end that first detects the failure takes on the tail-end role, initiating the switch. This is illustrated in the protection agent state machine of Figure 10.7 by the 'working trail #n fail' transition from the normal state. The head end performs the complementary role illustrated by the 'head bridge to #n' transition from normal. Figure 10.8 illustrates the operation of the process from the detection of a failure through the protection switch and the eventual reversion to normal. Again, exceptional behaviors and maintenance activities are not illustrated in the interests of simplicity.

With the very high speed line systems available today, values of N greater than 1 are less often justified. The 1:1 process operates in exactly the same way as the 1:N process. 1:N, 1:1, and 1 + 1 systems have been standardized in G.783 for the multiplex section layer. In this case, the end-to-end signaling is performed in the K1 and K2-byte channels of the MS overhead. The protection channel in 1:1 and 1:N systems may also be used for low-priority traffic, which is discarded in favor of the high priority traffic when a failure occurs. The system has also been used for fast facility protection in meshed VC-4 networks, but this has not been standardized.

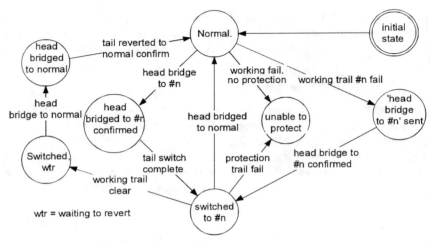

Figure 10.7   1:n protection agent state machine.

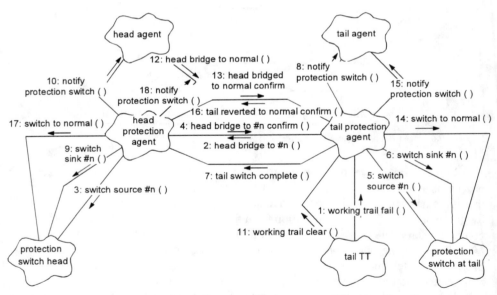

Figure 10.8   Operation and reversion of a 1 for N protection switching architecture.

10.2   PROTECTED OR SELF-HEALING SUBNETWORKS AND RINGS

Linear protection architectures with physical diversity become more difficult to justify economically as the bandwidth capability of fiber transport increases. Applying the planning constraints of minimum fiber use and physical diversity, a minimally connected flat mesh at the fiber level will generally result. Rings are particularly interesting because they represent the lowest possible physical connectivity between a set of nodes that still guarantees a minimum of two disjoint connections between any node pair. We have chosen to include rings under the more generic classification of protected subnetworks because some processes originally developed in a simple ring context are not restricted by this and can also be used in more highly connected subnetworks. Also, there are some protected subnetwork architectures that share similar characteristics to the rings but have some interesting additional advantages that may make them attractive in the future. Several of these mechanisms have been standardized or are being considered for standardization. Others are either still in a research phase or are not considered sufficiently interesting to pursue. Figure 10.9 attempts a classification based on this wider concept of the protected subnetwork. Currently standardized procedures are shown in shadowed boxes.

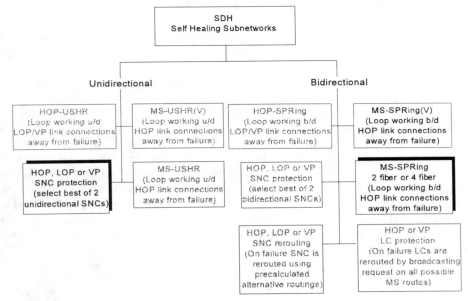

Figure 10.9   Classification of self-healing subnetworks.

10.2.1   Unidirectional Rings

In unidirectional rings, the switched entity is unidirectional. In the case of a multiplex section unidirectional self-healing ring (MS-USHR), the switched entity is a HOP layer unidirectional link connection group formed from the MS payload. In the case of unidirectional connection protection in HO or LO path layers, the switched entity is a unidirectional subnetwork connection (SNC).

The unidirectional self-healing ring (USHR) principle is illustrated in Figure 10.10. Bidirectional working connections are formed by allocating the send and receive components to a pair of unidirectional connections in a downstream and upstream server trail, respectively, such that the information flow from both components is in the same direction relative to the ring in the server layer. Unidirectional protection connections are available from the counterrotating server ring provided by the remaining upstream and downstream

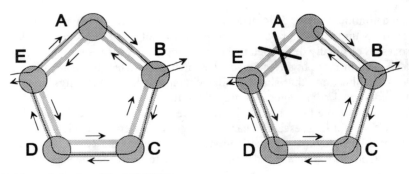

Figure 10.10   Operation of the MS-USHR.

trails. Failure as notified by the supporting trail termination causes the interrupted unidirectional connection to be looped to the protection ring away from the failure, as illustrated.

In the case of the MS-USHR, the server trail is the multiplex section. Signal failure and degraded signal conditions are derived from the MS sink termination. In the MS-USHR(Ph), all the HOP rings are constrained to be routed on the same physical ring topology, whereas the MS-USHR(V) allows overlapping virtual rings to be formed on an underlying flat mesh, each node of which implements the same MS-USHR mechanism. The protection response at the level of the affected link is the same in both cases; however, the virtual ring case allows greater efficiency to be achieved by exploiting the greater connectivity available in the flat mesh topology. This is illustrated in Figure 10.11. Two bidirectional paths, AB and HE, are shown. AB is allocated to virtual ring ABFH on channel x and HE is allocated to virtual ring HFEG on channel y. Both virtual rings share link HF in different channel slots. When link HF fails, both delimiting nodes loop away from the failure and take up the protected configuration illustrated.

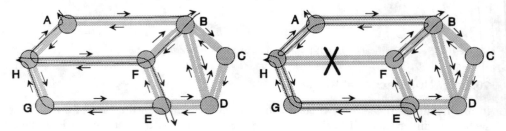

Figure 10.11   Virtual USHRs on a flat mesh topology.

HOP-USHR is based on a ring formed by a cyclic set of HO paths. It uses the same generic architecture illustrated in Figure 10.10 but operated one layer higher than the MS-USHR. The switched entity is the group of LOP link connections derived from the HO path. The protection switch trigger is given by the HOP path fail condition.

10.2.2   Subnetwork Connection Protection

Subnetwork connection protection in a unidirectional path switched ring involves bridging at the node where the connection enters the ring such that the signal is transmitted simultaneously in both directions round the ring. Selection is made at the exit node for the best of the two connections based on the normal connection fail criteria (AIS or LOP). To distinguish between AIS originating inside the protected subnetwork due to a subnetwork connection failure and AIS originating from a failure outside the protected subnetwork, the failure detection mechanism allows for a short delay greater than the delay difference of the two connection routings. If both SNCs indicate AIS, then no switching is performed. If only one SNC indicates AIS then the tail end switches to the remaining good SNC. The SNCP architecture is illustrated in Figure 10.12.

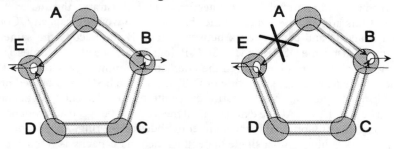

Figure 10.12   The subnetwork connection protection architecture applied on a ring.

The SNCP mechanism is not restricted to physical rings but may be used in any meshed subnetwork that provides two disjoint routings between each node pair, thus providing a bandwidth efficiency comparable to the virtual USHR usage described in Figure 10.11. The SNCP ring is "dedicated," having a fixed capacity independent of the traffic pattern with no possibility to share bandwidth. However, because the granularity of the protection mechanism is a single connection rather than a link connection group, it is normally possible to select a combination of protected and unprotected connections. The unprotected connections for a single service are diversely routed to ensure that no more than half fail at any one time. This hybrid scenario is often regarded as a good compromise for mixed PSTN/leased line traffic, particularly in access where the traffic patterns are anyway predominantly hubbed.

10.2.3   Bidirectional Rings

BSHRs are constructed from a cyclic set of trails whose client link connections are switched individually or as a set in response to a failure in a supporting trail. The generic architecture is illustrated in Figure 10.13.

Figure 10.13 may look superficially similar to Figure 10.10, but note the absence of arrows. The arrows in Figure 10.10 denote that the connections and links are unidirectional. The absence of arrows in Figure 10.13 indicates bidirectional connections. Also, in Figure 10.10 the link capacity is divided according to directionality, while in Figure 10.13 the link is divided into a set of working bidirectional link connections and a complementary

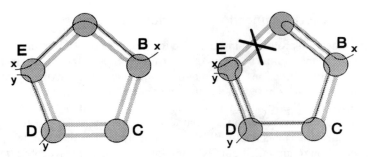

Figure 10.13   The bidirectional self-healing ring or shared protection ring.

set of protection bidirectional link connections. This illustrates the major perceived advantage of this architecture, which is the sharing of the working capacity. Connection y between E and D and connection x between E and B can occupy the same channel timeslot, in contrast to the USHR and the SNC-P rings where one path occupies a complete virtual ring. This property gives rise to the alternative terminology where the USHR is referred to as a dedicated protection ring or DPRing and the BSHR as a shared protection ring or SPRing. Depending on the traffic distribution, the efficiency advantage of the SPRing can be significant. At one extreme, if all the traffic is between adjacent nodes, then the total capacity is proportional to the number of nodes on the ring. At the other extreme, if all the traffic is hubbed onto a single node, then the total capacity is fixed and equal to the link capacity. For uniform traffic, the efficiency falls between these extremes, as illustrated in the graph of Figure 10.14.

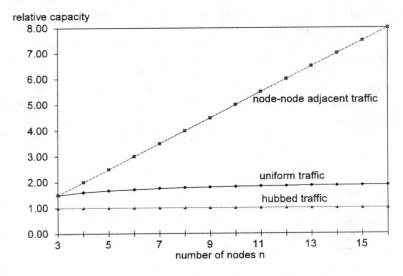

Figure 10.14   Relative capacity of SPRings with different traffic distribution.

10.2.4 Shared Protection BSHR Based on the Multiplex Section (MS-SPRing)

The MS-SPRing is based on the internode HOP links provided by a cyclic set of disjoint multiplex sections. The capacity is divided equally between working connections and protection connections, as illustrated in Figure 10.13. The working capacity in each span (the link between each node on the ring containing the working and protection connections is termed a span in G.841) is protected by the link formed by the protection capacity of all spans on the long route round the ring, as illustrated in Figure 10.13. Failure and degraded signal conditions are detected by the MS termination and used to trigger the protection-switching response. The protection-switching agents in adjacent nodes communicate using a message set encoded in the K1 and K2 bytes of the MS overhead. The nodes on the ring are numbered up to a maximum of 16 so that head and tail message destinations may be identified in the messages. A simplified state machine for the generic BSHR protection-switching agent is illustrated in Figure 10.15.

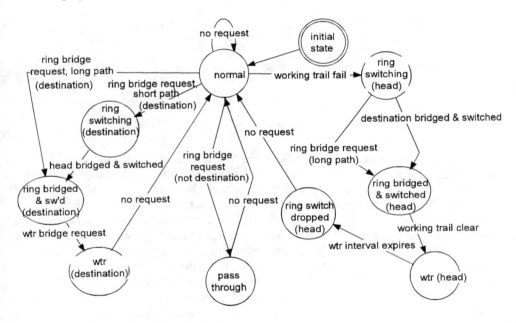

Figure 10.15   Behavior of a generic SPRing protection agent.

The agent that detects the failure and initiates the protection response acts as the head and the agent at the other end of the failed span acts as the tail. In the case of a bidirectional failure, both ends play both roles. The intermediate nodes on the ring go into pass-through mode and merely relay messages between head and tail round the ring. The protection response to a bidirectional failure between nodes A and E of Figure 10.13 is illustrated in the process scenario of Figure 10.16.

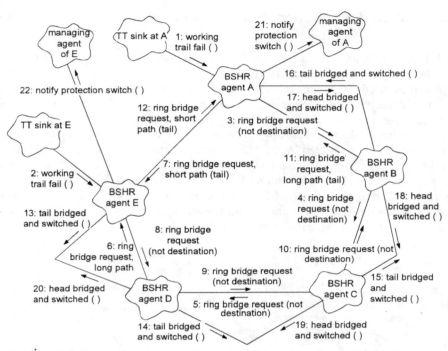

Figure 10.16   Protection response to a bidirectional failure between nodes A and E.

In this case, both nodes A and E detect the MS fail condition and enter the switching state, both acting as heads. Bridge requests are sent to the corresponding destination tails, both the long way round the ring and the short way. Naturally, as this is bidirectional failure, the short path message will not be received, but the heads do not know this. Each of the intermediate nodes enter the pass-through state and relay all messages. Eventually, the bridge requests reach their destination tails, which respond by executing a ring bridge and switch. The change of state is signaled back round the ring again to the corresponding heads, which then issue the protection-switch event notification.

In the case of a unidirectional failure detected at node A only, the short path bridge request reaches the destination tail at E first and stimulates the request to bridge around the long path. Node E does not itself bridge or switch at this stage, but returns a bridge request round the long path to the head, which then bridges and switches and acknowledges this to the tail. Meanwhile, the original bridge request round the long path from head to tail will reach the tail, which then bridges and switches and acknowledges this to the head. On bridging and switching, both head and tail notify the protection-switching event to their managing agent.

10.2.4.1   *Potential for Misconnection*

The simple SPRing process suffers from one major problem, which shows itself when a node fails. The effect is illustrated in Figure 10.17. In the event of a node failure (or certain

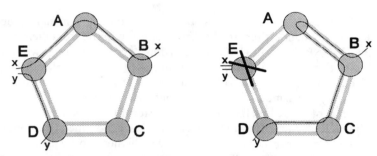

Figure 10.17   Potential misconnection due to node failure in a MSSPRing (BSHR).

double link failures), paths allocated on the same channel slot in opposite directions from the failed node, like x and y in Figure 10.17, get misconnected to each other. For this reason, the simple process is extended to include a mechanism to detect such node or double link failures. If such failures are detected, the misconnected connection ends are "squelched." That is to say, AIS is transmitted to the affected clients instead of the misconnection between them.

When a failure is detected, the detecting agent cannot know at this stage whether the adjacent link or the adjacent node has failed. Assuming a link failure, the initiating node identifies the destination node as the node known to be on the other side of the assumed failed link. The identifier of the destination node is included in the K1 byte message. First, this allows all other nodes except the destination node to relay the message in pass-through mode. When the relayed message reaches the other end of the failure, it will either find itself at the nominated destination node, in which case the failure was indeed a link failure as assumed, or it will find itself at a node one short of the destination identified, in which case it can be deduced that the unreachable destination is in fact a failed node. This information is returned round the ring to the originating agent adjacent to the failure.

In the case where a node failure has been diagnosed, the potentially misconnected paths that were previously dropped at the now failed node may be squelched. Squelching is performed in the node adjacent to the failure. Only when squelching is complete is the message sent back round the ring to eventually execute the switching. In this way, the protection switch is held back until it can be certain that the squelching is complete. The result is no misconnected traffic. Naturally, the traffic exiting at the failed node is not protected.

To squelch effectively using this method, each node in the ring must carry client layer connectivity information. Each node must know, channel by channel, which channels are dropped and which through-connected at each node. This information must be updated every time a connection is added, removed, or otherwise modified in the ring. The manner in which connectivity information is updated is not currently specified, but it can be assumed that the squelch table update will eventually be standardized as a TMN function in the NE information model.

Double link failures have a similar potential for misconnection. In this case, the ring segments into two subnetworks and the same mechanism that enabled squelching to proceed in the single-node failure case described above also applies in the case of segmentation due to double link failure. As far as each isolated segment is concerned, all nodes on the other segment are regarded as failed and any traffic from them squelched.

Four fiber MS-SPRings are also defined, with each node having two bidirectional ports in each direction. All the link connections in one fiber pair are allocated to working channels and all those in the other fiber pair to protection. The protection-switching response is basically the same as for the two-fiber case for cable or node failure, but is extended in the case of single link failure to include a local 1:1 response like that described above under linear protection. Thus in the event of a failure confined to a single fiber pair in the four-fiber link, a span protection is executed with higher priority than the ring switch. This enhances the availability of the subnetwork because several simultaneous span failures can be protected simultaneously. Only a cable break or a node failure will initiate the full MS-SPRing protection response described above.

10.2.5   The Well-Connected Mesh (WCM)

Despite the relative efficiency of the MS-SPRing and the comparative maturity of its protection protocol, the restriction to a physically cyclic topology does not best exploit the physical connectability that may be available in many cases. Consider the comparison below of protected subnetworks constructed from, respectively, two MS-USHRs, two 2-fiber MS-BSHRs, one 4-fiber MS-BSHR, and a well-connected mesh (WCM). The qualification "well connected" in this context implies a connectivity greater than that provided by a ring but less than full connectivity (i.e., every node connected to every other node).

Each example is similar in the resources it requires as each has five nodes (each with four bidirectional optical ports), as well as ten links (each with a fiber pair).

10.2.5.1   *Comparison of Efficiency of Some Five-Node Protection Topologies*

For each case, the total traffic carrying capacity has been calculated based on a uniform distribution of traffic between the nodes. The probability that some traffic is lost and the average capacity lost has been calculated for single link, double link, and triple link failure scenarios. In each case, this is based on half the capacity on each link reserved for protection. The results are illustrated in Figure 10.18.

We can see that a WCM can carry more traffic for a given node size and network connectivity than an equivalent physical ring, which implies that the fiber usage will typically be higher, in the range 20–50% more than the physical ring. There are a large number of cases, particularly in high density urban environments, where this translates directly into a lower cost per HO path (or connection) supported. WCMs based on transport nodes with four optical ports can be easily compared with their ring equivalents as shown above. These are functionally similar to the ADMs from a four-fiber ring except that a full connectivity VC-4 matrix at each transport node is assumed for the WCM, where the more limited VC-4 ADM connectivity is sufficient for the four-fiber ring. However, to achieve the potential efficiencies, the physical supporting topology must be more highly connected than is required for the ring.

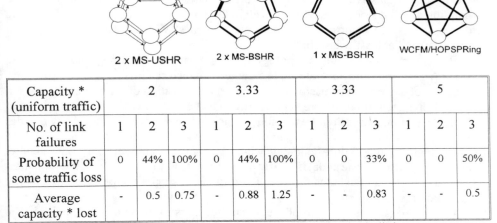

| 2 x MS-USHR | 2 x MS-BSHR | 1 x MS-BSHR | WCFM/HOPSPRing |

Capacity * (uniform traffic)	2			3.33			3.33			5		
No. of link failures	1	2	3	1	2	3	1	2	3	1	2	3
Probability of some traffic loss	0	44%	100%	0	44%	100%	0	0	33%	0	0	50%
Average capacity * lost	-	0.5	0.75	-	0.88	1.25	-	-	0.83	-	-	0.5

* capacity measured as a proportion of the capacity of a single link

Figure 10.18   Protection characteristics of four 5-node networks compared.

10.2.5.2   *Physical Topology for a Protected WCM*

Physical topologies of existing ducts and buildings generally have restricted connectivity. While it does not follow as an absolute rule, the physical topology of ducts and buildings will form a flat (or planar) mesh. (A flat mesh is one which can be drawn on a flat sheet of paper without any links crossing other than at a node.) It is possible for two duct routes to cross without coming into contact with each other, however, it is highly unlikely, as most planning procedures would place the crossing point either in a building or at a duct junction box. A more likely example of a physical topology which may not form a flat mesh, is a network which makes extensive use of radio. With these two caveats, it is reasonable to treat the physical topology as a flat mesh.

The example of a well-connected mesh used in Figure 10.18 is especially interesting as it is one of the simplest well-connected meshes which cannot be drawn as a flat mesh (Kuratowski's theorem); it must have at least one point where two links cross. Such a point we call a passive node. It is a point of common mode failure between the two links; however, it is not a node at which switching and routing can take place. Therefore, in order to correctly understand the resilience of a well-connected mesh, we must map it onto an appropriate flat mesh where the extra passive nodes which are introduced represent the points of common mode failures between sets of links. This flat mesh is called the common mode failure topology and was briefly introduced in Chapter 3 in the discussion of the management and control of disjointedness.

10.2.5.3   *Properties of Common Mode Failure Topologies*

The existence of passive nodes will reduce the resilience of a WCM compared to an assumption of full disjointedness between all links. The example in Figure 10.18 needs to be drawn, for resilience purposes, mapped onto a flat mesh with at least six nodes (at least one passive) and at least twelve links.

This common mode failure topology gives a means of examining the merits of a WCM, however, there are not any simple parameters which can give easy figures of merit. An analysis of cutsets can be useful and this gives an indication of the number of links in the WCM that need to fail before some traffic must be lost. A cutset is a set of links or nodes in the common mode failure topology, which, when removed, isolate the two nodes of a node pair from each other. Any pair of nodes may well have several cutsets. A WCM with a common mode failure topology which has any cutsets with only one member element (i.e. one link or one node) is clearly not resilient to any single failure. If the smallest cutsets have two members, then the WCM is resilient to single link or node failures, but not dual failures, and so on.

Another measure is the ratio of the number of links to the number of nodes in the common mode failure topology. While this gives no information on the robustness to multiple link failures, it will normally give an indication as to the efficiency of the WCM in terms of its ability to carry traffic for a given amount of node and link resource. As a simple ring has the same number of nodes as links, it has a link to node ratio of unity. As this ratio increases, the efficiency of the WCM will normally also increase. However, this cannot be increased without limit. It is a standard result of graph theory, that, if there is no more than one link between any pair of nodes, then the maximum value for this ratio is $(3 - 6/n)$ where n is the number of nodes in the common mode failure topology (derived from Euler's theorem of planar graphs).

We can therefore conclude that a physical ring requires the lowest connectivity among any set of nodes while still guaranteeing complete resilience to single link failures and this is the main property that has made it so attractive for many practical applications. It is possible to achieve this same resilience to single physical link failure but with greater bandwidth efficiency using a WCM, providing that when it is mapped onto its common mode failure topology, the links of the WCM achieve a suitable level of disjointedness.

Finally we can also identify a class of WCM which is already a flat mesh with no common mode failures between any links. This class we call well connected flat meshes (WCFMs).

10.2.5.4   *MS-SPRing(V)s on a Well-Connected Flat Mesh*

In the MS-SPRing described above, the cyclic link connection groups used to construct the physical ring are chosen from a single disjoint cyclic set of multiplex sections. An MS-SPRing(V), on the other hand, is implemented in a flat mesh of multiplex sections by independently allocating virtual HOP rings, each constructed from a link connection pair (one working, one protection) on a cyclic subset of the multiplex sections available.

This architecture is illustrated in Figure 10.19. Two paths, G-F and G-E, are illustrated, allocated to a virtual ring constructed on HFEG. The link connection on the channel slot E-F of this virtual ring is still available for traffic. Path H-B is also shown

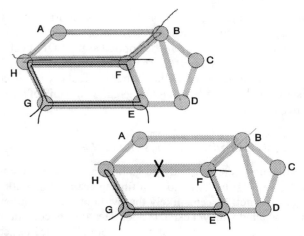

Figure 10.19   Virtual MS shared protection rings on a flat mesh.

allocated to the virtual ring on HABF, which still has two link connections available for traffic. The common link HF supports the two virtual rings on different channel timeslots. The simple protection action on failure is the same as for MS-SPRings as illustrated in Figure 10.13.

The same potential for misconnection exists as with the basic MS-SPRing, and therefore it will be necessary to implement a process capable of detecting misconnecting situations and squelching accordingly. An extension of the K-byte protocol as described for MS-SPRing is one possibility. The bridge request to the head may be broadcast on all possible routes, including those supporting virtual rings. When the bridge request reaches the destination node or another node terminating a failed link to the destination node, the same conclusion as to node or link failure can be drawn as for the physical ring. Squelching, if required, is achieved adjacent to the failure and the acknowledgment rebroadcast back to the tail. A simpler mechanism exploiting the VC-4 path trace mismatch may also be used. HOP terminations should be allocated unique identifiers that are transmitted in the path trace octet J1. The path trace mismatch indication may then be used to suppress the misconnected signal and replace it with AIS. This is particularly appropriate for rings that are administered at the LOP level because all the HOP terminations are within the same management domain.

10.2.5.5   The HOP-SPRing

The last BSHR option to be described has the same architecture as illustrated in Figure 10.13, but everything is moved up one layer in the client-server layer network hierarchy. Thus, it is based on HOP paths rather than multiplex sections and the protected entities are LOP link connection groups. The HO paths may be routed in a HOP network, with each path traversing several MSs, although this would not be visible to the protection sublayer. The same state machine behavior is appropriate for each HOP protection agent and the same potential for misconnection exists. If capacity were to be made available in the HOP overhead, the same K-byte message repertoire may be supported with squelching of the LO paths. The path trace misconnection suppression mechanism is also appropriate to this virtual ring architecture, but this must now be applied at the LOP level.

There is as yet no standard support for the HOP-SPRing or the MS-SPRing(V), but either may be considered for inclusion in the future, particularly in the context of LOP or ATM-VP administered rings or mixed service rings carrying a mix of LOPs and ATM-VPs, which are still under study.

10.2.5.6   *Practical Comparison of Rings and WCMs*

Although useful efficiencies can certainly be gained in well-connected mesh protection architectures, great care must be taken to plan fiber placement and allocate demands so that the objective can be met. It seems likely that without suitable automated design and planning algorithms which do not yet exist, the complex design constraints will be broken from time to time, either as a calculated risk to get more efficiency or accidentally during the normal churn of demands and network evolution. The process is in this sense vulnerable to accumulating what are in effect soft failures traceable to human error that will impair protection capability at some time in the future. The ring architectures, while not necessarily achieving the ultimate utilization efficiency in a well-connected physical mesh, are nevertheless more reliable in that they do not depend on the quality of the planning and management processes for their protection effectiveness.

10.2.6   Distributed Protection Processes in Meshed Subnetworks

The final two architectures to be described were originally conceived as service restoration techniques applicable in quite large meshed networks. We have chosen to include them here as protection processes because they are not service-driven and no service-driven renegotiation for resources takes place with the network. The distributed link protection process protects failed link connections individually like the MS-SPRing(V), while the distributed subnetwork protection process protects subnetwork connections individually between the subnetwork boundaries. However, given that the protected subnetwork is dimensioned to provide complete protection to single or double link failures, a simple deterministic route choice algorithm and no preemption of the perceived behavior and performance levels put both of these in the same class as the protected subnetworks just discussed. Subnetwork protection will generally be more efficient than link protection, as more efficient alternative routes may be chosen if they are available. This is illustrated in Figure 10.20.

10.2.6.1   *Distributed Link Protection in a Well-Connected Mesh*

The link protection mechanism "discovers" spare capacity by "flooding" failure information on all possible routes linking the ends of a failed link, with the objective of discovering all possible alternatives. On detection of link failure, one end of the failed link becomes the sender and the other end becomes the chooser. All other nodes act as transits. The protection agent at the sender emits a notification, termed a signature, containing informa-

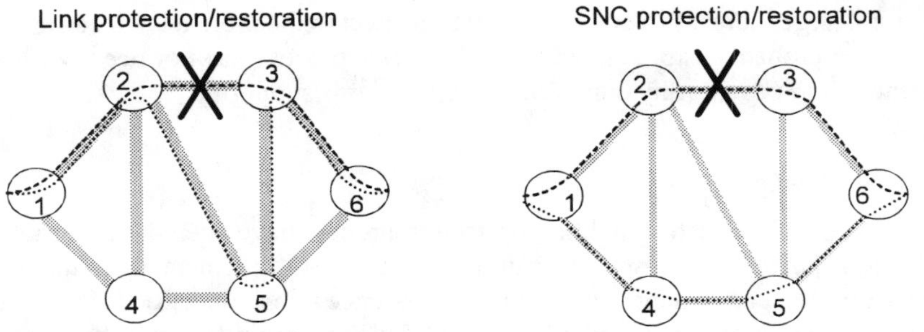

Link protection/restoration   SNC protection/restoration

SNC	Original SNC	Restored SNC		SNC	Original SNC	Restored SNC
(1,6)	1-2-3-6	1-2-5-3-6		(1,6)	1-2-3-6	1-4-5-6
(4,6)	4-2-3-6	4-2-5-3-6		(4,6)	4-2-3-6	4-5-6

Figure 10.20   Comparison of link protection and SNC protection in a meshed network.

tion about the failure and the identity of the failed link. This signature is issued on all possible links with spare capacity using a signaling channel provided for the purpose. In SDH systems, it would be natural to provide a channel from the overhead capacity in the server layer or the ECC itself. In ATM VP systems, a maintenance OAM VCh can be allocated for the purpose.

At the first transit, the signature is updated with the identifier of the first link (the link on which it arrived) and its available capacity. The revised signature is further broadcast on all other available links (except the link on which it arrived). The process is repeated at each transit node passed, thus building up a list of links traversed and their available capacity. Signatures arriving at nodes having no exit links with available capacity and that have not already been traversed are discarded. It is normal also to provide an upper limit on the hop count, depending on the size of the protected subnetwork, to avoid excessively long protection routes. This also gives the algorithm a measure of stability as the number of flooding messages can grow exponentially with the number of nodes in the network.

Eventually in a sufficiently well-connected network, a set of signatures will arrive at the chooser destination. The chooser now possesses a self-selected list of routes and capacities on the constituent links. It can therefore choose the set of routes best suited to achieve the desired level of protection, generating a set of return messages directed along the chosen routes in the reverse direction. The appropriate matrix reconfiguration is executed at each transit and, eventually, at the sender.

If the network is provisioned to have sufficient short path capacity to guarantee recovery from single and double link failures (as for the virtual SPRing), then the configuration before and after protection switching will probably be indistinguishable from that obtained by the MS-SPRing(V) applied to the same topology, despite the apparently nondeterministic nature of the route search mechanism. The same potential for misconnection exists when responding to node failure. The information to derive the failure mode is available in the received signature at the nodes adjacent to the failure, therefore squelching may be exercised accordingly. The comparatively small improvement in efficiency for a large increase in complexity with respect to the more advanced ring types discussed above probably explains why this particular process has not been considered attractive enough to provide any comprehensive standards support.

10.2.6.2   VP Link Protection

ATM VP versions of the distributed link protection process have been demonstrated (ref to IEEE article). There are two special features of the ATM implementations that can be exploited to improve performance. In ATM, the existence of a connection is independent of the bandwidth allocated to it, and ATM OAM cells can provide faster, more reliable messaging along the connections of a protection group for the communications required between the terminating agents.

A set of ATM VP connections can be established for protection between the nodes of each connected pair routed disjointly from the direct link between them. The set of disjoint alternative link connections has no traffic allocated initially, but sufficient capacity is reserved on the supporting links to be used for protection when needed. The capacity reserved for protection may be reused for protection of other links, but may not be used for working traffic.

On failure of a link, the termination processes signal to each other using ATM OAM cells on the standby VP connections. The criteria for choice of a suitable set of disjoint routes and how much traffic to allocate to each must also take account of the ATM traffic parameters of the VPs. The required capacity is "captured" on selected routes by further ATM messages from the chooser. Capturing the capacity implies allocating the required capacity to one of the VP link connections created as an alternative route. Naturally, once the capacity has been captured it is no longer available for subsequent use to protect against further failures.

A further refinement provides for detection of a collision between signatures generated from each end of a failed connection at a transit node, which will be topologically located approximately midway between the extremities of the protection connection. All the information pertaining to its capacity capture can be aggregated at the collision node and transmitted to both extremities, where the chooser can proceed as before. This reduces the maximum hop count necessary for messaging and consequently contributes to a reduction in response time.

10.2.6.3   Bidirectional Subnetwork Connection Protection in a Well-Connected Flat Mesh

The link connection protection process just described and the bidirectional ring options described previously all achieve efficiency levels limited by the need to "keep it simple," which generally precludes any attempt to seek more optimum protection routings in real time. Bidirectional subnetwork connection protection aims to achieve greater efficiency in a well-connected mesh within the same general constraints of response time and manageability.

Subnetwork connections are established in response to service demands in a normal manner, but in addition, for each working connection established, one or more disjoint alternatives are sought and reserved. That is to say, in the general case, the alternative routing is identified and a corresponding route table entry made in each node involved. Capacity is reserved in the links participating in the route, which may be reused for other protection connections but not for other working connections. Each new demand is handled in the same way.

Failure of a link or transit node is detected at the node adjacent to the failure, which then signals the failure to the connection terminations. In the SDH case, the existence of failure is known by the presence of AIS, but the location of the failure must also be signaled and this may be carried in the ECC or by means of a further refinement of the K-byte protocol, which was used to distinguish node failure in the SPRing architecture. In the case of ATM, described below, AIS OAM cells generated in the remaining ends of the actual failed connection may be used.

The simplest strategy is to ensure that the first choice alternative route is fully disjoint from the allocated one, thus maximizing the probability of successful protection with a simple response. For any particular single failure, more efficient protection routes will often be available that are only partially disjoint. In principle, all the possible alternatives may be identified and tried one at a time, starting with the most efficient, until a working alternative is found. If the process supports failure location, then this information may be used by the chooser to select the most efficient alternative from those that do not use the failed resource, hence achieving efficient protection within the shortest response time.

Naturally, a single network component failure can potentially interrupt multiple working connections that transit the failed component. The first choice alternatives for some of these may contend for the same protection resource. The bandwidth capture procedure is normally "first come first served," so that an attempt to capture bandwidth may be refused because the capacity, or part of it, has already been captured. In this case, a further attempt may be made to capture bandwidth on a second alternative, and so on until sufficient alternative connection bandwidth is successfully captured. The possibility to share the protection bandwidth makes for relatively high protection efficiency. On the other hand, the necessity to obtain a high probability of protection against any single, and some double, failures puts a conflicting constraint on the connection admission control, which must manage the network resources such that the aggregated demands generated by all the connections interrupted by a single failure may be met—if not by the first alternative, then by some combination of the other alternatives identified.

After recovery from a single failure, the remaining resources are not necessarily optimally chosen to deal with subsequent failures. Therefore, the spare alternate routings must be revised while the failure persists to allow optimal recovery from subsequent failures within the repair time.

10.2.6.4   VP Subnetwork Connection Protection

Connection/bandwidth independence and OAM cell signaling are also exploited here, but additionally, it is more easily possible in ATM to signal failure location information directly. Optimally routed VP connections are established in response to demand in the normal manner, but for each working connection established, a set of alternative VP connections are also established with partially disjoint routings to act as standby for the working connection. No capacity is actually allocated to the protection connections, but sufficient resource is reserved along the chosen route to enable each protection connection to carry its share of the displaced traffic. Fresh demands are similarly met by establishing working connections, each with one or more protection connections.

The establishment of unresourced VP connections in this way is similar in principle to marking routes in the STM case. The resource allocation and reservation algorithms are subject to the same rules, and the signaling of route selection is not so different in principle to the capturing of bandwidth, although the use of ATM OAM cells is more natural and likely to be faster than any equivalent scheme based on the use of SDH overheads.

The overcommitment of resources inherent in this approach raises the issue of scalability due to the possibility of exhaustion of VPIs resulting from the need to allocate many more VPIs per link than are actually used for working traffic. While this is generally acceptable in small subnetworks, the subnetwork size cannot be increased indefinitely without this eventually becoming a problem.

10.2.7   Interconnection of Protected Subnetworks

To ensure high availability for connections that transit multiple protected subnetworks, it is, in general, necessary to provide at least two independent interconnections between the two subnetworks. The architecture of such dual-node interconnections is greatly influenced by the protection strategy employed in the connected subnetworks. The simple interconnection of Figure 10.21 certainly protects against the single-node failure at the interconnection. Indeed, this is nothing more than the SNCP applied to a mesh network, as discussed above. However, independent failures in each of the protected subnetworks (EA and JH, for example) may still result in lost traffic. This in itself may be considered acceptable, but consider the general case where links in each subnetwork may share a single common failure mechanism; two fibers in the same cable or duct, for example. The two subnetworks of Figure 10.21 may share the same underlying topology, as illustrated by folding about the dual-node interconnect. In this case, a single cable failure may result in fiber links supporting AE and HG failing simultaneously. The connection from E to G will now fail due to a single cable failure. The recommended dual-node interconnect architecture is designed to avoid the effects of such common mode failure.

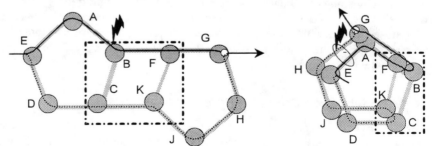

Figure 10.21   Impact of common mode failures in lower layers.

The guiding principle for interoperation of protected subnetworks is to preserve independence of operation by closing the protection at the boundary, thereby ensuring each is independently protected. Thus a long path traversing several such subnetworks may be protected against independent but concurrent failures in each transit subnetwork. Failure to ensure independence also leads to administrative complications, where each subnetwork may be in different administrative domains (even belonging to different operators). But the most important benefit associated with independence is seen where the protected subnetworks physically overlap each other, as illustrated in Figure 10.21.

Physical overlap frequently occurs in real networks. Although each of the subnetworks may itself be constructed from disjointed facilities as required by its own protection architecture, it is generally unacceptably restrictive to maintain disjointedness between the overlapping facilities of the two independent subnetworks. This is another example where even if the planning constraint were accepted in principle, the reliance on human decisions to not only optimize the design but to ensure the appropriate level of protection becomes a reliability hazard that will inevitably result in soft failures.

To achieve independence, the dual-node interconnect between two protected subnetworks must itself be a protected subnetwork, and can in principle be implemented by any of the mechanisms already discussed for protected subnetworks. Figure 10.22 illustrates two MS-USHRs with an interconnect architecture based on the MS-USHR, Figure 10.23 illustrates two MS-SPRings with an interconnect architecture based on the BSHR, and Figure 10.24 illustrates two SNCPs interconnected by an SNCP-based architecture. Note the inherent unidirectionality in this case. Complications occur when trying to interconnect protected subnetworks with different architectures and/or in different layers.

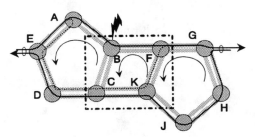

Figure 10.22   Dual node interconnection USHRs (interconnect architecture also based on USHR).

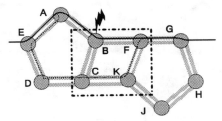

Figure 10.23   Dual node interconnect of BSHRs (interconnect architecture also based on BSHR).

10.2.7.1   *Dual-Node Interconnect Using Drop-and-Continue*

In fact, dual-node interconnect architectures have been standardized for SNCP HO to LO, HO SNCP to MS-SPRing, and MS-SPRing to MS-SPRing, all based on the SNCP principle of Figure 10.24. Figure 10.25 illustrates interworking between an MS-BSHR and a HOP SNC-P ring. In normal operation, both rings are required to present protected copies of the signal at the interconnect nodes. Even under failure conditions, at least one of these copies will be present. To achieve this, the signal is dropped to the primary node where the ring protection is closed. It is also continued to the secondary node to provide a duplicate. It is this feature that is responsible for the term "drop-and-continue," applied to this interconnect architecture. When accepting duplicated inputs, the SNC-P ring directs one to the east and one to the west. Only at egress from the ring is a choice made between the two copies of the signal, thus allowing for independent failure in this ring. The MS-BSHR makes a selection immediately at the primary node, thus restoring a single bidirectional connection routed as normal in an MS-BSHR.

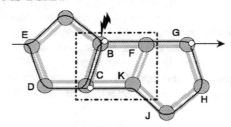

Figure 10.24   Dual node interconnect of SNCP rings.

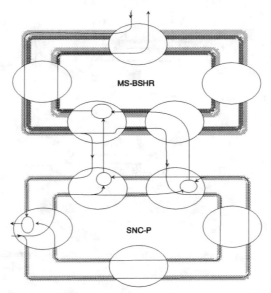

Figure 10.25   Dual node interworking between an MS BSHR and a HOP SNCP ring using drop and continue.

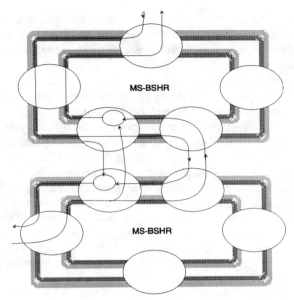

Figure 10.26   Dual node interworking between MS BSHRs using drop and continue.

Figure 10.26 illustrates two **MS-BSHRs** interconnected also using the drop-and-continue interconnect architecture.

10.2.7.2   *ATM Protection in SDH Rings*

Although the ring protection and the dual-node interconnect architectures described above were developed for use in the SDH transport layers, all may be used in ATM subnetworks also. In fact, hybrid SNCP and MS-SPRings carrying a mix of SDH LOP and ATM VP traffic are now commonly found in the planner's portfolio. In these cases, the ATM layer is implemented with unique VPIs allocated at the subnetwork boundary and no VPI translation is made at transit nodes. This simplifies the operation somewhat and contributes to faster protection response. The ATM protected subnetwork appears like a single NE to any superior ATM network levels. This also represents a size limitation, as the limited VPI resource must be shared by a potentially much larger access group, but this is usually acceptable as scalability of low-level protection subnetworks to large size is not considered a requirement.

10.2.8   **Service Restoration**

According to the terminology introduced at the beginning of this chapter service restoration is essentially service-driven with no visibility of the nature or location of the failure. But first we must describe a class of systems of which many examples have been implemented in recent years. These may be retrospectively applied to a legacy network infrastructure, which is not well endowed with service/failure alarms. They are generally based on a centralized control process, holding data on network configuration that can be used to deduce the services affected from some knowledge of the lower level service failure status of the network. The network may then be reconfigured in real time to restore as much service as possible.

There has been no standardization in this area and many ad hoc solutions exist, typically applied to a meshed configuration of network nodes with crossconnecting capability where each node will, in general, act as a transit node and as a transport access node. The objective is to restore connections between the boundary or access points after failure. The legacy transport network, typically consisting of point-to-point line systems and static multiplexers, often does not provide any open visibility of its service state or the operability of its components. This information may, however, be deduced from the lower level service alarms available at the crossconnect ports, either as signal fail or as AIS. Great care must be taken to design a high availability data communications network to ensure accuracy and completeness of the event reports from different parts of the network. Failures in communications within the process can lead to gross errors in the restoration plans generated. Where the infrastructure is already endowed with comprehensive surveillance, then more explicit and perhaps more effective failure information may be supplied with the help of some custom engineering.

After a suitable time interval chosen to ensure that all the information relevant to a single initiating event has been collected, the centralized control system will analyze the alarm data to determine which resources have failed, which connections have been interrupted, and hence which services have been affected. The next step is to reestablish interrupted connections using the remaining resources, starting with those of highest priority. If the remaining resources are insufficient to carry all the interrupted connections, it may then be possible to reestablish connections, allowing the additional use of resources currently occupied by lower priority services. If suitable resources are found, then the currently allocated low-priority services may be preempted to free their resources for use by the failed, higher priority services. The process is illustrated in Figure 10.27.

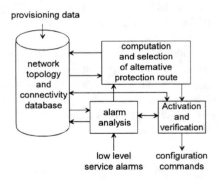

Figure 10.27   Functional block diagram for restoration system.

Such restoration systems were common in latter-day PDH networks, but the principles have also been used in SDH implementations based on ADMs, crossconnects, and line systems. Such systems have often formed part of the justification for an SDH introduction scenario in the sort of hybrid SDH-PDH network in which SDH crossconnects deployed at network nodes are used to manage a combination of legacy PDH facilities with new SDH facilities deployed for new growth. In this case, the end-to-end protected connections are a mix of SDH and PDH connections.

Response times from occurrence of failure to restoration of all failed connections is limited by the time to collect and analyze the network service alarm events, correlation with network connectivity information to determine which services have been interrupted, and then by the processor-intensive search for alternative routes. The route search may be streamlined by precomputing for each node a set of alternative routings to every other node in the network. The failure analysis will quickly eliminate those routes that are no longer available and an optimum choice may quickly be made from the remainder, taking account of the precomputed routing parameters such as implied cost (a measure of the efficiency and scarcity of capacity), quality, or delay, as appropriate.

Such systems have worked quite well, but the centralized architecture and the vulnerability of the communications network on which it relies ultimately limits the performance that may be achieved. Moreover, it does not scale well so that performance deteriorates in all respects as the network is scaled to larger size. Another issue is the positioning of the service restoration functionality within the overall management and control architecture of a large multilayer network.

10.2.8.1   Centralized and Distributed Architectures

The main perceived limitations of the type of process described above are associated with its centralized architecture. Here, we describe the functional decomposition required to support a distributed version based on the same principles. Figure 10.28 illustrates the essential differences between a centralized and a distributed control architecture. On the one hand, there is a highly available, high-performance centralized control system linked to the NEs by protected communications channels, and on the other, relatively compact control systems located at each node communicating with one another over a secure communications network. The OS plays no part in the real-time activities of event notification analysis or route search processes, although static configuration parameters of the NEs and of the restoration system itself, (implied costs, maximum hop counts, timeouts, etc.) are naturally still under TMN configuration control.

The process divides into two distinct parts: a routing information exchange process operating asynchronously in the background and a connection setup process acting in real time between cooperating nodes. Each node contains a route table containing a set of alternative routes from itself to every other node in the subnetwork partition. These tables contain essentially the same set of routes, sorted according to implied cost, as contained in the precomputed tables mentioned above, as a mechanism to speed the route search in the centralized architecture. However, the tables are computed using a version of the Dijkstra algorithm whereby each node asynchronously requests the route tables of each of its neighbors as described in Chapter 4. From this and the knowledge within a node that its neighbor may be reached by the single link between them whose parameters are known, every node can compute a set of routes to all other nodes in the subnetwork via each of its neighbors. In this way, it can regenerate its own route table to provide more current

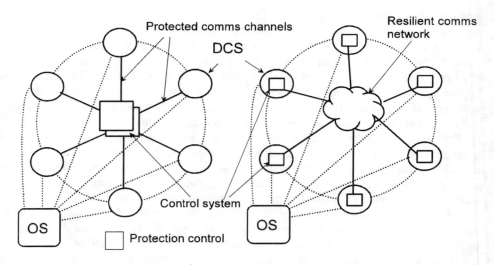

Figure 10.28   Centralised and distributed architectures for restoration.

information of the surrounding subnetwork topology. Each route table in turn must be disclosed to topologically adjacent neighbors on request.

From all the possible routes available, a minimum implied cost set of disjoint routes may be selected from the routing table. The concept of implied cost is based on the proportion of resources used, modified by their relative scarcity. Thus, while a link has lots of spare capacity, it shows a cost of use proportional to length, number of server hops, or other suitable measure of resources used. As this resource becomes used up, its implied cost increases to reflect the impending scarcity, thus the algorithm urges the selecting agent to seek other routes, which may be topologically longer but have more spare capacity. The onset of scarcity may additionally be used as a trigger to the server process to provision more capacity, thus restoring the lower implied cost for future demands. The parameters defining the implied cost, both its fixed and scarcity-dependent components, may be provisionable and may be used to fine-tune the algorithm. The mechanism works in a manner similar to a market economy, which may nevertheless be conditioned by adjusting monetary and fiscal fundamentals. This is described in more detail in Chapter 13.

When a node receives a request to establish a connection to another node, it merely selects the most efficient route available according to its ranking in the route table as defined above.

The connection setup protocol, which is replicated in each node, takes the selected route, in the form of a list of nodes and/or links, and signals forward along the selected route setting crossconnections at each step until the whole connection is made. When a connection subsequently fails, this is detected at one or both end points and the layer termination process alerts the far end in the case of unidirectional failure. The local process at the source end point has a record of the destination and the route used, and is therefore in a position to choose an alternative route from the same set but disjoint from the first. In this way, the probability of choosing a route that is still available is maximized, but without having any knowledge of the location of the failure. Naturally, in a situation where no alternative is completely disjoint, this high probability cannot be achieved and it becomes possible that the first alternative route chosen will also fail because of the finite probability that the original failure was common to both the original route and the first alternative. The result is that it will take a little longer on average to restore because several alternatives may need to be tried.

The new connection is set up in the same way as the original by signaling forward along the selected route, reserving resources at each step. Again, if no working alternative connection can be found, it may be possible to seek resources among those already allocated to active connections of lower priority. If such alternatives exist, they may be preempted and used to restore the failed high-priority connection.

Although distributed systems of this type may be implemented in any layer, the P-NNI network protocol of the ATM Forum described in Chapter 6 is the first standardized protocol that may be used in this way. The P-NNI protocol includes a routing information distribution part and also a connection control part based on source routing. Thus each node can keep a record of preferred route choices together with a set of disjoint alternatives. Failure may be signaled back to the source either by means of the tandem connection termination process or using the backwards clear mechanism of the connection control part.

10.2.9   Interlayer Dependencies in a Multilayer Network

The protection and restoration processes described above all act within a single layer network and effectively form a protection sublayer, offering protected subnetwork services, which effectively hides the details of the underlying protection processes, presenting instead an abstraction of the protected resources to the client. They also rely on certain capabilities or services from supporting layers, although good architectural design seeks to minimize interlayer dependencies.

10.2.9.1   Topological Visibility

A minimum requirement is visibility of the available topology, which is derived from the connectivity of the server layer. This is necessary to configure the protection topology by creating working and protection links between the protection-switching subnetworks. The problem of expressing the disjointedness in a layer topology to the protection client was discussed in Chapter 3.

10.2.9.2   Notification of Server Failure

All the protection mechanisms described rely on failure detection and notification mechanisms in supporting layers that may be used to initiate the protection response. Even where this is not explicit, the AIS indications that can be correlated to infer a layer connection failure event are themselves generated explicitly by failure detection processes in a serving layer.

10.2.9.3   Interaction of Nominally Independent Processes in Different Layers

Where the serving layer itself incorporates a protection or restoration process, care must be taken over interactions between the processes. There are no hard and fast rules. In general, the safest approach is to ensure the lower layer mechanism is fast and autonomous while constraining the client layer to delay its response long enough to allow the server to operate first. Only if the failure persists after the server has nominally had enough time, but nevertheless failed to recover, should the client response be initiated.

If the server layer process is already rather slow and the client standby configuration is equivalent to the working one (1 + 1 protection switching, for example), then it may be acceptable to allow the client to respond before the server. The main problem with this is that the client is then left in a switched configuration, which is not its normal state in the absence of failures within the layer. This, in itself, may be considered acceptable. For example, in the case of 1 + 1 client protection, the protection-switched and unswitched states are functionally equivalent and there is no loss of efficiency, whichever operating state is regarded as normal. There are operational difficulties, however, in that closer cooperation between the two layer processes is necessary than would otherwise be required. Two responses will have occurred as a result of a single failure: this must at least be acknowledged for auditing purposes and it may also be necessary to coordinate reversion to normal in the two domains after the failure has been cleared. In the same way that dual-node drop-and-continue was justified primarily on the basis of maintaining independence between domains to minimize operational difficulties, so with nested protection in layer networks, independence of operation is the target. If this target is compromised, then special steps must be taken to preserve operational integrity. Client layer processes based on shared protection resources cannot generally be operated this way. Normally, the subnetwork's protected state is less efficient than its normal state and its behavior in the presence of subsequent failures genuinely within the same layer (i.e., not protected by action of the server) will be at least suboptimal and probably ineffective.

10.2.10   Service Restoration in Existing Networks

Transport networks provide service in the form of leased lines that are used for contribution and distribution of broadcast programmed material or interconnecting LANs and PBXs to form private networks. In these cases, depending on the value placed upon availability by the client service, restoration may be applied using variations on the techniques described in Section 10.2.8, in addition to whatever protection may already be included within the network fabric. The same transport layers will also be generally used to support PSTN/ISDN or IP data networks, and these have their own methods for enhancing service availability.

10.2.10.1   Service Restoration in the PSTN/ISDN

The global PSTN/ISDN today achieves a very high availability, largely due to the high degree of automation of call control processes and the resilient mechanisms contained at all levels within the components, the switches, and the transmission facilities from which it is constructed. However, when a call in progress is interrupted—whether it be due to failure of an unprotected component, a coincidental failure combination beyond the scope of the protection systems involved, or simply a failure of a protection system itself—the caller can attempt to reestablish the call by redialing. Today "redial the last number" is typically an automatic function available in the telephone instrument itself. The new call attempt will proceed in a step-by-step manner but will avoid the failed resource, taking alternative surviving resources instead. Adaptive routing information processes in modern switched networks ensure that new routes can be made available to route around congested or failed resources. This rather crude mechanism has proved surprisingly effective. Of course, if such events became commonplace, expecting the subscriber to redial would not be acceptable; but in an environment where the frequency of occurrence of service affecting failures has been reduced to very low levels by providing protection of underlying facilities, it is a simple and effective backstop.

Private networks based on leased lines also use PSTN/ISDN as backup. That is, when the direct line fails, alternate routes may be set up by automatically dialing through the switched network. Although quite effective from the client viewpoint, such usage can be problematic for design of a network predicated on the statistics of voice telephony. The "calling pattern" of an automatically dialed standby connection will have a distribution very different to the Erlang statistics used to dimension switched networks.

10.2.10.2   Resilience of the Internet

The global public data network based on the Internetworking Protocol (IP) and known as the Internet is especially good at providing error-free transmission of data. The special needs for data integrity in an unreliable environment have been explicitly safeguarded right from the start. The transmission control protocol (TCP), operating at the transport layer in data protocol terms, essentially controls this aspect. A TCP connection is created as an association between two communicating host computers and essentially forms a trail between them according to the definitions in Chapter 3. The TCP trail terminations communicate using the data in the packet header to coordinate the parameter settings required to complete error-free packet transmissions. The TCP packets are entrusted to the network layer with the IP destination address. The IP is responsible for passing them from node to node until they reach their destination. If the source does not receive an acknowledgment from the destination sink before some preset timeout, it will retransmit part of this data, requesting a smaller window size. The window size determines the amount of data to be buffered before an acknowledgment. An acknowledgment is only given after a complete packet has been received without error. Failure to acknowledge a packet may result from various causes, including congestion, facilities failure, misrouting, or transmission corruption causing errors.

Packets are routed according to routing information at IP nodes, which is computed from similar routing information obtained from neighboring IP nodes. Changes in the network topology, whether due to failures, equipment temporarily being withdrawn from service, or new facilities being introduced to service, are discovered by the nearest neighbor. The change is notified to all the other network nodes by the routing information exchange protocol operating among neighbors, and the nodal routing information at each location is updated accordingly. Normally, the retransmitted packet will find an alternative route through the network at one of the subsequent attempts to retransmit. In this way, the Internet contrives to ensure that data is transmitted eventually without error, despite failures and misoperation in the underlying network.

10.2.10.3   *Service Restoration as a Feature of the Layer Process*

Each of the mechanisms described in this section is an integral part of its layer process. That is to say, the connection control process is essentially the same for a new connection as it is for a connection that is being restored, and the routing process supports the requirements of both aspects. The P-NNI protocol from the ATM Forum also shares this same characteristic, whereby the features of connection control and restoration may be effectively integrated in a single process, while the routing information exchange part ensures rapid adaptability to the changing state of the network as a result of failures, repairs, or new extensions.

10.2.11   Efficiency and Performance of Nested Processes

It is a fundamental feature of these network-level processes that in protecting or restoring a connection in a particular layer, account is also taken of failures occurring in all of the serving layers. What, then, is the value of retaining resilience mechanisms operating independently in the various layers of a layered network? The justifications fall under three headings: scalability, selectivity, and cost efficiency.

In a moderately large resilient subnetwork, the number of network components (i.e., network elements and physical connections) may be very large, and there are practical limits as well as economic limits as to how high an availability can be achieved in these components individually. Partitioning the components into internally redundant protection groups or subnetworks has the effect of reducing the number of components to be collectively protected at the higher level and simultaneously improving their individual availabilities. For a given overall redundancy, partitioning the resilience problem in this way will provide better overall availability, and is thus a considerable benefit as networks grow to large size.

The higher layer mechanisms are relatively more fine-grained in the service bandwidths available and, conversely, the volume of connections supported is much greater in the higher network layers. While there is a general need to improve the general availability of all services to all users, there is also a need to provide additional cover over and above this to a subset of high-priority services or users. The higher layer mechanisms provide the selectivity required to efficiently meet a wide range of expectations.

It is dangerous to generalize about cost-efficiency, but it is normally the case that network switching and control mechanisms are more costly per unit bandwidth in higher layers. This is a direct result of the finer granularity and a degree of service dependence found higher up the layer hierarchy. Of course, the onward march of silicon technology tends to erode such differentials over time. It is still the case today that protection of high-capacity transport is significantly cheaper in the SDH layer than in the ATM layer, and this makes it possible to get more value out of the ATM switch fabric rather than wasting ATM capacity on the rather mundane task of protecting against fiber breaks.

References

The following standards are the main source documents for the material in this chapter:

ITU-T Recommendation G.841, *Types and Characteristics of SDH Protection Architectures.*

The following contain useful reference material:

"Integrity of Public Telecommunication Networks," *IEEE Journal on Selected Areas in Communications*, Vol. 12 No. 1, Jan 1994.

Wu, T.-H., *Fibre Network Service Survivability*, Artech House, 1992.

Thulasiraman, K. and Swamy, M. N. S., *Graphs: Theory and Algorithms*, Wiley-Interscience, 1992.

Synchronization and Timing 11

Timing aspects of telecommunications networks are often regarded as a subject for specialists. With the public switched telephone networks (PSTN), to a large extent, the subject has been reduced to a set of pragmatic parameters and associated limits applied to network problems, while the rationale for these parameters and limits has not been normally considered a part of a general telecommunications training. The introduction of synchronous digital hierarchy (SDH) and asynchronous transfer mode (ATM), however, changes many of the original assumptions used in deriving these pragmatic rules, and it is necessary to reexamine some of the basic starting points of timing to successfully design and build broadband networks. However, the result of such an examination yields a coherent and uniform picture of timing that was previously obscured by some of the more pragmatic simplifications used in the past. In addition, some of the required solutions to timing problems turn out to be more complex than would have been expected from a naive analysis based only on the historical pragmatic rules.

A good understanding of timing may be gained by first considering the general concepts of phase variation, slip, and the phase control loop. Phase variation and slip are the basic measures of timing impairments, while the phase control loop is the basic element used in the control of timing. Depending on its precise characteristics, the phase control loop may generate phase variation, attenuate phase variation, generate slip, or a mixture of all of these.

This chapter first explains general characteristics and sources of phase variation and slip. Secondly, the chapter deals with the way phase control loops defined in ATM and SDH affect phase variation and slip. Thirdly, we deal with the effects as they occur at specific interface points. Finally, there is a section on the distribution of synchronization reference information to network elements in the broadband network. However, we first of all establish the terms that are used in this chapter and relate them to the historical parameters of jitter and wander.

11.1 TERMS AND DEFINITIONS USED IN TIMING

11.1.1   Basic Timing Signals

A digital signal may be regarded as binary data with each bit occurring at a discrete point in time: the characteristic information of the digital signal comprises the data itself and the time at which it occurs. The timing determines the discrete instances at which the data is defined and is often referred to as the clock, as illustrated in Figure 11.1. There are several types of clock signals that need to be considered in broadband networks.

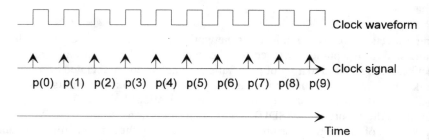

Figure 11.1   Representation of a regular clock.

11.1.1.1   *Regular Clock Signal*

A regular clock signal has all its expected discrete instants evenly spaced in time. It is defined in terms of its phase parameter $p(n)$ (n denotes the nth clock pulse), which is given as a proportion of the basic clock period. That is:

$$p(n) = 2\pi n t_0$$

where t_0 is the clock period. In general, physical media layers, including all STM-N optical/coax layers and all plesiochronous digital hierarchy (PDH) layers, use regular clocks. Shared media protocols such as Ethernet use a regular clock signal, even if it only lasts for the duration of a data frame. In addition, almost all integrated circuitry used to implement nodes use regular clocks and derive internal clocks such as gapped clocks, frame phase clocks, and embedded clocks from this.

11.1.1.2   *Gapped Clock Signal*

A gapped clock is derived from a regular signal and is of lower average frequency, yet has the same basic clock period t_0. The frequency is reduced by leaving "gaps" in the clock signal where the regular clock signal would have a pulse. The gapped clock is used extensively in SDH. Payloads within the SDH frame are timed using gapped clocks. For example, the VC-4 uses a gapped clock derived from the STM-*N* regular clock in which it is carried. The VC-12 uses a gapped clock derived from the VC-4 clock, which has already itself been gapped. In digital technology, it is very easy to generate a gapped clock by gating out the unwanted transitions, as shown in Figure 11.2.

STM-1 Regular clock

VC-4 gapped clock
72 missing STM-1 regular clock pulses followed by 2088 included STM-1 regular clock pulses

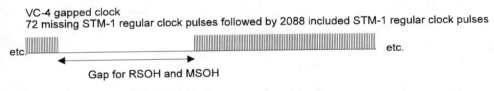

Gap for RSOH and MSOH

Figure 11.2   Example of a gapped clock.

11.1.1.3   *Frame Phase Clock Signal*

In addition to the basic clock, whether regular or gapped, a digital signal also includes some form of framing as a reference for the binary data format. This allows the receiving end to establish the ordering of the bits in such a way that those belonging to each client layer connection can be delivered correctly. Normally, an additional clock is derived from the main server layer clock that marks the start of each frame and is referred to as the frame phase. The way in which frame phase is established depends on the layer type:

- In circuit-based layers (e.g., SDH layers, PDH layers, and 64 kbit/s), frame phase is established by a regular counter; for example, 125-μsec frame phase is derived by counting 2,430 bytes of an STM-1 regular clock;
- In cell-based layers (e.g., ATM-VP and ATM-VCh), frame phase is established both by cell delineation and by the adaptation layers;
- In packet-based layers (e.g., FR, IP, SMDS, etc.), frame phase is established for each packet of data and a frame is formed around each packet of data.

11.1.1.4   *Embedded Clock Signal*

An embedded clock signal occurs when a layer is carried within another layer; for example, a VC-4 using a gapped clock is always carried within an STM-*N*, which uses a regular clock. Almost without exception, any layer that is to be switched using electronic time switching will have an embedded clock signal, as this makes for economic switching equipment. All HVCs, LVCs, VP, VCh, FR, IP, SMDS, and 64 kbit/s have embedded clock signals.

The fact that these have embedded clock signals has profound implications when they are used to transport layers that are associated with regular clock signals, for example, when a VC-12 is used to carry a PDH 2048 kbit/s signal. This arises because the embedded clock signal of the VC-12 will produce impairments on the regular clock signal of the PDH 2048 kbit/s signal. This is illustrated in Figure 11.3.

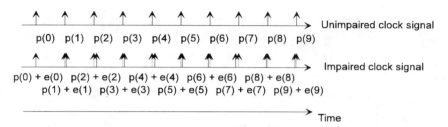

Figure 11.3   A clock signal with phase variation.

11.1.2   Timing Impairments

Most of our concern in network architecture is understanding and controlling the impairments to which the various forms of timing are subjected. We start with a review of the standard parameters that have been used historically to define timing impairments.

11.1.2.1   *Historical Parameters: Jitter, Wander, and Slip*

In the past, jitter, wander, and slip have been the main parameters used to quantify timing impairments and are defined in ITU-T Recommendations G.822, G.823, and G.824. Jitter and wander are both types of phase impairment, and there has never been a well-defined distinction between them. The distinction has been based primarily on the source of the impairment: jitter is produced by regenerators and multiplexer justification schemes; wander is produced by phenomena operating on a much longer timescale, such as temperature cycling effects in cables.

Because ATM and SDH use embedded clocks, they are capable of producing impairments that are not easy to classify as either jitter or wander according to this distinction. As a result, this loosely defined distinction is no longer adequate. The more general concept of phase variation, which covers both jitter and wander, takes its place. Phase variation is used in this chapter, and it will become clear to the reader who is familiar with jitter and wander that this is not simply a matter of convenience, but gives a clearer insight into the nature of timing impairments.

The terms jitter and wander are still used, however, and a new distinction is used to separate them based not on the cause of the phase variation, but on the way in which it is dealt with in the network.

Wherever timing impairments may be generated in a layer network, it is important to specify how a sink function handles the phase variation. The sink function will implement some form of phase control loop. When the phase control loop can adjust the outgoing clock some of the incoming phase variation is tracked and therefore passed on, while some is buffered in the phase control loop. Phase variation that is buffered is jitter, while phase variation that is passed on is wander. This understanding has allowed the clarification of the definition and limits of jitter and wander for all the PDH layers in ITU-T Recommendations G.823 and G.824. More detailed explanations of jitter and wander are given in Appendix A.

The distinction is therefore drawn on the basis of a high-pass filter, low-pass filter combination, each of the same fundamental pole frequency. The pole frequency thus determines the cutoff between jitter and wander, as illustrated in Figure 11.4.

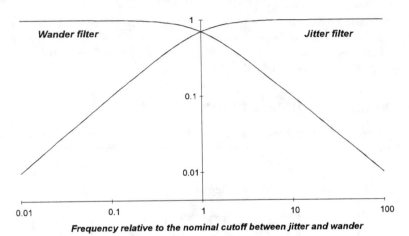

Figure 11.4   Distinction between jitter and wander based on filters.

11.1.2.2   Definition of Phase Variation

There are many mechanisms that can affect the quality of a clock signal, whether regular or embedded. These all have the result of altering the transition times $p(n)$ from their expected positions. For the transported timing signal, the actual times $p'(n)$ can be described as the expected time $t(n)$ with the addition of an error $e(n)$. The actual timing signal can therefore be written

$$p'(n) = p(n) + e(n)$$

It is the error $e(n)$ that is of particular interest when considering the quality of timing. This is the phase variation as illustrated in Figure 11.3. It is normal to show the phase variation $e(n)$ as a graph of $e(n)$ against time. This is the representation used throughout this chapter.

11.1.2.3   Slip

Slip occurs when pulses are added to or deleted from a clock signal. This occurs in phase control loops that cannot (or do not) fully adapt to the phase variation on their input and respond by slipping the output clock signal. When the buffer of a phase control loop either completely empties or completely fills, and there is no other mechanism available, the control must either drop information out of the buffer (if it is full) or add some in (if it is empty) to satisfy the requirements of the output clock signal.

In the PSTN, slip was only intended to occur in the 64 kbit/s layer frame phase and was constrained to be very infrequent. The 64 kbit/s slip performance is specified in ITU-T Recommendation G.822. (In fact, slip can also occur, spuriously, in PDH layers as a result of error burst corrupting the justification control bits in server PDH layers.) In cell- and packet-based layers, slip is normally the only realistic way to respond to the timing impairments inherent in these layers.

Under normal circumstances, slip only occurs as the response of a phase control loop to excessive phase variation. When considering sources of timing impairments, we note that they all cause phase variation first, and only produce slip if the buffering capability of the phase control loop is exceeded.

11.2 SOURCES OF PHASE VARIATION

Consider the arrangement in Figure 11.5. This shows a number of network elements and timing sources, and illustrates where different sources of timing impairments occur in the network.

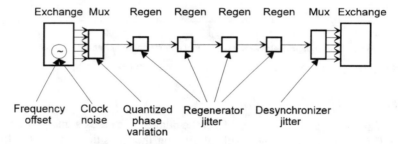

Figure 11.5   Example of sources of phase variation in the PDH transport network.

There are two principle types of impairment. First, there are impairments from the "physical world" that arise through imperfections in components and their environmental sensitivity. These are normally "analog" in nature. Second, there are imperfections that arise as a result of the design of digital systems, normally associated with embedded clocks, which are normally more "digital" in nature.

11.2.1   Analog Sources of Phase Variation

11.2.1.1   *Frequency Offset*

Any clock source is based on some form of resonant oscillator and so has a "natural" frequency at which it oscillates. In telecommunications, resonant oscillators can be made from quartz crystal, which has many different modes in which it can resonate using the piezoelectric effect of the crystal, or from an atomic resonance such as that of rubidium, cesium, or hydrogen. In general, atomic resonance is extremely accurate, although rubidium is less so than cesium or hydrogen. Table 11.1 gives an indication of the expected accuracy of clocks based on different resonant oscillators.

Table 11.1

Accuracy of Various Reference Clocks

Cesium	10^{-11} to 10^{-13}
Hydrogen	10^{-11} to 10^{-13}
Rubidium	10^{-9} to 10^{-10}
Double oven crystal	$\sim 10^{-9}$
Oven crystal	$\sim 10^{-8}$
Temperature compensated crystal	$\sim 10^{-6}$
Crystal	$\sim 10^{-5}$

Clocks based on cesium and hydrogen are rather expensive and delicate to maintain, and tend therefore to be restricted to applications such as primary reference for a complete telecommunications operator's network. The timing reference from such clocks is distributed to other clocks on the network, which are "slaved" to the "master" clock reference. Under normal circumstances, slave clocks operate as phase locked loops, so there should be no frequency offset between the output of the slave clock and the output of the primary reference clock. However, frequency offset does occur in real networks for two reasons:

- When there is a boundary between domains timed from different primary reference clocks;
- When a slave clock loses its connection to its primary reference clock, its frequency will start to drift from that of the primary reference clock at a rate dependent on the quality of the resonant oscillator in the slave clock; the slave clock is then said to be in a "holdover" condition.

ITU-T Recommendation G.811 defines the standards for primary reference clocks and is based on the expected performance of a free-running cesium oscillator. ITU-T Recommendation G.812 and G.813 define a range of standards for slave clocks. These are optimized around different quality requirements and are discussed later in the chapter.

If for either of the two above reasons there is a frequency difference between clocks, then (as phase is the integral of the frequency) a frequency offset is represented on a phase variation graph over time as the integral over time of the frequency offset. If the frequency offset is constant, then this is a straight line, as shown in Figure 11.6.

11.2.1.2   *Clock Noise*

The resonant oscillator, as well as the general circuitry of a clock, adds noise to the clock signal. The level and the characteristics of the noise depends on the type of the resonant oscillator and the design of the clock circuitry. In addition, a slave clock can attenuate some aspects of clock noise, again depending on the design. In general, the power spectrum

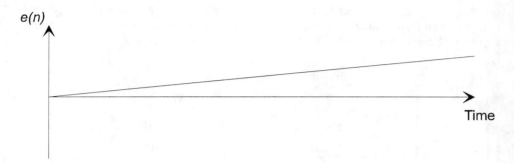

Figure 11.6   Phase variation from a constant frequency offset.

of clock noise is biased towards lower frequencies, and the sort of phase characteristic of this noise is illustrated in Figure 11.7. A number of parameters are used to measure and control its level, including *maximum relative time interval error* (MRTIE), *time deviation* (T_{dev}) as well as general plots of phase variation and its power spectral density. All of these parameters attempt to give statistical estimates of $e(n)$ in the units of time, normally *nanoseconds* (ns). There has been extensive debate within ITU-T, ANSI/T1, and ETSI on the relative merits of these parameters and they are discussed in more detail in Appendix A.

Figure 11.7   Phase variation from clock noise.

There is not always a straight correlation between the accuracy of a resonant oscillator for free-run or holdover operation and its clock noise performance. For example, a cesium clock does not have particularly good clock noise performance in the megahertz to hertz part of the noise spectrum and so is normally combined with a quartz crystal slave clock to reduce this clock noise.

11.2.1.3   Regenerator Jitter

In line system regenerators, simple circuitry is used to recover the clock from the incoming signal for regenerating the outgoing signal, and imperfections in this process will add to the error component of the phase variation. The variations in the data pattern of the regenerator input signal can cause pattern-dependent phase variations in the recovered timing information, and since all regenerators will be subject to the same tendency, such imperfections can accumulate through multiple regenerators (generally according to a systematic addition law). Most line systems scramble the data, which ensures that there are no low-frequency components in this pattern-dependent jitter. Regenerators also generate their own clock noise, which accumulates (generally according to a random addition law). This is described in more detail in Appendix A. The characteristics of regenerator jitter are similar to that of clock noise, however, it is normally measured in *unit intervals* (UI) of the line bit rate.

11.2.1.4   Diurnal Wander

The physical transmission medium (optical fiber, copper pair, coaxial cable, or air) will exhibit variations in electromagnetic propagation speed and physical length as a result of temperature variations. These effects have a diurnal and a seasonal component, and give rise to phase variations observable at the end of a transmission link. The number of clock pulses effectively "stored" in the media when pulse n emerges, $C_p(n)$, can be determined by the equation:

$$C_p(n) = \frac{b_r L}{v}$$

The way this changes with temperature ΔT, mainly a diurnal variation, can be calculated by taking partial derivatives

$$p(n) = 2\pi \frac{dC_p(n)}{dT} \Delta T(n) = \frac{2\pi b_r}{v} \frac{\partial L}{\partial T} \Delta T(n) - \frac{2\pi b_r L}{v^2} \frac{\partial v}{\partial T} \Delta T(n)$$

where b_r is the bit rate, L is the media length, and v is the propagation speed, and $\Delta T(n)$ is the temperature change when pulse n emerges. The phase variation resulting from diurnal wander is illustrated in Figure 11.8.

This change in the number of stored bits in the media is seen as phase variation, and is called wander. Wander is normally measured by a parameter that is independent of line bit rate whose dimensions are pure time (e.g., microseconds, $\mu\Phi255\sigma$):

$$wander = \frac{p(n)}{2\pi b_r}$$

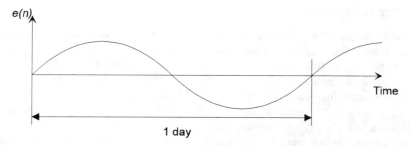

Figure 11.8   Phase variation from diurnal wander.

11.2.2   Digital Sources of Phase Variation

The clock and the data itself are integral parts of the characteristic information in a layer network. Both must be faithfully transported by a server, including the timing impairments described above. The analog phase variation must be sampled and quantized for transportation through a digital channel provided by the server.

11.2.2.1   *Justification Phase Variation From Quantization of Phase*

Quantization of phase occurs in any time division multiplexing system, whether it's a fixed or statistical multiplex. A server layer will make phase available to a client connection at discrete intervals, according to the server layer frame structure. For example, SDH makes a number of bytes of phase available every 125 ms, while ATM make cells available from a regular stream of cells. This has the effect of *quantizing* the phase of the client signal as illustrated in Figure 11.9. In addition, a client signal will be generated at a different geographic location to a server signal that may carry it. In this case, the reference clock for the client signal may differ from the reference clock for the server signal, and the server signal must take this into account in the quantization scheme.

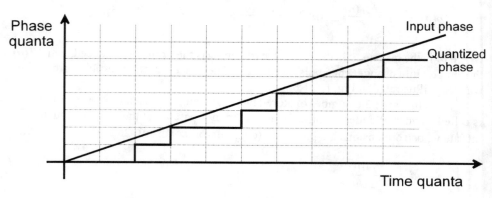

Figure 11.9   Quantization of phase variation in a digital multiplexer.

The combination of these two effects, phase quantization and frequency differences, gives rise to a phenomenon known as justification of phase. This originates from the fact that bits from the client are "justified" to align with the frame phase of the server, and this gives rise to a source of phase variation that is inherent in the definition of a particular client-server adaptation function.

Justification phase variation is hidden, or "virtual," until a regular clock signal for the client signal is required. While some attempts have been made in standards to specify parameters and limits for justification phase variation in its quantized form (i.e., in the server signal), these have been, as yet, unsuccessful. Their standardization is important, however, as this is the only way to properly specify the synchronizing end.

The characteristics of the quantization used in 64 kbit/s, PDH, SDH, and ATM are given in Table 11.2. As indicated in the table, the "quanta" used in ATM are cells, and

Table 11.2

Quantization of Phase

	Phase Quanta	Time Quanta
64 kbit/s->1544 or 2048 kbit/s	8 bits	125 µsec
64 kbit/s->VC11 or VC-12	8 bits	125 µsec
64 kbit/s->ATM	n × 8 bits*	125 µsec
PDH->PDH	1 bit	deliberately irregular - equal to PDH frame period eg 100.379 µs for 2/8 multiplexer
PDH->SDH	1 bit	500/n µsec where n = 1 for primary rates n = 12 for 34368 kbit/s n = 36 for 44736 kbit/s n = 36 for 139264 kbit/s
PDH->ATM	376 bits	none
ATM->SDH	424 bits	2.83/N µsec N = 1 for STM-1 N = 4 for for STM-4 N = 16 for STM-16
ATM->PDH	424 bits	6625/n µs n = 24 for 1544 kbit/s n = 30 for 2048 kbit/s n = 530 for 34368 kbit/s n = 636 for 44736 kbit/s n = 2160 for 139264 kbit/s

very large compared to other technologies, and as a result, the justification phase variation from ATM is proportionately larger.

11.2.2.2   Desynchronized Justification Phase Variation

Justification phase variation, while it remains embedded in a server signal, is not normally a problem. However, if it is necessary to produce a regular clock signal for the client signal, the client signal must be "desynchronized" and a regular clock signal reconstructed. At this point, the justification phase variation changes from being "virtual" into real phase variation. The exact characteristics of the real phase variation depends on the justification phase variation and the design of the desynchronizer, but will be similar in form to that illustrated in Figure 11.10.

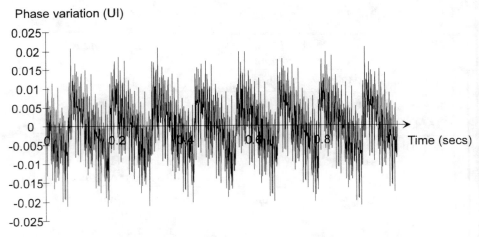

Figure 11.10   Phase variation from the desynchroniser of an 8448 kbit/s to 2048 kbit/s demultiplexer.

The spectral characteristics of justification phase variation are very sensitive to a number of design parameters and are discussed later in the chapter. In PDH multiplexers, careful specification ensured that the jitter part of the phase variation was the only part of interest and so, historically, the PDH interface specifications concentrated on the control of justification jitter. SDH, and to an even greater extent ATM, produce justification wander as well as jitter, and now all the PDH interface specifications in G.823 and G.824 have been enhanced to control justification wander as well as justification jitter.

11.2.2.3   Statistical Multiplexing Phase Variation

PDH, 64 kbit/s, and SDH layers are all circuit-switched layers, so when they are multiplexed into servers, they can be assembled using fixed (if programmable) multiplexes. The VCh and VP layers, however, are based on the idea that they are statistically multiplexed into their server layers.

The buffering delay inherent in the statistical multiplex is discussed in detail in Chapter 9. This buffering delay varies stochastically with time, and it is the *change* in buffering delay with time that causes phase variation. The total delay on an ATM connection, of which buffering is the time-dependent component, is given by the parameter cell transfer delay (CTD), so:

$$p(n) = \frac{d}{dt} CTD(t)$$

The characteristics of statistical multiplexing phase variation will depend on the queuing techniques, the size of the links in the ATM network, and the dynamics of traffic loading. It is also more sensitive to changes in traffic loading than to steady-state statistics. In any case, it is likely to be large in absolute terms, having an amplitude of many 100 sec of UI. In ATM networks, this source of phase variation is likely to dominate over all others.

11.2.3   Spectral Characteristics of Phase Variation

All the sources of phase variation discussed above have different characteristics. It is useful to show these characteristics in a way that can be easily interpreted, with the different sources easily identifiable.

Gapped clocks are subject to the phase variation of the regular clock from which they are derived as well as the quantized phase variation inherent in the gapping. Normally, only regular clocks are directly measured for phase variation, and a generally useful way of characterizing the phase variation is to calculate its spectrum from a measurement over a suitably long period of time. The different sources show significantly different spectra and can be readily identified. Figure 11.11 illustrates a typical phase variation spectrum with all the major sources shown on a single diagram.

11.3   PHASE CONTROL LOOPS

As part of the adaptation source function, there is a process that controls the timing performance by comparing the timing of the client connection with that of the server trail, which is normally generated from a local reference clock. As part of the adaptation sink function, a clock must be reconstructed for the client connection, and this also requires a special timing process in the adaptation function. These timing processes we call phase control loops, and they are illustrated in Figure 11.12.

In this chapter, we use a very general definition of the term "phase control loop," and include within it phase locked loops like a traditional voltage controlled oscillator (VCO) controlled by a linear analog control loop. In SDH and ATM, the control loop circuitry and even the VCO can be implemented digitally, and may also be highly nonlinear. The control loop may not be phase locked, as it may generate slips instead. In addition, each of the input clock signals, output clock signals, and the local reference clocks could be regular clocks or gapped clocks. While we use phase control loop as the general term, we still use the term phase locked loop for any phase control loop that does not produce slip.

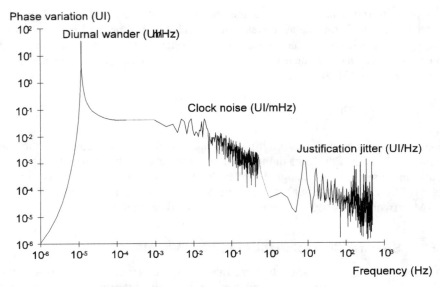

Figure 11.11   Spectral characteristic of phase variation from diurnal wander, clock noise and desynchronizer jitter on a 2048 kbit/s signal.

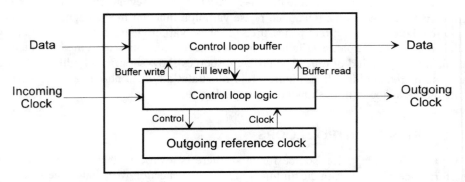

Figure 11.12   A phase control loop.

11.3.1   Line Rate Clock Generation and Recovery Phase Locked Loops

At the source of an optical/coax section, an optical or electrical signal is modulated with the digital signal of the regenerator section, and the role of the modulation scheme is to successfully carry both the signal and its clock over the physical medium. The modulation schemes vary depending on the media, and are discussed in more detail in Chapter 5. The clock used to drive the modulator will be that of the regenerator section so that the clock carried by the modulation scheme is synchronous with the regenerator section. The stability of the clock modulation, and more particularly, the stability of the local clock driving the regenerator section, is specified as part of the line system timing performance in ITU-T Recommendation G.958.

At the sink end of the optical/coax section, the clock must be recovered from the modulated signal before the data can be extracted from the modulated signal. The clock extraction circuit is normally an analog phase locked loop and must extract the clock signal from the presence of frequency offset, clock noise, diurnal wander, and regenerator jitter. It is not sensitive to frequency offset and diurnal wander, as it is simply tracking the incoming modulated signal and these will be well below the filter bandwidth. Both ends are illustrated in Figure 11.13.

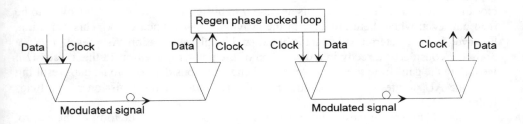

Figure 11.13   Combination of clock and data in line transmission systems.

The performance of this phase locked loop is also specified in ITU-T Recommendation G.958; however, two types have been allowed in the recommendation. This was not the intention in the original standardization program; however, it emerged that two policies of controlling timing through a long series of regenerators was possible and the two types reflect this (each was also strongly supported by a manufacturer). The first policy, assumed in regenerator type A, is to use low-noise, wide-bandwidth circuitry associated with the direct excitation of a crystal resonator with the modulated signal. The second policy, associated with regenerator type B, is to use low-bandwidth, but higher noise circuitry associated with a voltage controlled oscillator phase locked loop.

11.3.2   The AU Pointer Mechanism

An HVC signal has a gapped clock. If we consider VC-4 as an example, it has a gapped clock and all the VC-4s within equipment that have a VC-4 connection function need to be aligned to the same basic gapped clock (this makes it possible to design an efficient time switch for VC-4). When a VC-4 is generated within the equipment, then it will be generated with the basic gapped clock sourced from the local equipment clock. However, if the VC-4 was generated in distant equipment, then its gapped clock must be aligned to that of the VC-4 connection function.

There are two differences between the gapped clock of an arriving VC-4 and the VC-4 connection function basic gapped clock: first, the frame phase of each need not be aligned (i.e., the position of the J1 byte of an arriving VC-4 can occur anywhere in frame phase of the VC-4 connection function basic gapped clock) and second, the regular clocks from which an arriving VC-4 gapped clock and the VC-4 connection function basic gapped clock are derived are independently subject to phase variation, which means that the two clocks will move randomly with respect to each other. Both of these differences are accommodated in the MS-N to VC-4 adaptation sink function, specifically, in the AU pointer processor. The AU pointer process is described in more detail in Chapters 5 and 12; from the timing viewpoint, it is a phase locked loop.

At the heart of the AU pointer processor is a buffer. This buffer must take into account both the gaps from the section overhead in the VC-4 gapped clock as well as the more random gaps from AU pointer adjustments that may have been generated when the VC-4 was carried in previous MSs. As the gaps from the section overhead in the arriving VC-4 gapped clock and the corresponding gaps in the VC-4 connection function basic gapped clock may be in different places, it is necessary to buffer the payload to allow bytes to be read out, even when there are gaps in the arriving VC-4 gapped clock. This buffering, however, is fully deterministic, and is not altered by phase variation. As a result, it is not normally considered directly in a discussion of the pointer processor. In this discussion, we ignore the gaps from the section overhead and only consider occasional gaps resulting from the AU pointer (i.e., the inclusion of the H3 bytes or the omission of the bytes following H3).

The AU pointer processor is a nonlinear phase locked loop. The form of the control loop is similar to the "bang-bang with dead-band" control loop used in many control applications. The arriving VC-4 gapped clock is subject to a number of sources of phase variation: frequency offset, clock noise, diurnal wander, regenerator jitter, and justification phase variation (i.e., incoming AU pointer adjustments). The VC-4 connection function basic gapped clock can have an independent frequency offset and will definitely have independent clock noise. The pointer processor accommodates these differences by buffering to a given limit and then encoding any difference beyond this as pointer adjustments while resetting the buffer. This is illustrated in Figure 11.14. Overall, the pointer processor will attenuate the phase variation, as the buffer has a "dead-band" effect and the size of the buffer is critical to the performance of the pointer processor. The size of the dead band is specified in ITU-T Recommendation G.783 to be four times the size of the AU pointer adjustment, and so is 4 bytes for VC-3, 12 bytes for VC-4, and $n \times 12$ bytes for VC-4-nc.

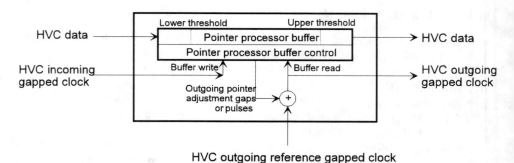

Figure 11.14   The AU pointer processor phase locked loop.

In the other direction (that is, in the VC-4 to MS-16 source adaptation function), the MS-16 will be formed using the basic gapped clock of the VC-4 connection function. The frame phase of each VC-4 together with any pointer adjustments that were produced by the pointer processor are now coded in the AU pointer values in the MS-OH.

11.3.3 The TU Pointer Mechanism

LVCs, like HVCs, have gapped clocks, and similarly, when an LVC is cross-connected in an LVC connection function, it needs to have a gapped clock that is aligned to the basic gapped clock of the LVC connection function. This is illustrated in Figure 11.15.

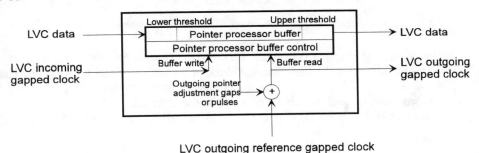

Figure 11.15   The TU pointer processor phase locked loop.

In any situation other than when the LVC is generated locally, there will be a timing difference between the LVC connection function basic gapped clock and the gapped clock of the incoming LVC. The HVC-to-LVC adaptation sink function must accommodate this difference, and this is done by the TU pointer processor.

The TU pointer processor is essentially similar to the AU pointer processor. However, the only incoming phase variation applicable to the TU pointer processor is the justification phase variation from the incoming AU pointer adjustments of the HVC that was carrying the LVC, as well as any TU pointer adjustments already associated with the LVC. There will not normally be any clock noise or frequency offset, as the clock that produced the HVC basic gapped clock (i.e., not including the AU pointer adjustments) is the local reference clock, which will also be producing the basic gapped clock of the LVC connection function. The way in which the HVC to LVC adaptation function will pass AU pointer adjustments onto LVCs is not precisely defined by the standards. However, logic and economic constraints are likely to mean that valid implementations will pass successive AU pointer adjustments to each TU in turn, forwards or backwards depending on the sign of the pointer adjustment, as illustrated in Figure 11.16.

The TU pointer processor has a buffer at its heart in which incoming TU pointer adjustment and any AU pointer adjustments that are passed on to the LVC are buffered. If the buffer reaches the end of its dead-band region, then an outgoing pointer adjustment is initiated and the buffer reset. The size of the TU pointer processor dead band is also specified in ITU-T Recommendation G.783 and is twice the size of the TU pointer adjustment (i.e., 2 bytes for all basic sizes of LVC and $n \times 2$ bytes for the VC-2-nc).

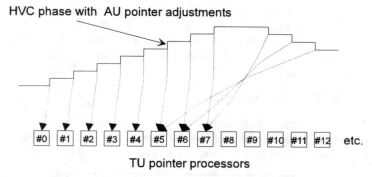

Figure 11.16   The passing of AU pointer adjustments to TU pointer processors by the HVC adaptation.

In the other direction, that is from the LVC to HVC source adaptation function, the HVC will be formed using the basic gapped clock of the LVC connection function. The frame phase of each LVC, together with any pointer adjustments that were produced by the pointer processor through which a LVC went on the other side of the LVC connection function, are now coded in the TU pointer values as defined in the LVC to HVC mapping.

11.3.4   Inserting and Extracting ATM Cells From a Cell Stream

Functionally, the cells of a VCh or a VP can be considered to exist at an arbitrary time. They are only aligned to the cells of a cell stream when the VP is multiplexed into its server layer (for example, a VC-4). This requires that the adaptation source function has a phase control loop that will align the cells of the client VPs with the cell stream of the server layer. The sink function, on the other hand, can deliver the VP cells directly to the VP client without an explicit phase control loop, as the cells of the VP can exist at an arbitrary time. The adaptation source function phase control loop operates a "first in first out" (FIFO) queue, as illustrated in Figure 11.17, and gives rise to two types of phase variation:

- Justification phase variation as the cells of the server cell stream occur at explicit, regular time intervals, which means that a cell in the buffer must wait until the start of the next cell before it is inserted into the cell stream;
- Multiplexing delay, which arises when the FIFO queue starts to fill up as a number of cells from different VPs all arrive in rapid succession and must be buffered until enough cells have passed out in the outgoing cell stream.

Stochastic models for the statistical multiplexing delay is described in Chapter 9 and the size of the buffer required in this phase control loop is also discussed in that chapter. If the buffer fills to its limit, a slip of one cell is inserted in one of the VP streams to restore the buffer. This is the phenomenon of cell loss. This may seem a surprising statement; the buffers in ATM switches are functionally slip-generating phase control loops.

The size of the buffer will depend on the buffering policy, mix of services, the policy toward cell loss, and the rate of the cell stream in the server layer. The buffer size will normally be expressed in a number of cells (for example, 200 cells). The maximum

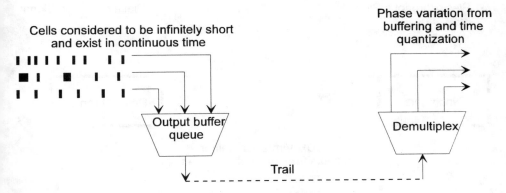

Figure 11.17 Adaptation between ATM VPs and VC-4.

buffering delay is therefore dependent on this size and the cell rate of the cell stream in the server layer. For example, if the server layer is a VC-4, the cell rate is around 350k cells/s, and if the buffer can hold 200 cells, the maximum buffering delay will be 570 ms.

In Chapter 12, the various possible designs of VCh and VP connection functions are discussed, along with the relative merits of input and output buffering. However, functionally, it is still possible to consider this as described above with all the buffering after the connection function.

11.3.5 PDH/SDH Justification Mechanism

In this section, consider the example of a 139264 kbit/s signal carried in a VC-4; however, the characteristics are basically the same for all the PDH signals when carried using their intended mapping. The 139264 kbit/s example has some of the most challenging performance constraints and is taken up later in the chapter.

When SDH is used to transport the 139264 kbit/s in a VC-4, the adaptation must be capable of accommodating the full range of the phase variation allowed on a 139264 kbit/s PDH interface and the most onerous is the frequency tolerance; in the case of 139264 kbit/s, this is +/-15 ppm. A possible technique might be to synchronize the VC-4 to the 139264 kbit/s signal and use the AU-4 pointer to encode any timing difference. However, there is a limit to the frequency difference that the pointer mechanism can effectively manage. While there is no hard bound to this limit, it has been agreed internationally that it should be +/-4.6 ppm. The 139264 kbit/s to VC-4 adaptation function must therefore incorporate more than the AU pointer mechanism to accommodate the +/-15 ppm.

A further complication is that the VC-4 may have pointer adjustments in its gapped clock. The adaptation has to produce a regular clock for the 139264 kbit/s interface and it must smooth out these pointer adjustments to produce a clock that meets the 139264 kbit/s jitter and wander specifications. This is illustrated in Figure 11.18.

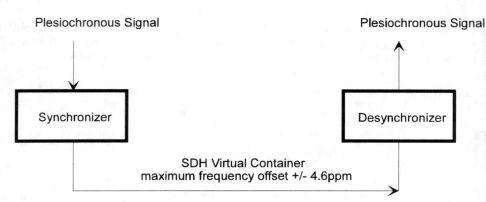

Figure 11.18   Synchronizing and desynchronizing plesiochronous signals carried in SDH containers.

11.3.5.1   *Synchronizing Plesiochronous Signals in the Adaptation Source Function*

The method used to synchronize is very similar to that used by existing plesiochronous multiplexers. A nominal capacity approximately equal to that of the plesiochronous signal is allocated in the VC and bits are passed into this capacity via a buffer. The fill level on the buffer is used to control the exact amount of capacity allocated in the VC, and as such, the synchronizer is just another nonlinear phase locked loop like the pointer processor. The basic principles of the mechanism are very similar to that of the pointer mechanism, with two main differences: The amount by which the VC capacity is adjusted is 1 bit and the capacity adjustment is signaled by sending the change in phase.

The VC capacity is divided into blocks, and the number of bits used in each block given to the input signal is encoded and sent forward (in the C bits). If this information on the capacity allocation is corrupted for any reason, it is impossible to recover it from the information passed in future blocks, which means that bits are added or deleted from the signal permanently, resulting in the need to reframe the transported PDH signal downstream.

Two basic schemes are used:

- *Positive justification*—The nominal capacity allocated in the VC is more than the maximum required by the plesiochronous signal. The justification scheme reduces the capacity when the buffer empties below a threshold, as shown in Figure 11.19 (this is used for DS-3 and 139264 kbit/s);
- *Positive-zero-negative justification*—The nominal capacity allocated in the VC is equal to the nominal of the plesiochronous signal. If the plesiochronous signal is above the nominal, extra capacity is allocated when the buffer fills above a threshold; if the plesiochronous signal is below the nominal, less capacity is allocated when the buffer empties below a threshold, as shown in Figure 11.20 (this is used for DS-1, DS-1C, 2048 kbit/s, DS-2, and 34368 kbit/s).

The first scheme is very similar to those used by existing plesiochronous multiplexers and the scheme is completely specified by the mapping format. All the SDH mapping formats are described in Chapter 5. The buffer only empties, and there is only one threshold,

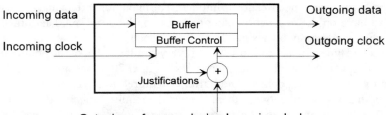

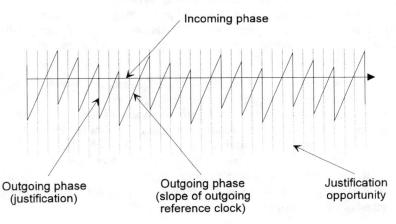

Figure 11.19   Synchronizer phase locked loop using positive justification.

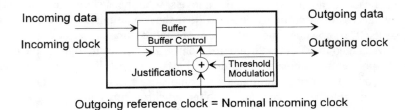

Figure 11.20   Synchronizer phase locked loop using positive-negative justification.

so the jitter performance specification is implicit. The maximum effect that the justification scheme can have on the desynchronizing end is also, therefore, predetermined, and the desynchronizing ends may be designed with confidence.

The second type of justification scheme is very similar to that of the pointer processor. However, the specification of the mapping formats in G.707 does not completely define the operation of the synchronizer. The buffer is required to have an upper and a lower threshold with a dead band between them to meet the required specification, but its size is not specified by the mapping. As there are several ways to implement this type of synchronizer, the desynchronizer cannot be designed without some further specification. It was agreed at an early stage in T1X1 that no particular mechanism should be specified, but the performance should be constrained by specification. Several implementations have been suggested, some of which have been claimed to reduce the amount of jitter encoded in the justification scheme, such as:

- Implement a two-threshold buffer with one UI spacing between the thresholds and justifications made when the thresholds are crossed.
- Lock some of the negative justifications in a fixed pattern so that they are forced to carry data, which means that the buffer and threshold control may now operate as a positive justification scheme.
- Add an additional phase signal to the buffer that has a long term of zero, but with sufficient amplitude and frequency to generate frequent justifications (this effectively modulates the level of the thresholds, and this scheme is referred to as threshold modulation).

The second scheme is rather simple to implement, but it does not give a significant improvement over a simple implementation. The success of the third scheme depends on the signal used to modulate the threshold levels. It has been shown that such schemes can be very successful at reducing encoded jitter that come purely from the synchronizing process itself. If, however, there is jitter on the input signal, its effectiveness is reduced.

It was noted in ETSI that if the desynchronizer has to smooth out the eight UI steps from the pointer adjustments, the additional affect of one UI steps from the justification scheme should present no additional extra complexity burden. In light of this, there seemed little point in adding extra complexity to the synchronizing end.

ANSI has worked on an assumption of the second method (locking some negative justifications), although this has since been demonstrated to be of little benefit, and ETSI has worked on an assumption of the first method (simple positive-zero-negative with two buffer thresholds spaced by one UI), although these assumptions are not part of the specification.

To guarantee transverse compatibility so that any synchronizer can work into any desynchronizer and still meet the jitter limits, it is necessary to specify the required performance of the synchronizer. This can be done by specifying the maximum level of the signal recovered in a defined low-pass filter operating on the output of the synchronizer justification channel. In ANSI T.105 is a simple description of the performance without any defined measurement method, while ETSI specifies the use of a digital filter that uses the justification control bits as the input signal. The digital filter then models a low-pass filter representing a desynchronizer and a high-pass filter representing the jitter measurement filter. This is illustrated in Figure 11.21. The filter frequencies for the different bit rates are given in Table 11.3.

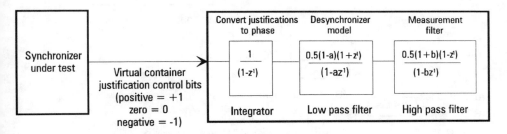

Figure 11.21   Measurement method of a synchronizer.

Table 11.3

Levels of Synchronizer Justification Jitter

	LPF	*HPF*	*Limit*
1544 kbit/s	350 Hz	10 Hz	1.5 UI
2048 kbit/s	40 Hz	20 Hz	0.3 UI
34368 kbit/s	200 Hz	100 Hz	0.3 UI
44736 kbit/s	1000 Hz	10 Hz	1.5 UI
139264 kbit/s	500 Hz	200 Hz	0.3 UI

11.3.5.2   *Desynchronizing Plesiochronous Signals*

The STM-N arriving at the final desynchronizing element must be adapted several times between intermediate layers before the plesiochronous signal can be recovered. Each adaptation is carried out using a gapping process so that the gapped clock signal for each layer is produced by additionally gapping the gapped clock signal from its server layer. These multiple interlayer adaptations are illustrated in Figure 11.22.

The recovered plesiochronous signal that appears in the final adaptation function is not a regular clock signal, but contains the accumulated gapping components from all the intermediate layers. The desynchronizer takes this gapped clock signal and produces a regular clock signal with a specification such that the plesiochronous interface can meet the jitter and wander requirements of existing PDH transport networks, while at the same time not producing slip. The plesiochronous signal is clocked into a buffer using the gapped clock signal and clocked out using a regular clock signal, while the fill of the buffer is used to control the exact frequency of the regular clock signal; however, it must not adjust the regular clock in such a way as to exceed the existing PDH jitter and wander limits. The desynchronizer is, yet again, an example of a phase locked loop. However, this time, unlike the pointer processor and synchronizer, its control loop cannot be too nonlinear if the jitter

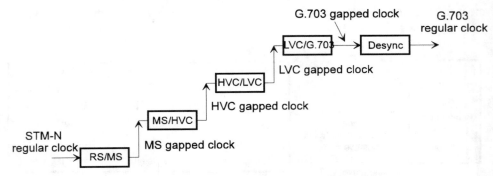

Figure 11.22   Position of AU pointer processor and TU pointer processor.

and wander criteria are to be met. The desynchronizer at the plesiochronous boundary is one of the rather more complex aspects of the SDH because of the following:

- The effective time constant of the control loop must be long if jitter and wander limits are to be met, generally longer than one second (this is especially true of the higher bit rates DS-3, 34368 kbit/s, and 139264 kbit/s); the pointer adjustments may be encoding a frequency offset of up to 4.6 ppm.
- The size of the buffer should be small, both to reduce the network delay and the complexity of the desynchronizer.

Much meeting time in T1X1 was devoted to establishing that simple implementations existed for the desynchronizer that could meet these requirements. Early in the discussions it was recognized that conventional linear analog control loops would be very difficult to implement and would require very large valued components, so this was not considered to be a viable implementation. It was concluded that some form of digital circuit was required before the analog circuit to reduce the effect of the step from the pointer adjustment that produced the most troublesome gaps. Two techniques are used, and these are discussed in more detail in Chapter 12.

The first technique uses a control mechanism that buffers the pointer adjustment and passes the phase out of the buffer one bit at a time (or even a fraction of a bit at a time) with a fixed interval between the bits. These single "leaked" bits are then fed to a conventional phase locked loop. The limitation of fixed bit leaking is that the maximum rate at which bits can be passed out is predetermined, and if the pointer adjustments represent a frequency offset of greater than this, then the buffer will overflow or underflow. This limit for bit rates other than DS-1 and 2048 kbit/s is generally below 4.6 ppm, so the scheme cannot meet all the requirements for these higher bit rates.

The second technique uses a linear digital phase locked loop for the pointer adjustments sampled at a relatively slow rate (e.g., 125 μs) to produce a simple desynchronizer that will cope with any input in a stable way. The output from the linear digital phase locked loop is a digital control signal. This can be encoded in single bit justifications using a +0- justification scheme. A possible implementation of a desynchronizer based on this principle is described in Chapter 12.

11.3.6   CBR/ATM Justification Mechanism

There are many existing layers that we would wish to carry on ATM that are not network synchronous, like 64 kbit/s, but at constant bit rate, and included in this are all the existing PDH layers. The ATM Forum circuit emulation service (CES) defines two ways of carrying CBR clients—adaptive cell desynchronization and SRTS—both within AAL1.

11.3.6.1   *Adaptive Cell Desynchronization*

Adaptive cell desynchronization is a simple scheme whereby no external timing reference is required and no explicit timing information is created in the adaptation source. The adaptation source buffers the incoming signal until there is sufficient data to fill a cell, at which point the cell is launched into the VCh. At the adaptation sink, the cells arrive and the long-term average arrival rate is used to reconstruct the regular clock of the client signal, which requires a nonslipping phase locked loop in the adaptation sink function. This overall mechanism is illustrated in Figure 11.23.

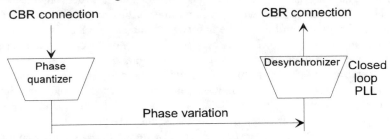

Figure 11.23   Adaptive cell desynchronization of ATM cell streams.

Many early workers on ATM assumed there would be no difficulty in designing a simple circuit that could average the arrival times of cells carrying a CBR client and reconstruct the regular clock. However, we noted that designing a suitable circuit for pointer adjustments is not trivial and the phase variation from cell justification and statistical multiplexing is many times greater than that of pointer adjustments.

This early work on adaptive cell desynchronization centered on building examples and testing them, and little effort was put into any theoretical analysis. The early tests showed promising results; for example, a 2048 kbit/s signal carried in a VCh with the cells carried in a VC-4, when passed through a prototype ATM crossconnect, could be desynchronized while keeping within the limits of G.823. The crossconnect was not subject to the dynamic variations in load and the example of 2048 kbit/s in a VCh in a VC-4 was only one of many possible examples, and other examples may behave differently.

More recent theoretical work and practical tests based on this theoretical work suggest that adaptive cell desynchronization should not be used for the transport of PDH signals, as many circumstances will arise where G.823 cannot be met. For other client layers, where the specification of the regular clock signal is much looser, this scheme may be of use. The general conclusion, which has now been agreed upon in ATM Forum for the circuit emulation service, is that for nonsynchronous CBR clients, SRTS should be used unless a network reference clock is unavailable, in which case adaptive cell desynchronization can be used.

11.3.6.2   *The SRTS ATM Justification Mechanism*

AAL1 defines a synchronous residual timestamp (SRTS) mechanism, which is described in Chapter 5. This is a mechanism that allows CBR clients to be carried and their regular clock recovered in the presence of phase variation from cell justification and statistical multiplexing. While it has many of the features of an adaptive scheme, it is actually a slipping scheme and relies on a network clock. If there is any difference between the network reference clock at the adaptation source and network reference clock at the corresponding adaptation sink, slip will occur in the adaptation sink.

The adaptation source works in the same way as adaptive cell desynchronization in that it has a buffer and a cell is launched when the buffer has filled with sufficient bytes to fill a cell. However, in addition, it has a mechanism for dynamically measuring the rate of the CBR signal. It creates a local cyclical counter that is synchronized to the local reference clock, and this is read every given number of bits of the CBR signal (3,008 in the case of PDH signals), and this value is forwarded across the 8 cells required to carry the 3,008 bits (in fact, it is the difference between the value of the counter and an expected value established from the nominal rate of the CBR signal that is forwarded).

The adaptation sink function must have a desynchronizing buffer that is big enough to hold all phase variation arising between the source and the sink, including phase variation from diurnal wander, cell justification, and statistical multiplexing. The buffer is filled with bytes from cells as they arrive; however, they are played out by realigning the timestamp counter values from the incoming cells with a parallel cyclical counter in the adaptation sink function. This will work as long as the two cyclical counters are working at the same rate. If one is running faster than the other as a result of a difference between the two local reference clocks, the buffer will either empty or completely fill, and the adaptation sink function will have to slip the client signal to reset the buffer. This is illustrated in Figure 11.24.

Desynchronizers based on the SRTS mechanism are fully capable of meeting the requirements of G.823 and G.824 when there is an accurate network reference clock available at both ends. This is increasingly the case, as the technology for distributing accurate synchronization is improving.

The phase variation from the statistical multiplexing will depend on the design of the ATM network, and in particular the buffering strategy in the switches and the capacity of the links. Both the ATM Forum and the ITU-T have declined giving a figure for a required size of desynchronizing buffer, stating it is an implementation decision. This is not the case, as the phase variation accumulates across the whole network, making it an interworking standards issue like the buffer spacing in SDH pointer processors.

The reason has its origins in the two approaches for designing ATM networks described in Chapter 9. "Thin" ATM networks based on low-capacity links and large switch buffers may require large buffers in the desynchronizers, and figures of around 6 ms have been quoted. On the other hand, "thick" ATM networks based on high-capacity links and small switch buffers may only require buffers of 500 μs to 1 ms. Neither side

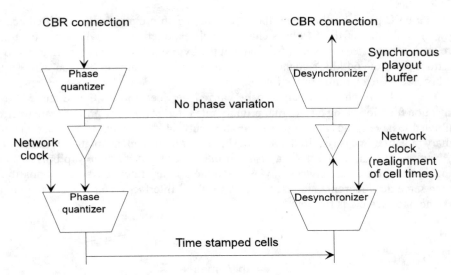

Figure 11.24   Adaptation using SRTS.

liked the cost of accepting compatibility with the other, and this is another example of where network operators and manufacturers need to be careful because ATM standards are not as complete as the large quantity of paper would suggest. Early transverse compatibility issues are highly likely where ATM is used to carry time-critical information.

11.3.7   Transport of the 64 kbit/s Layer

11.3.7.1   *Transport of 64 kbit/s on SDH*

The transport of 64 kbit/s on SDH is defined using the byte-synchronous mapping. This mapping, as described in Chapter 9, makes each 64 kbit/s channel observable as a byte in the LVC mapping.

When the 64 kbit/s channels are being delivered directly to a 64 kbit/s connection function with the SDH equipment, then the adaptation would naturally integrate the slip buffer that is required to align all the 64 kbit/s channels to the synchronous rate of the connection function. There are several different ways in which this could be implemented, each of which requires a different amount of buffering. One method would have a separate wander buffer, slip buffer, channel concatenation buffer, and time switch alignment buffer, each according to their specification. It is also possible to combine these buffers into a single buffer between the LVC termination and the 64 kbit/s connection function. This mechanism is a single slip buffer in the adaptation sink function, and when a slip occurs, all 24 or 30 64 kbit/s channels (depending on the primary rate) will slip at the same time.

It is possible to reduce this buffering further by using the pointer information from the LVC to directly control the first time switch of the connection function. Rather than aligning the channels at the 64 kbit/s connection function in timeslot order, they can be lined up as they arrive and the time switch reprogrammed to take account of the order. If there is a LVC pointer adjustment, then the time-switch programming is shifted accordingly and only one of the 64 kbit/s channels will experience a slip. Few manufacturers have, as yet, exploited this simplification that the byte synchronous mapping gives to the construction of the 64 kbit/s connection function, and it remains to be seen whether this will become more common in the future.

The 64 kbit/s channels can also be carried between the LVC and the 64 kbit/s connection function using an external primary rate interface. As the primary rate interface is allowed to have some level of phase variation, any 64 kbit/s connection function with a primary rate interface is required to have a slip buffer in the sink adaptation function. This is illustrated in Figure 11.25. As a result, it is not necessary to have slip buffers in the adaptation sink function between the LVC and the primary rate interface, it is possible to use the same desynchronization circuitry as for PDH interfaces, and it is possible to use the same interface cards for both.

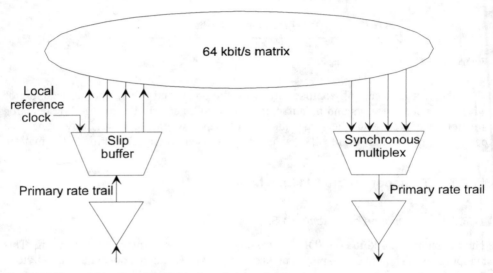

Figure 11.25   Adaptation of 64 kbit/s channels in a primary rate server.

There is, however, a specific requirement in the primary rate to LVC adaptation source function that needs to find the 125-ms frame alignment of the primary rate signal and generate a frame reference to which the LVC payload may be aligned.

11.3.7.2   *Transport of 64 kbit/s on ATM*

The circumstances under which ATM might be carrying 64 kbit/s client channels are complex. Unlike SDH, ATM is not specifically designed to carry 64 kbit/s channels in a way that is directly compatible with the existing 64 kbit/s switches, and the ideal assumption has been that ATM will carry every 64 kbit/s trail directly as an ATM VCh trail. Any explicit 64 kbit/s switches are then not required. However, there comes the point at which 64 kbit/s connections on ATM do need to interwork with existing 64 kbit/s switches, generally as an evolutionary arrangement. Architecturally, we can consider two options, even if practical examples include hybrids of both:

- The 64 kbit/s layer is embedded as a sublayer in the adaptation function and all switching is done by ATM VCh connection functions according to the ideal assumption described above;
- ATM is used to transport 64 kbit/s channels between 64 kbit/s connection functions where the ATM is only providing a link transport service to the 64 kbit/s client layer network.

In the first case, one ATM VCh must be used for each 64 kbit/s connection, and AAL1 has been standardized for this application. In the second example, a number of options are available, as it is possible to multiplex several 64 kbit/s link connections into one ATM VCh. One option is to multiplex the 64 kbit/s link connections into a primary rate signal and then carry the primary rate signal on the ATM VCh. Another is to use the frame structure that the ATM Forum has agreed to as part of AAL1 for the transport of a programmable number of 64 kbit/s link connections on a single VCh.

All three of these basic options—single 64 kbit/s link connection per VCh, use of a primary rate, and $n \times 64$ kbit/s frame structure—use the same timing technique. In the 64 kbit/s to VCh adaptation source function, the 64 kbit/s channels are multiplexed and segmented into cells, which requires an adaptive phase locked loop. The multiplexed 64 kbit/s channels are fed into a buffer, and when there are sufficient bytes to fill a cell, they are loaded into the cell.

This phase locked loop introduces a delay, as there is a delay in collecting sufficient bytes to fill a cell. This is governed by the number of connections that are multiplexed into a single VCh. For a single 64 kbit/s connection, this delay is just under 6 ms (i.e., 47 times 125 ms). For an E1 primary rate signal, the delay is reduced to around 185 ms, and it is fixed and predictable.

Since 64 kbit/s is already defined as a synchronous layer, the adaptation sink function must include a slip buffer that will produce byte slips if the timing of the 64 kbit/s that has been carried on the ATM VCh is different than that of the timing of the local 64 kbit/s connection function. However, the ATM VCh layer and its servers produce more phase variation than diurnal wander and frequency offset assumed when the existing slip performance standards were set in ITU-T Recommendation G.822.

The ATM layers will add cell justification phase variation and statistical multiplexing phase variation to the VCh, and this must be taken into account in the adaptation source function that recovers the 64 kbit/s signal. The amount of phase variation arising from these sources will vary greatly depending on the design and size of the ATM network, so it is hard to predict at this stage. If the adaptation function is to meet the existing requirements of G.822, then it must completely buffer the phase variation from these sources. The size of this buffer is the same as the one described in the section on SRTS, and a value has not yet been agreed to as a standardized interworking parameter. Worst case analysis suggests that at least 6 ms of buffer is needed to meet the requirements.

While such a buffer can, in principle, eliminate the phase variation caused by cell justification and statistical multiplexing, the 6 ms of buffer creates a new difficulty. We noted that when a single 64 kbit/s is mapped into an ATM VCh, this created 6 ms of delay, and now when it is recovered from the VCh, there is an additional delay of up to 6 ms, so 12 ms of delay has been added to the 64 kbit/s connection in the adaptation functions. This delay is new and is not envisaged in the delay budget that has been established over many years for the control of echo in the PSTN and specified in ITU-T Recommendation G.114. Adding such a delay, at minimum, requires that echo control needs to be considered in the network design of the interworking between 64 kbit/s carried on an ATM VCh and the existing PSTN/ISDN. The network situations in which this interworking is taking place will vary. The ATM VCh to 64 kbit/s adaptation sink function is likely to require all of the following:

- A 6 ms slip buffer;
- Some means of detecting whether the 64 kbit/s connection is carrying a PSTN connection or a purely digital connection (this could be derived from signaling information and be passed to the adaptation function from the connection control processor);
- An echo control device that can be activated if the 64 kbit/s is carrying a PSTN connection.

Such complexity makes direct interworking economically unattractive, and it seems more likely that alternatives will be found wherever possible. The transport of multiple channels reduces the delay, but is not consistent with the goal of replacing 64 kbit/s switches. The alternative is to move very rapidly towards end-to-end VChs. This suggests that broadband terminals will have independent narrowband and broadband voice capabilities. If the required destination is only capable of PSTN, then the broadband terminal will use PSTN capability that is fully compatible with the PSTN and all its delay requirements. On the other hand, if the destination also has broadband capability, a direct VCh can be assigned for the voice communication.

11.4   CHARACTERISTICS OF IMPAIRMENTS AT DIFFERENT LAYERS

11.4.1   Regenerator Section Jitter

In the early days of digital transmission, with regenerators every 1–2 km along the line system, control of regenerator jitter was a serious problem. Today, with optical transmission systems with regenerator spacings of more than 50 km, the problem is markedly reduced. This is even more the case when optical amplifiers are used. The subject of regenerator jitter requires complex analysis and would require more space to adequately describe than is warranted by the subject when taken in the context of all the other sources of phase variation.

11.4.2   The Statistics of AU Pointer Adjustments

In attempting to assess the nature of the pointer statistics, it is necessary to form a good model of the phase variations that cause AU pointer adjustments. Such a model has to take into account the four main sources of phase variation that cause AU pointer adjustments;

that is, frequency offset, diurnal wander, clock noise, and incoming AU pointer adjustments. In principle, regenerator jitter could cause pointer adjustments, however, much of this is filtered in the clock recovery circuit and any residual jitter is small compared to the size of the AU pointer processor buffer.

11.4.2.1   AU Pointer Adjustments From a Long-Term Frequency Difference

Consider the example where there is a difference between an incoming VC-4 gapped clock and the VC-4 connection function basic gapped clock of +0.1 ppm. All other sources of phase variation are assumed to be zero. The outgoing VC-4 connection function basic gapped clock is 150336 kHz while that of the incoming VC-4 is 150366.015 kHz; that is, a difference of 15 Hz. In terms of gradient on the graph, this is 15 *unit intervals* (UI) per second.

Every second, the outgoing reference gapped clock will clock 15 fewer bits out of the AU pointer processor buffer than are clocked in. For the VC-4, the pointer adjustment is 24 bits, so the pointer processor will have to make a negative pointer adjustment every 1.6 seconds to ensure that the same number of bits are clocked out as are clocked in and that the pointer processor buffer does not overflow.

In practice, the pointer processor decisions to make the pointer adjustments are based on when the buffer fill crosses the threshold. This is shown in Figure 11.26, and it can be seen that pointer adjustments are produced every 1.6 seconds and that the full VC-4 gapped clock, including the AU pointer adjustments, is 150,366.015 kHz. The frequency offset has effectively been quantized and encoded in the AU pointer adjustments.

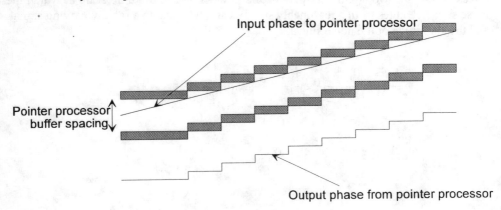

Figure 11.26   Response of the AU pointer processor to a frequency offset.

11.4.2.2   AU Pointer Adjustments From Diurnal Wander

The phase variation can be represented as a sinusoidal signal of periodicity 86,400 seconds (equivalent to one day). A long line system might give rise to a diurnal wander amplitude of 2 μs peak to peak, and this is illustrated in Figure 11.27.

Again, considering the example of the VC-4, the VC-4 connection function basic gapped clock is 150,366 kHz and the 2 μs of wander is equal to 300 UI. The pointer processor buffer has 96 bits between the thresholds, and it may be expected that the wander will result in approximately two pointer adjustments in each direction. The way the thresholds will determine the pointer adjustments is illustrated in Figure 11.27.

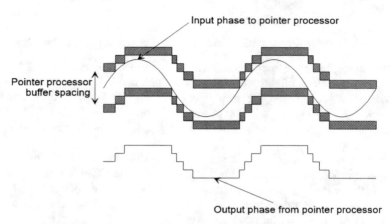

Figure 11.27   Response of AU pointer to diurnal wander.

11.4.2.3   AU Pointer Adjustments Resulting From Clock noise

Figure 11.28 illustrates a typical, noise-like phase error signal viewed over different time scales. There are two significant parameters in clock noise: the time scale at which the noise signal changes from a reasonably slowly varying signal to a rapidly varying signal, and the amplitude of the noise signal.

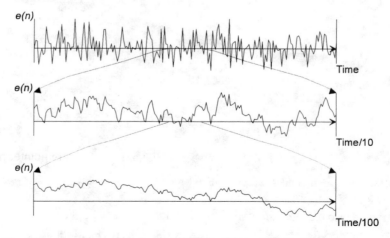

Figure 11.28   The appearance of clock noise when viewed over different timescales.

If the amplitude of the noise signal is sufficiently small, then it will be absorbed in the pointer processor buffers. Most noise processes result in a signal that, if sampled at random, has a phase error that fits a normal (gaussian) distribution. The normal distribution has the property that large phase errors are less likely than small ones. For example, the pointer processor buffer of the VC-4 has 96 UI hysteresis. If a local reference clock produces noise with a standard deviation of less than 50 nanoseconds, it is unlikely to produce pointer adjustments.

If the clock noise is of sufficient amplitude to produce pointer adjustments, the time scale at which the noise signal changes (from slowly changing to rapidly changing) determines the time between the pointer adjustments.

The interrelationship between these two parameters can be seen in Figure 11.29. The way in which the AU pointer processor responds to clock noise is illustrated in Figure 11.30.

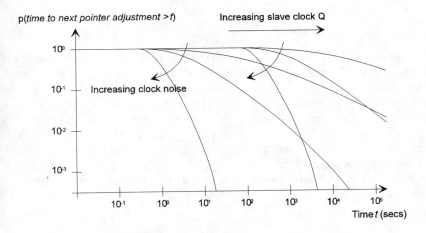

Figure 11.29   The effect of noise level and noise spectrum on pointer adjustments.

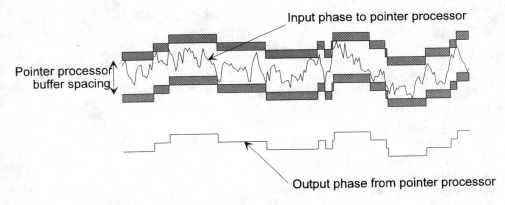

Figure 11.30   Response of Au pointer processor to clock noise.

11.4.2.4   General Case of AU Pointer Statistics for a Single Pointer Processor

The general case has to consider the effects of all three sources of phase variation. In most practical situations, one source is likely to dominate. For example, in the case where a local slave clock has lost reference to its master and has entered a holdover condition, the frequency offset is likely to dominate the other two mechanisms as the main source of pointer adjustments. However, in normal synchronized operation, it will be the level of the clock noise that is likely to determine the AU pointer statistics. This has been a largely unknown quantity, as slave clocks designed for 64 kbit/s digital exchanges are not generally tested for this parameter since the 64 kbit/s switches are not sensitive to it. The design of synchronization networks that control clock noise is discussed later in this chapter. For the general case, an aggregate phase variation signal is considered, which is the addition of the phase variations from the individual sources.

11.4.2.5   Accumulation of AU Pointer Adjustments

The case where the incoming VC-4 gapped clock has no incoming pointer adjustments has now to be considered. When the incoming gapped clock does contain pointer adjustments, this will result in steps on the phase variation signal of the incoming gapped clock. The phase variation signal will be the sum of the stepped phase signal from the previous pointer processor, a slope from any frequency difference between the incoming reference gapped clock and the outgoing reference gapped clock, any wander from the line system, and the difference between the slave clock noise signals on the incoming reference gapped clock and the outgoing reference gapped clock. The effect of the combined phase signal is illustrated in Figure 11.31. A path with several connections and hence, pointer processors, is illustrated in Figure 11.32.

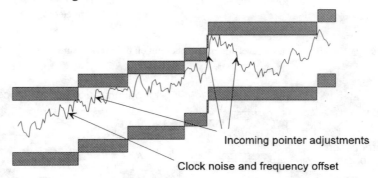

Incoming pointer adjustments

Clock noise and frequency offset

Figure 11.31   Response of AU pointer processor to combined clock noise, frequency offset and incoming AU pointer adjustments.

Analytical prediction of accumulated pointer activity is very difficult as the basic pointer processor mechanism is nonlinear. To achieve some level of confidence in the accumulation properties on pointer adjustments, extensive simulations have been undertaken within standards bodies. These have shown that the accumulation is inherently stable and that after a certain number of tandem pointer processors, pointer activity does not increase significantly despite more phase variation added at each node. Figure 11.33 shows

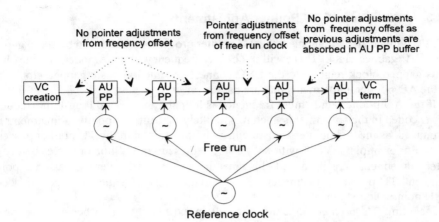

Figure 11.32   Cascade of AU pointer processors on an HVC trail and their response when one pointer processor is timed to a clock with frequency offset.

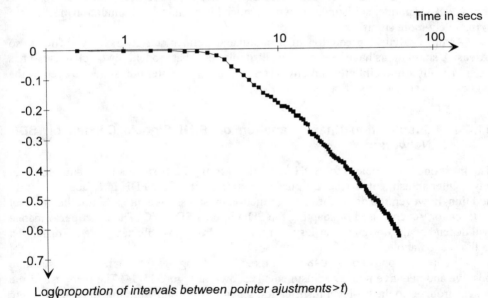

Figure 11.33   Simulated results of the distribution of the time between AU pointer adjustments following four cascaded AU pointer processors each with the maximum allowable clock noise.

the results of a simulation of four tandem AU-4 pointer processors, each with independent clock noise added to its input at the maximum allowable level.

11.4.3   Statistics of TU Pointer Adjustments

The LVC gapped clock that enters the TU pointer processor is produced by fixed gapping of the HVC gapped clock. There will not be any frequency offset, wander, or clock noise on this gapped clock relative to the LVC connection function basic gapped clock other than the AU pointer adjustments that are passed on.

If the AU pointer adjustment frequency is low and there is no long-term frequency offset encoded in the adjustments, then it is unlikely that any TU pointer adjustments will be produced, as the AU pointer adjustments will be absorbed in the TU pointer processor buffers. For example, a VC-4 with 84 VC-11s represents 1,344 bits of buffer before any pointer adjustments are produced. In general, other than the very occasional pointer adjustment, TU pointer adjustments are only produced by a frequency offset encoded in the AU pointer adjustments.

Incoming TU pointer adjustments result in exactly the same behavior in the TU pointer processor as incoming AU pointer adjustments cause in the AU pointer processor. Relative to the pointer adjustment, the TU pointer processor buffer spacing is less than half that of the AU pointer processor buffer (two as opposed to four adjustments). Thus, the overall TU pointer adjustment activity is much lower and the accumulation mechanism is relatively more significant.

Again, analytical prediction of accumulated pointer activity is very difficult, so extensive simulations have been run. In all simulations that model only clock noise, it has been virtually impossible to find any TU-11 or TU-12 pointer adjustments, even at the highest allowance levels of clock noise.

11.4.4   Justification Phase Variation on PDH Signals Carried on SDH Networks

The most onerous problem for SDH networks carrying PDH signals is the smoothing of any pointer adjustments so that the phase variation limits of the PDH interface are met. In addition, however, the desynchronizing function must also take into account the bit-level justification scheme used in adapting the PDH into its SDH VC. The exact performance will depend on the design of the desynchronizer, and two possible designs were discussed in the section above.

The jitter response of a 2048 kbit/s fixed bit leaking desynchronizer to consecutive positive and negative pointer adjustments is shown in Figure 11.34. The response of the desynchronizer to the bit-level justification scheme is not shown in this diagram as there are several ways in which this could be incorporated into the design. As can be seen from the diagram, the jitter transient from each leaked bit is well below the limit specified within ITU-T Recommendation G.783 of 0.175UI.

The response of an adaptive desynchronizer designed for the 139264 kbit/s signal to a single isolated pointer adjustment is shown in Figure 11.35. In this case, the design produced the high-frequency jitter associated with a PDH bit-level justification scheme, as this is the way the phase information on the pointer transient is passed to the clock generating circuit. The response is therefore a single jitter transient from the pointer adjustment with high-frequency justification jitter superimposed on the top. Again, the

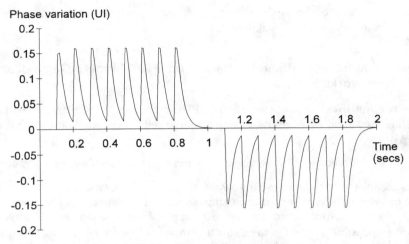

Figure 11.34   Response of a 2Mbit/s fixed bit-leaking desynchronizer to a positive and a negative pointer adjustment.

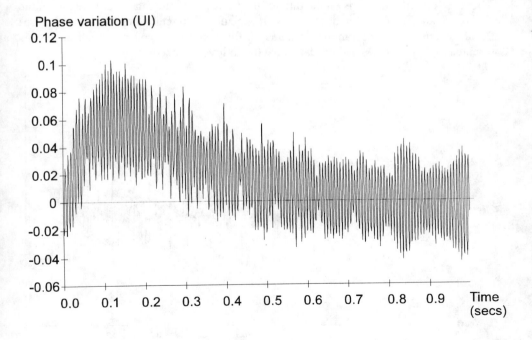

Figure 11.35   Response of a 140Mbit/s linear digital control desynchronizer to a single positive pointer adjustment.

amplitude as a result of the single pointer adjustment is well within the limit specified in ITU-T Recommendation G.783 of 0.175UI.

11.4.5   Justification Phase Variation on PDH Signals Carried on ATM Networks

The nature of justification phase variation on PDH carried on ATM networks will depend on whether they are desynchronized using adaptive cell desynchronization or SRTS.

11.4.5.1   *Justification Phase Variation From Adaptive Cell Desynchronization*

Justification phase variation resulting from adaptive cell desynchronization is hard to analyze other than for very particular circumstances. This is one reason why there has been extended discussion in standards bodies on the applicability of adaptive cell desynchronization. It is possible to state some general principles that frame the size of the phase variation coming from cell justification and statistical multiplexing. From this, the results of desynchronization can then also be framed.

Figure 11.36 shows the effects of cell justification. The example includes a desynchronizer and is measuring jitter in a jitter measuring set. As can be seen from the diagram, the level of the jitter depends on the ratio of the CBR rate to the total available rate in the cell stream, and is a complex dependency. It is actually a fractal pattern that has features of both deterministic and random systems. A full analysis of cell justification jitter is presented in Appendix A, and we can make the following observations:

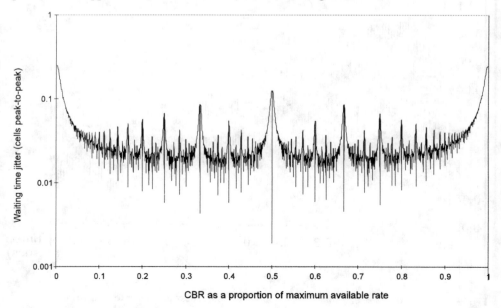

Figure 11.36   Waiting time jitter in an adaptive cell desynchronizer.

- Justification of a CBR signal into a cell stream results in a complex, repeating pattern of filled and unfilled cells.
- When the pattern is very regular with very occasional resets to the pattern, a single transient of phase variation is produced that enters the desynchronizer.
- The amplitude of transients is inversely proportional to the length of the repeating pattern, and the maximum amplitude is therefore one cell, or 376UI.
- The width of the jitter peaks, in Figure 11.36, and hence the chance of encountering a peak, depends on the rate of the cell stream and the jitter measuring set and not on the desynchronizer.
- Low bit rate access to an ATM network is likely to generate significant cell justification jitter.

In some ways, the final conclusion is reasonable. The time taken to send a cell on a low bit rate cell stream is long, in relative terms, and the justification time is therefore proportionately longer. A lightly loaded, high bit core ATM network is unlikely to affect this cell justification phase variation, and it will arrive at the desynchronizer. This dependency on the cell stream bit rate makes cell justification jitter hard to control, other than accepting it may happen. Designing a desynchronizer that can handle an isolated transient of 376 UI is a very difficult design problem!

The multiplexing phase variation may well be larger than the cell justification, and its characteristics are poorly understood. While a significant amount of analysis has been carried out on statistical multiplexing, this has been oriented at gauging the size of buffers required in ATM equipment. To do this, it is possible to use models with static traffic loading. However, adaptive cell desynchronization is sensitive to the *rate of change* of buffer fill in ATM equipment, and this depends on the dynamics of the traffic loading. A sudden surge in traffic could cause a very large transient in phase variation, potentially of many cells in size. At the moment, it is safe to say that an adaptive cell desynchronizer will need to cope with transients of several cells (i.e., ~1,000 UI). This is not easy, and if adaptive cell desynchronization is going to be used at all, it is likely to be in the context of new applications that are not sensitive to these effects.

11.4.5.2 *Justification Phase Variation From SRTS*

The resolution of SRTS as defined for PDH signals depends on the CBR rate, and will be between half a bit and one bit, and this determines the performance of the STRS scheme. The effect is the same as the cell justification scheme described above in that a justification pattern is set up that will have peaks at particular rates of CBR signal. In this case, however, the amplitude of the highest peaks is between 0.5 UI and 1 UI. In this case, the worst case is isolated transient on 1 UI, which must be filtered in a simple desynchronizer to meet the requirements of G.823 or G.824.

11.5 SYNCHRONIZATION NETWORKS

A synchronization network is a set of clock nodes that are maintained in synchronization with one another. To achieve this, it is necessary to accurately transfer synchronization reference information between nodes so that their relative synchronization may be monitored and maintained.

11.5.1   Existing Synchronization Networks

The PSTN/ISDN is based on the 64 kbit/s layer network, which has no means of accommodating long-term frequency difference between clocks at different nodes other than by introducing slips. This has meant that the development of the 64 kbit/s layer network went hand in hand with the development of a synchronization network that ensured that all digital exchanges were ultimately all slaved to the same master clock. For speech, this is not a problem, as the speech coder (A-law or μ-law) is not sensitive to slips in the data if they are controlled to be complete bytes. The loss of a whole byte or the insertion of a repeated byte is scarcely audible once decoded. However, this is not the case for data services carried on N-ISDN, which are very sensitive to such slips.

Internationally, it was recognized that it is difficult to produce a single international synchronization network with all clocks throughout the world slaved to one master clock, and so it was agreed in the ITU-T that each national operator should have a master clock of the highest possible standard using a cesium beam (ITU-T Recommendation G.811). In this way, slip would, in theory, only occur at international boundaries, and even then at a very low rate. It was possible that under severe network failure conditions, any one slave clock could lose its reference to its master, and in this case, the slave clock would enter a tightly specified holdover condition where the rate at which the slave may drift from the reference is kept within strict but looser limits and, again, keep the resulting slip within well-controlled limits.

Existing synchronization networks were designed for the requirements of the 64 kbit/s layer network, where the primary parameter is the absolute accuracy of the frequency in order to prevent slips. The bit rate is low, and when wander is translated into a number of bits at 64 kbit/s, wander is very small (indeed, normally being only a fraction of a bit). As a result, the 64 kbit/s layer network is designed to completely buffer any network wander, up to a network maximum of 18 μs (ITU-T Recommendations G.823 and G.834), which is just over one bit at 64 kbit/s. Slave clock noise is even smaller in amplitude than wander, and so is easily absorbed in the wander buffers. As a result, existing synchronization networks have not been engineered to minimize slave clock noise, which is the key parameter for the SDH path layer networks.

11.5.1.1   *Existing Synchronization Network Topology*

The existing synchronization networks are generally based on a strict tree-type network topology, normally with a strict hierarchy of slave clocks. This is illustrated in Figure 11.37. The quality of the synchronization at the end of the network will depend on the performance of the slave clocks and the performance of the transmission links between the slave clocks, including both the situation when the slave clocks are correctly locked to their master clock and also the situation when a slave clock loses its connection to its master clock and enters a state called "holdover."

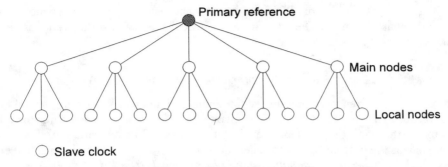

Figure 11.37 Example of the existing synchronization network topology.

11.5.1.2 Existing Slave Clocks

The existing slave clocks are designed for good holdover performance, with clocks higher up the hierarchy having the better specifications. If a clock higher up the tree were to lose its reference to its master, it would start to drift and so would all the slave clocks below it in the tree hierarchy that are slaved to it. To preserve the synchronization quality of the network, the slave clock must have an oscillator with good long-term stability and, generally, a very long time constant control loop. While no absolute rule can be made, especially as technology has advanced considerably since most operators specified and deployed their existing slave clocks, optimizing a slave clock for holdover performance often means the clock will not have minimal clock noise. The latest version of ITU-T Recommendation G.812 divided slave clocks into those that are only suitable to synchronize 64 kbit/s networks (reflecting existing clocks which are not designed for minimal clock noise), from those which are designed to synchronize any network (including SDH and ATM networks as well as 64 kbit/s networks). These are specified to have good holdover performance and low clock noise.

11.5.1.3 Existing Synchronization Links

Different network operators have chosen to transport the synchronization reference information between clock nodes in different ways. The most popular has been to transport the synchronization as part of the primary rate traffic signal. Both the North American primary rate of 1544 kbit/s and the European primary rate of 2048 kbit/s are directly locked to the timing of the 64 kbit/s data they are carrying since there is no frequency justification in the adaptation between the 64 kbit/s layer network and either the 1544 kbit/s layer network or the 2048 kbit/s layer network. This means that the synchronization information is carried automatically with the traffic in a primary rate signal.

The plesiochronous multiplexers mainly produce high-frequency phase variation that can be filtered out by a slave clock. Jitter from the line system regenerators can sometimes have lower frequency components, but the majority of the phase variation is of high frequency and so can also be filtered. However, the wander from PDH line system moves so slowly that a slave clock will track it (the time constant of a slave clock is likely to be substantially less than one day). This gives a band of frequency between diurnal wander and high-frequency jitter from PDH multiplexers and regenerators, within which there is very little phase variation. The existing slave clocks have assumed they can work in this band where there is little variation.

11.5.1.4   Carrying Existing Synchronization Links on SDH or ATM

If SDH is used to transport the primary rate signal, it will be subject to substantial phase quantization. The spectral content of the phase variation due to a pointer adjustment falls exactly in the band that was previously largely free of phase variation. This is not surprising, as the time constant of the desynchronizer will be comparable with that of the slave clock. *In general, this situation should be studiously avoided.*

The existence of the dead band in the pointer processor gives rise to the following situation:

- If there is no pointer activity on the line, then, by definition, the slave clock is not receiving any synchronization information from the clock to which it is slaved (in fact, the timing is coming from the local timing reference of the last SDH node).
- If there is pointer activity, then synchronization information is being passed between to the slave clock from the clock to which it is slaved; however, the slave clock is unlikely to be able to track it.

The slave clock cannot win! An even worse situation arises if the last SDH node is itself timed from the slave clock, which is highly likely. In this case, within the deadband of the pointer processor, there is a free running timing loop. This will amplify the clock noise until pointer adjustments occur. This is illustrated in Figure 11.38.

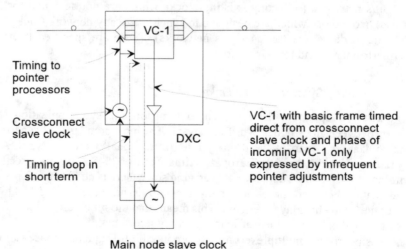

Figure 11.38   Illustration of short term timing loops created by transport of synchronization on VC-1x.

If the primary rate were to be carried on ATM, the situation would be even worse. In theory, the adaptive scheme could be used; however, with phase quantization of 376 bits, the phase variation generated is many orders of magnitude higher than that desired for these links. SRTS needs a synchronization signal to work, and if that signal is available, then there is no need to carry one! *If SRTS is used to carry the synchronization reference signal, the slave clock will never receive any synchronization information, as a free running timing loop has been established* in exactly the same way as SDH; however, now there are no pointer threshold buffers to limit the free run other than slips, which guarantees the synchronization information is lost.

11.5.2 Synchronization Networks Suitable for Broadband Networks

Standards bodies have been working on the basic components of a general synchronization network incorporating the requirements for the existing 64 kbit/s layer network, the ATM layer networks, and SDH path layer networks. The work defines a completely new synchronization network covering the following basic requirements:

- Functionality of many layer networks and many pieces of equipment require high-quality networks synchronization.
- Holdover, clock noise, wander, and slip are all important parameters to control in a synchronization network.
- There are now often many operators in one country, and often operators are active in many countries.
- Synchronization should be distributed by a managed network where the timing performance at all points of the network is monitored and observable at a central management center.

These are all in marked contrast to existing synchronization networks, and this fundamental change has raised the profile of synchronization as a major networking issue and many major operators have either already changed their synchronization network or are currently considering evolutionary strategies to achieve this.

11.5.2.1 Synchronization Supply Unit

At the heart of a new synchronization network is the synchronization supply unit (SSU). This is an independent synchronization utility that is not associated directly with any one technology and is normally implemented as a stand-alone unit. It has four main functional areas: timing input and measurement, internal oscillator, timing output generation and distribution, and management. An example is illustrated in Figure 11.39.

The input functions can take a timing signal from a wide variety of sources, check the source quality and integrity of the signal, extract the timing information, and then carry out real-time phase variation measurements of this input signal relative to other input signals and also relative to the local oscillator. The local oscillator provides the SSU with its "fly wheel," which is used to improve the quality of the timing output and also provide the timing source if the SSU enters a holdover condition.

The timing output generator uses the timing information for the input signals and for the local oscillator to produce the highest quality timing signal covering all the parameters of clock noise, holdover, wander, and slip. It can do this by using the phase variation

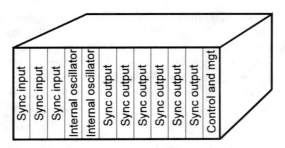

Figure 11.39   Synchronization supply unit (SSU).

measurements and also a knowledge of the characteristics of phase variation for each source. For example, by setting a control loop time constant, clock noise on an input signal can be filtered out; however, clock noise from the local oscillator will be inserted in its place. If the characteristics of each are known, then the control loop time constant can be set to give the lowest possible clock noise. It should be noted as well that if there are no inputs and the local oscillator is cesium or hydrogen, then the SSU can become a stand-alone primary reference clock.

The SSU management can record and report all the performance information as well as allow the configuration to be set up remotely.

The ITU-T has recently agreed to a complete revision of Recommendation G.812 to include a range of three SSU specifications (in addition, there are three older slave clocks specified, but these can only be used for 64 kbit/s networks). The principal details of each are given in Table 11.4. The different clock types have the following origins: type I is oriented towards ETSI based networks while types II and II are oriented towards ANSI based networks, with type II designed for large central exchange buildings and type III for smaller local exchange buildings. Type IV reflects a large embedded base of clocks in North America while types V and VI are the transit node clock and local node clock from the previous version of G.812.

Table 11.4

G.812 SSU Specifications

Clock Type	MTIE	T_{dev}	Holdover drift	Long-term precision
Type I	160 ns	12 ns	2×10^{-9} / day	not specified
Type II	100 ns	10 ns	1×10^{-9} / day*	1.6×10^{-8}
Type III	100 ns	10 ns	1×10^{-8} / day	4.6×10^{-6}
Type IV	100 ns	10 ns	4×10^{-8} / day	4.6×10^{-6}
Type V	1000 ns	under study	1×10^{-9} / day	not specified
Type VI	1000 ns	under study	2×10^{-8} / day	not specified

* suggested figure

Clock types IV, V, and VI are not suitable for SDH and ATM networks

11.5.2.2 Primary Reference Clocks and "Off Air" Distribution

"Off air" distribution is a major new opportunity in synchronization networks. There are a growing number of radio signals that can be tracked that are timed from clocks of primary reference quality. Most notable of these is the global positioning satellite system (GPS), which is a large constellation of medium earth orbit satellites that can be used to determine any position on earth to within a few meters. To do this, each satellite carries its own cesium clock. The receivers are inexpensive and can be used as an input to an SSU. The link between the "primary reference clock" on the satellites and the SSU is a direct, reliable radio link, and at any one time there are normally at least three satellites visible (the minimum needed for positioning). This means that an SSU timed from GPS is, effectively, a primary reference clock.

In addition, the exact offset of the clock on each satellite relative to universal coordinated time (UTC) is also carefully tracked, so the SSU can have better than primary reference clock accuracy and can track UTC.

The general architecture of GPS-based distribution of timing is illustrated in Figure 11.40. It has many advantages:

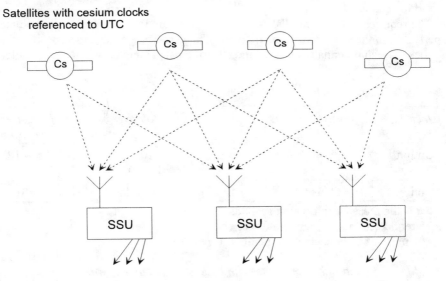

Figure 11.40 Off air synchronization with GPS.

- GPS is global and therefore is not constrained by national boundaries, and any issue arising between multiple operators in the same geographic region and pan-national operators disappears.
- The timing link is simple and robust, and not subject to protection switching and other transmission network phenomena that cause "glitches" in synchronization links.

- It enables a flat distribution tree and the build-up of clock noise across a synchronization network is easily controlled.

A number of operators have already included GPS as a major part of the synchronization strategy, and this is sure to increase as more SDH and ATM networks are deployed.

11.5.2.3   SDH Slave Clocks

All SDH equipment must contain a slave clock to synchronize all the outgoing STM-N line signals, any pointer processors, and any SDH layer matrixes. Because some SDH equipment, such as regenerators, line terminals, and ADMs, are small and may be located in remote sites, it may not be appropriate to include an SSU in these sites. SDH slave clocks fall into two distinct categories: those associated with line system regenerators, and those containing an SDH path layer matrix, including add/drop multiplexers. This latter slave clock is normally referred to as an SDH equipment clock.

The requirements of the SDH regenerator slave clock are given in ITU-T Recommendation G.958 as part of the general specification of line systems, and two types are specified as noted in section 11.3.1.

The requirements for the SDH equipment clock are that it be low-cost and that it fit properly within the synchronization network. The ITU-T has agreed to a specification for this clock in Recommendation G.813, and the principal parameters are given in Table 11.5.

Table 11.5

G.813 SDH Equipment Clock Specifications

Clock Type	MTIE	T_{dev}	Holdover drift	Long-term precision
Option 1	$25.25\ \tau^{0.2}$ ns * $+ 50$ ns **	6.4 ns ***	1×10^{-8} / day $+ 1 \times 10^{-6}$ **	4.6×10^{-6}
Option 2	60 ns ***	10 ns ***	5×10^{-7} / day ***	2×10^{-5}

* τ is the measure interval in seconds
** term accounts for temperature effects
*** measured at constant temperature (+/- 1°K)

11.5.2.4   Synchronization Links Using the SDH Section Layers

The SDH optical (or electrical) signal is timed from the local reference synchronization, and so can automatically carry synchronization reference information to the remote sink. This gives, compared to PDH primary rates, high-quality transport of synchronization information and is the natural way to transport synchronization reference information in broadband networks (other than using "off air" synchronization).

11.5.2.5   Synchronization Network Topology

As before, the general synchronization network topology is based on a tree structure, but now it is possible to have synchronized SDH equipment in the branches of the tree, as illustrated in Figure 11.41. While SDH regenerator slave clocks do not add much noise, an SDH equipment slave clock will filter some clock noise, but it will also add some clock noise. An SSU similarly, as specified in G.812, is likely to filter out a certain amount of clock noise, but it can also add a certain amount of clock noise and so it is important to limit the maximum number of SDH equipment clocks and SSUs that can be chained together in the synchronization tree. ITU-T Recommendation G.803 proposes a maximum of 10 slave clock SSUs with a maximum of 20 SDH equipment clocks between each pair of SSUs subject to a maximum of 60 SDH equipment clocks.

The general topological structure is such that regenerator slave clocks come between SDH equipment slave clocks and SDH equipment slave clocks come between SSUs, as illustrated in Figure 11.41. The most important new feature of this topology is that the SDH equipment slave clock must signal to the SSU when it enters holdover mode; otherwise, the SSU could take its timing reference from the inferior holdover timing of the SDH equipment slave clock. This is signaled in the synchronization status message defined in the multiplex section overhead byte S1.

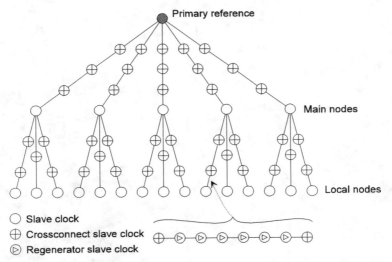

Figure 11.41   General tree synchronization network topology including SDH network elements.

Such signaling is not required between the regenerator slave clocks and an SDH equipment slave clock, as there is no circumstance where MS AIS is not generated, automatically signaling the loss of timing reference.

References

The following standards are the main source documents for the material in this chapter:

ITU-T Recommendation G.810, *Definitions and Terminology for Synchronization Networks*.
ITU-T Recommendation G.811, *Timing Requirements at the Outputs of Primary Reference Clocks Suitable for Plesiochronous Operation of International Digital Links*.

ITU-T Recommendation G.812, *Timing Requirements at the Outputs of Slave Clocks Suitable for Plesiochronous Operation of International Digital Links*.

ITU-T Recommendation G.813, *Timing Characteristics of SDH Equipment Slave Clocks (SEC)*.

ITU-T Recommendation G.822, *Controlled Slip Rate Objectives on an International Digital Connection*.

ITU-T Recommendation G.823, *The Control of Jitter and Wander Within Digital Networks Which are Based on the 2048 kbit/s Hierarchy*.

ITU-T Recommendation G.824, *The Control of Jitter and Wander Within Digital Networks Which are Based on the 1544 kbit/s Hierarchy*.

ITU-T Recommendation G.825, *The Control of Jitter and Wander Within Digital Networks Which are Based on the Synchronous Digital Hierarchy (SDH)*.

ETSI ETS DE/TM 03017, *Transmission and Multiplexing; Generic Requirements for Synchronization Networks*.

In this chapter, we consider some of the issues that face implementers and specifiers in the realization of equipment with which to build broadband transport networks. From a hardware designer's viewpoint, much of the implementation technology for synchronous digital hierarchy (SDH) and asynchronous transfer mode (ATM) will be logical extensions of techniques already familiar from the plesiochronous digital hierarchy (PDH) worlds of transmission and switching. In SDH, new concepts in synchronizing information transfer and the demands of flexible bandwidth management at higher data rates raise challenging problems for the designer that require a reappraisal of old assumptions. ATM designers take their inspiration primarily from the packet-switching world, but the special problems associated with monitoring and guaranteeing traffic flow in a multiservice network with widely varying traffic characteristics raise many new problems for implementers to solve.

The deployment of the SDH has coincided with heightened expectations of management functionality, and this has resulted in much larger software loads in the transmission equipment. Recently matured object-oriented design methodologies bring the promise of more straightforward software implementations with a degree of reuse comparable to that achieved in integrated silicon design. ATM deployment has coincided with the rise of the Internet, and the original broadband integrated services digital network (B-ISDN) vision is being hastily adapted to fit more easily into the connectionless Internet Protocol (IP) environment. Some specific implementation issues in both SDH and ATM will be discussed in this chapter, but first we will consider the subject of specification and requirements capture as it has been applied to transport equipment, and to SDH equipment in particular.

12.1   GENERAL REQUIREMENTS OF BROADBAND TRANSPORT EQUIPMENT

The general requirements for BB transport equipment are no different in principle from the general requirements for other telecommunication equipment. The physical equipment practices on which telecommunication equipments are delivered vary around the world. There is no fundamental reason for this other than the fact that the various markets have developed largely in isolation, and at each new stage the constraints of backward compatibility have been paramount, tempered only by local attempts at optimization and rationalization. In the United States, de facto standards have existed for some time, driven by the large single market. These are quite rudimentary, being based on the 19-inch rack, and provide lots of scope for variation and differentiation at the shelf and card level. In Europe, there has traditionally been a greater degree of physical specification. Major national operators have traditionally sponsored the development and tooling to support national equipment practices that not surprisingly have been different in different countries. Recent advances in the European Telecommunications Standards Institute (ETSI), driven by the single European market, have resulted in harmonized requirements at the physical level, leading to a uniformity of physical design across Europe comparable to that which already existed in the United States.

There has been a general tightening of limits on environmental specifications in recent years, particularly in the area of electromagnetic compatibility (EMC), that is reflected in the detail of the packaging design and in the functional partitioning. The predominantly optical interconnect in SDH and BB environments neatly solves one of the more difficult problems traditionally faced by the designer of interference-proof packaging, namely that of conducted interference along electrical connecting leads.

Thermal performance with regard to cooling and ambient temperature introduces significant application and regional variations. These partly reflect prevailing climatic conditions in different regions, but are more directly related to environmental management policies in particular organizations. New levels of functional integration increasingly found in network equipment are putting greater strain on thermal management systems at all levels. The functional density achievable in ASIC technology, coupled with the high-speed signal-handling advantages that come with small physical size, is driving the watts per unit volume generated in SDH and ATM equipment ever upward. For this reason, forced cooling is now commonplace in much of the higher functionality transport node equipment. The power density at which forced cooling becomes necessary within equipment is, of course, largely dependent on external ambient conditions and local thermal management policy. Many operators, recognizing this trend, are installing forced air circulation systems in their buildings, with ducts contained in false floors and ceilings. Thermal management policy is also the determinant as to the density at which equipment can be installed in racks and suites.

None of these issues is specific to the technologies we are discussing, and therefore we will not dwell on these aspects of equipment specification. However, the specification of functional requirements for equipment supply in a multivendor environment without unduly constraining implementation options is providing a considerable challenge for the equipment specifier in the SDH and ATM context because of the high level of functional integration that is expected.

12.2   FUNCTIONAL SPECIFICATION METHODOLOGY

In Chapter 3, we analyzed the architecture of transport networks in general and SDH and ATM layers in particular. We developed functional definitions and diagrammatic conventions with which we can describe all the functions of these networks necessary for their construction and management. Figure 3.10 and Figure 3.11 in particular were concerned with the intersection between the network view and the network element view. The network element (NE) view was presented as a conical surface intersecting a number of parallel planes that represented the layer networks. The NE view is reproduced in Figure 12.1 for the example of an NE capable of flexibly allocating any one of six DS3 signals to either of two STM-1 (OC-3) ports. If we now project the connection points (CPs) contained within the cone onto the conical surface, show the termination and adaptation functions explicitly for all of the network layers, and then project this conical surface onto a plane, we produce the equivalent view illustrated to the right in Figure 12.1. We use this view predominantly in this chapter to describe NE functions. The functional requirements of a piece of SDH equipment whose NE function can be derived from such an intersection can therefore be seen as the aggregation of the functional requirements of all the atomic functions generated by the intersection.

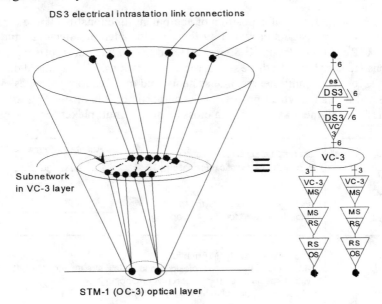

Figure 12.1   Intersection of the network view and the network element view and the functional representation of equipment functions.

To make the specification complete, it is necessary to provide full formal definitions of all the specific atomic functions required for the complete set of NE intersections to be specified. This approach was first attempted in ITU Recommendations G.782 and G.783, which describe SDH multiplexers, and G.958, which describes SDH line systems. However, the resulting documents are a compromise between this more formal methodology

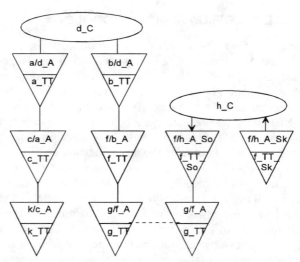

Figure 12.2   Combining atomic functions to form compounds by interconnecting at matching
reference points.

and a more traditional informal exposition of equipment functionality. These recommendations predated new insights provided by the more detailed architectural study documented in ITU Recommendations G.805, G.803 and I. 326. It is not surprising, therefore, to find inconsistencies between these sets of recommendations. The supposed distinctions between crossconnects, multiplexers, line systems, and other equipment types evaporate when subjected to the analytical process of decomposition. It is misleading to classify a particular nodal equipment as a switch, a crossconnect, a multiplexer, or a line terminal

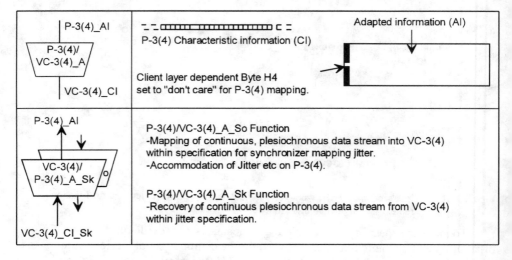

Figure 12.3   Atomic function specification components for P-3(4)/VC-3(4)_Adaptation.

when several of these functions may be integrated within a single unit. We may use the terms crossconnect or multiplexer when referring to the use of such functions in the network, but there is no benefit in attempting a formal classification of equipment along these lines for specification purposes.

The notion of generic transport requirements (TRs) that apply across a wide range of transport applications is well-established. Bellcore TRs make extensive use of generic specifications and their European counterparts, the European telecommunication specifications (ETSs) produced by ETSI, are doing the same. We use this method in the discussion that follows, as it provides definitions that are concise but sufficiently precise for the purpose: the precise definition of functionality in an implementation-independent manner.

12.2.1 Atomic Functions and Their Combination to Form Compound Functions

It is not our purpose here to provide a complete set of formal specifications for all the atomic functions. This is the role of the ITU and the regional standards bodies themselves, and to do so would involve a degree of replication of, and some supposition about, the

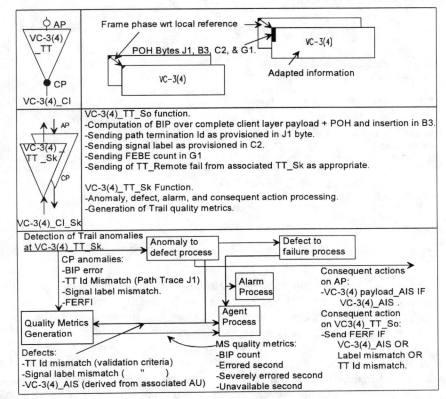

Figure 12.4   Atomic function specification for VC-3(4)_TT.

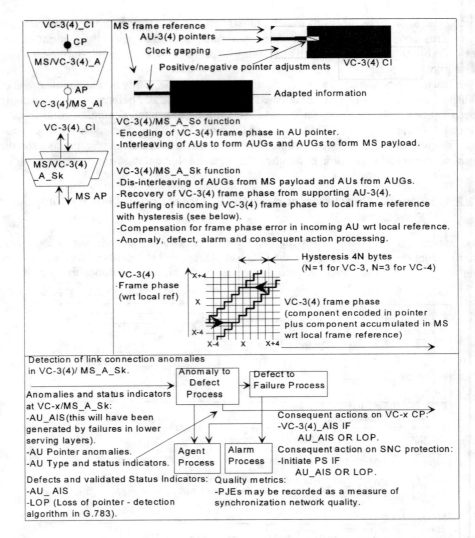

Figure 12.5   Atomic function specification components for VC-3(4)/MS_Adaptation.

precise content of the emerging documents. However, the following paragraphs indicate the nature of the information required to define specific atomic functions.

Transport networks are constructed from three generic function types. These are the trail termination function and the connecting (or subnetwork) function within a network layer, and the adaptation function between layers. When defining unidirectional NE functions it is necessary to distinguish between the source and sink components of the trail termination and adaptation functions and to specify the directionality of the connections in the connecting function. However, the discussion that follows concentrates on the bidirectional combination of sink and source in the interests of simplicity and because of

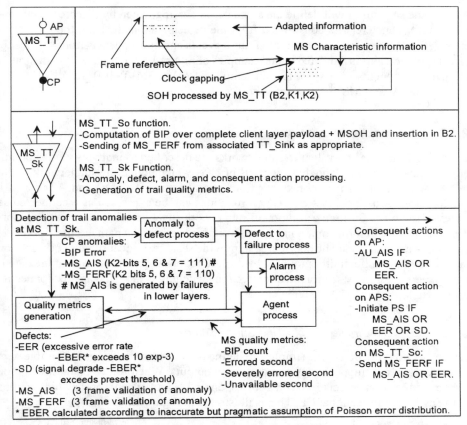

Figure 12.6 Atomic function specification components for MS_TT.

its predominance in transport equipment except in cases such as synchronization distribution where unidirectional behavior is implicit.

Atomic functions are combined by binding inputs to outputs at matching reference points to ensure information continuity. A trail termination function that is bound to one of its possible adaptation functions at their mutual *access point* (AP) forms a basic compound from which the larger compounds are formed by binding at like *connection points* (CPs), in the same way that dominoes are laid alongside each other with matching numbers. This is illustrated in Figure 12.2. The adapters are represented textually as X/Y_A where X is the layer designator of the server layer and Y is the layer designator of the client layer to which the adaptation function applies. X/Y_A defines the information processing between a server layer X CP and its client layer Y CP(s). The trail termination functions are designated X_TT where X is again the layer designator. These are the same layer designators as used in the diagrammatic representations, and are listed in *appendix B*.

A complete specification for an atomic function will define the information processing that occurs between the reference points, anomaly detection algorithms, interface levels (if relevant), process performance in terms of delay variation, phase error, regeneration margins, and so forth. It will define the anomaly processing, whereby defects are inferred and/or performance reports constructed from the detected anomalies and the process by which defects in turn are converted into failure events and consequent actions. The scope of such specifications is indicated in the examples of Figure 12.3 to Figure 12.6 illustrating the main components of a specification template for VC-3(4)/P-3(4)_A, VC-3(4)_TT, VC-3(4)/MS_A, and MS_TT respectively.

12.2.2   Connection Functions

Connection functions equivalent to subnetworks are described as matrices with explicitly defined connectivity available between sets of CPs, which bound the subnetwork within the NE. A range of atomic subnetwork functions of interest in SDH and ATM NEs is defined below. Most practical connection functions can be expressed as a connected set of these basic types forming a compound subnetwork function. They may be used alone or in combination to form compound subnetwork functions in the path or circuit layers. A single physical matrix may be configured to provide separate logical subnetworks in more than one layer. The flexibility provided by such functions is realized in one of the path or circuit layers. Multiplex section layer protection switching is modeled as link connection group switching in the HOP layer.

12.2.2.1   The Simple Nonblocking Square Matrix

This connection function is the simplest to define but the most challenging to implement. It connects any of n inputs to any one or more n outputs. When folded so that each input forms a bidirectional association with a corresponding output, it connects any bidirectional CP to any other bidirectional CP. This is illustrated in Figure 12.7.

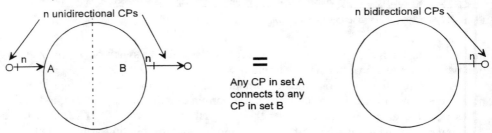

Any CP in set A
connects to any
CP in set B

Figure 12.7   Nonblocking square matrix.

The CPs are of course characteristic of the layer in which the matrix is operating and will be defined by the configuration of the adaptation and termination functions to which they belong.

12.2.2.2   The m × n Grooming/Consolidation Matrix

This matrix connects any CP in one set A of *m* inputs to any one or more CPs in another set B of *n* outputs. This is generally used in combination with a similar matrix that connects any CP in one set B of *n* inputs to any one or more CPs in another set A of *m* outputs. The combination in which the A inputs are associated with the A outputs and the B inputs are associated with the B outputs provides *m/n* grooming, or consolidation between the sets A and B of bidirectional channels. These are illustrated in Figure 12.8. The 1 to *n* selection matrix is a special case of the *m* × *n* matrix in which *m* = 1. A further special case in which *n* = 2 is used to generate the compounds defined below.

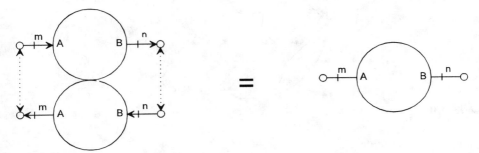

Figure 12.8   The m:n grooming/consolidation matrix.

12.2.2.3   *Add/Drop Matrices*

The unidirectional add/drop matrix is a compound consisting of two unidirectional 1:2 selection matrices connected as illustrated in Figure 12.9(a). It will either pass unidirectional traffic directly between E and W or it will drop traffic from E and add traffic at W. The simple, four-port bidirectional add/drop matrix is a compound consisting of two 1:2 bidirectional selection matrices connected as illustrated in Figure 12.9(b). It can provide bidirectional connections east-west or east-drop east and/or west-drop west.

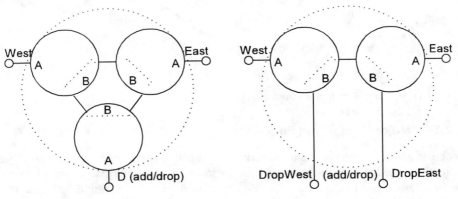

Figure 12.9   Add/drop matrices.

12.2.2.4   *The Test/Break Matrix*

This is a compound consisting of two unidirectional 1:2 selection matrices. It connects the input of an existing subnetwork connection to a nonintrusive monitor by setting up a new multicast subnetwork connection for test purposes. It simultaneously breaks the associated contradirectional subnetwork connection and connects its output to an input to which a test source can be attached. The compound function is illustrated in Figure 12.10.

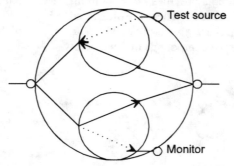

Figure 12.10   Test/break matrix.

Simple 1 + 1 Protection Switch

This is another compound of two unidirectional 1:2 selection matrices, illustrated in Figure 12.11. It connects one subnetwork connection input to two outputs and connects the corresponding subnetwork connection output to the best of two inputs. The switching criteria are generally based on status information derived from the protected entity. Protection-switching models are discussed in more detail in Chapter 10.

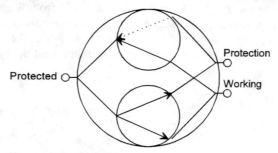

Figure 12.11   The 1+1 protection switch.

12.2.3   Major Compound Functions

Combinations of the atomic and basic compound functions described above exhibit levels of functionality that qualify them as major components of an NE that are significant for network planning and design purposes. At lower levels of functional integration, a whole equipment unit may have been dedicated to such major compound functions, with only the addition of some interfacing functions. In an environment characterized by high functional integration, they still form useful modules of specification.

Certain functional groupings are emerging as more important than others in the minds of planners and implementers. Some of these are presented below. A short informal description is presented for each, together with a diagrammatic representation which, taken with reference to a complete set of atomic function definitions, represents a quite formal functional specification for the new compound complete with process description, performance limits, interface requirements, failure and performance reporting, consequent actions, and any specified autonomous behavior.

12.2.3.1 Regenerated Line Termination

The regenerated line termination (RLT) function is implemented where a piece of SDH line terminating equipment is required to interface directly to a regenerated line. It is required to implement OS, RS, and MS network layer functions as illustrated in Figure 12.12.

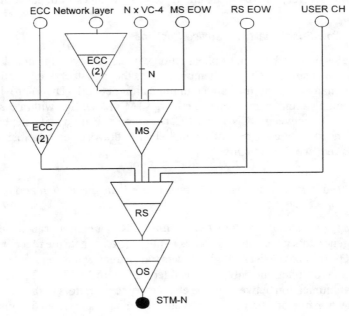

Figure 12.12   Regenerated line termination.

The function is generally required to provide management access to remote regenerators by means of the RS-DCC for the transfer of NE management messages. The messages themselves are defined by the information model that was discussed in Chapter 7. It will generally be associated with a message-switching function for this purpose capable of routing messages between the Q-interface on the NE, a local managing process within the line terminating NE, and the agent process in each dependent regenerator.

The multiplexing process whereby the different components are aggregated into the STM-N signal is defined by the STM-N format definitions of G.707 and illustrated in Figure 5.5. Performance constraints are placed on the physical interfaces, the synchronization, and alignment recovery. The adaptation to the VC-3/4 layers is defined by the pointer process also, described in Chapter 3. This implies that the data be buffered to adapt the frame phase of the incoming signal to the local reference. A constraint is placed on the threshold spacing of this implied buffer to introduce some hysteresis into the pointer process, thus limiting unnecessary pointer adjustments due to short-term phase noise.

Functional specifications of this type allow flexibility as to the placement of the associated management processes. In the case of a regenerated line system, there will be a management process associated with its supervision and that of the remote regenerators of which it is comprised. Such processes will often be located in the terminating equipment to optimize messaging efficiency and resilience to operating system (OS) failures.

The provision of voice communication capabilities between terminals and remote regenerators via engineering order wire (EOW) facilities is still expected in many line system applications. There is, however, an increasing tendency to rely more on the normal mobile telephony service, where it is available, to provide this sort of capability.

12.2.3.2   *Low-Functionality Intraoffice Sections*

The intraoffice section is a basic multiplex section with no RS layer functionality. The MS embedded control channel (ECC) is often omitted, as the intrastation application is likely to be served in many cases by the data communication network (DCN) provided as part of the telecommunications management network (TMN) facilities within a station. The economic tradeoffs between DCN and ECC for communication with dependent SDH equipment on the same site are not sufficiently clear to draw strong conclusions as to the benefits of this reduced functionality.

12.2.3.3   *Elementary Regenerator Section Terminating Function and Regenerated Line Systems*

The *elementary regenerator section terminating* (ERT) function can be expected in elementary regenerators where an SDH NE is required to terminate the regenerator section overhead (RSOH) but pass MS and higher layers transparently. It is required to implement the OS and RS layer functions only, as illustrated in Figure 12.13. ERTs are required to provide for communication between all the elementary regenerators in the same multiplex section for the transfer of regenerator management and supervisory information. The

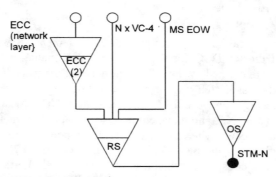

Figure 12.13   Elementary regenerator section.

information exchanged and the message set will be defined by the associated information model. The ERT function will normally be associated with a message-switching function for this purpose capable of routing messages to and from the local agent process and relaying messages between upstream and downstream ports.

Synchronization and frame reference information is required to be transferred transparently from the RS/MS_A_Sk to the associated RS/MS_A_So in the same direction of transmission for each direction of transmission independently. In other words, regenerators must be through-timed. The ERT function may also be required to support an RS EOW channel and/or the 64 kbit/s user channel.

Elementary regenerators achieve STM-N frame alignment independently in the two directions of transmission. After aligning to the incoming frame from the upstream direction in the OS/RS_A_Sk function, the alignment information in the form of the A1 and A2 bytes is regenerated and inserted in the OS/RS_A_So function in the downstream direction. This cannot be done until correct alignment has been validated, which implies that realignment of a long chain of regenerators requires each regenerator to align in sequence. Alternatively, G.958 allows for simple relaying of the alignment information to avoid this sequential buildup and thus reduce the alignment time.

An important set of parameters for a regenerator is that associated with timing recovery from the optical (or electrical) line signal in the OS/RS_A_Sk function. It is necessary within a regenerated line system to constrain variations in the bandwidth of the timing recovery filter in a series of regenerators, to limit the relative phase error between the received signal and the recovered timing, so that adequate buffering can be provided. The higher the Q factor, the narrower the bandwidth of the recovery filter and the smaller the phase error. This topic was discussed briefly in Chapter 11. Current technology offers two popular implementations: the phase locked loop capable of providing very high Q factors and consequently low phase error, and the surface acoustic wave (SAW) filter capable of providing moderately high Q factors and hence, not such low phase error. The ITU recommendations cannot exclude any viable technology, so both of these possibilities have been specified in the form of the type A and type B regenerator jitter tolerance mask specification of G.798. Either can be used, but the two types should not be mixed within a single line system.

12.2.3.4   SDH NE Timing Function

All SDH NEs (except the elementary regenerator) are required to provide an SDH NE timing function that generates synchronization and frame phase reference information derived from a selected reference input to the NE. This is used for the generation of all the synchronous containers that originate at the NE. If more than one input reference is available, selection may be made according to the priorities accorded each potential source by the managing system. Its performance in terms of long-term stability (holdover) and short-term stability were discussed in Chapter 11 and are defined by ITU-T in the new Recommendation G.813. The typical functional representation is shown in Figure 12.14.

The characteristic information of the synchronization distribution layer consists of the frame phase plus validity and hierarchical status information. The NE timing function illustrated provides for selection from three potential reference sources. The status and

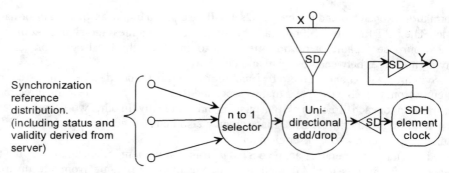

Figure 12.14   SDH NE timing function.

validity that is used together with any preassigned priority to determine the choice of reference must be derived from the associated server trail. The unidirectional add/drop function provides the capability to present synchronization information at output X as a reference for an external clock source such as a building integrated timing supply (BITS). This can in turn be used as a superior reference for the SDH NE itself. Y is the reference point at which the new system frame phase is available for onward distribution and local timing.

A variation of the timing function for use in rings has been described that automatically switches reference source while protecting against the possibility of timing loops. Schemes have also been described that transmit synchronization reference status information, which can be used as an enhanced form of synchronization trail termination, and to validate synchronization configuration or, in conjunction with a suitable algorithm, as a mechanism to recover a valid reference on failure of the primary reference.

12.2.3.5   *Message-Switching Function and Q-Interface*

Message switching is required in association with the embedded communication channels (ECC) provided in the regenerator section and/or multiplex section overheads. Each managing process and agent process on the data communications network acts as an end system (ES) as defined in the X.200 series Recommendations. Each ES must be allocated a network service access point (NSAP) with an agreed upon standardized format. The message-switching function acts as an intermediate system (IS) as defined in the X.200 series. The function is illustrated in Figure 12.15.

The message-switching function is required to inspect the network layer destination address information field of all messages arriving from SDH section terminating ports. The messages must then be routed accordingly using network routing information held locally. Local routing information may be built up in a NE by exchanging routing information between elements topologically adjacent in the same ECC network layer. The IS-IS routing exchange protocol (ISO 9542) has recently been standardized and will generally be used for this purpose. Elements used in subnetworks whose topologies are simple, static, or otherwise constrained may use manual or semimanual methods according to local discretion.

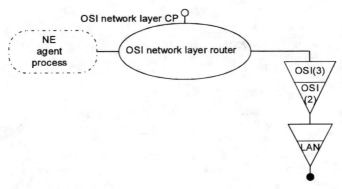

Figure 12.15 The OSI message switching function.

12.2.3.6 *HO Path Layer Access Functions*

SDH equipment providing VC-4 layer access functions may be optimized to provide access for DS4 paths (139264 kbit/s), ATM virtual paths, or in Europe where VC-4 is used as an administrative layer for the transport of tributary unit (TU) structured lower order (LO) paths. In North America, the VC-3 is used in a similar way to provide access for DS3 (44736 kbit/s) and TU structured lower order paths (LOPs). These compound functions are illustrated in Figure 12.16 for the VC-4 and Figure 12.17 for the VC-3.

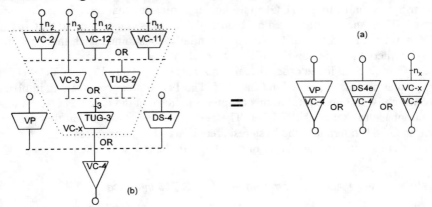

Figure 12.16 Access to the VC-4 layer.

The different mechanisms used for adapting these signals into the VC-3 and VC-4 payload are defined by the mapping procedures described in Chapter 5. These effectively define the information process implemented by the adaptation functions. The performance required of the DS3 and DS4 mapping processes is defined by the jitter specifications of G.783, which have been derived to be consistent with existing PDH requirements. The corresponding performance specifications for ATM cell streams will be defined in terms of cell arrival statistics and acceptable cell loss ratio.

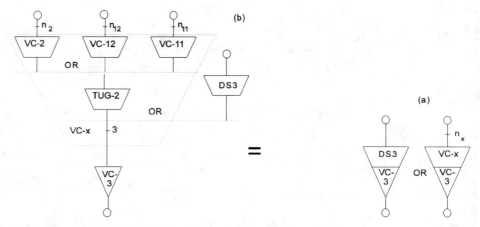

Figure 12.17   Access to the VC-3 layer.

It is common to provide flexible access for any combination of the LOP layers. Selection may be made as part of the initial configuration at the time of commissioning, or there is provision for single-ended control using the signal labels in the VC-4 overhead together with the labels in the TU pointers to interpret the required setting at the remote end. In the case of TU structured payloads, the valid combinations are only constrained by the tributary unit group (TUG) multiplexing structures that have been discussed in Chapter 5. Thus, the TU structured payload of the VC-4 is constrained so that $7n_3 + n_2 + n_{12}/3 + n_{11}/4 = 21$ and that of the VC-3 is similarly constrained such that $n_2 + n_{12}/3 + n_{11}/4 = 7$ (where n_x is the number of VCs at level x).

The adaptation of lower order virtual containers (LVCs) into VC-3 or VC-4 is defined by the pointer process described in Chapter 5. The TU pointer processor, an implementation of which is described later in this chapter, implies that the data be buffered to adapt its frame phase to that of the VC-3/4. The performance constraints on this process are defined in G.783 in terms of the hysteresis required to limit unnecessary pointer activity resulting from short-term instability of the clocks in the reference chain.

12.2.3.7   *VC-1 Access From Primary Rate G.703 Interface*

SDH NEs providing access for PDH primary and basic rate signals onto the VC-11 or VC-12 layers from PDH ports conforming to G.703 and/or G.704 have to take into account a number of options. The applicability of the various options is discussed in Chapter 13 on deployment. The implementation issues are the same for the 2048 kbit/s signals of the ETSI PDH as for the 1544 kbit/s signals of the ANSI PDH. In both cases, the mapping options available are constrained by the composition and timing of the offered signals.

If the incident G.703 signal is not synchronous with the network, but merely constrained within the +/-50 ppm limits associated with such locally timed signals, then the only option available is the asynchronous mapping. This is illustrated in the functional diagram of Figure 12.18. The extension "Asy" is added to the layer designator P1 to indicate that although conforming to the P1 layer definition that only constrains the

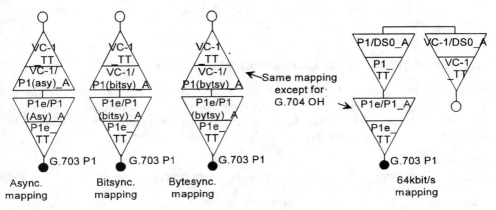

Figure 12.18   VC-1x access from the plesiochronous interface.

nominal rate of the information stream according to G.702, it is in fact not synchronous with the local reference. This puts it outside the range for which the pointer mechanism was designed. The asynchronous mapping option described in detail in Chapter 5 has been designed for this situation. The adaptation source and sink functions between P1(Asy) and VC-1 will define the jitter limits associated with the synchronizing and desynchronizing processes, respectively.

If the incident G.703 signal is synchronous with the network, then it becomes possible to use the bit- or byte-synchronous options. The byte-synchronous option can only be used if the incident synchronous signal is indeed byte-structured according to G.704, which defines the PDH primary frame structures. If the incident signal is synchronous but unstructured, then the bit-synchronous option may be used. The byte-synchronous option has significant implementation advantages in 64 kbit/s machines that interface directly via the SDH, and this can be expected to be the norm in the future. The bit-synchronous option appears to offer no user advantages, but is a simple subset of both the asynchronous and the byte-synchronous options and as such may have a role as a common mode of exchange between implementations that have been cost-optimized around one or the other.

There are further options to be considered in relation to byte synchronous working. In an integrated SDH network, $n \times 64$ kbit/s information will be transported by mapping directly into the VC-1 containers. In such a scenario, there will be no requirement to support a vestigial P1 layer and hence no requirement to generate or terminate G.704 overhead or framing information. A versatile multiplexer with primary rate tributaries might therefore be required to terminate the G.704 overhead and frame indicator or, alternatively, pass this information transparently as defined in the byte-synchronous mappings of G.709. If the G.704 frame is terminated, it may be necessary to support the channel-associated signaling superframe within the SDH container. The mechanism for doing this is defined as an extension of the pointer multiframe format and is described in Chapter 5.

All the options discussed above have some defined role in the evolution of the SDH transport network. This is discussed more fully in Chapter 13. The implementor must make a choice between a general purpose implementation that can be configured in all or some of the roles described above or an implementation that may be cost-optimized but supports only a restricted subset of these roles.

12.3   PRACTICAL DESIGN AND IMPLEMENTATION

In the following sections, we consider some of the equipment design considerations in practical terms, relating the functional models to implementation models that are described in terms of practical realizations. The design examples are not claimed to be unique nor even optimum, but have been chosen merely to illustrate the principles involved.

12.3.1   Adaptation Between Section and HO Path Layers

Adaptation between section and HO path layers involves multiplexing and demultiplexing of payload data and frame phase information. The section derives its frame phase from the master clock within the NE in which it is implemented. Under normal circumstances, this is traceable to a high-stability network reference that determines its frequency. The 125μsec frame boundary is derived arbitrarily from this by simple division. This becomes the frame reference for the NE and therefore defines the frame phase of all sections originating within it.

   If the higher order virtual containers (HVCs) are also originated within the same element, it is likely they will have bit rate and frame phase derived from the same reference, although this is not a requirement. More generally, HVCs will have been originated

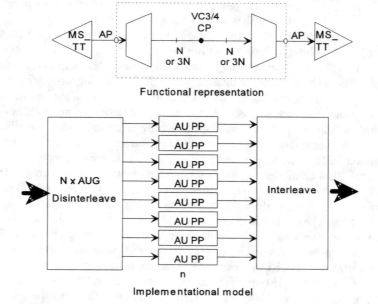

Figure 12.19   Section adaptation and AU pointer processors.

remotely with bit rate and frame phase, which will be encoded in the pointer, derived from that of the remote element. Moreover, as explained in Chapter 11, further phase variations may have accumulated in transmission, and these will also be encoded in the pointer.

Figure 12.19 illustrates the functional representation and an implementational model of a unidirectional combination of an MS sink connected back to back with an MS source. HVCs pass transparently between them across a VC-3/4 layer CP. This involves disinterleaving the AUs from the STM-*N* payload, recovering the HVCs, realigning them to the outgoing frame reference, and reinterleaving them in the outgoing STM-*N* payload with their new frame phase alignments. This particular functional combination is likely to be quite rare in practice, but it is described first as it separates the pointer process from the frequently associated process of *timeslot interchange* (TSI). The disinterleaving and reinterleaving processes may be logical, in which case the time division multiplexed sequence of bytes is presented to a time-shared pointer processor together with a sequence of AU timeslot identification labels. The level to which such a stream is physically disinterleaved is a design compromise and will very much depend on the capability of the technology.

12.3.1.1   The AU Pointer Processor

Realignment of the frame phase to that of the outgoing MS is a function of the AU pointer processor, and this is illustrated in more detail in the implementational model of Figure 12.20. This diagram shows a single AU pointer processor. It is presented with the MS frame reference and gapped information stream, which is characteristic of an MS access point. The gapping wave form and the MS frame reference are shown for AU-4 and AU-3 in the inset.

From this timing input, the AU sequencer generates byte strobes to identify pointers and justification opportunities. The pointer value recovered from the data stream corresponding to the current frame is held in register PV_N. The pointer value from the previous frame is held in register PV_{N-1}. The current receive pointer value is held in register RPV. This is updated from time to time according to the pointer interpretation rules. These are defined in G.709 and were also described in Chapter 5. The pointer interpreter (PI) compares RPV with PV_N and adds or subtracts one clock-enabling pulse from the VC clock in the positions corresponding to the justification opportunities whenever the increment or decrement pointer conditions are detected, while simultaneously incrementing or decrementing the value in RPV. The gapping signal corresponding to the dejustified stream is shown inset as the VC clock. The "NDF or X3" detector sets RPV directly if the new data flag is detected or, following a discrepancy between RPV and PV_N, if identical pointers in three successive frames are subsequently detected. Either of these conditions results in RPV assuming the value held in PV_N.

The write address generator takes the form of a counter that counts the bytes of the VC and is driven by the justified VC clock. Its zero state corresponds to the first byte of the VC and is set when the value in RPV is equal to the AU byte location as generated by the AU sequencer. Data is written into buffer locations identified by the least significant bits of the address field. The whole address number is used to drive the buffer fill comparator.

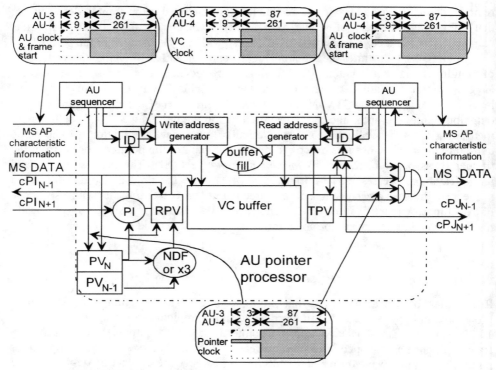

Figure 12.20 An AU pointer processor.

A similar AU sequencer is driven from the outgoing section clock, which will have been derived from an element clock synchronized to the network but whose frame reference will be independent of the incoming frame. This determines the structure and frame phase of the outgoing AU and drives the read address generator in the same manner its counterpart in the transmitter drives the write address generator.

The read address generator has the same structure as the write address generator. The buffer fill comparator continuously monitors the state of fill of the buffer by subtracting the read address from the write address. It will generate outputs when the fill crosses its upper (nearly full) or lower (nearly empty) threshold values. These threshold crossing events trigger increments or decrements in the read address counter, simultaneously enabling pointer justification events (PJEs) in the transmitted data stream. The value of the transmit pointer held in TPV is taken as the byte location number in the AU generated by the AU sequencer that is coincident with the zero value of the read address counter.

The TPV (transmit pointer value) register holds the value of the current transmit pointer, which is incorporated into the AU to be transmitted. The buffer fill comparator continuously monitors the state of fill of the buffer by subtracting the read address from the write address. It will generate outputs when the fill crosses its upper (nearly full) or lower (nearly empty) threshold values. These threshold crossing events trigger increments or decrements in the value of TPV and transmit VC clock. They simultaneously enable PJEs in the transmitted data stream.

12.3.1.2   AU Buffer Thresholds and PJEs

Figure 12.21 illustrates the four types of PJEs and their effect on the buffer fill. The buffer read and write clock gapping wave forms are reproduced as a reference. These are equivalent linear representations of the VC clocks represented two-dimensionally in Figure 12.20. Positive or negative PJEs resulting from upstream phase variations are detected in the input stream and used to control the writing of data to the buffer. The state of the buffer fill is used to generate PJEs in the output stream subject to hysteresis built into the pointer process. Note that a positive PJE at the input has the same net effect on the buffer fill as a negative PJE at the output and, conversely, a negative PJE at the input has the same effect (i.e., a decrease) as a positive PJE at the output.

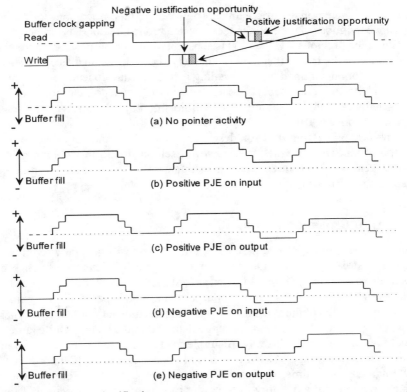

Figure 12.21   AU pointer justification events.

The buffer fill thresholds are the main parameters in pointer processor design. The threshold spacing must be large enough to accept, without error, the maximum differential phase resulting from fixed justification due to the SOH gapping and the possibility of either a positive or negative PJE. Examination of Figure 12.21 will demonstrate that this is equivalent to $(3 + 1 + 1 = 5)$ bytes for AU-3 and $(9 + 3 + 3 = 15)$ bytes for AU-4. In addition, it is necessary to provide some hysteresis in the buffer threshold spacing to reduce the pointer activity associated primarily with accumulated phase variations due to short-term instability limitations of transit reference clocks. This value has been set at 4 bytes for AU-3 and, correspondingly, 12 bytes for AU-4.

Allowance must also be made for practical limitations connected with buffer headroom, phase sampling at plesiochronous boundaries and waiting time between threshold crossing and justification opportunity. All these considerations combine to set a minimum value on the threshold spacing. The maximum threshold values, efficiency considerations aside, are only limited by the need to confine the absolute and differential transit delay within reasonable bounds. Values in the range 3–5 μsec might be considered reasonable.

12.3.1.3   Concatenation

Contiguous concatenation requires a linkage between pointer processors corresponding to AUs that are contiguous in time. When the concatenation indicator (CI) is detected in PVN, the PI state of the previous AU (cPI_{N-1}) is substituted for the purposes of incrementing or decrementing. On the transmit side, AUs that have been designated as concatenated take the PJ output from the immediately preceding member of the concatenated group and act on it as though it were spontaneously generated within its own pointer processor. Concatenated VCs will therefore shift as a rectangular block through the three-dimensional frame structure in response to PJEs.

The pointer processing function described above is a mandatory feature of any MS termination in an SDH network. It is however, seldom found in isolation as described above. It is generally used in conjunction with an HVC crossconnection function or an HVC termination, or both.

12.3.2   HVC Crossconnecting

Elements that perform an HVC crossconnecting function will generally terminate two or more STM-N streams. These must first be aligned to the element reference clock as described above and their pointers adjusted accordingly. If disinterleaved down to the AU level, they can be synchronously crossconnected in a single space-switching matrix. The synchronous alignment of the streams is an important feature not available in asynchronous space switches, as it enables hitless rearrangement within the matrix. Quite large VC-3 or VC-4 matrices can be built in this way. An implementational model of such an arrangement with only four terminating STM-N streams is illustrated in Figure 12.22.

Alternatively, in a time multiplexed architecture, bytes from each of the incident streams are written into a memory in the order in which they are received. They can be read out in an order that has been transposed according to a TSI pattern supplied by a controller. The TSI RAM must be large enough to hold one complete pattern repeat from each of the incident streams. In principle, a single multiport RAM can be used efficiently to implement all the pointer buffers and the TSI. Single-stage time switches of this type for the HVCs are only feasible for the smallest of matrix sizes such as those found in certain

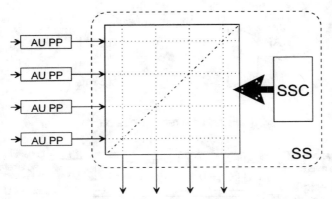

Figure 12.22   Implementational model of a four port HO path layer crossconnect implemented as single stage space switch.

classes of flexible multiplexers. They may be used to implement one of a group of first-stage matrices in a three-stage T-S-T structure, but this again remains something of a design challenge for HVCs.

12.3.3   Adaptation Between the HO and LO Path Layer

The mechanism for transporting LVCs within an HVC is the tributary unit (TU). The structure of the TU and its location within the HVC are described in Chapter 5. The TU has the same flexible alignment relationship with the HVC in which it is transported as the AU has with the STM in which it is transported. In the general case, LVCs recovered from an incoming HVC must be demultiplexed and realigned before remultiplexing in the outgoing HVC. The realignment function is performed by the TU pointer processor.

12.3.3.1   *The TU Pointer Processor*

Multiple TU pointer processors are invariably time-shared. In other words, the processing logic is the same for each TU in the implemented group and the process repeats in each TU channel timeslot. Buffer storage and the registers required to store pointer values and other TU parameters may be provided by the logical partitioning of a single RAM. The logical equivalent of disinterleaving in the time domain is the generation of a sequence of channel strobes and address labels that activate the registers and RAM in timeslots corresponding to a particular TU. Functional representation and implementational model are illustrated in Figure 12.23.

The TU channel sequencer is complicated somewhat by the necessity to allow for mixed payloads in the HVC. Such a mixed payload is illustrated in Figure 5.20. Any one of the three 84-column groups interleaved in this payload is equivalent to a mixed VC-3 payload. A TU channel sequencer for the VC-4 payload is illustrated in Figure 12.24.

It consists of three tandem counters corresponding to the three levels of TUG. The first counter divides by three and identifies the TUG-3s within the VC-4 structure. The second counter advances once for each cycle of the first counter and itself has a count of seven, corresponding to the seven TUG-2s in a TUG-3 (or equivalently, in a VC-3) The

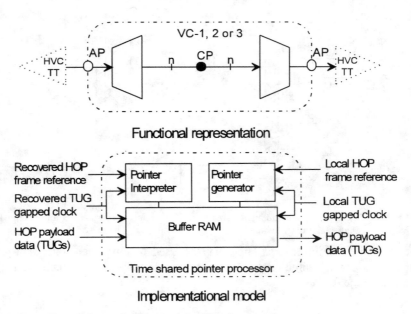

Functional representation

Implementational model

Figure 12.23   HO path layer adaptation and the TU pointer processor.

third has a count of 12 corresponding to the 12-column group in TUG-2. This last counter operates as a counter to three followed by a counter to four or, alternatively, as a counter to four followed by a counter to three. The choice of count mode is determined on a per TUG-2 basis according to whether it is organized as TU-12s or TU-11s. The counter-to-12 mode selection is made under the control of the 21-bit TU-type register (TU-11/12) that can be loaded from an OS system or by an automatic process responding to the TU-type fields in the pointers themselves.

The sequencer is driven by the VC-4 clock corresponding to the container from which the TU payload is being recovered. This will have been already gapped to allow for the VC-4 POH and fixed justification. The counter sequence is set to zero by the HVC frame start flag. It thus carries all structural information pertaining to the TUG mapping in the HVC, including its frame start and the TU channel allocation of each byte. This is coded in the 9-bit sequencer output that is presented to the time-shared pointer processor to identify TUs in the TUG.

The pointer processor must have a TU format generator for each TU type supported; four in the example considered here. These are driven by the TU sequencer, with the TU-3 format generator using the output from: the first (divide by three) counter, the TU-2 format generator using the output from the second (divide by seven) counter, and TU-11 and TU-12 format generators using the output from the divide by four and divide by three counters, respectively. The TU format generators are programmed to count the number of bytes in their respective TUs, identifying in the process the location of TU pointer bytes and justification opportunities. Of course, these are located with reference to the HVC frame and the TU multiframe marked by the H4 byte of the HVC POH.

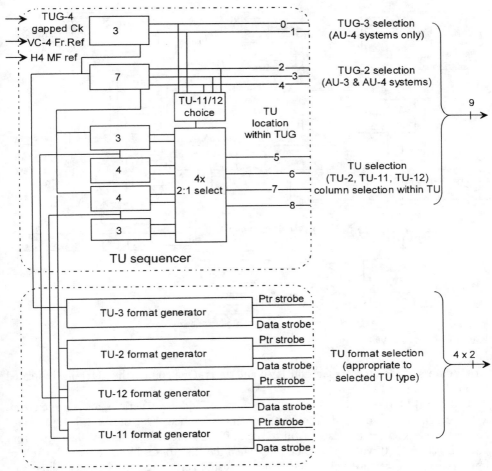

Figure 12.24   A TU Sequencer.

Each TU pointer processor illustrated in Figure 12-25 will have been provisioned to represent a particular location within the TUG structure. Using the pointer byte strobe from the TU format generator and matching the location address to the sequencer output, the TU pointer in the received data stream is identified and placed in register PV_N. This is compared with the current receive pointer value in register RPV by the PI. If the increment or decrement condition is detected, the PI causes the *increment/decrement* (ID) function to insert one extra or one less pulse into the sequence in the locations corresponding to the positive or negative justification opportunities. If the value in RPV does not match the value in PV_N, then, provided PV_N is stable for three successive multiframes, the detector "NDF or x3" will cause RPV to take on the value of PV_N.

The write address generator for a particular TU takes the form of a counter that is advanced once for each byte of the corresponding justified VC, starting from the point where the TU format generator count is equal to the value in RPV. In other words, the

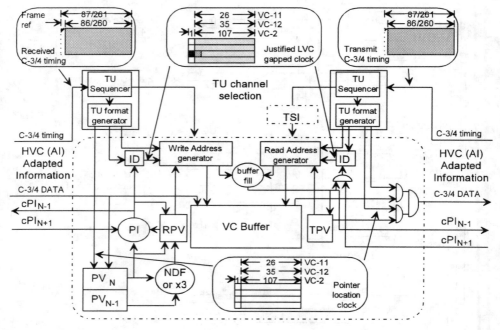

Figure 12-25   TU pointer processor.

write address generator represents the LVC and starts at the point in the TU frame indicated by the TU pointer.

The read address generator is similar to the write address generator, but is driven from a VC clock that is derived from the element reference. The buffer fill comparator continuously monitors the state of fill of the buffer by subtracting the read address from the write address. It will generate outputs when the fill crosses its upper (nearly full) or lower (nearly empty) threshold values. These threshold crossing events trigger increments or decrements in the read address counter, simultaneously enabling PJEs in the transmitted data stream. The value of the transmit pointer held in TPV is taken as the byte location number in the TU generated by the TU pointer of the TU frame.

12.3.3.2   TU Buffer Fill and Buffer Threshold

The fixed justification due to the TU pointer itself and the frame phase quantization increment resulting from a PJE are both equal to 1 byte. The fixed justification from the HVC POH and HVC fixed justification is scaled down to a small fraction of a byte phase offset in the LVC. The resulting behavior of the buffer fill parameter in the presence of PJEs is illustrated in Figure 12.26. The minimum buffer size therefore, to allow for justification and fixed stuff, is between 3 and 4 bytes.

PJEs are much less frequent in a TU pointer processor than in a similar AU pointer processor. This is because the hysteresis built into the AU pointer processor will have suppressed those PJEs resulting from short-term effects such as short-term instability in

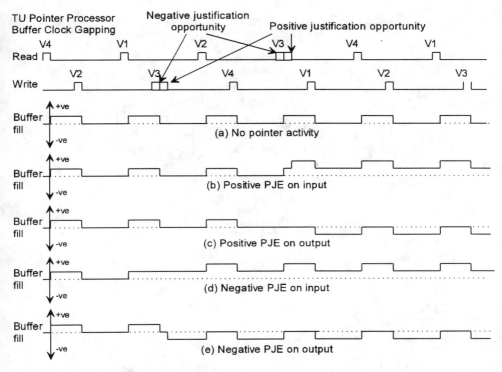

Figure 12.26   TU pointer processor buffer fill and PJEs.

oscillators and in many subnetworks even the diurnal variations. ITU-T has therefore set the threshold spacing hysteresis for TU pointer processors at 2 bytes. Hence, the overall minimum buffer capacity required for a TU-11/12 pointer processor is somewhere between 5 and 6 bytes, depending on implementation details.

Contiguous concatenation is defined for TU-2s transported in VC-3. In this case, the cPI and cPJ conditions corresponding to previous $(N + 1)$ and following $(N - 1)$ TUs are enabled as in the case of AU concatenation. The counter increments and decrements are driven by the PI and buffer fill conditions of the previous TU. Virtual concatenation as required in VC-4, for instance, does not require any hardware support and therefore is not shown in this implementation model

12.3.4   LVC Crossconnecting

A functional representation of an LVC crossconnect is shown in Figure 12.27. The harmonic structure of the TUGs is well-suited to time-space-time (TST) crossconnecting architectures. An implementational model is illustrated in Figure 12.28. The unidirectional structure of Figure 12.28 can be turned into a bidirectional structure by folding about the diagonal.

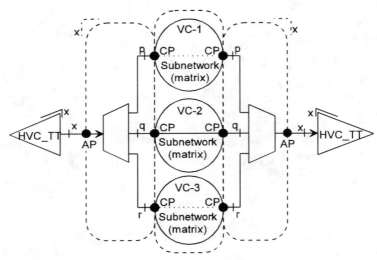

Figure 12.27   LO path layer crossconnect.

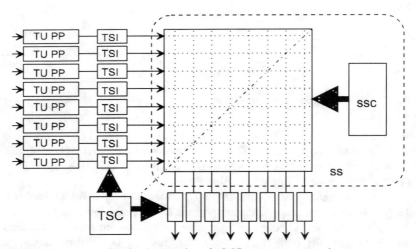

Figure 12.28   Time-space-time implementation of a LOP crossconnect matrix.

The TSI may be implemented together with the pointer processor sharing the same RAM and much of the control circuitry. The timeslot interchanging function may be achieved by translating the sequential read addresses supplied by the TU sequencer that selects the TU-specific RAM segment operated on by the read address generator according to a lookup table supplied by an operations support system (OSS) and held locally. The memory capacity allocated to TSI must be large enough that bytes written near the beginning of an incoming frame can be read near the end of an outgoing frame. This is equivalent to 1 byte per TU in addition to the capacity mentioned above for the pointer processor alone.

The *space switching matrix* (SS) in a TST architecture is time-shared. That is to say, it takes up a different switch pattern in each timeslot. The sequence of switch patterns is supplied by a space-switch control register SSC. It is, of course, essential that the TUG structures of all TSIs be aligned before presentation to the SS. This requires additional buffer delay in each stream equivalent to one row of the relevant TUG structure. In the general case of a mixed capacity payload, as illustrated in Figure 5.20, this implies additional buffer capacity of 3 bytes per TU-11 and f bytes per TU-12. If TU-12 is excluded, as will often be the case in products aimed at North American applications, then an additional 4 bytes per TU-2 and 1 byte per TU-11 is sufficient. If TU-11 is excluded, as will generally be the case in Europe, then an additional buffer capacity of 3 bytes per TU-2 and 1 byte per TU-12 is required. Implementations that can be configured for either market will need to provide sufficient data memory for the worst case, which is TU-11.

The buffer store requirements discussed above may be taken as the minimum possible. Practical realizations will inevitably carry some additional implementation burden that will be dependent on features particular to the realization. Nevertheless, they may usefully be taken as the lower bound on the required storage, corresponding to a minimum value for the consequent transit delay that can be used in the derivation of delay specification targets.

12.3.4.1 Blocking Characteristics

Crossconnects are often required to be nonblocking for many applications. In the case of the multistage switching architectures discussed above, this has generally required expanding the size and connectivity of the center stages. An unconditionally nonblocking three-stage switch requires almost twice ($2n - 1$) the number of center stages as a minimal conditionally nonblocking equivalent. The latter is more hardware efficient, but will, in general, require rearrangement of existing connections to be nonblocking. Traditionally, there have been two problems with rearrangement. First, the difficulty of executing the rearrangement algorithm in real time, and, more importantly, the inevitable hit (on perhaps several traffic carrying paths) when rearrangement is made.

The first problem has been solved by the teletraffic community in recent years, and it is now possible to implement simple, fast, and reliable algorithms to rearrange a blocked matrix. The second problem is solved with the SDH because hitless rearrangement within a single fabric by switching on frame boundaries is made possible by the strict control of phase alignment required for switching SDH containers. A small overprovision in the center stage of about 5–6% can reduce both the frequency and the extent of the rearrangements required to acceptable levels.

12.3.5 Adaptation From SDH Path Layers to the PDH Layer

This adaptation function recovers PDH layer information from an SDH VC. In doing so, it must smooth out the discontinuities introduced by the intervening processes. These include the bit-level justification process by which PDH signals are mapped into VCs, the fixed justification associated with each level of multiplexing, and, most importantly, the 1-byte (or 3-byte in the case of VC-4) phase steps associated with pointer processing in normal and degraded modes. The function is called desynchronizing, and the implementation of desynchronizers has been the subject of intense debate in the standards community in relation to the feasibility and cost of meeting the various specifications.

12.3.5.1   Fixed Bit-Leaking Desynchronizers

An implementational model of a fixed bit-leaking desynchronizer is shown in Figure 12.29. There are two key elements: a digital control part and a linear, *analog phase locked loop* (APLL). The APLL must have a sufficiently long time constant, in which single-phase steps of one UI can be attenuated to a level within the existing recommendations for jitter at PDH interfaces. The digital control part detects the pointer adjustments, but only releases the phase step to the APLL one bit at a time. The time the digital circuit must wait before passing the next bit is determined by the time constant of the APLL. One bit must have been leaked out by the APLL before the next one is passed to it; hence the term fixed bit leaking.

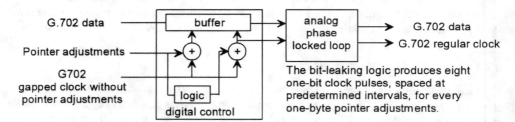

Figure 12.29   A fixed bit-leaking desynchronizer.

A variant on this theme uses a regular clock between the digital circuit and the APLL that is a multiple of the outgoing plesiochronous clock, which allows the pointer adjustment to be passed to the APLL in fractions of a bit. Indeed, it is possible to implement the whole APLL digitally by using such an oversampled clock. While this is relatively straightforward at DS-1 (1544 kbit/s) and 2048 kbit/s, it is not at all straight forward for the higher PDH rates (34368, 44736 and 139264 kbit/s). Tightening the degraded-mode, long-term frequency stability specification has been discussed, but was considered unacceptable due to cost of maintaining robustness in the presence of failures in the synchronization network where 4.6 ppm was well established as a stratum 3 reference in the United States. Refinements of the simple schemes described above were also considered using two and even three different leak rates. However, it proved impossible to prove that such schemes could provide acceptable jitter levels on the recovered signals.

12.3.5.2   Linear Digital Control Desynchronizers

The linear digital control desynchronizer is implemented as a digital low-pass filter, acting directly on the quantized phase information provided by the interlayer adaptation process as 8-kHz samples and a spectrum-spreading quantizer that makes linear the bit-quantized phase of the output presented to the APLL.

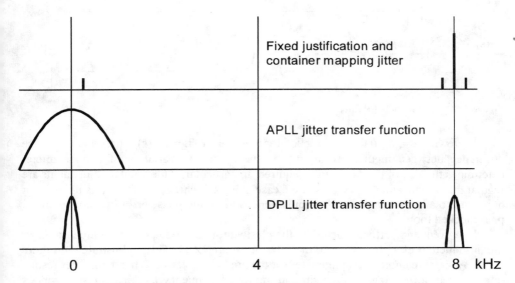

Figure 12.30   Spectral components of jitter and roles of APLL and DPLL.

The jitter presented to the desynchronizer at the end of an SDH path is made up of the following three components:

- Jitter due to fixed justification (SOH, POH, etc.);
- Jitter due to container justification, including waiting time jitter;
- Jitter due to pointer adjustments.

We can ignore the jitter due to system clocks.

The frequency domain representation illustrated in Figure 12.30 clearly demonstrates the main features. The conventional jitter components are concentrated at the frame rate of 8 kHz and its multiples. There is a strong line for the fixed component and for plesiochronous offsets, as small sidebands are produced that may fall inside the passband of the APLL. These sidebands constitute the well-known phenomenon of waiting time jitter associated with plesiochronous demultiplexing. The pointer jitter is not illustrated, but a single pointer transient produces a flat jitter spectrum from zero to infinity.

The digital *low-pass filter* (LPF) and the analog LPF represented by the APLL are illustrated with somewhat distorted frequency scales so that their main features are discernible on the same diagram. The digital LPF has a narrow passband but is replicated at all the harmonics of the sampling rate. It is a well-known feature of such digital filters that their characteristics may be defined in a range equivalent to half the sampling frequency.

A two-pole filter of the type illustrated in Figure 12.31 can provide adequate performance with 40 dB per decade rolloff and the characteristics can readily be tailored to combine adequate low-frequency tracking performance with sufficient attenuation of the pointer phase transient.

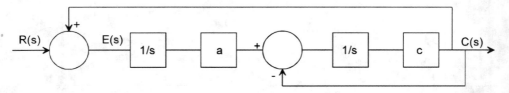

Figure 12.31   Two pole LP filter.

If viewed as a digital phase locked loop, the error signal E(s) represents the phase difference between input and output, and the maximum variation of this parameter determines the size of the data buffer required. In particular, if the pointer adjustments are regular monotonic and frequent, as in the case of a plesiochronous offset, this phase error builds up to a large constant value equivalent to the static phase error in a conventional phase locked loop.

Low-frequency effects, including the plesiochronous offset, can be compensated by the introduction of a predictor into the filter of Figure 12.31. This is illustrated in Figure 12.32. A suitable choice of a parameter for the predictor relative to the main filter results in a reduced buffer size requirement and improved frequency tracking. Plesiochronous adaptation can therefore be accommodated without excessive buffering. Figure 12.33 shows the time domain response to a single transient and the fixed phase offset that is produced at the output by regular adjustments resulting from a plesiochronous offset between the APs of the VC path. The equivalent phase offset within the loop, which would otherwise have required excessive data buffering but is canceled by the predictor, is of the order of 250 times the offset produced at the output. Figure 12.34 illustrates the frequency domain response of the double pole filter with predictor and typical parameter values. Figure 12.28 shows the transient response due to a single isolated pointer adjustment on the output phase of a 139264 kbit/s signal as measured in the 200-Hz high-pass filter specified in G.823. This represents the calculated value based on the linear model just described and compares well with the response of Figure 11.35, which represents the result of a precise simulation of the digital implementation described below in Section 12.3.5.3.

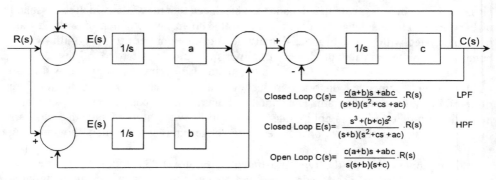

Closed Loop $C(s) = \dfrac{c(a+b)s + abc}{(s+b)(s^2 + cs + ac)} \cdot R(s)$ LPF

Closed Loop $E(s) = \dfrac{s^3 + (b+c)s^2}{(s+b)(s^2 + cs + ac)} \cdot R(s)$ HPF

Open Loop $C(s) = \dfrac{c(a+b)s + abc}{s(s+b)(s+c)} \cdot R(s)$

Figure 12.32   S-domain diagram of double pole desynchronizer with predictor loop.

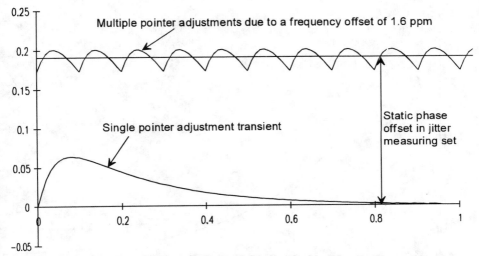

Figure 12.33   Time domain response to pointer adjustments.

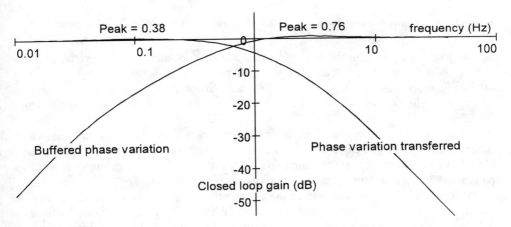

Figure 12.34   Frequency domain response.

12.3.5.3   *Digital Implementation*

Filters of this type, when transformed to the *z*-domain, have a straightforward digital implementation. The approximation in transforming to the *z*-domain is very good. The linear adders of the *s*-domain transform to digital adders; the perfect integrators of the *s*-domain transform to digital accumulators and the linear multiplier coefficients a, b, and c transform to digital multipliers or dividers. If the coefficients are constrained to be powers of 2, then this function reduces to a simple *n* bit shift for a multiplier coefficient of $2n$.

Figure 12.36 illustrates an implementation model of a desynchronizer based on the transformation of Figure 12.32. The interface between the binary representation of the SDH format and the digital signal-processing domain is implemented by some clock

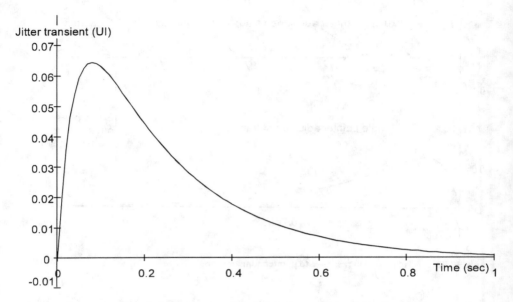

Figure 12.35   Time domain response of a 139.264kbt/s recovered signal to a single pointer transient.

control logic such that the filter is presented with a sequence of numbers at intervals corresponding to the pointer frame that will either be 0 for no phase change, 1 for a single bit justification, +24 for a positive pointer adjustment or -24 for a negative pointer adjustment. The pointer adjustment values are ±8 in the case of recovering 44736 kbit/s from VC-3.

The output of the digital filter is quantized and will introduce or suppress clock transitions in the output clock stream so as to accurately reconstruct the original plesio-chronous frequency. The rate at which this clock modifying process operates is a design parameter. The frame rate or the row rate are convenient values. The role of the "dither" signal in modulating the threshold of the quantizing process serves to make the system linear. A sawtooth waveform is easy to generate and has the required amplitude distribution to achieve this purpose, while a frequency between 3 kHz and 4 kHz is appropriate so that any artifacts that may be produced can easily be attenuated by the APLL.

The time domain response of the phase variation of a recovered 139264 kbit/s signal due to a single pointer adjustment is illustrated in Figure 11.35. The effect of a frequency step (i.e., a sudden change from one frequency encoded in the pointer adjustments to another frequency) and the conditions on the relationship between the time constants of the main loop and the predictor loop to guarantee stability are left as an exercise for the interested reader.

The comparative simplicity of desynchronizers of this type are such that they are now being applied at all bit rates and not just for 44736, 34368, and 139264 kbit/s where fixed bit leaking cannot meet all the requirements.

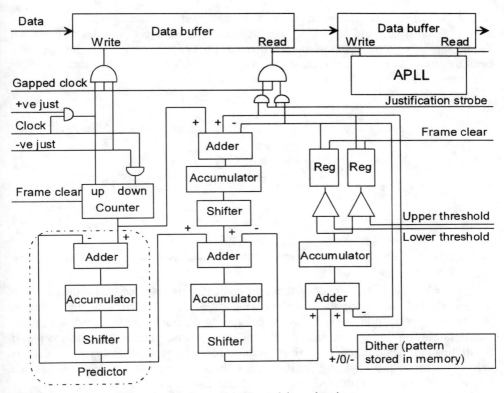

Figure 12.36 Implementation of a linear digital control desynchronizer.

12.3.6 ATM Cell Switching

High-capacity STM switches exploit the synchronous time multiplexing of the synchronized data streams in combinations of time- and space-switching elements. A simple time-space-time (TST) system was described in Section 12.3.4 to illustrate SDH LVC crossconnecting. ATM virtual channels are asynchronously multiplexed in a cell stream, with each channel identified by the label in the cell header rather than a numbered timeslot in a periodic frame. The analogous function to timeslot interchanging in this case is cell label interchanging. Similar basic switching mechanisms between ports can be implemented, but the statistical nature of the cell traffic means that more than one cell can be directed to a particular output port in the same timeslot. Contention for output slots is resolved by queuing cells in a buffer until a free slot is available.

Large switching fabrics are constructed by interconnecting a large array of basic switching elements (BSEs) in a multistage interconnection network (MIN). Each BSE comprises a switching part capable of routing between a set of input ports and a set of output ports and the queuing buffers required to resolve port contention. The BSE behavior is strongly dependent on the distribution of the buffers in relation to the switching part, and the queuing discipline employed in the buffers.

ATM cell switching is one of the most extensively studied topics in telecommunications in recent years, and there is a huge literature on the subject. The descriptions that follow attempt to classify some common themes in the huge variety of designs that have been described. A functional representation of a VP/VCh switch with associated VC-4 terminations is shown in Figure 12.37.

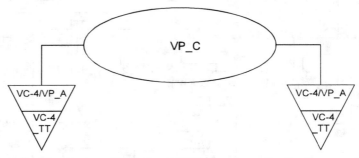

Figure 12.37   Functional representation of an ATM VP crossconnecting function on an SDH platform.

It has been already mentioned that port contention is resolved by queuing cells in buffers. The buffers may be placed at the input ports, at the output ports, or, depending on the switching technique used, it may be possible to distribute the buffering within the switching part. In the input queuing case, a cell is only passed across the switching part if the destination output port is free. This has the unfortunate side effect that while a cell is held in the input queue waiting for its outlet to become free, it may block another cell whose output port is already free. The blocked cell may be allowed to jump the queue by departing from a simple first in first out (FIFO) rule, but this adds significant complexity to the buffer control. The effect is known as head of line blocking and will seriously limit the performance of an input queuing implementation. Input queuing is illustrated in the implementation model of Figure 12.38(a).

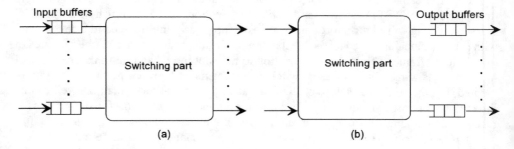

Figure 12.38   Switching and queuing components in an ATM switch.

Output queuing is illustrated in the implementation model of Figure 12.38(b). In this case, there is no head of line blocking and the queue serving rule is simple FIFO. Cells arriving simultaneously at an output port are simply queued until a slot is available. The output queue must be fast enough to read in $N + 1$ cells in one cell period. In the worst case, it may have to accept a cell from each of the N inputs and also write one cell to the output, all within one cell period. Doubling the word size at which the buffer is organized relaxes the access time requirement by a factor of 2. The required memory access time for output queued systems is given by the relation $W/\{(N + 1) \times F\}$ where W is the word width and F is the cell stream bit rate. For example, for $N = 16$, $W = 64$, and $F = 622$ Mbit/s, an access time on the order of 6 nsec is required.

We can assume that the cell arrival process at each inlet is based on independent and identical Bernoulli processes. This means that in any given timeslot, the probability that a cell will arrive on a particular inlet is p $(0 < p < 1)$. On average, each inlet is utilized at a level p. With N inlets and N outlets, we also assume that a particular cell has equal probability $1/N$ of being addressed to any given output, so that the probability of a cell arriving at one inlet for a given outlet is p/N. A BSE with output queuing has a very well-defined behavior and cell loss probability better than 10^{-9} up to $p = 0.8$, equivalent to a traffic load of 80%, while for input queuing, only 59% is achievable.

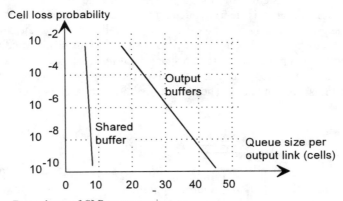

Figure 12.39   Dependence of CLP on queue size.

The queue must be dimensioned to hold maximal length bursts of closely spaced cells. Given the statistical nature of the cell traffic, this implies some probability of cell loss in any finite queue. The load on the separate port buffers is also statistical, and considerable savings in buffer size can be achieved by sharing the buffering requirements of all ports. For the same traffic assumptions above, for example, a gain factor of about 5 can be obtained for $N = 16$, increasing to about 7 for $N = 32$. Figure 12.39 illustrates the achievable cell loss ratios as a function of queue size for $N = 16$ with traffic loading of 80% for separate and shared output queues.

A shared output buffer clearly has a significant advantage in size; however, it is now necessary to allow for each input and each output to access the memory simultaneously. The access time required is thus given by $W/\{2 \times N \times F\}$. This is close to a factor of 2 smaller than for the simple output buffer. Use of dual-port RAM restores the factor of 2 but the dual-port RAM is a more complex function, which implies a greater area silicon.

There are design choices to make for the switching part also. The switching part may be implemented as a shared memory. In this case, cells from each input port are written sequentially into the shared memory and read out sequentially to the selected output ports. The shared memory is similar at this level to a timeslot interchanger. When used in conjunction with shared buffer output queues, the switching and queuing functions effectively take place within the same RAM, which must be enlarged slightly to allow for the sequencing time to connect a low-numbered input port to a high-numbered output port. The combined memory control logic is now significantly more complex. Having to deal with switching and buffering the access time for the combination is the same as that for the shared output buffer alone. The shared memory switch can be quite efficient, providing the required access speed can be achieved; but multicasting and broadcasting is not natural, and to achieve these functions implies considerable extra complexity.

Conceptually, the simplest switching part is a shared medium connecting all inputs to all outputs via a TDM bus. Thus cells at input ports are gated onto the bus in time sequence and the appropriate output ports are also selected in each timeslot. The TDM bus must operate at the full speed so that there is no contention for timeslots at the input side. Some compromise is possible whereby the bus may be operated at reduced speed, provided some input buffering is included to avoid contention on the bus. Queuing to avoid bus contention is a quite separate issue from queuing to avoid output port contention. Although broadcasting and multicasting may be easily and naturally achieved, the severe physical limitations of high-speed buses for large switches has made the shared medium approach less popular today. Output queues may be shared, but this makes the solution somewhat redundant, combining the access time limitations of the shared memory with the speed/size limitations of the shared medium. The shared medium approach is illustrated in Figure 12.40

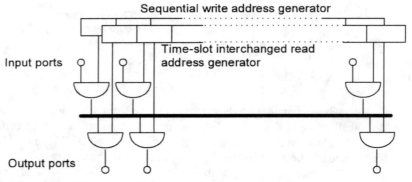

Figure 12.40   Shared medium switching.

A third alternative for the switching part is provided by effectively broadcasting each input to all outputs and selecting appropriately at the output port. Naturally, there is no contention among the N^2 independent paths between inputs and outputs; hence, all queuing occurs at the outputs. Like the shared medium technique, broadcasting and multicasting is natural and straightforward. Unlike the shared medium case, buffers and address filters operate at the port speed, and there is therefore considerable scope to increase throughput by increasing speed. The main limitation is in the quadratic dependence on N, which

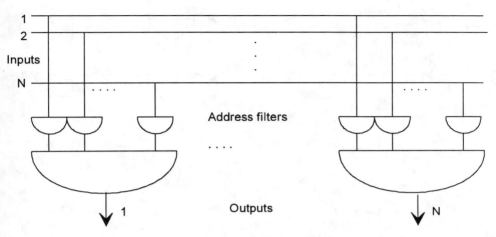

Figure 12.41   Sending all inputs to all outputs and selecting at output.

ultimately determines the maximum physical size that can be achieved. This approach is illustrated in Figure 12.41.

A fourth switching part alternative uses space switching in the form of multistage interconnection networks (MINs). MINs have themselves been studied extensively for many years, initially in the search for solutions to the N^2 crosspoints problem in crossbar switches and later in packet switching. An important subset of these, known as banyan networks, have attracted the attention of the ATM community—first, because they are constructed by simple replication of a simple 2×2 module, and second, because each basic module is controlled by a single bit, which allows for a simple implementation of self-routing cells based on a routing tag attached to the cell indicating the output port to be connected.

The basic principle is illustrated in Figure 12.42. Each 2×2 switching element routes an incoming cell to one or the other output according to the binary value of a single control bit. If the control bit is a 0, the cell is routed to the lower output; if it is a 1, it is routed to the upper output. A 4×4 element is shown constructed from four 2×2s and an 8×8 element from twelve 2×2s. If the destination output address is attached as a binary tag in front of the cell at the input to the network, each bit of the address is read by the corresponding switching stage and used as the control bit. This self-routing property allows considerable simplification of the switch control system. The example of Figure 12.42 can easily be extended to generate an $N \times N$ switch. If $N = 2^n$, the banyan will consist of $n = \log 2N$ stages, each consisting of $N/2$ switching elements. All stages are clocked at the same speed as the port speed, and there are no systematic limitations to speed or size

The simple banyan as described is not nonblocking; thus, in its simple form, only reduced throughput would be available. A popular remedy is to add a batcher sorting network in front of the banyan as illustrated in Figure 12.43. The sort network sorts cells according to their addresses directing those with larger addresses towards the output indicated by the arrow. The combination is internally nonblocking because cells are presented to the banyan in sorted order. Naturally, output contention will still occur, but

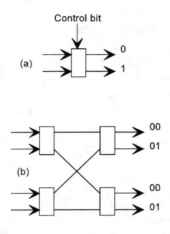

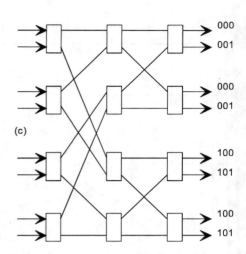

Figure 12.42   The banyan switch.

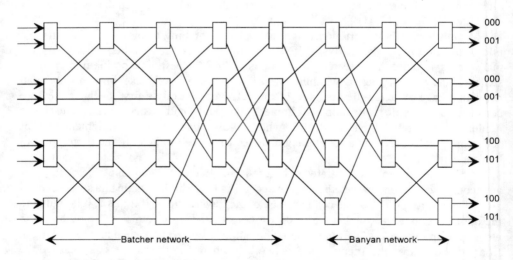

Figure 12.43   A batcher-banyan switch network.

may be resolved by buffering. Input buffering may be applied as discussed above, but this limits throughput due to head of line blocking. The internal structure of the network allows the buffering to be distributed within the network. When contention occurs, one cell is buffered internally at one of the basic elements. This is signaled backwards to apply backwards pressure so that further contention causes buffering at an earlier stage. In extreme cases, the back pressure reaches the first switching stage. Internal buffering limits throughput by the same head of line blocking mechanism described for input buffering.

Again, output buffering is considered the best compromise. However, output buffering is not straightforward in such space-switching networks. A banyan network can deliver only one cell per cell period to any output port. Output queuing requires that the network deliver multiple contending cells to the selected output in the same cell period. In other words, it is necessary to provide multiple paths through the network by means of multiple parallel switch planes, by adding extra switching stages (e.g., multiple banyans in tandem), or by increasing capacity (by increasing speed) of interstage links.

12.3.6.1 Interconnecting Basic Switching Elements to Form Large Cell-Switching Fabrics

BSEs generated by any of the mechanisms discussed above may themselves be intercon-nected in an MIN to form the cell switching fabric and very large switches may be built in this way. A three-stage Clos structure is illustrated in Figure 12.44. In this example, the first and third stages consist of (N/n) BSEs of dimensions $n \times m$ and $m \times n$ respectively. The first and third stages are interconnected via m second-stage BSEs, which are square matrices of dimensions (N/n) $\times$ (N/n). This structure is nonblocking and provides m possible paths between any pair of outputs. Nominally blocking MIN dimensions can be rendered nonblocking by speeding up the center stages relative to the outer stages.

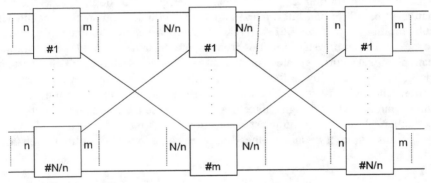

Figure 12.44   A Clos (N,n,m) network.

Multipath MINs have several advantages in addition to their contribution to the output queuing problem. A randomizing element introduced in the routing algorithm has the effect of distributing traffic more uniformly in the network, thus rendering its performance less dependent on external traffic statistics. Multipath networks are also more resilient. Partial failures can be taken into account at the routing stage and alternative routes set up around the failure. Of course, such self-healing capability requires that connections across the switch are supervised by introduction of test patterns, cell sequence numbering, error detecting codes, or some such performance-monitoring overhead.

Where cells from the same virtual channel are allowed to take different paths through the network, there is the probability of differential delays, causing cells to arrive out of sequence at the output ports. One solution is to pass all cells belonging to the same virtual channel on the same path through the switch, but this increases the blocking probability and hence reduces the throughput. Another solution is to extend the self-routing tag to include a time marker or sequence number that can be used to resequence cells at the output.

The concept of blocking is more complicated for cell switching than for circuit switching because of the statistical nature of the virtual channel cell streams. In most of the preceding discussion, a simplified and more conservative view of blocking is taken and the virtual channels are assumed to be of constant bit rate and defined by this single parameter.

12.3.6.2   *Copying and Multicasting in an ATM Switching Fabric*

To support multicasting in an ATM fabric, all the BCEs that make up the fabric must individually support multicasting or, alternatively, a separate copy network can be added to the routing fabric. Ease of multicasting was mentioned in connection with each of the BSE options above. Multicasting in a banyan space-switching element is essentially quite simple, but an extra control bit is required for each 2 × 2 element to distinguish the broadcast condition from the other two switched states. There is a further difficulty that applies to self-routing networks in that a cell duplicated to both outputs also has a duplicated routing tag and both copies would be routed to the same output.

A popular solution to this difficulty is to distinguish the routing mode as simple or multicast in the self-routing tag. When a cell is designated to be multicast, the normal destination Id-based routing information is replaced with a unique number identifying a particular multicast tree. Each BSE will read the multicast connection identifier and use it to read the specific multicast information from an internal table that holds a list of multicast destinations for each current multicast connection. This means that each BSE must hold a table large enough to hold the few extra bits of multicast information for each current connection. The multicast connection identifier is used as a key to access the table. There is obviously an important design choice as to how big the table should be to cope with a target number of concurrent multicast connections. This turns out to be a very elegant solution because it preserves all the other desirable properties of the multipath self-routing structure.

12.4   SOFTWARE SYSTEMS EMBEDDED IN NETWORK ELEMENTS

We have referred before to the trend towards management automation in the transport network and the consequent increase in the level of functionality in transport NEs. The proportion of software to hardware in SDH NEs can therefore be expected to be much higher than was traditionally the case in transport equipment. It is now quite normal to find quite modest transport equipment such as multiplexers and line terminals with multiprocessor control systems, while the larger crossconnect nodes can be expected to have several layers of control below the level of the open management interface. The design and implementation of such large complex systems in a form that balances high functionality and performance with robustness and cost efficiency is not an easy task. A generalized multiprocessor system is illustrated in Figure 12.45.

The complexity of large systems, which must also be extensible and evolvable in service, has stretched traditional software design methodologies to the limit. Today object-oriented (OO) technology seems to hold many of the answers and is used extensively in modern systems. The power of the OO approach in managing complexity derives

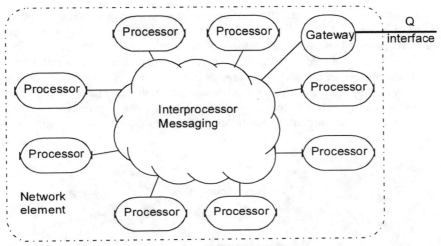

Figure 12.45   Network element containing multiple communicating processors.

mainly from the properties of encapsulation of functionality within objects, in such a way that the operations and data encapsulated may only be accessed by defined interface protocols. Object access security can render more difficult the sort of ad hoc patches to "fix" problems common in traditional software systems, but it also places a limit on the sort of arbitrary complexity towards which such development practices inevitably lead.

12.4.0.1   The Object Support Environment

An effective OO system must operate within a supportive OO software environment where the implementational complexity of providing OO capability is hidden from the designer. The difficulty of providing such object support environments was often underestimated, and early attempts met with mixed success. However, the technology is now more mature, and most of the components to implement a robust framework for objects are now available in true OO style—off the shelf. Several important features are required from an OO framework:

- *Concurrency*—Objects are considered to act independently within the constraints of the relationships in the object model. In particular, this implies that a system is capable of maintaining many concurrent, independent threads of control.
- *Distribution*—The object behaviors are essentially independent of their location on a multiprocessor platform. This implies the maintenance of a unique naming system over the whole multiprocessor platform, a name to location referencing mechanism, and an effective interobject communication network.
- *Robustness*—Objects should have a strong "will to survive" and should be able to take corrective action to recover from errors and failure situations.

Interobject communication within a process can be orders of magnitude more efficient than communication between objects in different processes. Partitioning of large systems into subsystem processes where strongly bound objects are placed closer together, in communication terms, than loosely bound objects is perhaps the most important architectural decision facing the designer. The closeness of coupling among objects, illustrated in Figure 12-46 , is not a well-defined concept. It may be based on the frequency of interobject references, the amount of information shared or interchanged, or the speed of interaction demanded by the application. Difficult choices must invariably be made. An object cannot be in two places at once, but a local copy may suffice in some circumstances, providing account is taken of the burden of maintaining consistency between the copies and the original.

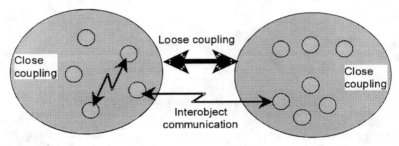

Figure 12-46   Relative coupling between objects.

Whatever the partitioning decision, care must be taken to dimension the processes, the power of the processors, their local memory requirements, and the interobject traffic bandwidth required of the interobject communication network. Closely coupled objects, which for other reasons may not be placed in the same process module, require special attention to the corresponding interprocess communication channels.

12.4.0.2   *Memory Management*

In a live, dynamic system, objects are being created and deleted all the time. So long as an object exists, it is taking space in memory. NE software is intended to run nonstop for very long periods, and great care must be exercised to eliminate the possibility of "memory leaks." That is to say, objects should be deleted as soon as they are no longer needed (a process known as "garbage collection"); otherwise, the address space, even a virtual one, will eventually become exhausted and force a software restart. This was a common software bug in early C++ systems implemented ad hoc without the support of automatic memory management provided in the environment.

12.4.0.3   *Persistency*

An object exists if it is present in memory. Nonvolatile records of persistent objects must be kept that transcend the association with the process that created them. Such an image can be used not only for restart or recovery purposes, but also to support economic use of program workspace. Objects can be "swapped out" of resident memory to be recalled (with an inevitably longer delay) when referenced later by a message from another object.

12.4.0.4   Encapsulation

Arguably the most important feature of OO methodology is the ability to encapsulate the operations, the behaviors, and the data pertaining to an object. This implies strict access control so that not only is access confined to the defined operations, but can also be denied if the accessing agent is not cleared. This is particularly important where one administrative domain may be responsible for service configuration while another may be responsible for operational maintenance and repair. It should not be permissible for an object in the maintenance domain responsible for running an intrusive diagnostic test, for example, to do this while the target object is "in service." Only the service administration can take it out of service to enable the test to run. Apart from the obvious security from hackers in an open communications system, it is the robustness provided in the face of human error and system bugs that is often the most important feature in complex systems.

12.4.0.5   Interobject Communication

The freedom to locate objects to optimize aspects of performance or resilience while maintaining a desirable level of encapsulation implies that the operations and notifications supported by the object are the same whether executed locally between collocated objects or remotely between objects on different subsystems. The former may execute via procedure calls within a single processor system, while the latter may use a backplane bus, an intershelf LAN, or even a station LAN. It is an important aspect of system design to ensure that the interobject communications network has enough bandwidth to handle peak loads and, more importantly, that it reacts gracefully to peak load congestion, either by flow control or queuing messages for later execution. System "crashes" due to congestion of the interobject communication network must be avoided at all costs.

12.4.1   NE Control Architecture

The control architecture within the NE is shaped by similar pressures to those acting on the architecture of the external management network with respect to the placement and relative coupling of components. Processes responsible for collecting event and perform-ance information are often inactive for much of the time, but when something happens they are expected to handle large volumes of data with speed and accuracy. The objective of such systems is generally to classify, filter, correlate, and otherwise process the data before reporting it in condensed form to a higher management level. Alternatively, the processed data may be held in the NE until requested. Both procedures are well-adapted to a hierarchical architecture in which the lower levels are implemented on a multiplicity of processors correlating and filtering data with local scope before passing the consolidated data results up the hierarchy.

Processes responsible for the dynamic control of internal connectivity, whether in response to service requests or other stimuli such as protection requests, are similarly required to exhibit uniformly fast response, even in periods of high activity with bursty inputs. Connection control processes are generally required to coordinate the behavior of two or more entities within the NE to change its connectivity. In large systems, such processes are best distributed such that connectivity changes are effected by direct peer-to-peer communication between the entities themselves. The effective parallel processing can significantly improve speed of response during periods of high activity, and additionally, the reduced dependence on a single processor provides resilience to failures.

There is no conflict in choosing within the same multiprocessor system a hierarchical architecture for one application, a distributed one for another, and hybrid of the two for a third. The OO framework will allow objects to communicate freely with one another, within the NEs embedded system according to the demands of the architecture. Communication outside the system with other components of the management architectures are governed by the requirements of the management applications as expressed through the information model.

12.4.2   Control of Connectivity in a Three-Stage Crossconnecting Network

The practical implementation of LO path crossconnect equipment was discussed earlier in this chapter. Figure 12.27 illustrates the manner in which all the LO path layers may be supported simultaneously within one digital crossconnect (DXC) equipment unit. A crossconnect fabric is partitioned among the various layers that it is capable of supporting by configuring the HOP adapters. This is normally considered to be a form of static configuration, and is done in advance of setting up crossconnections. The creation of TUG3, TUG2, and TU CTPs determines how much resource is allocated to each layer and hence the size of the matrix available. The matrix is further defined by allocating the TPs that have been created to TP pools associated with specific links that terminate on the matrix. When this static aspect of configuration is complete, matrices will have been created in each of the layers, with TP pools to identify groups of TPs with routing significance.

The view of connectivity presented across the interface shows no internal detail of the supporting three-stage network. However, the same topological decomposition procedure that has been applied to the network at large in Chapter 3 can be applied to the subnetwork implemented by the three-stage switching network. This is illustrated in Figure 12.47 for the case of a 16 × STM-4 port DXC. In this example, the links between switch stages have STM-1 equivalent capacity, there are eight center stage switches, and there is a 1:2 expansion between the outer stages and the center stages. Each outer stage switch has a single STM-1 equivalent link to each center stage. We refer to these as intermatrix links (IMLinks).

Before crossconnections can be made between outer stages across one of the center stages, the internal adaptation processes associated with the IMLinks must also be set, thereby selecting the TU format on the IMLinks themselves and by implication the layer-type characteristics of the matrices to which they are bound. The simplest method to understand is one based on using the same adaptation process internally on the IMLinks as is used on the external links; that is to say, one based on the TUG structures defined at the interface. In this case, the capacity presented to the center stages by each outer stage must proportionally match that being presented to the interface by that outer stage. However, there is a choice as to how that capacity is best assigned among the center stages while still meeting this basic constraint.

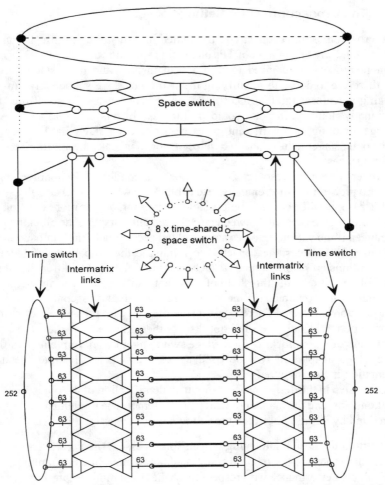

Figure 12.47   Functional decomposition of a three stage T-ST crossconnect with 16 STM 4 ports.

It is generally desirable to minimize payload fragmentation as far as possible by attempting to configure each IMLink and each center stage uniquely to a single layer type. In a general case of mixed payloads, the relatively coarse capacity quantization of the TUG-4s (or equivalently 3 × TUG-3) will make this an inefficient procedure. Where it is necessary to mix capacity on a single IMLink, the next lowest TUG capacity should be kept unfragmented and so on until a sufficiently close match to the predicted traffic is achieved. At any point in time, it is sufficient to configure only enough center-stage capacity to maintain adequate blocking performance to the next crossconnection request in each layer. As this capacity is used up by setting new crossconnections, more resource must be configured to maintain the level of blocking performance.

12.4.3   The Crossconnection Setup Process

A typical crossconnection request will be framed in terms of a specific TP on a specific outer stage X and any one of a set of TPs on another specific outer stage Y. The connection management process is required to find a path through the three-stage network. The problem therefore, reduces to identifying the two outer stage subnetworks involved and then seeking a center stage that has free capacity between these two. If the structure is unconditionally nonblocking, as would be the case in this example as a result of the center-stage doubling, it is sufficient to search for a center stage that has free outlets on both of the outer stages to be connected, select it, and then make the appropriate connection on the outer stages.

However, where the resource is scarce (as in a conditionally nonblocking arrangement), such a procedure will eventually lead to blocking when there is actually still capacity left. In this case, it will generally be possible by moving one or more existing crossconnections to create free capacity between the selected outer stages. The blocking probability can be reduced if the various choices are weighted according to their impact on the scarce resource, namely center stage capacity, to minimize the probability of blocking later. Simple algorithms exist for computing such weighting factors. Additionally, if the blocking probability reaches some threshold of unacceptability, then the same weighted allocation procedure can be operated offline to reassign the current crossconnections in a manner that releases more capacity. There is extensive literature on the subject of rearrangeable nonblocking networks, and we shall not discuss these further here. The crossconnection control process of a nonblocking three-stage switching network implemented with standard center-stage adaptation will be used as an example to illustrate the design and implementation issues involved.

The first step in the design process requires identification of the objects in the system, their relationships, and the manner in which they interact. The implementation architecture is then derived by iterating between the following two steps:

- Partition the design allocating objects to processors (in a multiprocessor system), taking account of the relative coupling and the interobject communication needs.
- Evaluate performance with respect to a behavioral model representing the external stimuli acting on the system.

12.4.3.1   Modeling the Crossconnection Setup Process

In large systems, objects with close functional relationships can be grouped together in application modules. The externally visible crossconnection fragment of the information model is extended in Figure 12.48 to include an object-oriented representation of the three-stage switching matrix, which is not visible externally across the open interface. The *outer stage matrix objects* (OSM) represent the first and third stages of the three-stage network. They are characterized by their link relationships to each of the center stages. The center stage subnetworks themselves are represented by the *center-stage matrix objects* (CSM) characterized by their complementary link relationships via the IMLinks with each of the OSM objects. Each OSM controls the connectivity between a set of externally visible TP objects and a set of internal TP (ITP) objects representing the ends

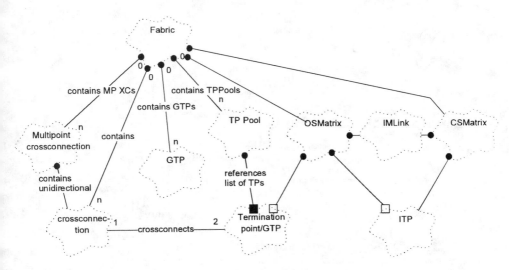

Figure 12.48   Class diagram for a crossconnect including some internal objects.

of the internal link connections belonging to the IMLinks between OSMs and CSMs. The
TPs and ITPs are all from the same layer, but only the TPs are visible at the open interface.
Similarly, each CSM controls the connectivity between sets of ITP objects that represent
the other ends of the internal link connections belonging to the same IMLinks between
OSMs and CSMs.

In the example given, there will be up to 16 instances of OSM and 8 instances of CSM
for the unconditionally nonblocking case, with double capacity in the center stage. The
LOP layer network topology is created by a separate external network building application
in response to predicted demand by setting up HOP layer trails between selected nodes
and by selecting interlayer adaptation options sufficient to create enough TPs in each layer
type associated with the fabric in each of the crossconnecting elements in the subnetwork.
The internal object instances can then be instantiated, taking account of considerations of
matrix fragmentation and blocking probability as outlined above.

The sequence of actions following a connect request to the fabric is illustrated in
Figure 12.49. A crossconnection request is received by the fabric in terms of the entering
TP and a destination TPPool. After creating a crossconnection object instance (at this stage
nonoperational, as the supporting hardware is not yet configured), the fabric then requests
from the destination TPPool the identity of an attached OSMatrix object having sufficient
capacity to complete the crossconnection. The TPPool then passes on the crossconnection
request to the selected OSMatrix, referencing the identity of the entering TP and OSMatrix.
The destination (i.e., the exiting) OSMatrix then requests the entering OSMatrix to list the
set of CSMatrices towards which it has spare capacity. The selected exiting OSMatrix
selects from this list a CSMatrix towards which it also has spare capacity and then sends
a request to interconnect the entering and exiting OSMatrices towards the selected
CSMatrix. The CSMatrix selects suitable ITPs, makes the requested center-stage connection, and returns the identity of the selected ITPs to the OSMatrices. The OSMatrix objects

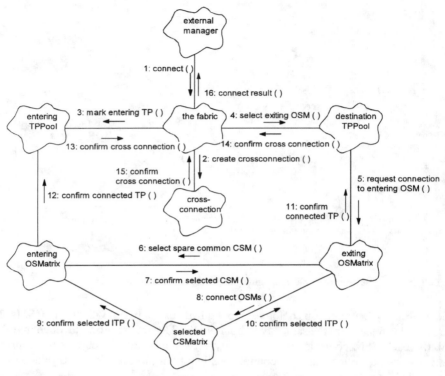

Figure 12.49   Object instance diagram illustrating the setting up of a crossconnection.

each make their appropriate connections between the selected ITPs and the TPs. All three participating matrix objects issue the appropriate configuration instructions to the configurable hardware. The OSMatrix objects return crossconnection confirmed. If validity and integrity checking are included in the design, these should now read true indicating that the crossconnection object has become operational. The crossconnection object notifies its new state to the fabric and the operation is complete.

The hardware objects are derived from an internal specialization of the equipment object and represent actual chips, registers, and so forth that are mounted on some superior equipment object within the NE. The objects controlling the crossconnecting function may be distributed in the NE in different ways. At one extreme, the whole process except for the hardware objects may be implemented on a single processor. In this case, the crossconnection setup algorithm proceeds by procedure calls between objects on the same processor system, but finishes by passing messages down to each of the hardware objects. At the other extreme, each OSM and each CSM may be implemented together with their associated TP, ITP, and hardware objects on separate processors. In this case, the algorithm proceeds by messaging between participating objects using an interprocessor communication network within the NE, with the final instruction to the hardware being a local procedure.

In the former example, the single processor must be capable of handling the worst case total processing load even with high crossconnection request arrival rates, while in the latter case the load is spread over the multiple processors. This load-sharing property is typical of multiprocessor systems applied in each of the application areas. There is no unique best solution. The design compromise may be struck differently according to the density and distribution of the external event generating mechanisms and the different weights given to cost, performance, and resilience in different situations.

References

The sections on specification methodology draw heavily on the work in ETSI recorded in:

ETSI ETS DE/TM-1015, *Transmission and Multiplexing: Generic Requirements for SDH Equipment.*

Reference is also made to:

ITU Recommendation G.783, *Characteristics of SDH Equipment Functional Blocks.*
ITU Recommendation I.732, *Functional Characteristics of ATM Equipment.*

Additional reading on ATM design and performance issue in ATM switching equipment can be found in:

de Prycker, M., *Asynchronous Transfer Mode: Solution for Broadband ISDN*, Ellis Horwood Ltd, 1991.
Chen, T. M., and Liu, S. S., *ATM Switching Systems*, Artech House, 1995.

There are a large number of books on analysis, design and construction of object oriented software systems. The following have been found instructive and informative:

Booch, G., *Object Oriented Analysis and Design*, Second Edition, Benjamin/Cummings Publishing, 1994.
Meyer, B., *Object-Oriented Software Construction*, Prentice Hall International series in computer science. 1988.

Optimal Design and Planning of Transport Networks

In this final chapter, we bring together all the pieces and describe how to set about designing and building a broadband network. There are many factors that make the process of designing broadband networks different from the traditional networks characterized by the public switched telephone networks (PSTN) and the plesiochronous digital hierarchy (PDH). It is possible to conceive of a big, single, monolithic broadband network and to use most of the design and planning methodology of the past. Some of the earlier work on broadband integrated services digital network (B-ISDN) tended to this approach; however, as broadband technology has moved towards reality, networks are no longer attempting to build a massive, ubiquitous, single network to cater for all services and all customers. The design process has to take account of the changed environment for broadband networks. The factors that most significantly influence the design process include the following:

- The coexistence of narrowband networks alongside new broadband networks for some considerable time.
- Rapidly increasing number of telecommunications services.
- The dynamic nature of the market for broadband services makes forecasting of services and traffic difficult and hence service-specific investment risky.
- The capabilities of technology are developing very fast, which makes network equipment obsolete much faster than was the case for the PSTN.
- The technology is influenced by both the telecommunications and the information technology industries, which have very different approaches to design and planning.
- The marketplace for telecommunications is increasingly governed by the competitive forces of many players rather than by national, monopolistic planning and investment.

- ATM, especially, brings together the different worlds of Internet, private networks, data networks, and large public networks, which have had different design constraints and different attitudes to design.
- The general expectations for reliability have risen significantly as more companies trust "mission-critical" processes to depend on telecommunications services;
- Modern algorithms and processing power enable a much higher level of network automation.

Asynchronous transfer mode (ATM) and synchronous digital hierarchy (SDH) technologies were developed around these expectations. Simply recognizing this provides a basic rationale for deployment of ATM and SDH; however, the design methodology will largely determine the extent to which these factors may be satisfactorily taken into account. For example, a single, national ATM network intended for all services is very likely to be obsolete by the time it is finished and incapable of supporting many currently unforeseen applications. Deploying SDH without due regard to the planning implications of flexible path management and simply because it is the next transmission technology is likely to result in a network that will be difficult to migrate, which is also expensive, and cannot meet market expectations for quality of service. Indeed, it is in considering these factors that it becomes clear that an effective broadband network requires both ATM and SDH technologies.

In this chapter, we present a general methodology for the design of broadband networks and a description of the specific implications for different parts of the network.

13.1   GENERAL APPROACH TO BROADBAND NETWORK DESIGN

The design of a telecommunications network is not a simple task, and a modern broadband network is generally more difficult than its predecessors. There are now many factors to be taken into account, and an optimal design will be a calculated compromise between many apparently conflicting requirements such as high functionality and low cost. The objective of the design exercise is to arrive at a design that achieves the best balance between all requirements.

In general, we try to develop a formal and systematic way of representing all these requirements so that the design problem can be analyzed rationally and a design solution selected for clear, objective reasons. Formal, systematic representation of requirements and their rational analysis is a key input to the design process; however, there are often other factors that make this difficult to achieve. Where the design problem is complex or the methods for rational analysis are not well developed, the design is often left to the "intuitive feel" of an experienced designer. Similarly, systematic information is often hard to collect, so the results of any rational analysis based on this can be no more than partially correct. In either case, network design effectively becomes a subjective process; it is often costly or fails to meet all requirements, and frequently both.

It has been difficult to design large, public networks by means of clear and objective analysis. Private network designers have found it easier to apply more rigorous processes, partly because private networks are significantly smaller than public networks and partly because they tend to be well-bounded by commercial interfaces. In network design, larger problems are disproportionately harder than small problems. The design of the topology of a network is a classic example of what mathematicians euphemistically call a "hard" problem. The network design problem gets hugely more complex the greater the number of nodes and links. We must consequently divide up most design problems into smaller parts to reduce the computational complexity. In addition to limiting the complexity, there are other reasons for partitioning the design problem, most particularly because layer networks are owned and operated by many independent organizations, each of which has independent control of the design of its part.

13.1.1 Network Design and Commercial Interfaces

Commercial interfaces are an inevitable feature of multioperator telecommunications networks. Many organizations are involved in running a network, and supporting this "federation" of parts from these many independent organizations is a major constraint on the design and planning process. The overall network design and planning process must ensure the essential independence of the participants to plan their own part of the network; however, there must be sufficient commonality in the separate processes to ensure effectiveness and efficiency overall. In the past, the existence of the national monopolies for PSTN simplified matters somewhat, as the common design and planning was limited to international gateway and transit networks.

The advent of competitive network operators in many countries, and also pan-national network operators, severely limits the applicability of this model. An improved model is needed within which to fit the shared aspects of design and planning, and understanding the nature of the commercial interface is central to this.

The Internet offers an alternative model, but this is also limited in its current applicability. While it addresses the competitive multioperator aspect of network design, the current Internet model does not address the more commercial aspects of service guarantees and, in particular, differentiated quality of service. Layering and partitioning are central to the framework for design and planning to allow maximum freedom of design within an operator's domain while ensuring that the global federation is still effective and near optimal.

If we consider a reasonably self-contained private network, we can see at least two very important relationships with other networks, which are also commercial interfaces, as illustrated in Figure 13.1:

- The private network is unlikely to be a completely isolated network and so will have one or more points at which connections can be onward-connected to the wider network, often referred to as gateways or points of interconnection.
- A private network will often resource its links from another network operator in the form of private circuits.

In the first case, the gateways can be seen as part of a partitioning of the wider, global layer network, which is formed by the existence of separate organizations that own and operate different parts of the layer network. In the second case, the lease of the private circuits is a fully commercialized implementation of an interlayer client-server relationship.

The cost of interconnect to other networks and the cost of private circuits are normally critical factors in the design of the private network, and from this we draw the basic

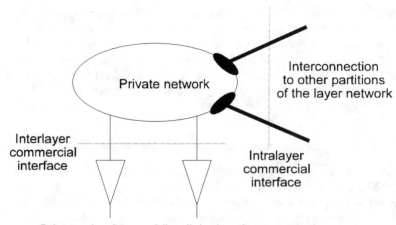

Figure 13.1   Layering and partitioning by commercial boundary.

paradigm for global design and planning: *the effect of network design on one side of the commercial interface is passed to the other in the form of a price for an interconnect or link transport service.* This is illustrated in Figure 13.2.

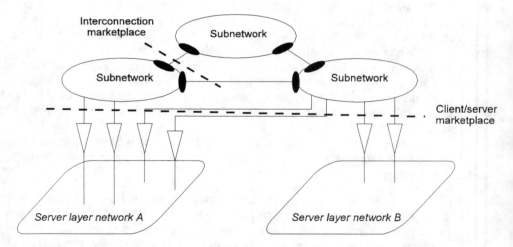

Figure 13.2   Commercial interfaces exist in marketplaces.

Using this as a paradigm for global design and planning is fully workable as it is based on the commercial reality that already exists and is imposing no technical constraints on the commercial operation of these interfaces. However, it also has two profound and important properties that we can exploit in the global design and planning exercise:

- The commercial interfaces between the layers and partitions of the global network operate in a dynamic marketplace, which is itself an optimizing mechanism that will act to optimize the global network.
- We can use models of a marketplace to construct network design tools that will work across a broad range of layering and partitioning to find optimal layering and partitioning structures, as well as predicting likely effects of price structures and levels.

As we will see later in the chapter, the best class of optimization that can successfully optimize complex and ill-behaved problems are statistical algorithms, and a dynamic marketplace effectively operates, in real time, a statistical optimization algorithm. Indeed, we can see a convergence of theories in optimal network design and economic progress through competition and the market mechanism. We can understand that the environment within which we are carrying out a particular network design is not static but is dynamic and is, itself, self-optimizing continuously. We can, in a very real way, relate the large scale mechanisms that the market applies to the global network to the small-scale design optimization mechanisms we apply when designing small parts of the global network.

13.1.2　General Methodology of Network Design

We now state the network design problem as a commercial problem. *An optimal subnetwork design is one that gives the maximum profit by maximizing the revenue from client layer and interconnect revenues and minimizing the costs of the network needed to achieve the revenues.* This is a brief outline of the process, and some of the stages will be discussed in more detail in the following sections. Broadly, this falls into the general methodology of:

- Framing and bounding the design problem;
- Characterizing the potential sources of revenue;
- Characterizing all the cost elements (transport, control, and management) from which the design algorithm can choose in order to build a solution;
- Imposing additional constraints, such as performance, as either extra costs or lost revenue if a design does not meet the constraint;
- Running an optimization algorithm that finds the solution that yields the most profit.

The framing of the design problem for a subnetwork is illustrated in Figure 13.3. This design exercise is essentially a functional design as it is assembling the optimal set of functions required in the network to profitably meet client requirements. As a functional design, it is not directly concerned with equipment and other implementational issues. However, these have a profound bearing on the functional design, so we need to establish the relationship between the functional design and its implementation. The functions of a subnetwork design can be provided by either:

- The services of another subnetwork across a commercial interface;
- Equipment and other implementations that are exclusive to the subnetwork;

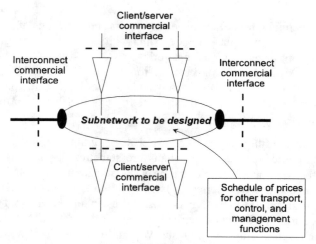

Figure 13.3   Interfaces surrounding the design of a subnetwork.

- Equipment and other implementations that provide functions to more than one subnetwork (normally in different layer networks).

This last case is another example of a common resource that is providing functionality to several clients, and this points to the way of handling this case in the design exercise. *Where an implementation is providing functionality to more than one subnetwork, it is modeled by a commercial interface between the implementation and its client subnetworks.* This is illustrated in Figure 13.4.

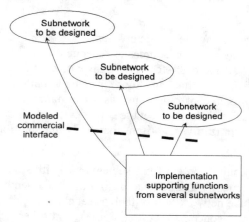

Figure 13.4   Cost apportionment of implementations is established by a modeled commercial interface.

The main body of this chapter is divided into three principle sections. The next section discusses many of the important areas involved in the design and planning of networks and effectively describes a "toolbox" of features and techniques. The second section discusses the features of the functional design of different types of subnetworks. The final section looks at the relationships between the functional design of subnetworks and their implementation.

13.2 SPECIFIC AREAS IN NETWORK DESIGN

The number of factors that potentially impact on the design and planning of a network are potentially huge, and this section only describes some of the more important. Moreover, in any network design exercise, there is a law of diminishing returns whereby the detailed consideration of certain low-significance factors is simply not economic, as the cost of analysis is not justified by the potential improvement in the network design. The areas described below are not exhaustive; however, we do intend that they should cover, at least for many practical network design situations, a sufficient set to deliver effective network designs.

13.2.1 General Aspects of Managing With Uncertainty

There are now few aspects of the network design process that are unaffected by some form of uncertainty requiring a statistical analysis. This was true to a lesser extent with the PSTN/ISDN when the network was designed using Erlang statistics. The use of Erlang statistics had became so automatic in the design of the PSTN/ISDN that the very particular nature of telephony traffic for which the Erlang model was developed was frequently taken for granted. The use of Erlang statistics has been reduced to simple rules of thumb that have successfully guided network planners for several generations.

Broadband traffic is very different in kind to that of telephony, and the simple rules of the Erlang model can no longer be applied. A fundamental redevelopment of statistical descriptors is required for broadband network design. For example, four features that are implicit in Erlang statistics that are clearly not true in a broadband network are as follows:

- The Erlang model is based on simple circuit switching and cannot describe the variety of traffic sources that can exist in broadband networks.
- In a planning timeframe, the telephony traffic statistics are always in "steady state." (This means that the statistics of the traffic at the forecast date will be fully independent of the traffic existing on the subnetwork at the time the planning decision is taken. This is true when the holding time of a connection is very much shorter than the planning lead time.) This is not generally the case where there are private connections that have long lead times, and it is especially not the case where automated interlayer processes (for example, between the VCh and VP layers) can add or reduce link capacity on a minute by minute basis, making the effective planning lead time for link capacity minutes, not years.
- With Erlang statistics, the standard deviation in the traffic will always equal the mean traffic. When dealing with forecasts, especially of services in early stages of growth, the standard deviation can be many times the mean.
- When two Erlang traffic sources are added together on a link, the statistics of the two traffic sources are assumed to be fully independent. With the wide variety of broadband services envisaged, the chances of traffic being generated at different

points in the network related to the same cause is much higher, so the traffic cannot be added on the basis of full independence.

There are now many circumstances where the uncertainty in a parameter is more important than the absolute level of the forecast. For example, if the lead time for introducing equipment into a network is one year, we need to design to a traffic forecast for at least one year's time. There are now many services where the level of uncertainty in such a forecast is very high, and the network design will owe more to the attitude adopted to the uncertainty than to the actual forecast level. It would be too expensive to build the network to meet the maximum likely demand, as this is more likely to result in a heavily underutilized network, and so a policy is required that balances the risk of not meeting demand with the risk of an underutilized network.

In general, the strategy for managing uncertainty will include at least the following policies:

- Changing some parameters to reduce the uncertainty in others, for example, reducing lead time times to reduce the uncertainty in the forecasts;
- Adding independent risks to reduce overall risk, for example, combining several services onto the same network so that the uncertainty in the total traffic forecast is proportionately less than in any one individual forecast.

13.2.2   Uncertainty in Forecast Traffic

The relevant timescale of telephony traffic was reasonably well constrained to be that of the few minutes of hold time. Here, there are no statistical characteristics in any shorter timeframe and those of the longer timeframe are reasonably well understood—normally, there is a deterministic daily pattern of traffic with a peak in the "busy hour" and over a planning timeframe, a reasonably steady growth rate.

We noted in Chapter 9 that broadband traffic exists in many timescales ranging from the cell scale, burst or packet scale, call scale, human activity scale, and at the long-term equipment planning scale, and that each scale has its own statistical characteristics. The effects of these different timescales is to overlap with the previously deterministic daily and growth timescales, so we must use statistical techniques at these timescales as well as at the call scale. In addition, as noted above, the forecast growth is not the relatively certain, steady rate that telephony planners were able to assume.

13.2.2.1   *Probability Distribution of the Traffic Forecast*

While it may be possible to see the general nature of the probability distribution for the traffic forecast, including the number of parameters required to describe it, it would be easier to work with a "standard" model of the traffic forecast that could be completely characterized by a mean and standard deviation (or mean and variance) only. The distribution itself must be constrained such that the forecast traffic takes only positive values. The gamma distribution would frequently work well as illustrated in Figure 13.5.

In a stochastic traffic model, each unit of traffic is considered to be the result of stochastic birth process and a stochastic death process. The birth event gives rises to a new

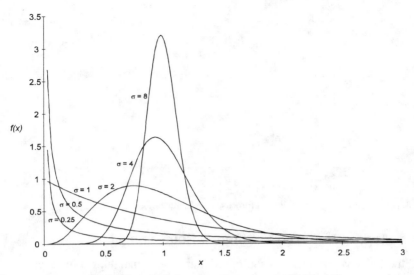

Figure 13.5   Gamma distribution with unit mean and varying standard deviation.

instance of traffic (e.g., a call), while a death event causes one of the current instances of traffic to cease (the death rate is normally proportionate to the number of traffic instances current at the time). From the analysis in Chapter 9, we can conclude a number of points:

- The mean level of traffic is governed by the ratio of the birth rate and the death rate.
- The rate at which the uncertainty in the level of traffic increases with time depends on the death rate (i.e., the current state of traffic is known and so the uncertainty increases as these die).
- If all the birth events are statistically independent and all the death events are also statistically independent, then, in the steady state, the variance of the distribution will equal the mean as in the Poisson distribution.
- If the birth events and/or the death events are bursty (i.e., the existence of one event is likely to increase the probability of another), then the variance will be larger than would be the case for statistical independence.
- If the birth events and death events are scheduled or queued, then the variance will be reduced.

13.2.2.2   Effects of Long Holding Times

In general, the effect of long holding times is that the forecast will depend on the state of the traffic as it is currently known. That is, the standard deviation of the forecast depends on the ratio of the average holding time to the expected lead time. When this ratio is of the order of unity or greater, then it is significant; when it is small, it is not significant. When the forecast does depend on the traffic as it is currently known, it will reduce the standard deviation in the forecast. This is illustrated in Figure 13.6.

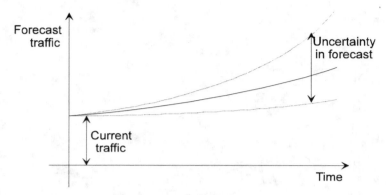

Figure 13.6   Increasing uncertainty in forecast with increasing lead times.

13.2.3   Adding Traffic From Different Routes on a Link

While fatter links will normally cost proportionately less than thinner links, and designing topology to exploit this is advantageous, there are also good traffic reasons for trying to consolidate on fatter links. Combining different traffic flows onto the same link has the effect of reducing the overall level of uncertainty for the traffic on the link. By reducing the level of uncertainty, the amount of capacity that must be allocated to give a required grade of service can be reduced, and this is particularly important for the thinner routes where, proportionately, the levels of uncertainty are higher. A network that exploits this ability to reduce uncertainty by combining traffic flows will naturally have fewer fatter links, and will therefore be less meshed than those that have not exploited this feature. This tendency does have to be balanced with the risks of exposure to failures in the network, which, when the network is less meshed, can have more significant effects.

13.2.4   Planning and Equipment Lead Times

Generally, the shorter the planning lead time, the more accurate the traffic forecasts can be and the less risks in designing the network. This suggests that there are many advantages in reducing the planning lead times for building and/or augmenting a network to a minimum.

However, the planning lead times for new equipment are themselves notoriously uncertain, especially when new technology is involved. This requires that the designer must understand the way in which changes in the planning lead times will affect the level and uncertainty of the forecast and be able to translate the uncertainty in the planning lead times to uncertainty in the forecasts. And this in itself is not simple. Traffic demand may be waiting for the technology to be available and forecast may not be that sensitive to delays in the technology. On the other hand, there are many occasions where there is only a relatively short "window of opportunity" for a particular network; if it is not available at the start of that window, there may be no demand at all. In general, there can be no simple rules for relating planning lead times to traffic forecasts other than recognizing that there is a relationship. This is a very important insight.

13.2.5   Performance and Failures in the Network

Many performance factors influence subnetwork design. These include the following:

- Blocking probabilities;
- Cell loss probabilities in ATM layers;
- Probability of a network failure affecting a given connection;
- Probability of successful restoration following a network failure;
- Ability to offer fully disjoint routings.

It is less obvious that error performance is a major factor in the design of a subnetwork, as significant errors are normally indicative of substandard equipment and have nothing to do with the design of the subnetwork. If there are aspects of error performance that need to be taken into account (for example, separation of connections that can tolerate radio/satellite from those that demand optical transmission), then these must be characterized and included, generally as an equivalent cost. In general, the primary network design factor relating to performance is the effect of equipment failures on the network.

As we know, all equipment and their associated systems are liable to failure without warning. Such failure mechanisms give another uncertainty that must be taken into account in the network design. The network can be designed to be robust to many failure mechanisms, often at little additional cost. However, there is normally a basic relationship between the resilience of a network to failures and the cost of the network, and the network design should give the best balance between resilience and cost. This balance will be affected by:

- The probability that the failure mechanism will occur;
- The consequence on the network of the failure;
- The importance the users of the network attach to the effects of the failure mechanism;
- The time and costs of effecting a full repair;
- The costs associated with restoring the connections affected by the failure mechanism, including the cost of capacity held spare for such an eventuality.

Achieving a sensible balance between all these has tended to require a subjective judgment as the complexity of the trade-offs have not easily yielded to mathematical analysis.

Now that many network design techniques are sufficiently powerful to produce robust designs that can be implemented directly, the failure mechanisms need to be considered carefully as inputs to the design process rather than modifications to the design once it has been produced. This generally means making a pragmatic judgment on the way the failures are described and parameterized, and this would normally take into account the control mechanisms used to restore connections and/or repair the faulty equipment.

13.2.6   Modeling the Commercial Aspects at Boundary Interfaces

The relationship between a client layer subnetwork and a server layer is essentially a commercial interface and shows most of the features of a commercial interface. There are many occasions when there is a real-market commercial relationship (for example, when the links of a subnetwork are purchased as private circuits). However, there is also strong commercial flavor to all client-server relationships and also to the interconnect relationships between subnetworks in the same layer network. These relationships are illustrated in Figure 13.3. It is important, therefore, to understand the way in which we can model such a commercial relationship.

We consider a server network offering a link transport service to a client subnetwork. The client subnetwork sees the amount paid for each link as a cost, which the design algorithm will add together with all the other costs to arrive at a total cost for the subnetwork.

When the subnetwork offers service to its clients (interconnect of client layer), it has this total of all its costs as the basis by which it charges its clients. However, there is no prescriptive rule by which this total cost can be "allocated" among the client connections using the subnetwork. The subnetwork can apportion costs and must charge a *price* for two reasons:

- The amount the client subnetwork is prepared to pay will depend on the usefulness of the link within the subnetwork, and also on the availability of other server networks that can offer a comparable link but with a different cost structure;
- The great majority of the costs must be incurred in the server network *before* a client subnetwork asks for any connection, while the marginal cost incurred as a direct consequence of the request for a connection is normally very small indeed.

The first reason implies that a market mechanism should operate for the supply of connection services, and the price will be established by the balance of supply and demand for the service, including all possible servers and clients.

The second reason implies that this is a "difficult" market, as all costs, at least in the shorter term, are fixed. This usually means it is worthwhile for a network operator to price in such a way as to extract as much profit from the marketplace as possible by balancing demands from different clients. This is already a well-established practice in the PSTN, where connections during the night, when demand is low, are priced lower than during the day. When all costs are fixed, if resource is otherwise idle, any revenue is better than no revenue. *Establishing the effective price for capacity is central to network design.* In practice, prices are set either by market forces or by regulation.

Assuming there is a market mechanism between a client subnetwork and its servers, the competitive forces will forge a price. This is classical marketplace economics, and many economics modeling techniques for marketplaces are readily usable.

Where there are not competitive forces but regulation and/or monopoly, the price is set by a formula that must be arbitrary. This was what was happening frequently in telecommunications. The tariffing structures of monopolistic telecommunications marketplaces are frequently based around a general formula of time × distance × bandwidth. The long-run costs of today's technology do not justify this formula and suggest instead heavy volume discounts in all dimensions, which is borne out in more competitive marketplaces such as the long-distance U.S. market. The fact that it is often still so slavishly applied is firm proof of its arbitrariness!

The price at which a server network sells its services cannot be established a priori by calculation. It is either by competitive market forces or by an arbitrary regulatory formula, or, as is frequently the case today, an evolving combination of both.

13.2.7   Interacting With the Operational Environment

The target operational environment also has an important impact on network design. In this section, we look at the way design information is carried forward into the operational environment. This transfer forward of information should attempt to meet the two following requirements, which sometimes seem to be in conflict:

- It should ensure that in the operational environment the network is used in accordance with the way in which it was designed.
- It should not unduly restrict the capability of the operational environment to adapt to meet unforeseen demands and to take whatever action is necessary to maintain service under stress.

Historically, designers of switched networks tended to be overrestrictive on the operational environment, resisting the beneficial use of, for example, dynamic rerouting, which can allow connections to follow different routes to those preplanned in route tables. Conversely, designers of transmission networks have often allowed total freedom in the choice of routing, which can easily result in long and circuitous routings that would have been better refused in the interest of maintaining the blocking performance to future connection requests.

13.2.7.1   Understanding Network Utilization

There is always a temptation to view 100% utilization as the target and that anything less is wasting resources. This is dangerous as almost every network has a "foldover" effect, where the number of connections supported on a subnetwork actually declines above about 70–80% network utilization, as illustrated in Figure 13.7. If an attempt is made to use a

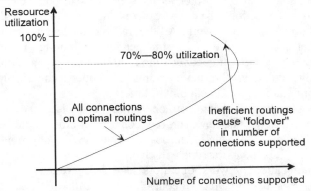

Figure 13.7   The foldover effect in network utilization as a result of inefficient routings.

subnetwork above this figure, it will undoubtedly contain connections with inefficient routings. The art of operating a network at its peak utilization is to be able to block connections that can only be routed inefficiently, and accept connections that can be routed efficiently. Such a policy will generally hold the utilization at about the 80% level.

This gives the necessary "headroom" to accommodate the general change rate in the subnetwork as some connections are taken down and new ones are set up. This is often referred to as "churn." In the past, this was achieved in switched networks by using very rigid routings and blocking a connection when its prescribed route is blocked. There are many examples where more flexible routing algorithms have been applied to try and marginally increase utilization only to have the whole subnetwork "seize up."

In addition, long planning lead times in a growing market will reduce the maximum utilization levels still further as the network must hold capacity against larger demand in the future, which must be planned into the network in advance.

13.2.7.2   Characterization and Costs of Control and Management Functions

Increasingly, control and management form a very large proportion of the costs of a subnetwork; however, they are not always reflected in the network design exercise. They are often difficult to characterize and quantify, but it is still important that they are included in the overall costs of a subnetwork. This is not simply because of the size of their contribution to the cost burden; they can also be significant in the optimization process itself. For example, it is often possible to simplify and cost reduce the control and management by placing restrictions on other aspects of the subnetwork, such as its topology. If the network capacity is relatively cheap while the control required to do dynamic network restoration is expensive, it will often be cheaper to employ intralayer protection processes such as automatic protection switching (APS) or subnetwork protection. This is reflected in the different costs, but also in the greater benefit to the client.

13.2.7.3   Routing Exchange Algorithms and Protocols

Routing information exchange protocols have been used extensively in the Internet; however, as it is a connectionless network, there is no opportunity to block a connection before the data arrives in the network. All the traffic management has to take place on a packet-by-packet basis and is therefore independent of routing. The routing exchange protocol runs in the background between the routers of the connectionless network and implements a dynamic algorithm that attempts to produce a route table for each router, which is near optimal for any given point in time.

The use of routing exchange protocols has been discussed over the years for connection-oriented networks; however, few connection-oriented networks actually implement them. Now that the P-NNI specification from the ATM Forum includes a routing exchange protocol, it is likely that this will change.

Telecommunication network operators have been reluctant to implement such autonomous systems, fearing a loss of control; however, there is nothing fundamental to suggest that the use of a routing exchange protocol may cause a subnetwork to behave badly or "seize up." However, the autonomous intralayer routing algorithms must be constrained to prevent unacceptably inefficient routings from causing blocking. One technique for doing this is the use of "implied costs," where a notional cost is attached to using a link for a connection, which increases dramatically as the link resource becomes scarce (i.e., when it approaches 80% utilization). In any event, the network design algorithm must take account of the routing exchange algorithm to effectively design a network to use it on.

13.2.7.4   *Priority and Preempt Schemes*

Many networks, especially the transmission parts of private networks, use prioritization and preemption. In such schemes, every connection is assigned a priority level that can be referenced when the network resources come under pressure, for example, under failure conditions. If there is a network failure and insufficient capacity to restore all the affected connections, the subnetwork controller will attempt to fully rearrange the subnetwork so that no connection of a higher priority remains unrestored while a lower priority connection remains on the network.

The aim of priority and preempt schemes is laudable. However, in practice, they often prove very difficult to get to work effectively, even on relatively small networks. The principle reason for this is they normally require a total view of the topology of the subnetwork and so are not easily partitioned (i.e., distributed). This means they scale poorly both in the absolute limits of the number of nodes the algorithms can accommodate and also in the time taken to recalculate all the new routings.

Much of the difficulty arises from the precise definition of priority. In a subnetwork, some connections, even if they are of lower priority, are not causing immediate congestion to other connections. If they are thrown off to place higher priority connections on the network, the high priority connections often have circuitous routings. The priority and preempt scheme is actually taking the subnetwork utilization well into the danger zone above 80%. This means that large numbers of connections are being frequently rearranged in urgent attempts to stop it from "seizing up."

It is possible to conceive of more stable priority and preempt schemes; however, they are softer in operation. In general, they will contain routings, even of high priority connections to efficient routings, and if none is available, the connection is dropped. This may mean that some higher priority traffic is dropped while lower priority connections are maintained and/or there is apparently spare capacity in the network. By restricting routing, the subnetwork can be kept within the bounds of stable operation. Another way of looking at this is that the prioritization problem is so complex that a statistical approach is very much more efficient than a prescriptive approach, even if it is not scrupulously fair.

13.2.7.5   *General Model of Information Flow From Design to Operations*

Figure 13.8 gives a model of the different stages between forecasting, network design, and network operations. In addition, the network operations have been divided between those associated with "normal conditions" and those associated with "disaster conditions." The distinction between these two is largely determined by the processes involved. Normal conditions, which should include many failure scenarios, would describe the condition for

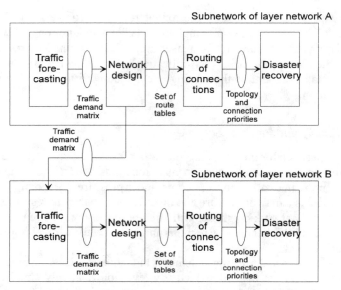

Figure 13.8   General flow of information from network design to network operations.

which the network is designed and for which there are clear operational processes. Disaster conditions are those that go beyond what it is sensible to design for but have prescribed operational responses. These could include situations of simultaneous multiple failures in the network or of unprecedented demand, for example, from the press agencies following a major incident at a remote location. Under disaster conditions, ad hoc actions have to be taken to accommodate the particular circumstances that, by definition, have not been planned in advance.

The information passed between the traffic forecasting stage and the topology design stage is the traffic matrix for the subnetwork that is to be designed. The information passed between the topology design and the connection control can vary; however, this would normally be a set of route tables for the nodes in the subnetwork. The information passed under disaster conditions will normally be specific to the particular situation that occurs.

Under many circumstances, the passing of route tables to the connection control process achieves a good balance between constraining the connection control to routing anticipated in the topology design while allowing some freedom to the connection control process. This is particularly true if the route table allows a level of alternate routing, which is taken into account in the design. Alternate routing can successfully take into account a large number of failure scenarios and maintain a stable utilization at all times.

13.2.7.6   *Time Lag Between Designed Routings and Actual Routings*

From time to time it is usually desirable to update the route tables in nodes. This can be either as part of an automatic routing exchange protocol or as part of a topology design exercise to accommodate anticipated growth in traffic.

However, once new route tables have been implemented, only new connections will use the new routing because existing connections, unless explicitly rerouted, will remain on the old routing. The significance of this will depend on the holding time of the existing connections. If holding times are relatively short, then there will be no need to take any explicit action since as soon as the connections are ceased, any new connections in their place will use the new routings. However, if the holding times are relatively long (that is, in relation to the timescale for growth increments in the network), then such connections could start reducing the effective utilization levels below that anticipated in the design. Under such circumstances, it becomes desirable to take explicit action to reroute the inefficient connections. This can be done by actively ceasing the connection and immediately setting it up again on a new routing.

In general, there should not be any major concern in having a proportion of the connections in a subnetwork using old routing as long as the proportion is kept low. This will manifest itself as a time lag between the anticipated routing and the actual routing, with the time constant being the same as the mean holding time of the connections.

13.2.8   Parameterizing Commercial Relationships [1]

In this section, we bring together a number of network design areas already discussed and look at their impact on the commercial interfaces, be they real or modeled commercial interfaces formed to divide up the network design problem into partitions of a manageable size. The result is a set of parameters with which a commercial negotiation can take place, as shown in Figure 13.9. We noted above that the client-server and interconnect relationship are fundamentally commercial relationships. The client-server relationship is the more demanding of these two and the parameters are principally described for this case; however, the interconnect relationship follows exactly the same principle.

13.2.8.1   Future Demand (Passed From Client to Server)

As part of a subnetwork's forward planning process, it will start forming a view of the services that it will require in the future, and they would like to know likely pricing for these services. In the same way, the forward planning process of each server is equally keen to form a view on the likely demand for its services into the future. In general, it suits neither the subnetwork nor its servers to be working with unnecessarily high levels of uncertainty. A server must compensate for high levels of uncertainty with either overprovision of capacity or accepting that it will not be able to meet all demands. This affects the

---

[1] There is a strong analogy with the mechanisms used by the spot, futures, and options markets for financial instruments and commodities. While the futures and options markets are now dominated by speculators, their origins are in derisking against an uncertain future. A common way of looking at these markets is that they "discover" a best estimate for the price of a commodity for some point in the future. The futures market gives an estimate for the mean of the price in the future, while the options market gives an estimate of the variance about this mean. The spot market is analogous to the actual contract between the client and server for explicit provision of trails.

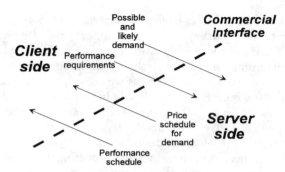

Figure 13.9   Parameters in a commercial relationship.

client as either higher prices or a risk of long lead times for link capacity. It therefore should suit both sides to cooperate in estimating future demand. The client can provide two forms of information:

- The likely trails it will use at points in the future ahead of the firm demands for trails that the client will ask for at the last possible moment to reduce its risk;
- The possible trails that form the full set from which the client's design process will select the ones that it will actually use.

The demand schedule can be passed in two forms: possible demand and likely demand. The possible demand is the complete set of services the subnetwork may need to procure to meet its own commitments, and the likely demand is the actual forecast based on whatever information is available (prediction algorithms, growth expectations, etc.). Both are of interest to the servers.

13.2.8.2   *Performance Requirements (Passed From Client to Server)*

Again, the subnetwork needs to know the differential cost it must bear for performance, and its servers must know the estimated performance requirements.

While for some performance parameters this is relatively straightforward, other performance requirements are not well-behaved. One example is availability performance. In general, changes in availability requirements as set by the subnetwork are "ill-behaved" in that they may well necessitate disproportionate changes in the server's network in order to meet the request. Often, a server's network will show major cost breaks, so that a small change in availability performance necessitates major changes in the design of the server's network (for example, the introduction of guaranteed restoration). Generally with these parameters, the subnetwork and its servers need to dialogue. The client can pass an approximate set of requirements to the server and the server can respond with a more precise price and performance schedule. This can then be refined over a number of repetitions of the dialogue.

In either case, the client acquires an understanding of the server's capabilities and the server acquires an understanding of the client's requirements. These requirements and capabilities are framed in statistical terms.

13.2.8.3   Price Schedule (Passed From Server to Client)

When the subnetwork asks for price information on either its possible or likely demand, the server will respond in a format that both can understand. In some countries, the structure of the price schedule is established by regulation to encourage an open marketplace; however, where the client and server are fully internal to one company, the form of the price schedule will take the form of an internal transfer charging mechanism. There is undoubtedly increasing pressure to form more flexible pricing schedules for all client-server relationships, and as this happens, it should make the process of network design that much easier. The formation of an effective price schedule has been dogged historically by two factors:

- The "time × distance × bandwidth" myth;
- The treatment by public network operators of their networks as a single monolithic entity, which does not recognize the client-server relationships within.

As competition, regulation, and technology break down these two legacies, we should see more effective price schedules emerge. Modeling the commercial relationships in the network design process will help considerably in understanding effective price schedules.

13.2.8.4   Performance Schedule (Passed From Server to Client)

As with price schedule, the server needs to pass back information to the subnetwork in a way both can understand. For many performance parameters, this is reasonably straightforward. Common mode failure has always been an area of difficulty; however, the use of the topology of common mode failures described in Chapter 3 appears to meet this requirement.

13.2.9   Well-Behaved and Ill-Behaved Topologies

We noted that some performance parameters have well-behaved effects while some have ill-behaved effects. The same phenomenon is also true of the sensitivity of different topologies to growth. We define a well-behaved topology and an ill-behaved topology as follows:

- A well-behaved topology is one where the optimal topology changes predictably and proportionately with changes in the traffic matrix for which it is designed.
- An ill-behaved topology is one where the optimal topology changes unpredictably and disproportionately with small changes in the demand matrix.

Clearly, it would be desirable to only use well-behaved topologies as these can grow and evolve easily. For example, with a well-behaved topology, existing nodes, links, and connections would not need significant rearrangement if the traffic matrix were to grow by 50%. The extra capacity required for the new demand would be added by adding new nodes and new links and/or augmenting old ones. Conversely, in an ill-behaved topology, the existing nodes, links, and connections would require significant rearrangement to implement the new optimal topology, which includes the new traffic. In this situation, the rearrangements required to grow the network would be both costly and disruptive. A good test for a well-behaved or ill-behaved topology is a cost comparison between a design for the new traffic that uses the existing nodes and links as currently configured, and a design where the optimization has freedom to rearrange them. If the two results yield the same cost, the topology is well-behaved; if the former is more expensive, then it is ill-behaved (if the latter is more expensive, then the design optimizer is not working!).

The difficulty arises that there are many occasions when-ill behaved topologies are cost-optimal for a given traffic matrix. Most notably, when there is a major cost advantage in using an add/drop multiplexer switch matrix rather than a full crossconnect switch matrix, the resulting topology is likely to be ill-behaved. In such situations, optimal topologies are likely to be based on interlocking rings, where the rings are based on careful balancing of traffic to get a high level of reuse of the capacity around the ring. However, small changes to the traffic matrix can suggest very different sets of interlocking rings, but it will be difficult to change from one set of rings to the other. This has been a subject of much discussion in the design of SDH networks, especially in Europe where the demographics frequently point towards such topologies. This is further complicated by the fact that interlocking rings also have the advantage of inherent resilience because of their "flat" nature.

There is no simple answer to this dilemma. The use of both mean and variance in the traffic matrix for the design will help considerably in producing better behaved topologies as the design algorithm will now attempt to derisk against uncertainty in the forecast.

13.2.10 Evolution and Interworking

The introduction of new technology often results in the introduction of new layer networks that seek to meet the requirements of a set of client layers that (at the time of introduction) are met by existing layer networks. In such a situation, an evolution and interworking policy is required between the old layer networks and the new layer networks. This is illustrated in Figure 13.10 and applies particularly to the evolution from PDH to SDH and interworking of ATM with PDH and the PSTN.

There are three possible ways of interworking, as illustrated in Figure 13.11:

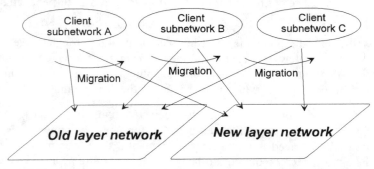

Figure 13.10   Migration and evolution from a layer network of old technology to a layer network of new technology.

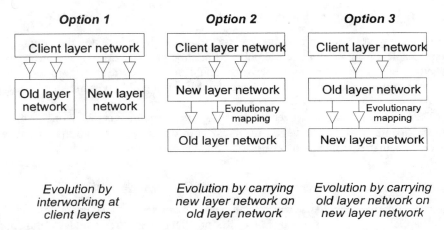

Figure 13.11   The three basic classes of evolutionary interworking.

- New layer networks and old layer networks are mutually independent and all interworking is at the common client layer networks;
- A new layer network using an old layer network as a server;
- An old layer network using a new layer network as a server.

These are shown in more detail in Figure 13.12 by considering whether node technology or link technology is to be deployed first.

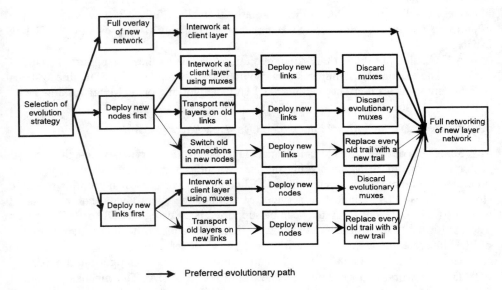

Figure 13.12   Possible evolutionary paths.

13.2.10.1   Mutually Independent Evolution

This is the easiest to design and manage, as it does not introduce any new client-server hierarchies. If it were possible to use only this method of interworking, evolution from an old layer network to a new layer network would be simple, involving no more than a gradual migration of the links of the client layer from the old layer network to the new layer network.

There are occasions, however, when it may be advantageous to use one of the other methods of interworking; however, they will both create new transitional client-server relationships that complicate the design and management of the network.

13.2.10.2   New Layer Network Carried Over Old Layer Network

A new layer network could use a new layer network as a server. Two examples where this might happen are the transport of SDH VCs on PDH line systems and the transport of ATM on PDH line systems.

An SDH path layer could use a PDH path layer as a server to make use of existing line system capacity, yet providing all the features of the SDH path layer network. The ITU-T has defined mapping of LVCs in the 139264 kbit/s, 97728 kbit/s (the fourth Japanese PDH line rate), 44736 kbit/s, and 34368 kbit/s layer networks as described in Chapter 5. The procedure is most useful where there is a well-established PDH infrastructure in place with many years of service left, as is the case on many submarine cables where these mappings are used. These mappings allow for the SDH connectivity required to realize the specific SDH benefits of flexibility and also efficiency by using an existing PDH line capacity.

In a similar way, mappings of ATM cells have been defined into the same set of PDH line rates as well as into primary rates. Similarly, they allow fully ATM networking to be established without the need for SDH transport by making use of existing PDH line capacity.

As indicated in Figure 13.12, final evolution from using these mappings is straightforward. The existing PDH line systems can be gradually replaced by SDH line systems without any significant effect to either SDH LVC networking or ATM networking.

13.2.10.3   Old Layer Network Carried Over New Layer Network

Some network operators offer PDH path layer *connection* services, where the external customer is at liberty to supply his or her own nonstandard path termination using proprietary equipment on the end of the PDH connection with its own proprietary frame structure. These PDH connection services cannot see the client layers that are actually being carried, so the only way they can be supported is by using an SDH or ATM trail as a server for the PDH path layer connection. This is done using the appropriate mapping of PDH line rate into an SDH VC or into an ATM CBR service.

These mappings can also be used as a transitional means of supporting the existing PDH layer networks on a growing SDH network (or even ATM). This mechanism should be used sparingly and avoided whenever possible, as it inhibits the evolution of automated SDH and ATM networking as indicated in Figure 13.12. An existing PDH layer network supports PDH clients using existing PDH multiplexers and manual management, and if these mappings are used, the lower order PDH paths and the ultimate client layer signals that they support remain locked up and inaccessible within the higher capacity PDH path format, thus effectively preventing flexible access to this capacity. Furthermore, it becomes very difficult to evolve from this situation, as any upgrade to full SDH or ATM networking must be coordinated across large parts of the network, presenting an enormously difficult operational challenge.

13.2.11 Selection Algorithms

The design of topology, as we noted earlier in the chapter, is, in mathematical terms, a "hard" problem. We noted earlier that we solve the topology design problem to an optimization procedure, where we are trying to select the optimal set of links and nodes from a wider set of candidate links and nodes; that is, the topology that achieves the best balance between features and cost. One way of looking at this is that the optimization algorithm is looking for the most profitable topology. The general process is illustrated in Figure 13.13.

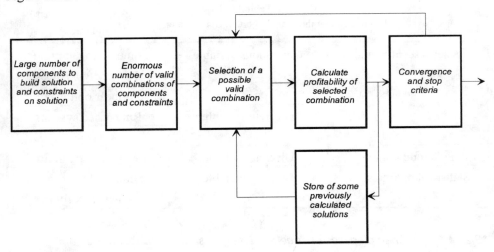

Figure 13.13   The general process of optimization.

Many of the various parameters that may be included in the calculation of the profit have been discussed in the preceding sections; however, this list is certainly not exhaustive and some of the features described in these sections may be irrelevant for many subnetworks. The selection of the parameters for inclusion in the optimization will depend on the particular subnetwork.

In recent years, we have seen significant advances in the mathematical understanding of optimization algorithms as well as huge increases in the computing power available to process them. These algorithms fall into three main categories, which we have called steepest ascent methods, full enumeration methods, and statistical methods. Each of these has advantages, however, it is the statistical methods that have attracted more recent interest and show a great deal of promise. The effectiveness of an algorithm can be judged in several different ways, and three of the most important (particularly for designing networks) are the confidence in the result, the speed of calculation, and whether it can run autonomously in real time. We shall see that for the design of network topology, selecting an algorithm implies balancing these properties.

Before describing some of the algorithms, it is useful to classify optimization problems according to the nature of the parameters in the profit function. In many optimization problems, the parameters in the profit function that the optimizer is trying to select can vary continuously over a given range of values, while in other problems, the parameters in the cost function can only take on a finite set of discrete values. These are continuous variable optimization and discrete variable optimization, respectively. The latter is often called combinatorial optimization as the number of possible solutions is directly related to the number of possible, valid combinations of values of each variable.

Most algorithms are oriented to one class of problem or the other, however, it is always possible to develop an approximate technique where an algorithm developed for one class of problem can be used on the other, with varying degrees of success. Various optimization algorithms are listed in Table 13.1.

Table 13.1

List of Some of the More Important Optimization Techniques

Steepest ascent methods	Continuous variable	Linear progamming
		Sensitivity analysis
		Non-linear optimization
	Discrete variable	Rule based systems
		Relaxation
Full enumeration	Discrete variable	Integer programming
Statistical optimization	Discrete variable	Simulated annealing
		Genetic algorithms

13.2.11.1   Characteristics of Topology Design as an Optimization Problem

In general, there are three sets of parameters that form the external inputs to the optimization problem: the traffic matrix, the cost profiles for resources, and performance constraints. The actual parameters that represent the topology are less clear. We noted earlier that to map the traffic matrix onto a topology, it is necessary to make some assumption on routing mechanisms. Irrespective of whether the same routings will be used in the operational environment, the optimizer must route each element of the traffic matrix across the candidate topology. This indicates that the optimization variables should not be just the topology itself, but also the complete set of route tables. The revenue is determined from the traffic matrix and the achieved performance. Subtracting the cost from the revenue gives the profit that the optimizer is seeking to maximize. In short, the optimizer is seeking to select the set of route tables for a given topology that maximizes profit. This is illustrated in Figure 13.14.

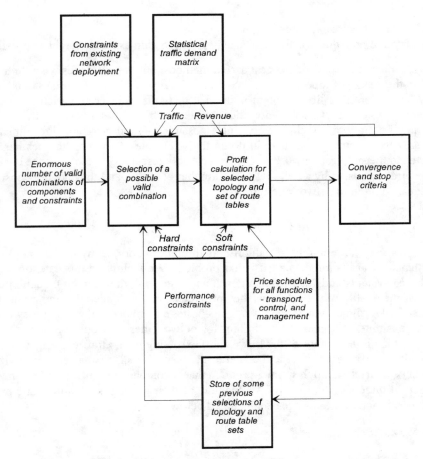

Figure 13.14   Process of optimizing the design of a subnetwork.

Network design is essentially a discrete variable problem, as components are generally included in the design or not. Few variables of the design are genuinely continuous, and even when they are, they can be approximated by a discrete variable.

13.2.11.2   *Optimization Techniques for Topology Design*

In general, optimization algorithms that are in widespread use today are steepest ascent methods. While, they can appear quite different, they all work by the same basic principle. This is easily explained by the analogy of climbing to the top of a mountain, where the top of the mountain represents the optimal solution. The algorithms seek to climb the mountain in the shortest number of steps (the analogy is a little weak here as the size of each step is not limited, unlike our human steps, nor is it any problem to climb cliff faces!). The number of steps represents the time taken in the calculation. These algorithms are generally characterized by the following two rules:

- Always ensure each step is uphill.
- Try and pick the direction and size of a step to gain as much height as possible.

There are many examples of steepest ascent methods, but the only ones we note that are useful in the design of topology are relaxation techniques and rule-based systems.

When an optimization problem is defined in terms of discrete variables, the total number of possible solutions is finite, and when the problem is relatively small, a possible technique is simply to calculate every single possibility—and there are algorithms, such as integer programming, that help in doing this. The role of such techniques has to be restricted to the simplest network problems, probably less than 10 nodes, as the computation required increases by a factorial of the number of nodes, and they are not considered a practical solution to network design.

13.2.11.3   *Rule-Based Algorithms*

There are a very large number of algorithms in this category, as by definition, they are algorithms that are tailored to a particular problem. We include in this category knowledge-based systems and "heuristic" algorithms. In general, they can be made to be fast; however, the quality of the solutions they provide is, at best, uncertain, and often poorer than a skilled designer would do by using his or her experience and good guesswork.

An example of a heuristic algorithm for network design would be one where the topology, the dimensioning of the links, and the routing of the traffic are all carried out independently from each other and one after the other. This is fast, as each optimization problem is, by itself, a much simpler problem than a complete optimization, which would consider all three simultaneously. An example of the rules that such a heuristic process might use at each of these stages is illustrated below.

Design of topology:

- Provide a direct link between two nodes if the direct traffic (i.e., the single entry in the traffic matrix) between the two nodes is greater than a threshold.
- If any nodes are left isolated, connect them to the nearest connected node.
- Take all nodes with more than a threshold number of links after steps 1 and 2 and make them "hubs," and then provide additional links to fully interconnect all hubs.
- Look at the results of applying the above rules and make any changes where the rules have inserted "obviously inappropriate" links or missed "obviously appropriate" links.

Dimensioning of links:

- Apply a simple routing algorithm (e.g., shortest path).
- Route traffic from traffic matrix and dimension links accordingly.
- Look at the results and "rebalance" traffic from significantly oversized links onto undersized links.

Establishment of route tables:

- Apply a simulated routing exchange protocol to the dimensioned network (see below).

This process, if taken in very broad terms, is possibly the most frequently used method of network design today. The design process is generally computable in reasonable time and is reasonably flexible in that changes can be made if the requirements change. However, there are a number of drawbacks.

First, by separating the processes, the resulting network cannot be optimal, as the routing, dimensioning, and selection of topology are all mutually interdependent. Separating them is bound to miss more effective solutions. This is especially true where topology changes can significantly affect the number of efficient routings.

Second, the process relies on a human "sanity check." As a consequence, the process cannot be fully automated and the design process cannot be implemented in a way that responds dynamically and automatically to changes in requirements, such as growth in the traffic matrix. In addition, even skilled designers are prone to subjective prejudice. In larger networks, it becomes necessary to have more than one designer, and differences in their subjective judgments can also lead to significant nonoptimality.

13.2.11.4   Relaxation Techniques (Continuous or Discrete Variable)

These are examples of a particular type of nonlinear optimization technique that can be implemented as a parallel process. This can be seen by taking the example of the Dijkstra algorithm, and those derived from it, used in routing exchange protocols. This is a very important type of algorithm in telecommunications, and is already standardized for use in the Internet, the OSI layer three protocol (IS-IS protocol), and the ATM Forum P-NNI routing exchange protocol. They have been in use in the Internet for around 15 years, so at least in the context of connectionless networks, are very well tried and tested.

The algorithm takes both the topology and the capacity available on the links as a given. It attempts to find the optimal set of route tables for this give topology and set of resources. This problem is a reasonably well-behaved, discrete variable problem. The success of the Dijkstra and related algorithms is that the profit function can be broken down into a number of independent profit functions, each of which can be optimized in independence. It performs a local optimization at each node, using its current local route table, available resources on adjacent links, and the route tables passed to it from the nodes on the ends of the adjacent links. Once the node has calculated its new route table, it passes it to all its nearest neighbors for them to use in their calculation. If certain conditions are obeyed in the way the new route tables are updated, then this algorithm will converge on a maximum of the overall profit function. Since it is a steepest ascent method, there is no guarantee that this is a global maximum, but since the problem does not attempt to alter the topology, it is reasonably well-behaved and, in practice, yields very good results. The general process of a routing exchange protocol based optimization of route tables is illustrated in Figure 13.15.

The great power of this type of algorithm is that it can work in real time in a fully distributed system. There is no need for any central planning of routing other than

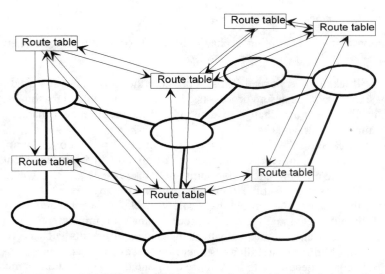

Figure 13.15   The use of the Dijkstra algorithm, a relaxation algorithm to optimize route tables.

agreement to the routing exchange protocol and the updating algorithm. If a new node is added to the network, then it simply connects to a number of neighboring nodes and, after a few automatic iterations, all the route tables throughout the world can then make use of new node.

However, routing exchange protocols only optimize the routing on a *given* topology and link capacity resource. Today, they do not attempt to optimize either the topology itself or make a recommendation to capacity management on optimal increases to link capacity, although this is obviously not excluded.

13.2.11.5   Statistical Techniques

There are several forms of these algorithms, and the most important two have their origins in analogies with processes observed in nature; however, they are all, at a basic mathematical level, based on the same principle.

13.2.11.6   Simulated Annealing and Annealing of Metals

The resoftening or annealing of metals by heating the metal to a critical temperature and then letting it cool slowly has been understood since man first started to work metal thousands of years ago. However, a more complete understanding of the crystalline process by which annealing happens is more recent. A metal becomes harder when it is bent and worked to form useful shapes, a process called strain-hardening. This makes the metal stronger, but it also makes it more brittle, and if a metal is to be worked further, it must be annealed.

The hardness of a metal is dependent on the number of crystal boundaries and "dislocations" in its crystal lattice. As the metal is worked, these increase and become interlocked, stopping the atoms from sliding over each other and stopping the metal from being worked further. Annealing releases the dislocations.

The metal can be considered to be in an energy state where the greater the number of crystal boundaries and dislocations, the higher the energy state. As the metal is worked, it is pushed into higher and higher energy states. The annealing process seeks to return the metal to its lowest energy state. And this gives the clue to designing an optimization technique. The annealing process is looking for the global minimum energy state. From a knowledge of the annealing of metals, we know the following:

- A strain-hardened metal, even at room temperature, will reduce its energy level over time; however, this process is very, very slow.
- When the metal is heated to a temperature normally referred to as the annealing temperature, which is normally well before it melts, it becomes malleable and is changing its energy state rapidly.
- If a metal at its annealing temperature is cooled very rapidly (e.g., by quenching it in cold water), it will be very brittle and is in a high energy state.
- If a metal at its annealing temperature is cooled gradually, but still in a reasonable time, it will settle into a low energy state.

In the simulated algorithm, every possible solution equates to an energy state and the level of the energy equates the profitability of the solution. The lower the energy state, the more profitable the solution it represents. The simulated annealing algorithm works by "heating up" the problem to a "temperature" at which all solutions can be potentially entered and then letting it cool at a rate slow enough for it to "freeze" into the optimum solution. We will see that the temperature relates to the likelihood that the algorithm will accept a worse solution. At high temperatures, the algorithm will accept most solutions, while at low temperatures, the algorithm is likely only to accept better solutions.

13.2.11.7   *Genetic Algorithms and Evolution Through Mutation and Natural Selection*

Natural selection was first persuasively argued and documented by Charles Darwin in his book *The Origin of Species*, while the discovery of genetics gave solid foundation to the theory and its use in biology. Since Darwin, it has been used (and abused) in many other areas such as, philosophy, politics and economics, but any philosophical position on evolution as it is applied in these areas is not relevant to our discussion here; we are simply taking the *theory* of evolution through mutation and natural selection as a model for which to design an optimization algorithm. Success depends not on the philosophical reality of evolution in any particular sphere, but on the ability to successfully optimize hard problems!

A number of individuals of a species exist in an environment, and they survive according to their fit to the environment. The longer an individual survives and the stronger it is, the more likely it is to pass on its genes to the next generation; on the other hand, the process will tend to kill off individuals that are ill-adapted, and they then cannot pass on their genes. However, the process is statistical, as those well adapted are more *likely* to pass on their genes while those ill adapted are less *likely*; those well adapted can still die without passing on their genes and those ill adapted can still live and pass on their genes.

Also, in creating the next generation, random genetic mutation can occur. These are random changes, which means that a next-generation individual may be different from its parent, and this change may mean it is better adapted to the environment or worse. Over time, successive generations of individuals become better adapted to their environment.

The genetic algorithms follow this analogy. The fit of an individual in the environment equates to the profitability of a solution, while the genes of the individual are an encoding of the controlling parameters in the optimization. The profitability then controls the number of children in the next generation; however, the passing of "genes" is enhanced by the mixing of genes between members of a generation and by "mutation," where small random changes are introduced into a few members of the generation.

13.2.11.8   Nature of Statistical Algorithms

At the heart of any of the statistical algorithms is that they progress in a nondeterministic way. There are two elements of randomness injected into the process. First, selecting a new solution from a previous solution has an element of randomness. Second, accepting this new solution has an element of randomness. This injection of randomness stops the algorithm getting locked at a local maxima, as there is always a probability the algorithm will leave this local maxima to investigate other maxima. If we return to the analogy of trying to find the top of the highest mountain, statistical algorithms are based on a random walk across the landscape with a general bias to climb rather than descend. The degree of this bias is the equivalent of the temperature in simulated annealing or the reproduction rate in genetic algorithms.

The importance of statistical algorithms is that they offer a technique that is effective with ill-behaved problems. As we noted earlier, for this class of problem, steepest ascent methods give unreliable answers while full enumeration techniques suffer from very poor scaling. For example, most statistical algorithms will scale according to a polynomial power (generally around 3 or 4) rather than the factorial scaling of full enumeration. This scaling generally will allow statistical algorithms to tackle many practical network design problems and yield reliable results. It is only in more recent years that these techniques have been tried outside a research environment and on real network design problems, as they still require large amount of computation time and power; however, those that have used them report that they have found them very effective.

Another feature of the mathematics of statistical algorithms is that it is possible to prove that they are the *only* way of reliably tackling these sorts of problems without full enumeration. They are also reasonably robust, as even when they are run more quickly than theory would suggest, the resulting solutions are still good, even if not a global maximum.

These algorithms are often tailored to particular problems, and frequently combine the features of simulated annealing with genetic mutation and selection. The following are the two basic rules that are at the heart of these algorithms:

- The selection of a new possible solution from an existing one should be such that if the process were left to run to long enough, all possible solutions would be tried (i.e., there are no sets of solutions disconnected from the "random walk").

- The decision as to whether to accept the new possible solution is based on the following:

$$p(accept) = \begin{cases} 1 & \text{if } profit_{new} - profit_{old} \geq 0 \\ e^{\frac{profit_{new} - profit_{old}}{kT}} & \text{if } profit_{new} - profit_{old} < 0 \end{cases}$$

where T is the "temperature" of the acceptance criterion and k is a scaling constant. This is the criterion for simulated annealing. Genetic algorithms normally use a "Monte Carlo" method for determining the number of the next generation's population according to profitability, and when run over a few generations, the two stochastic mechanisms are essentially similar.

In general, simulated annealing is easy to implement, as coding of the algorithm for any given problem can be relatively easy. Genetic algorithms have the advantage that for many problems, they will run more quickly; however, they are normally much harder to implement, especially if speed improvement is desired. The speed improvement comes about because the genetic algorithm is effectively running many optimizations in parallel, but the success of this depends on the gene coding.

13.2.12 Summary of Specific Areas in Network Design

In this section, we have described a number of considerations that must be taken into account in defining the design problem. Much of this amounts to strategic planning choices between transition strategies or basic technology choices. We have stressed the importance of the internal commercial interface in partitioning the network design problem. Finally, we have surveyed the formal optimization processes available to the designer, an area of network design which is currently very dynamic.

We take it as an objective to automate these processes as much as possible, minimizing the need for sophisticated and experienced human judgments and at the same time eliminating much of the human error. Of course, if we can succeed in automating the processes with well-characterized, quasi commercial interfaces between layers and partitions in the network, we can easily embed them as intralayer processes within the layer networks themselves and their partitions.

Such goals suffer from both the view that it is basically simple but operators make their networks unnecessarily complicated and also the view that it is much too complex even to start. However, it is now possible with the mathematical techniques and computation power that are available to realistically tackle large-scale networks with automated network design tools. A judicious selection of techniques is needed for each subnetwork, which allows the automated design process sufficient freedom to find good solutions while not giving so much freedom as to make the problem too large to solve in realistic time. To maintain integrity of overall design, the notional service prices generated as an output of a design in one subnetwork should be used as part of the cost base for its client subnetworks. These notional costs and prices must be held stable between components of the network, but must nevertheless evolve over time and respond to the quasi market forces on which the optimization depends.

13.3   FUNCTIONAL DESIGN CONSIDERATIONS FOR DIFFERENT SUBNETWORKS

In this section, we illustrate the general functional design considerations for a variety of subnetworks. The list of subnetworks that we describe is not exhaustive, but we hope it is illustrative of those currently relevant to broadband networks. Each subnetwork is considered from a number of viewpoints:

- Control mechanisms;
- Requirements of users;
- Requirements placed on server layers;
- Formal design considerations;
- Practical design outcomes.

This is intended to capture the primary user requirements, the primary constraints on the design, and the likely shape of optimal solutions. First, however, we consider possible client-server relationships.

13.3.1   Potential Client-Server Relationships

Each subnetwork must know the layer networks which may wish to buy capacity from it (i.e., its clients). It must also know the layer networks from which it can buy link capacity (i.e., its servers) and so we need to know which layer networks can have client-server relationships with other layer networks. For this, a complete map can be drawn up showing all the main client-server relationships between all the layer networks that a network operator may be required to support. Few real network scenarios will simultaneously contain all these possibilities. Any real network will, in general, be required to support a defined subset. A European example is given in Figure 13.16 and a North American example is given in Figure 13.17. Both cover the following:

- Circuit-switched layer networks,
- Connectionless packet-based layer networks,
- Connection-oriented packet based layer networks,
- The ATM layer networks,
- The PDH layer networks,
- The SDH layer networks,
- Transmission media layer networks.

Such a diagram gives the highest level of structure of an overall network operator's network and records the basic flexibilities that can be exploited by exercising these client-server relationships.

However, we must remember that the decision on client-server relationships cannot be made in full isolation, as the clients that a server network may support are constrained by the use of the agreed upon interface standards, the available mappings, and the desire to interwork with other operators. If one network operator chooses one hierarchy of client-server relationships and another network operator chooses another, then the two must agree on how they can interwork between the two hierarchies. In general, standards

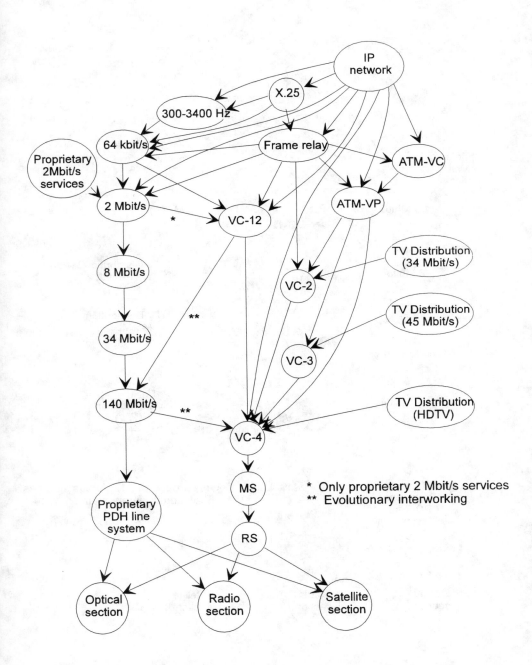

Figure 13.16   Possible client-server relationships for operators using the ETSI PDH hierarchy.

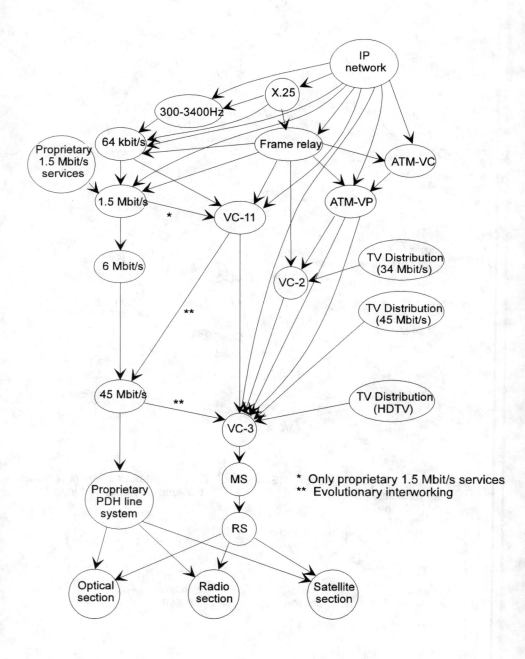

Figure 13.17   Possible client-server relationships for operators using the North American hierarchy.

bodies try to attain a good balance between a need for flexibility and the need to keep the costs of maintaining options to a minimum.

Having established all these possible client-server relationships, and taking into account the need to interwork with its associated costs, it is now possible to start designing each subnetwork in the network operator's network.

Although a multilayer network design will necessarily be iterative, it is normal to start with the design of the highest client layer, based on estimates of server charges from earlier designs, and, from the requirements they place on their servers, go on to develop a design for each server. However, two factors ensure that this cannot be a simple linear process:

- The precise tariff structure as generated by a server network may well depend on the level and type of demand (i.e., a definitive price schedule from the server may only evolve over several iterations);
- Many performance primitives depend on the detailed design of the transmission media layer networks, and the higher client layer networks cannot finalize their design until the server gives precise information on these performance primitives.

13.3.2 Internet Protocol (IP) Connectionless Packet Transit Subnetworks

There are many aspects of the IP network that are different, at a practical level, from many of the connection-oriented networks with which public telecommunications network operators are more familiar.

The extent to which ATM can be used as an IP server layer network to reduce cost or improve performance and features is currently very topical. There is in fact considerable scope for achieving this by exploiting the flexibility of the client-server relationships. An IP subnetwork may, in principle, use any set of servers that optimally meets its requirements and ATM can offer significant advantages to an IP subnetwork in several circumstances. ATM can carry IP, frame relay, X.25, SMDS, private leased CBR services, and so forth, and where the derisking associated with this multiservice capability is important, it is likely that ATM will be able to offer a price advantage over other layer networks in satisfying the requirements for link transport in the IP subnetwork.

At the end of the day, the marketplace will decide the extent to which IP and ATM are complementary or competitive, but in the meantime, the design methodology should take careful account of this market mechanism as expressed in the quasi commercial interface between the layers.

13.3.2.1 *Control Mechanisms*

Being connectionless, the IP packet header contains the full network address of its destination. Each router through which it is switched looks at this address and uses it with its route table to select an outgoing route. Each packet is effectively its own connection from the point of view of the control of connectivity.

IP transit subnetworks use one or more routing exchange protocols to build and update their route tables and thus adjust dynamically to changes in the topology and available capacity in the subnetwork.

13.3.2.2   *Performance Requirements of Users*

These are hard to define for any connectionless network. Effectively, no performance requirements are directly imposed on the design of the IP network by the IP connectionless service. Clearly, there is a relationship between the amount of capacity available, the number of concurrent users, and the performance experienced by the users; however, this is hard to translate into a performance versus price curve as it must be averaged across all users.

If the IP transit subnetwork is an "intranet" (i.e., it is internal to one organization), then it is easier to define the value of the performance, as the requirements of the users of the intranet will be less diverse.

13.3.2.3   *Performance Requirements on Server Layers*

As the packet size of an IP packet is variable and can potentially be very long, corruption of packets can start having a multiplication effect on the performance of the IP network. One small corruption in an IP packet will cause its full retransmission. For example, if the IP packet is being carried by ATM cells and the ATM network drops a cell through congestion, the ATM network may now receive around 10 to 100 new cells (based on an average IP packet size), as the packet is retransmitted, which will only add to the congestion in the ATM network.

This is a classic control loop stability problem where there is one control loop nested inside another. A practical condition for stability in this situation can be stated as

- The control loop responding to congestion in the server layer should respond with a time constant of at least 10 times longer than that used to control congestion in the client layer in order to avoid interaction between the control loops.

In this case, the control mechanisms in the client layer are not in at the IP network layer but in the termination band of the service protocols TCP and UDP. The user datagram protocol (UDP) is a fully connectionless protocol and has no congestion control mechanisms; however, the transmission control protocol (TCP), which is used for connection-oriented applications like file transfer, does. The TCP congestion control protocol uses a window mechanism that has grown up in the light of experience, as the Internet has been adapted to deal with catastrophic congestion oscillations in the past.

While the ATM UBR service was designed with IP transport very much in mind, there are still issues over its traffic stability. If the server is ATM VCh with an ABR service, then the congestion control is managed by the resource management (RM) protocol. This protocol does not currently specify any time constants, so the use of an ABR service is potentially dangerous for carrying IP. In principle, a CBR service would appear to be the safest and most suitable; however, an AAL5/CBR service is not one that is readily acknowledged as a standard.

13.3.2.4   *Formal Design Considerations*

Using the routing exchange protocols together with the connectionless nature of IP considerably complicates the network design process, as it makes it more difficult to decide how to augment an IP transit subnetwork to meet growth requirements. It is difficult to track traffic on a source/destination basis across the transit subnetwork, and even more difficult to spot which end-to-end routings are causing congestion. However, this information is essential to cost optimal network design. Historically, most links were provided on a bilateral basis between organizations, and the Internet was then able to "hitch a ride" by adding routing capability using these bilateral links. As commercialization has advanced in the Internet, optimal design of topology has become a major issue.

Bringing together the parameters for a formal design process is always likely to be more difficult in a connectionless layer network; however, it is not impossible, and there is a considerable commercial advantage in doing it. This is linked with the commercial partitioning of the IP network. It is, of course, much more attractive to an intranet, where the cost and performance advantages can be passed directly to the intranet users paying for the service, and not divided among all the Internet users in the world.

13.3.2.5   *Practical Design Outcomes*

The IP transit subnetwork is a collection of interconnected routers. Currently, the links are usually high-capacity leased lines (e.g., T1, T3, E1, etc.) and the costs of these are generally high and scale substantially less than proportionate with bandwidth (e.g., the price of a T3 is much less than 21 times the price of a T1). This ensures that the topology is relatively sparse, based on a few fat links and many relatively small routers.

The potential price structure, rather than any technical advantage, may well be the determining factor in the use of ATM in transport links between IP routers. The PVC service, which is normally suggested as the obvious solution, offers no special advantage over the current leased lines and will simply have to compete on price. However, the bandwidth on demand nature of an SVC service may well significantly change the topology of IP transit subnetworks. These should be able to allow the IP transit subnetwork to have many thinner links, which can be put up and taken down relatively rapidly, at a similar price to the existing leased lines, thus reducing significantly the number of routers through which an average packet must pass and hence reducing the total cost of the router network for a given amount of traffic carried.

One of the most attractive situations where this can be used is where a host system (i.e., a system with an access point to the IP layer network) can set up a temporary link directly to another host, bypassing all the transit network routers. The routers in each host system can detect and now use this link. This not merely reduces cost by not using the transit subnetwork routers, it now offers a level of performance guarantee, as the performance is controlled entirely by the two host systems and the performance of the ATM SVC. This would be especially attractive for applications such as the transfer of large files and real-time applications. This is illustrated in Figure 13.18.

A most attractive feature of such a solution is that it has not changed anything on the IP network as seen by its users; the only thing that has been adjusted is the topology of the IP network to give cost and performance advantages.

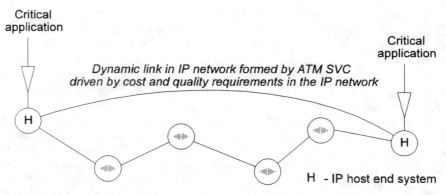

Figure 13.18   Use of ATM as a server to the IP layer network in order to enhance quality.

13.3.3   The 64 kbit/s PSTN/ISDN Transit Subnetworks

The major feature of these subnetworks is that they exist and the marketplace for the services they offer is very large, still growing strongly, and is maturing.

13.3.3.1   Control Mechanisms

Signaling protocols used for control in the PSTN/ISDN are generally fast and robust. However, the call controllers tend to be fully integrated into the switch, which means that services based on extensions to the signaling protocols often result in circuitous routings of the transport path as well as the signaling messages. For example, an "800" call may route the transport path to a switch that will carry out the number translation and then route the call onto the network address associated with the "800" number. In another example, when a GSM mobile telephone that has roamed onto another mobile network receives a call, the call will first route to the home network and then the call will be extended to the roamed network. This frequently results in a call to a mobile which is "in the next block" being made up of two international calls.

Such situations make the development of the traffic matrix more complicated, as the routing may well be constrained by the capabilities of the switches to separate number translation from routing. In principle, IN should solve this problem, however, since there is large overlap and basic incompatibility between ISDN signaling (ISUP), the IN messaging protocol (INAP), and mobile messaging protocol (MAP), it seems unlikely that IN will ever provide a complete solution simply because the installed base of equipment will persist.

13.3.3.2   Performance Requirements of Users

These are dictated not so much by the users of the PSTN/ISDN themselves as by the historic service specifications of the PSTN/ISDN. Terminal equipment and user expectations are largely built around the existing service, and any changes that do not result in full backwards compatibility are likely to be strongly opposed by the marketplace and regulators alike.

13.3.3.3   *Performance Requirements on Server Layers and Peer Interworking*

The support of the existing narrowband network by a new broadband network has been much discussed in standards and other fora and frequently misunderstood, most notably when voice delay was used as an argument in the selection of the ATM cell size. The confusion arises by mixing the multimedia voice service with interworking to the PSTN, as illustrated in Figure 13.19.

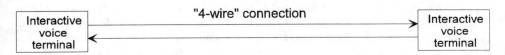

Multimedia voice - echo is purely a matter for terminal equipment design

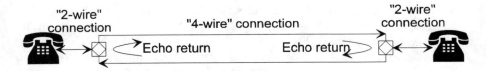

PSTN - has strict delay and echo planning rules

Figure 13.19   The different management of echo in PSTN and multimedia voice.

- A multimedia voice service supports a real-time stream of encoded voice, and the coding scheme and the characteristic of the stream can be tailored to suit the application. The transport is fully digital, and the terminal equipment is entirely responsible for encoding and the control of echo entering the microphone from the audio output device. The transport service will contribute impairments such as absolute one-way delay, delay variation, degraded error performance, and so forth, and the terminal equipment can be designed to be robust to these impairments. The definition of the stream transport service is independent of the voice application, and it is the responsibility of the terminal equipment to adapt the voice application to the transport service.
- In the PSTN, as the original design was analog, echo is a feature within the network and is controlled within the network. It is not simply a terminal equipment consideration, as is the case for multimedia voice. The single copper pair that is generally used in the access network carries both directions of the audio signal in the same band and must be converted to a duplex mode, or "four-wire," both in the telephone and in the local exchange. The conversion in the local exchange produces a significant echo return in addition to that produced by the terminal equipment,

and the PSTN has strict planning rules to control the effects of this echo. These planning rules must be adhered to any service that is to fully interwork with the PSTN.

These rules seek to keep the delay below a level at which echo becomes disturbing to the user (around 100-msec round-trip delay) for most PSTN calls, and to add echo control devices only to links that will cause calls to exceed these delays (normally intercontinental cable links and any satellite links). The ITU-T specifies echo control for all round-trip delays longer than 50 ms; conversely, most connections of less than 50 msec round-trip delay have no echo control.

This 50-msecs round-trip delay gives a 25-msec one-way delay, which must account for all transmission media delay (limited by the speed of light), switching delays, and any other buffering delays. The plain fact is that in most circumstances, adding 6 msec for the packetization of a PSTN channel into an ATM cell cannot be tolerated in existing network delay plans, and often adding any more than about 1 msec for packetization will cause instability in the operation of existing echo control procedures. This means that interworking any switched ATM voice service with the PSTN will cause echo difficulties. We must consider two separate scenarios:

- ATM is used as one of the set of server layer networks that can provide link transport for the PSTN and meeting all the performance requirements, including delay, of the current PSTN;
- An ATM multimedia voice service is interworking with the PSTN and special consideration must be given to the interaction of the two different approaches to echo control.

The latter scenario is basically similar to the interworking between the GSM mobile network and the PSTN, for which full interworking guidelines have been developed. The GSM mobile network has a 100-msec one-way delay between the mobile terminal equipment and the point of interworking with the PSTN.

13.3.3.4   Formal Design Considerations

The PSTN/ISDN is capable of being supported by many server layers. In the case of private networks, there is little restriction on the network changing from one server layer network to another. However, where switching and transmission for the network are owned by the network operator, the economic rationale for charging is very different. In cash flow terms, owned equipment costs nothing and it is unlikely that any new investment will cost less, unless there are very substantial operational savings. The way this is formulated into the subnetwork design problem is reflected in the very low variable cost of transmission passed as a price for server capacity to the subnetwork design process.

13.3.3.5   Practical Design Outcomes

Apart from private PSTN/ISDN subnetworks, a PSTN/ISDN transit subnetwork is only likely to be a strong influence on the design of a broadband network when it is either new or growing fast. In such a situation, the ability of SDH to carry VC-1s that are tailor-made for the PSTN/ISDN along with broadband payloads makes SDH very attractive for the support of links in the network. The extent to which PSTN/ISDN link capacity will be carried by ATM must still be in question and the marketplace will decide over the course of time.

Before fiber transmission systems reduced the importance of distance as a cost factor in many PSTN/ISDN transit subnetworks, they were designed in such a way that routing was kept as close as possible to a shortest path routing. This tended to result in many small switches and many thin routes concentrating onto a few larger switches and some very fat links. The routing would often be based on several levels of partitioning from a local transit level, to a regional transit level, to a national transit level.

However, for many transit subnetworks now, distance is not a major factor in the price of link capacity and the opportunity is presented to transform the topology. Generally, this results in fewer overall nodes and links, but with higher meshing as well as a greater consistency in the size of the noses and links. This is often true to such an extent that if many public network operators were to start again with the design of their PSTN/ISDN, they would look radically different from their existing networks. The extent to which a network operator with an existing network can perform this "node consolidation" will depend on the growth in the network and the financial attitude to the existing network.

13.3.4   Multiservice Access Subnetworks

A multiservice access subnetwork seeks to provide a number of features:

- Collect traffic from a number of transit groups (or access groups), which may be geographically dispersed, to one or more core transit groups;
- Distribute traffic to a number of transit groups (or access groups), which may be geographically dispersed, from one or more core transit groups;
- Concentrate/consolidate the traffic between the dispersed transit groups (or access groups) and the core transit group(s);
- Be able to support a large number of client layers, including those that may be defined in the future.

This is illustrated in Figure 13.20. The actual media used to support the multiservice access subnetwork may be point-to-point optical fiber, passive optical-fiber network (PON), coax tree/bus, point-to-point copper pair, and radio. On all these media, it makes little sense to dedicate technology to the supply of the services of only one layer network (although this is historically the case with PSTN on copper pairs), as there is a strong economic case based on flexibility and the reduction of risk for introducing an electronic transport layer that is tailored to the support of many client layers. We call these layer networks multiservice access layers (MSA layers).

In this section, we discuss the nature of this first electronic layer, which uses the media layer, be it media copper pair, coax, fiber, or radio. Current protocols include LANs, PONS, and some radio-based access systems.

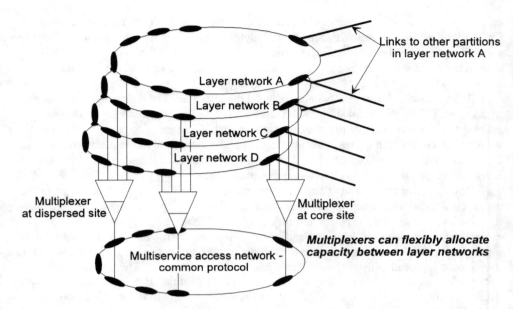

Figure 13.20   A multiservice access network supporting many subnetworks of many layer networks.

13.3.4.1   *Control Mechanisms*

Other than the usual management functions, the principal role of the control is to concentrate the traffic. The capability for concentration is set by the traffic characteristics of the client layer links, and, in general, the more the traffic on each client link varies over time, the more concentration can potentially take place.

- *Concentration at the attachment scale*—A level of concentration happens automatically when the MSA layer receives a request to provide a trail to support an access link.
- *Concentration at the call scale*—At call timeframe, this concentration is achieved by intercepting the signaling of the client layers to assign the capacity for the client layer's access link when required.
- *Concentration at the packet or burst scale*—Controlling concentration at the packet/burst scale in this environment can be difficult. This arises as the concentration is taking place at the same time that the geographical collection/distribution of the packets/bursts, and the precise place at which the statistical multiplexing takes place to achieve the concentration is very often distant from any buffer holding the packets/bursts awaiting permission to send. This means that a full protocol is required to message to the buffers when they can send, in other words, a flow control protocol. The rate at which the rate of packets/bursts can be changed is ultimately limited by the round-trip delay in the flow control protocol. Media access control (MAC) protocols for LANs are the most notable current examples and include Ethernet's TDMA/CD (IEEE 802.2), Token Ring (IEEE 802.3), and

DQDB (IEEE 802.6); however, in principle, the resource management (RM) protocol could also work here.

13.3.4.2 Performance Requirements of Users

Since there are potentially so many client layers, even those currently unknown, it is hard to give clear performance requirements. This does, however, make it more important that a multiservice access subnetwork is capable of very high performance. Two issues are worthy of special consideration:

- Dual-parented access;
- Transmission delay.

Some client layers are likely to require dual-parented access. This allows a dispersed transit group (or access) to connect simultaneously to two transit groups. Frequently, the request also includes that the two connections have no common mode failure other than the equipment providing the access group itself. This requirement can be a major factor in the design of the MSA layer protocol.

If the PSTN is a client layer, then it is likely to place very strict transmission delay restrictions on the MSA subnetwork and, again, is likely to be a major factor in the design of the MSA layer protocol.

13.3.4.3 Performance Requirements on Server Layers

This will depend on the media. The ducted media, copper pair, coax, and fiber will all have design issues in supporting the requirement of dual-parented access. Radio is likely to have design issues in reaching high availability and good error performance.

13.3.4.4 Formal Design Considerations

The exact parameters by which the traffic forecast for the MSA subnetwork should be specified will depend on the exact protocol. This will also govern the way statistical addition of traffic can take place.

When a ducted media is being used, the failure topology of the transmission media service must be declared for the design of dual-parented access.

13.3.4.5 Practical Design Outcomes

Achieving good traffic performance will depend on the size of the MSA subnetwork. For a given set of clients, there will be a critical size below which it will be hard to achieve good statistical gain without significantly affecting the performance. However, above this size, the statistical gain is unlikely to improve greatly. It is this factor of statistical gain that is likely to be the controlling factor on the optimal size of an MSA subnetwork.

The statistics from different timescales are likely to affect different parts of the MSA subnetwork. For example, all the investment associated with the MSA protocol is generic to all client layers, so it is derisked by aggregating the risks from all these clients. This means that it should be possible to make longer term and preprovided investment in these parts of the MSA subnetwork. However, the equipment providing the adaptation to the client layer is specific to that layer and so carries all the risk associated with that forecast. Since the dispersed transit groups (or access groups) are likely to be small in traffic terms, this risk will be relatively high. This makes it desirable to have this provided on a "just in time" basis and this gives a rationale to access equipment with plug-in cards. Even if for a given configuration it is more expensive than purpose-built equipment, the costs associated with the uncertainty in the client layer forecasts make the flexible equipment with plug-in cards much cheaper overall.

13.3.5   ATM Virtual Channel Transit Subnetworks

Most ATM services would normally expect to use the VCh layer. The fact that it is a ubiquitous multiservice layer means that it must be flexible and widely available. Traffic management and dimensioning are major issues in the design of this layer network, especially as forecasting will be difficult. The traffic aspects were discussed in Chapter 9, so they are only referred to in this section.

13.3.5.1   *Control Mechanisms*

The primary control mechanism for VCh transit subnetworks will be signaling, assuming SVCs form the majority service. In addition, the RM protocol will control the capacity used by any ABR connections.

13.3.5.2   *Performance Requirements of Users*

These depend on the application, but they are specified using the service parameters of each connection type. The service parameters offer the possibility for a wide variety of service and performance; however, the ability for an ATM service to justify price differentiation based on service parameters may be more restricted.

13.3.5.3   *Performance Requirements on Server Layers*

Currently, standards only define one server of the VCh layer, the VP layer. The links in an ATM VCh transit subnetwork will contain concentrated traffic and the statistics of the concentrated traffic of the VP are more slow-moving than those of any individual VCh connection within the VP. This means that there is little extra to be gained by the VP layer responding rapidly to capacity on the VCh link and an overrapid mechanism is potentially unstable. The capacity of the VP must therefore react slowly to any changes in the capacity in the VCh link. A safe way of managing this is in the adaptation by controling the bandwidth on the VP by making periodic changes to the capacity of a CBR service, or possibly an ABR service with long time constants.

13.3.5.4   *Formal Design Considerations*

This is likely to contain a classical network topology design problem. The routing across the network is either based on preallocated route tables or on the results of a routing exchange protocol. With this knowledge, and starting with the statistical traffic matrix for the transit subnetwork, an optimization can be undertaken to find a suitable network topology. In addition, a failure topology for the candidate links is required from the VP layer.

13.3.5.5   *Practical Design Outcomes*

The general topology is likely to be controlled by two factors:

- Where links are relatively expensive, the topology's degree of meshing will be less than when links are cheaper.

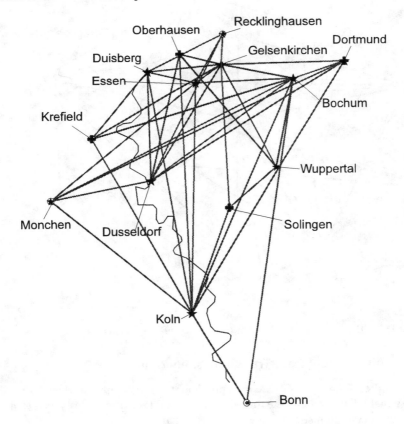

Figure 13.21   Example of an ATM VCh network in a large urbanized and industrialized area (reproduced by kind permission of Network Design House Ltd.).

- Since the VCh and the VP are likely to be switched in exactly the same switch fabric, the effective node costs are likely to be controlled by the cost of call control, and increased meshing will reduce, on average, the number of times a call controller is invoked.

Figure 13.21 and Figure 13.22 show two examples calculated using exactly the same traffic matrix but with different link-to-node cost ratios driven by a much larger geography in the second example. These were calculated using a statistical network design algorithm and demonstrate that at the VCh layer, distance is not a major factor in the topology. Both examples are well-meshed, however, there is a marginally greater level of meshing when the links are cheaper.

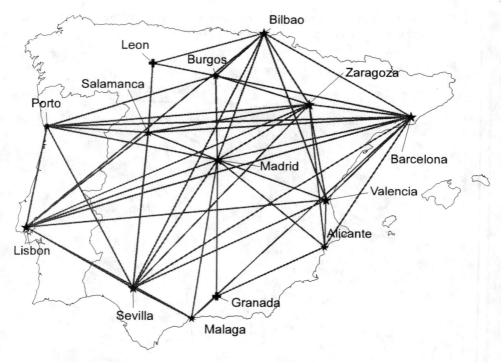

Figure 13.22   Example of an ATM VCh network in a country with long distances between centers (reproduced by kind permission of Network Design House Ltd.).

13.3.6   ATM Virtual Path Transit Subnetworks

The uses of the VP layer in ATM remain problematic. It will only be when ATM networks become widely deployed that the true value of the VP layer can be truly assessed. Among the objectives cited in justification were the following:

- A means of reducing the amount of call control by providing a links between VCh switches and ensuring a high degree of meshing in the VCh layer;

- A means by which a public network operator can provide private circuits between VCh switches in a private network;
- A means by which further statistical gain can be made by aggregating traffic from a number of VCh links.

To a large extent, there is little that could not be achieved by using the SDH LVC layers; however, as the VP layer now exists, it will no doubt be used for all these purposes.

13.3.6.1   Control Mechanisms

The expectation is that the main demands for service at the VP layer will be for PVCs. In principle, these are set up by network management; however, as experience with SDH has shown, network management solutions for connectivity control have proven to be costly and slow to implement. The SPVC, where the signaling protocol is used to establish the connection, does appear an attractive solution to this control problem.

13.3.6.2   Performance Requirements of Users

While there is nothing to stop other layers being clients of the VP layer, the VCh is the principle intended client. This layer is looking for simple robust connectivity between VCh switches and a failure topology for all the paths used by a particular client subnetwork.

13.3.6.3   Performance Requirements of Server Layers

The VP layer can use many client layers including any of the SDH VCs (not just VC-4), PDH 2048 kbit/s, 34368 kbit/s, 44736 kbit/s and 139264 kbit/s, or any multiservice access (MSA) layer. The SDH and PDH layers are circuit-switched layers and the VP is probably wise to select a CBR or circuit-switched service from any MSA layer. All of these should be capable of providing robust, high availability, effectively error-free service. As with other subnetworks, any VP transit subnetwork is likely to require a failure topology from its servers.

13.3.6.4   Formal Design Considerations

Once again, this is likely to contain a classical problem in network topology design. In this case, the routing used may be more open, as the network management protocol can allow almost any routing algorithm.

13.3.6.5   Practical Design Outcomes

Once again, the degree of meshing in the network will depend on the relative costs of the nodes and links. Since the principle client is the VCh layer, the nature of a VP transit subnetwork traffic matrix will depend on the topology of the subnetworks of the VCh layer. As we noted in the section on VCh transit subnetworks, this is likely to depend on the cost of call control relative to the cost of the switch. If this is high, the demand will be for more thinner VPs, while if it is low, then the demand will be for fewer but fatter VPs.

In either case, the total amount of capacity in the VP layer network should remain largely unaltered.

The networks shown in Figure 13.21 and Figure 13.22 could equally be for a VP layer network.

13.3.7   SDH Lower Order VC Transit Subnetworks

The LVC layers provide a general set of layers that can be used for consolidating traffic from different client layers into a common transmission infrastructure. Most notable, the LVC layers are capable of supporting links of the 64 kbit/s layer, and direct support of IP and frame relay, as well as the ATM VP layer.

As well as supporting their own client layer subnetworks, public network operators also provide private circuits that form the links in private networks. The variety of layer networks that are in use in private networks has grown considerably and includes 64 kbit/s, all the data layer networks discussed in this book, and a large number of proprietary layers created by private network multiplexing equipment. The LVC layers (even if supporting a PDH layer) are an effective way for a public network operator to meet the needs of all these private networks.

13.3.7.1   Control Mechanisms

The choice of connection control within a network operator's domain will depend on the size of the SDH path layer network, the desired level of reliability, and the speed of response to a client request. The automatic assumption of most network operators is that network management protocols should be used. However, these have proved difficult to produce, costly, relatively slow, and difficult to scale. An alternative that will become available as integrated ATM and SDH networks emerge is to use the SPVC part of the P-NNI signaling protocol.

An important part of the connection control is likely to be restoration and diverse routing. There are several ways in which this can be achieved. As discussed in Chapter 10, there is a need to ensure that any restoration mechanism at the LVC layer does not interact with any protection mechanisms at the lower layers. In addition, 1 + 1 network protection at the LVC layer could be considered as an option for the provision of very reliable service.

13.3.7.2   Performance Requirements of Users

These can vary, but demand often falls into a "low-cost" category or a "high-availability" category, reflecting the key parameter in a buying decision. Often, a client layer network will make its own arrangements for restoration and low cost becomes the critical parameter (assuming the failure topology yields no major pinch points). On the other hand, if the connection is to be used directly by an application, or forms part of a very sparse network, then the buying decision is often controlled by the value of the applications, as the LVC connection is now a critical component. In these cases, the availability of the connection becomes important.

In almost all cases, error performance is important, and there is a growing assumption among users that a transmission network will not produce errors. Where an SDH network is based largely, or exclusively, on optical fiber, this is becoming realistic as an LVC connection can run for days without observing any errors.

13.3.7.3   *Performance Requirements on Server Layers*

While an HVC layer is the usual server for the LVC layers, there are some circumstances where the LVCs are carried directly on a PDH layer, on a dedicated radio, or on a satellite section layer. There are a number of features that standards indicate any server layer of an LVC layer must support:

- Transfer of frame phase by a TU pointer mechanism;
- A path/section overhead to assure the integrity of the links in the LVC client networks, including error monitoring, path/section trace, and signal status.

The provision of these features will ensure the consistency of performance of the LVC layer networks across a variety of media.

13.3.7.4   *Formal Design Considerations*

The design of a transit network may appear to be the same as the design of other transit networks such as 64 kbit/s VCh and VP. However, the performance requirements may make this a more difficult exercise and the topology may be required to be resilient to any single failure mechanism in the network. This requirement adds considerable extra complexity to the design of the topology, but there is still no reason why simulated annealing and/or genetic algorithms cannot be used.

In addition, the frame structure of SDH makes add/drop multiplexer (ADM) equipment especially economic. The ADM places strong constraints on the development of the topology, which must be included in the way the optimization problem is constructed. This must reflect the economic advantage of ADM while also fully costing any resulting restriction in growth and flexibility.

13.3.7.5   *Practical Design Outcomes*

Most practical designs of LVC layer networks show a mixture of large crossconnects at a few key nodes with smaller nodes served by chains and/or rings of ADMs. There is, however, still considerable difference of opinion on the extent to which the network should be based on ADMs or crossconnects. Some operators show a strong preference for many crossconnects with relatively few ADMs feeding into these crossconnects, while others make extensive use of ADMs, as illustrated in Figure 13.23.

Networks based exclusively on ADMs have a number of advantages from the control point of view. The rings are controlled by a low-level protocol and large centralized network management-based configuration control is often not necessary. In addition, a resilient topology is effectively guaranteed, as all the rings are built on physically disjoint media layers and intersecting rings have full ability to switch connections between rings

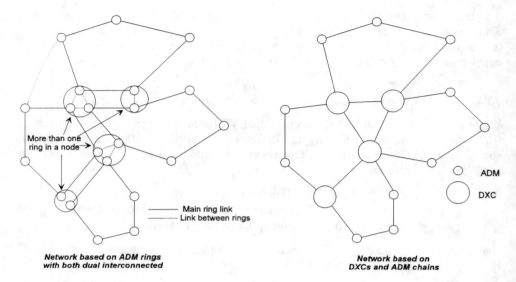

Network based on ADM rings
with both dual interconnected

Network based on
DXCs and ADM chains

Figure 13.23   Difference in design between a network based on ADM rings and one based on DXCs
with ADM chains.

at the two points of intersection. It is also possible to conceive of all the ADM at nodes at
which a number of rings intersect being replaced by crossconnects, but the topology and
control associated with the rings remains intact.

Noting these points about ADMs, once again the network designs illustrated in Figure
13.21 and Figure 13.22 could equally be that of an LVC network, as the network design
constraints, at least within the context of an example, are very similar.

13.3.8   SDH Higher Order VC Transit Subnetworks

The principal purpose of the HVC layer networks is to provide flexibility between the rigid
nature of the physical layer network topology which is expensive to alter and the more
dynamic nature of the topology of the HVC layers' clients. These clients are intended to
be the SDH LVC layers and the ATM VP layer; however, there are some layers that can
use the HVC layers directly, in particular, studio-quality TV. The HVC layers provide a
convenient way of transporting the very high bit rate TV signals between studios,
production centers, and distribution centers before they are compressed (e.g., using
MPEG2) for more general distribution.

The other purpose of the HVC layers is to provide an evolutionary path from PDH
transmission by carrying 139264 kbit/s and 44736 kbit/s PDH signals. For some network
operators, this can be the main volume for traffic on their HVC layer network.

13.3.8.1   *Control Mechanisms*

Connections at the HVC layer are likely to have a long holding time, often months to years.
This suggests that the control mechanisms need not be fast or sophisticated, and while this

is true for most aspects of the control mechanisms, two other factors suggest that some level of sophistication is needed:

- The control mechanisms must be able to offer resilient service, either through protection or by restoration.
- As the network grows, some old routing may be very inefficient as the topology is likely to be relatively sparse.

There has been much discussion in the more academic circles on the use of low-level dynamic restoration algorithms following an original paper by Grover; however, the efficacy of these algorithms over other mechanisms has not been universally accepted. In general, most network operators are using a mixture of autonomous protection (e.g., 1 + 1 or SPRING) and network management-based configuration management.

13.3.8.2   Performance Requirements of Users

These are basically the same as those defined for the users of LVC layer networks.

13.3.8.3   Performance Requirements on Server Layers

Currently, the SDH transmission media layers are the only ones defined in standards for the support of HVC layers. The requirements are basically the same as those described for the LVC server layers.

13.3.8.4   Formal Design Considerations

In the case of the HVC layers, the cost advantage of ADMs is even more marked, so inclusion of the constraints surrounding ADMs is even more important. Once again, the resilience of the topology is likely to be a major design objective.

An extra subtlety in the design is that for each of STM-4, STM-16, and STM-64, ADMs, line terminals (LTs), and regenerators are normally variations on the same basic equipment. This means that if a regenerator is needed in a line system, it is economic to make this an ADM in a chain, as there is normally little extra cost. This means that the costs of the transmission media layers need to be accurately included, and this may well be achieved most effectively by a single-stage optimization including the HVC layers and the transmission media layers.

13.3.8.5   Practical Design Outcomes

Once again, some network operators have made extensive use of ADMs while others have based their networks on crossconnects with only a few ADMs. The reason for this probably owes as much to the lack of good optimization tools and differing availability of capital for investment as to genuine differences in network requirements. Moreover, designs have often been subject to the general preferences within a network operator's planning group. In addition, if capital for investment is easily available, operators have tended to favor crossconnect-based networks as they appear to offer more flexibility (though, in practice, this is not always the case), while if capital for investment is tight, operators have tended to favor ADM-based solutions as they have the lowest up-front cost. Neither approach is probably optimal and in almost all cases, there could have been a better balance between crossconnects and ADMs.

Figure 13.24 and Figure 13-25 show the VC-4 subnetworks that are supporting the network in Figure 13.21 and Figure 13.22, respectively (be they a VP network or an LVC network). Again, this was derived using a statistical design algorithm and a modeled commercial interface between the layers. The VC-4 price structure was based purely on the number and size of the access groups to the VC-4 network and giving significant volume discount on both the number and the size. A nonlinear distance charge was included for the VC-4 connections. In this case, the effect of the distance is more marked. The optimal design in the network with long distance has many fewer links than the network in the densely urbanized and industrialized area. The VC-4 network is generally well-coupled to the costs of line systems, so the design reflects this. However, the VC-4 network

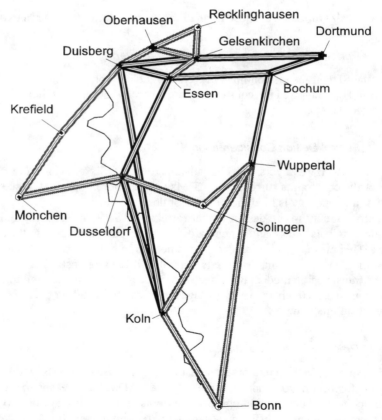

Figure 13.24   Example of a VC-4 network in a large urbanized and industrialized area (reproduced by kind permission of Network Design House Ltd.).

Figure 13-25   Example of a VC-4 network in a country with long distances between centers (reproduced by kind permission of Network Design House Ltd.).

is decoupling these effects from its client layers, which is why there is not a great deal of difference between the design of the VCh networks in the two geographies.

13.3.9   SDH Section Layer Subnetworks

13.3.9.1   Control Mechanisms

Multiplex section protection can provide for the protection of HVC links by any of the mechanisms described in Chapter 10. The decision whether to use MS protection is driven by the availability requirement of the particular part of the client layer.

13.3.9.2   Formal Design Considerations

The size of the MS depends on the total HVC demand placed on the subnetwork. This is the capacity between the two terminating nodes of the linear topology and the dependence of ring capacity on ring size and protection mechanism. In general, the cost of the line system capacity is now low compared to the overall network costs, so the cost of significant overprovision may be advantageous when compared to more frequent upgrading of line systems to higher bit rate as demand develops. This, again, suggests that the design of the transmission media section layer may be best integrated with the HVC layer optimization.

Finally, fiber is now generally expensive compared to optoelectronics. In principle, it is likely to be cheaper to increase the total bit rate on a fiber by replacing the line system electronics than to add more fiber. This means that the lifespan of the use of line system may be shortened not by its obsolescence, but because it is using a fiber that could be more profitably used at a higher bit rate. This can be modeled by including the cost of the fiber as an "opportunity" cost. This is a mechanism by which the fiber is costed by the extent to which it brings nearer the day when a new cable is required. This makes fiber in poorly used cables cheap while those in well-used cables very expensive.

13.4   PRICING AND RELIABILITY OF IMPLEMENTATIONS

The design exercise of the subnetworks described in the previous section has been a functional exercise, and the cost of each of the functional components was included in the design exercise. In this section, we complete the picture by looking at the relationships between these functions and their implementations. When we separate out the functional aspects, implementations are left with two properties: they cost money, and they can go faulty. These two properties are essential input to the network design exercise.

We noted at the beginning of the chapter that functions can be provided in one of three ways: services across a commercial interface, implementations exclusive to the subnetwork, and implementations that provide functions to more than one subnetwork.

The first category has already been discussed at some length, as it is trading in pure functionality. The second category is reasonably straightforward in that the implementational properties of cost and failure are the sole concern of the subnetwork. However, the third category is more difficult and is new in that it only applies when implementations have a significant level of functional integration, which has only been the case in more recent years. With this category, we must arrive at a price for the functionality provided to each subnetwork and provide information failures, especially as they impact common mode failures.

13.4.1   Pricing of Functions Provided by Implementations

The levels of functional integration in ATM and SDH are very high, and not all manufacturers package the same functional combination in one implementation package. Moreover, parts of the implementation can often be reassigned from one function to another. For example, an ATM switch fabric can either be a VCh matrix or a VP matrix, and, similarly, an SDH crossconnect fabric can normally be flexibly assigned as matrices in any of the LVC layers. Such functional integration is advantageous to both the operator and the manufacturer, as it provides substantial derisking against precisely which subnetwork will require to be resourced and to what level. One implementation can share the uncertainty across several subnetworks.

Some implementations are designed for exclusive use within one subnetwork, although this is decreasing in broadband networks, and here it is appropriate to include the cost of the implementation directly into the design optimization for that subnetwork.

However, where implementation is designed for more than one subnetwork, and especially where this is configurable, then we need a modeled commercial interface to fix a price for the functionality delivered to each subnetwork that will cover the cost of the implementation.

Such an implementation will have a number of client subnetworks with which it operates a notional marketplace, and this provides all the necessary mechanisms for optimizing the functionality in each subnetwork and also the configuration of the implementations, most notably determining what functionality is high risk and needs to be on a short lead time (supported by plug in units or software configurable) and what functionality is lower risk and can be on a longer lead time.

In return, the clients are providing forecasts and risk levels associated with the forecasts as well as an indication of the costs incurred when the equipment cannot supply the required functionality. This allows the equipment "enterprise" to plan and schedule the implementation. This is illustrated in Figure 13.26.

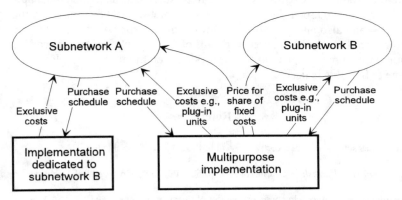

Figure 13.26   Implementation costs are a mixture of direct, exclusive costs to a subnetwork and a share of fixed costs, charged at a price, for multipurpose implementations.

13.4.2   Implementation Failure Topology

As equipment can go faulty, this information is critical to the performance aspects of subnetwork design. The major problem to be resolved is the way in which a single implementation failure can affect many functions and cause "common mode" failures within subnetworks. There are many examples:

- A cut in an optical fiber will affect all links in all layers that are supported on that fiber.
- A failure of a core component (e.g., a power supply unit) in functionally integrated equipment will affect many functions in the equipment, often in different subnetworks.
- The "bug" in the software of a control and management system may be replicated across many different equipments and many seemingly unrelated equipments can all fail in the same way.
- Other systematic implementation errors, either software or hardware, can create common mode failure and affect seemingly unrelated functions.

- Major building failures (e.g., a fire or a major power failure) can affect large numbers of functions, all implemented by equipment within the building.

Tracking all these common mode failure mechanisms can be very difficult, especially when more than one organization is involved. For example, a private network operator may deliberately buy private circuits from two public network operators expecting there will not be common mode failures between the two sets of private circuits. However, it is not infrequent that one public operator has bought a higher capacity private circuit from the other, so some of the links that the private network operator assumed to have no common mode failures are actually running on the same line system. The private network operator is then surprised and annoyed that two links it presumed were separated and had used to create resilience in the private network failed at the same time.

To a large extent, tracking common mode failure will remain difficult, especially when more than organization is involved. The failure topology mechanism provides a parameterized framework within which it can be solved, as described in Chapter 10. This allows the common mode failures to be accurately tracked across all the different subnetworks, and also encapsulates this information in such a way that the operator of a subnetwork is not divulging the precise details of the design of the subnetwork. It provides the minimum information necessary to track common mode failures. However, it may be some time before this is used extensively.

References

The following has several papers directly relevant to the subject of the chapter:

Internet Economics Workshop, Journal of Electronic Publishing, University of Michigan Press.

The following are relevant to delay and echo aspect of interworking with the PSTN:

ITU-T Recommendation G.114, *One-way transmission time.*
ITU-T Recommendation G.126, *Listener echo in telephone networks.*
ITU-T Recommendation G.131, *Control of talker echo.*

The following describe the operation of statistical optimization techniques:

Aarts, E. H. L., and Korst, J., *Simulated Annealing and Boltzman Machines*, John Wiley and Sons, 1989.
Goldberg, D. E., *Genetic Algorithms in Search, Optimization, and Machine Learning*, Addison-Wesley, 1989.

Appendix A
Mathematical Analysis of Phase Variation

Chapter 11 described the synchronization and timing features of SDH in largely qualitative terms. To be quantitative, it is necessary to use relatively advanced mathematical techniques and this appendix gives an introdution to their use. It covers the mathemetical modeling of phase variation and its measurement, a mathematical model of clock noise and its measurement, analysis of justification phase variation, and a stochastic model of pointer adjustment statistics from an AU pointer processor responding to clock noise.

A.1   MATHEMATICAL REPRESENTATION OF PHASE VARIATION

A.1.1   Representation of a Clock Signal

Phase variation is the difference in phase between a nominal reference clock signal and the clock signal under test. It is measured in radians, as illustrated in Figure A.1. A measurement of phase variation starts at a given point in time at which the phase variation is assumed to be zero. Thereafter, the phase variation is a function of time and is called $e(t)$. The clock signal is a periodic signal and is therefore a periodic function of time. Any periodic signal can be expressed as a Fourier series and so the clock signal can be expressed mathematically as:

$$clock\ signal = \sum_{n=0}^{\infty} A_n \cos(n\omega_{nom}t - \phi_n)$$

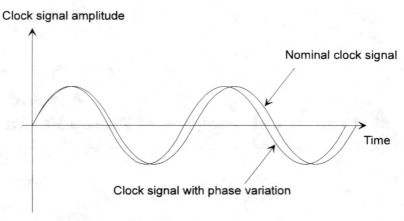

Figure A.1   Phase variation on a clock signal.

where ω_{nom} is the nominal clock radial frequency. The actual clock signal including phase variation is therefore:

$$clock\ signal = \sum_{n=0}^{\infty} A_n \cos(n\omega_{\text{nom}}t - \phi_n + e(t))$$

While the clock signal itself is of some interest, it is usually used to trigger clock pulses, often by a zero crossing on the clock signal (for example, when the clock signal crosses from a -ve value to a +ve value). This is determined solely by the phase of the clock signal. Therefore the clock signal can be expressed mathematically as a series of impulses, with an impulse every time the clock signal makes a positive going zero crossing:

$$clock\ signal = \sum_{m=-\infty}^{\infty} \delta(\omega_{\text{nom}}t - \phi + e(t) - m2\pi)$$

A.1.2   Bandwidth and Spectrum of a Clock Signal

The phase variation signal, $e(t)$ is an arbitrary function of time, and like other functions of time, can be described both in time and frequency domain. There are two practical limits on spectral bandwidth of $e(t)$, first the time taken for any measurement must be finite, that is $0 < t < T$, which will set a limit on the lowest frequency of $e(t)$ that can be measured. Second, the highest frequency of $e(t)$ that can be recorded is limited by the design of the phase variation measurement equipment. In many practical measurement equipment, the phase variation is sampled with a period τ_s. This sampling establishes the maximum frequency possible to measure by the Nyquist sampling theorem, that is, $f_{\text{max}} = 1/2\tau_s$. Sampling is modeled mathematically by multiplying $e(t)$ by a series of impulse functions.

$$e_s(t) = \sum_{n=-\infty}^{\infty} e(t)\, \delta(t - n\tau_s)$$

The phase variation $e(t)$ can also be expressed by its spectrum by taking the Fourier transform of $e(t)$ to give $E(\omega)$, where the lowest frequency of the Fourier transform is set by the measurement time T, and the maximum frequency is set by the sampling period τ_s. When $e(t)$ is a sampled signal, the Fourier transform becomes a *discrete Fourier transform* (DFT).

$$E(\omega) = \int_{-\infty}^{\infty} e(t)\, e^{-j\omega t}\, dt$$

$$E_s(\omega) = \int_{-\infty}^{\infty} \sum_{n=-\infty}^{\infty} e(t)\, e^{-j\omega t}\, \delta(t)\, dt$$

$$E_s(m\omega_s) = \sum_{n=0}^{N-1} E_s(n\tau_s)\, e^{-j2\pi nm/N}$$

where $N = T/\tau_s$ and is the total number of samples in the measurement. When N is an exact power of 2, then the computationally efficient *fast Fourier transform* (FFT) is used to obtain the spectrum of the phase variation.

More usefully, the *power spectral density* (PSD) can be calculated from this spectrum:

$$P(\omega) = E(\omega)\, E^*(\omega)$$

$$P_s(m\omega_s) = E_s(m\omega_s)\, E^*(m\omega_s)$$

where $E^*(\omega)$ is the complex conjugate of $E(\omega)$.

Remember that a spectrum analyzer is simply a real-time Fourier transformer and that modern test equipment actually use an FFT to implement the spectrum analyzer.

Therefore, there is no loss or gain of information in defining the spectrum of phase variation by the Fourier transform of $e(t)$.

A.1.3 Mathematical Representation of a Reference Phase Variation Signal

In many of the examples detailed below, it is useful to have a reference phase variation signal $e_r(t)$ to test the effect of the existing measurement parameters. There are five major mathematical forms by which phase variation can be represented, and most real variation signals can be modeled using a linear superposition of these five forms:

- $e_a(t)$ from a steady-state frequency offset in the clock radial frequency of $\omega_a = 2\pi f_a$;
- $e_b(t)$ from an independent set of sinusoidal components, each having a phase variation frequency;
- $e_c(t)$ from an independent set of frequency transients where there is frequency offset for a short period of time, each transient having a frequency offset and a time during which the frequency offset lasts;
- $e_d(t)$ from an independent set of phase transients where an amount of phase is instantaneously added or subtracted from the clock signal, normally as a result of justification of phase;
- $e_e(t)$ from a random component as a result of noise processes.

Mathematically,

$$e_r(t) = e_a(t) + e_b(t) + e_c(t) + e_d(t) + e_e(t)$$

$$e_a(t) = \omega_a t$$

$$e_b(t) = \sum_p B_p \cos(\omega_{bp} t - \phi_p)$$

$$e_c(t) = \sum_q \omega_{cq} \left[(t - t_{0q}) \, u(t - t_{0q}) - (t - t_{1q}) \, u(t - t_{1q}) \right]$$

$$e_d(t) = \sum_r A_{dr} \, u(t - t_r)$$

$$e_e(t) = X(t)$$

where $u(t)$ is the Heaviside unit step function and $X(t)$ is a random process. Most practical cases of phase variation are described by a combination of these components.

In practice $e_a(t)$ models a longer-term frequency offset, $e_b(t)$ models diurnal wander and is also used to model some form of desynchronized justification jitter, $e_c(t)$ models the effects of pointer adjustments and the effects caused by some digital slave clocks, $e_d(t)$ models most forms of justification phase variation including pointer adjustments, and $e_e(t)$ models clock noise from all types of clocks, including regenerators.

A.2   MEASUREMENT OF JITTER

Jitter is now regarded as the spectral components of phase variation above a filter cutoff frequency. The standard jitter measuring set is described in ITU-T recommendation O.171 and can be modeled by a phase-locked loop as shown in Figure A.2, where the measured jitter is proportional to the phase error signal in the phase lock loop. The local oscillator in the *voltage controlled oscillator* (VCO) part of the phase locked loop can either derive its timing from an external timing reference or from the clock signal being tested.

General Configuration of Jitter Measurement Equipment

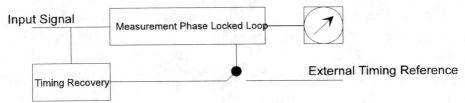

Major Components of Measurement Phase Locked Loop

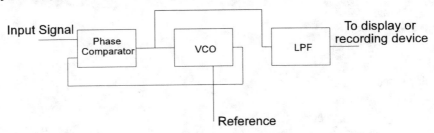

Laplace Transform Representation of Jitter Measuring Equipment

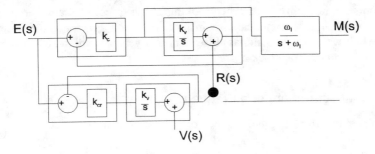

Figure A.2   Mathematical modeling of a jitter measuring set.

This test setup is best analyzed using the Laplace transform technique. The overall transfer characteristic function of the jitter measuring set is the combination of a high-pass and low-pass filter, which gives an overall bandpass filter

$$M(s) = \frac{k_c \omega_l s}{(s + k_c k_v)(s + \omega_l)} [E(s) - R(s)]$$

where $M(s)$ is the transform of signal recorded by the jitter measuring set, and $R(s)$ is the transform of the reference signal from the VCO. If the VCO derives its timing from an external reference, then $R(s)$ is used directly and

$$R(s) = \frac{\omega_r}{s^2}$$

however, if the VCO is timed from the incoming signal,

$$R(s) = \frac{k_{cr} k_{vr} E(s) - V(s)}{s + k_{cr} k_{vr}}$$

where $V(s)$ is the transform of the local oscillator free-run phase signal and will be of the form

$$V(s) = \frac{\omega_v}{s^2}$$

and in this case,

$$M(s) = \frac{k_c \omega_l s^2}{(s + k_{cr} k_{vr})(s + k_c k_r)(s + \omega_l)} [E(s) - V(s)]$$

which is the transfer function of a double pole high-pass filter and a single pole low-pass filter.

Random noise is not considered when looking at the effect of the different phase variation components. In most practical circumstances it is not a factor as the frequency components are below that of the bandpass filter. The analysis of random noise is covered under the section on measuring clock noise.

A.2.1   Jitter Measurement of a Frequency Offset

Taking the Laplace transform of the frequency offset component

$$E_a(s) = \frac{\omega_a}{s^2}$$

In the case when the measuring equipment is timed from an external reference

$$M_a(s) = \frac{k_c \omega_l (\omega_a - \omega_r)}{s(s + k_c k_v)(s + \omega_l)}$$

Taking the inverse transform

$$m_a(t) = k_c \omega_l (\omega_a - \omega_r) \left[\frac{1}{k_c k_v \omega_l} - \frac{e^{-k_c k_v t}}{k_c k_v (\omega_l - k_c k_v)} + \frac{e^{-\omega_l t}}{\omega_l (\omega_l - k_c k_v)} \right]$$

which is a constant term plus two transients. The constant term is often called the static phase offset and can be troublesome to jitter measurements; however, it usefully reflects a real phenomenon.

In the case when the measuring equipment is timed from the incoming signal

$$M_a(s) = \frac{k_c \omega_l (\omega_a - \omega_v)}{(s + k_{cr} k_{vr})(s + k_c k_v)(s + \omega_l)}$$

and taking the inverse transform

$$m_a(t) = k_c \omega_l (\omega_a - \omega_v) \left[\frac{e^{-k_{cr} k_{vr} t}}{(k_c k_v - k_{cr} k_{vr})(\omega_l - k_{cr} k_{vr})} + \frac{e^{-k_c k_v t}}{(k_{cr} k_{vr} - k_c k_v)(\omega_l - k_c k_v)} + \frac{e^{-\omega_l t}}{(k_{cr} k_{vr} - \omega_l)(k_c k_v - \omega_l)} \right]$$

which are all transient terms, and there is no static phase offset when jitter is measured this way.

A.2.2   Jitter Measurement of Sinusoidal Jitter Terms

The effect of the jitter measuring set on sinusoidal phase variation components depends on their frequency. If the frequency of the sinusoidal component is inside the frequency bounds of the bandpass filter, it will be unaffected and the measurement will be a true reflection of the original phase variation component. If the frequency of the sinusoidal component is outside the passband, then it will be attenuated and will not be properly measured. Historically, jitter measurement is designed for this type of phase variation component, and the bandpass filter has been set accordingly to ensure all major components are included in the measurement.

A.2.3   Jitter Measurement of Frequency Transient Terms

The Laplace transform of a single frequency transient is

$$E_{cq}(s) = \frac{\omega_{cq}\left(e^{-st_0} - e^{-st_1}\right)}{s^2}$$

In the case when the measuring equipment is timed from an external reference

$$M_{cq}(s) = \frac{k_c\omega_l\left(\omega_{cq} - \omega_r\right)\left(e^{-st_0} - e^{-st_1}\right)}{s(s + k_ck_v)(s + \omega_l)}$$

Taking the inverse transform

$$m_{cq}(t) = k_c\,\omega_l\,(\omega_c - \omega_r)\left[\begin{array}{l} \dfrac{u(t-t_0)}{k_ck_v\,\omega_l} - \dfrac{e^{-k_ck_v\,(t-t_0)}\,u(t-t_0)}{k_ck_v\,(\omega_l - k_ck_v)} + \dfrac{e^{-\omega_l\,(t-t_0)}\,u(t-t_0)}{\omega_l\,(\omega_l - k_ck_v)} \\[2ex] -\dfrac{u(t-t_1)}{k_ck_v\,\omega_l} + \dfrac{e^{-k_ck_v\,(t-t_1)}\,u(t-t_1)}{k_ck_v\,(\omega_l - k_ck_v)} - \dfrac{e^{-\omega_l\,(t-t_1)}\,u(t-t_1)}{\omega_l\,(\omega_l - k_ck_v)} \end{array}\right]$$

The measured effect of the frequency transient depends on its duration in relation to the time constants k_ck_v and ω_l, illustrated in Figure A.3.

In the case when the measuring equipment is timed from the incoming signal

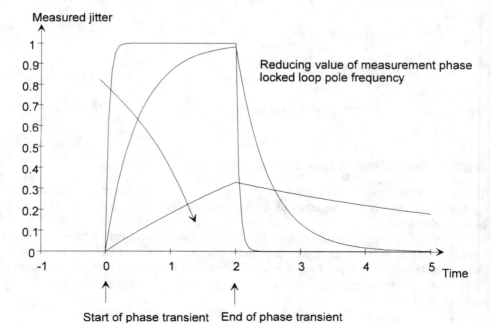

Figure A.3   Measured jitter from a frequency transient when the jitter measuring set is externally timed.

$$M_{cq}(s) = \frac{k_c\,\omega_l\,(\omega_c - \omega_v)(e^{-st_0} - e^{-st_1})}{(s + k_{cr}k_{vr})(s + k_ck_v)(s + \omega_l)}$$

and taking the inverse transform

$$m_{cq}(t) = k_c\,\omega_l\,(\omega_c - \omega_v)\left[\begin{array}{c} \dfrac{e^{-k_{cr}k_{vr}\,(t-t_0)}\,u(t-t_0)}{(k_ck_v - k_{cr}k_{vr})(\omega_l - k_{cr}k_{vr})} \\[2mm] + \dfrac{e^{-k_ck_v\,(t-t_0)}\,u(t-t_0)}{(k_{cr}k_{vr} - k_ck_v)(\omega_l - k_ck_v)} \\[2mm] + \dfrac{e^{-\omega_l\,(t-t_0)}\,u(t-t_0)}{(k_{cr}k_{vr} - \omega_l)(k_ck_v - \omega_l)} \\[2mm] \dfrac{e^{-k_{cr}k_{vr}\,(t-t_1)}\,u(t-t_1)}{(k_ck_v - k_{cr}k_{vr})(\omega_l - k_{cr}k_{vr})} \\[2mm] - \dfrac{e^{-k_ck_v\,(t-t_1)}\,u(t-t_1)}{(k_{cr}k_{vr} - k_ck_v)(\omega_l - k_ck_v)} \\[2mm] - \dfrac{e^{-\omega_l\,(t-t_1)}\,u(t-t_1)}{(k_{cr}k_{vr} - \omega_l)(k_ck_v - \omega_l)} \end{array}\right]$$

The measured effect again depends on the duration of the transient in relation to the time-constants; however, this time the effect is even more marked as all the terms have exponential decays. For example, if the duration is short enough, the measured effect may be negligible. This is illustrated in Figure A.4.

Jitter is usually measured in *unit intervals* (UIs), and this is simply related to phase variation by converting the radian measure so that it is based on cycles of the nominal clock signal

$$x \text{ radians } = \frac{x}{2\pi} \text{ unit intervals}$$

A.2.4   Jitter Measurement of Phase Transient Terms

Phase transients are normally caused by a phase justification process, and only truly exist as phase transients when the clock signal has been quantized and is being carried by a server layer. While the phase transient remains encoded as part of the quantized signal, it is unlikely to be of much direct interest; however, at some point the phase transient will be desynchronized. At this point, the phase transient is buffered in some form of control loop and will either give rise to slip or some form of frequency transient. Very often the frequency transient is not a linear rise in phase as indicated in the previous section, but an exponential rise, depending on the design of the desynchronizer. If the desynchronizer is equivalent to a single pole low pass, a single phase transient will appear as

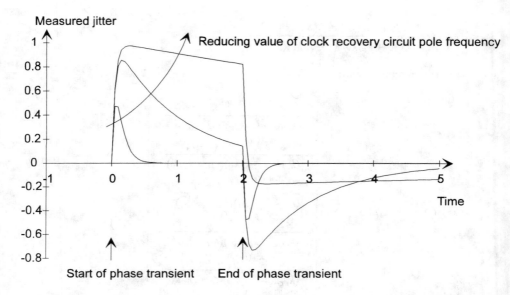

Figure A.4   Measured jitter from a frequency transient when the jitter measuring set is recovering its timing from the input signal.

$$E_{dr}(s) = \frac{A_{dr}\,\omega_d}{s(s+\omega_d)}$$

where ω_d is the pole frequency of the desynchronizer, and assuming the transient is at $t = 0$. The output of the jitter measuring set with an external frequency reference will therefore be:

$$M_{dr}(s) = \frac{k_c\,\omega_l\,(A_{dr}s - \omega_r)}{s(s+\omega_d)(s+k_ck_v)(s+\omega_l)}$$

Taking the inverse transform:

$$m_{dr}(t) = k_c\omega_l \left[\begin{array}{c} -\dfrac{\omega_r}{\omega_d k_c k_v \omega_l} \\[2ex] +\dfrac{(A_{dr}\omega_d + \omega_r)\,e^{-\omega_d t}}{\omega_d(k_ck_v - \omega_d)(\omega_l - \omega_d)} \\[2ex] +\dfrac{(A_{dr}k_ck_v + \omega_r)\,e^{-k_ck_v t}}{k_ck_v(\omega_d - k_ck_v)(\omega_l - k_ck_v)} \\[2ex] +\dfrac{(A_{dr}\omega_l + \omega_r)\,e^{-\omega_l t}}{\omega_l(\omega_d - \omega_l)(k_ck_v - \omega_l)} \end{array} \right]$$

When the jitter measuring set is recovering its timing from this input signal then the output will be:

$$M_{dr}(s) = \frac{k_c \, \omega l \, (A_{dr}s - \omega_v)}{(s + k_{cr}k_{vr})(s + \omega_d)(s + k_c k_r)(s + \omega l)}$$

and taking the inverse transform:

$$m_{dr}(t) = k_c \omega l \left[\begin{array}{c} -\dfrac{(A_{dr}k_{cr}k_{vr} + \omega_v) \, e^{-k_{cr}k_{vr}t}}{(\omega_d - k_{cr}k_{vr})(k_c k_v - k_{cr}k_{vr})(\omega l - k_{cr}k_{vr})} \\[2mm] -\dfrac{(A_{dr}\omega_d + \omega_r) \, e^{-\omega_d t}}{(k_{cr}k_{vr} - \omega_d)(k_c k_v - \omega_d)(\omega l - \omega_d)} \\[2mm] -\dfrac{(A_{dr}k_c k_v + \omega_r) \, e^{-k_c k_v t}}{(k_{cr}k_{vr} - k_c k_v)(\omega_d - k_c k_v)(\omega l - k_c k_v)} \\[2mm] -\dfrac{(A_{dr}\omega l + \omega_r) \, e^{-\omega l t}}{(k_{cr}k_{vr} - \omega l)(\omega_d - \omega l)(k_c k_v - \omega l)} \end{array}\right]$$

A.2.5   Jitter Measurement of Clock Noise

The frequency spectrum of interest for clock noise is below that of jitter measuring sets, and so its measurement using a jitter measuring set makes little sense. Measurement of clock noise is its own subject and is discussed later in this appendix. However, trying to measure jitter resulting from other phenomenon when there is a significant amount of clock noise on either the measurement signal or any external reference signal may well affect the results.

A.3   MEASUREMENT OF WANDER

A wander measurement is normally trying to characterize the diurnal effect of the heating up and cooling down of cables and equipment over 24 hours. The cycle period of the wander is therefore 24 hours, which corresponds to a frequency of 12 µHz. It is normally impractical to design a bandpass filter for 12 µHz so a low-pass filter is used. The low-pass filter will eliminate any higher frequency jitter and any random noise above the cutoff frequency of the low-pass filter. However, it will include any zero frequency component (i.e., a clock signal frequency offset). If the level of the random noise after the filter is significantly less than the level of the diurnal wander, then an accurate measure of the wander can be made despite a frequency offset by taking the average of the difference between maxima and minima, as illustrated in Figure A.5.

Wander is normally measured as a *time inteval error* (TIE) where the phase variation is converted into a time difference usually in microseconds. This is done by dividing the phase variation by the nominal clock radial frequency:

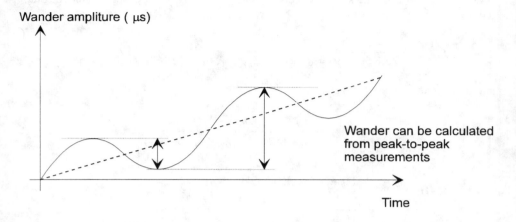

Figure A.5   Measurement of wander in the presence of a frequency offset.

$$TIE = \frac{e(t)}{\omega_{nom}}$$

A.4   CLOCK NOISE AND MEASUREMENT OF CLOCK NOISE

Clock noise is the major source of random, rather than deterministic, phase variation. The spectrum of clock noise potentially extends from zero frequency up to several megahertz. In practice, however, it is the lower frequency components that are of most interest.

A.4.1   Model of Clock Noise

Within broadband networks, slave clocks are the most numerous and the most important. It is useful to form a simple mathematical model of the phase variation on the output of the slave clock relative to the output of another slave clock. It should be assumed that the two slave clocks can have different master clocks. However, in the timescales of interest for clock noise, accumulated frequency differences between master clocks is normally small.

The slave clock illustrated in Figure A.6 generates random noise as well as low pass filtering any random noise on its synchronizing input. In general, the internal noise processes inside the slave clock produce a combination of "white" noise, "flicker" noise, and "random walk" noise. White noise is generated by thermal effects in resistive components and has a flat PSD (normally considered only up to some maximum practical frequency). Flicker noise is normally generated by chaotic quantum effects in semiconductor components and has a PSD proportional to $\omega\Delta^{-1}$. Random walk noise is caused when white noise is continuously integrated through some storage mechanism and has a PSD proportional to ω^{-2}. There are likely to be other noise-generating processes, for example, within the quartz crystal of the VCO but the characteristics are similar to one or more of the above noise types. The internally generated noise will therefore be a combination of white, flicker, and random walk noise that can be injected at all points in the slave clock. The following equation for PSDs is derived from the slave clock shown in Figure A.6:

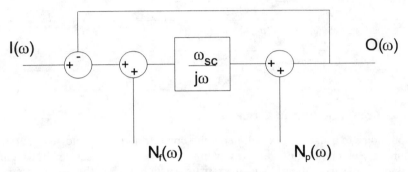

Figure A.6   Model of a slave clock passing on and producing clock noise.

$$O(\omega)O^*(\omega) = \frac{I(\omega)I^*(\omega)\omega_{sc}^2}{\omega^2 + \omega_{sc}^2} + \frac{N_f(\omega)N_f^*(\omega)\omega_{sc}^2}{\omega^2 + \omega_{sc}^2} + \frac{N_p(\omega)N_p^*(\omega)\omega^2}{\omega^2 + \omega_{sc}^2}$$

Noise injected before the VCO is therefore effectively subjected to a low pass filter while those after the VCO are subjected to a high-pass filter. When the time constant of the slave clock is long, which is normally the case, the timescale between the pole of the slave clock (maybe thousands of seconds) and where clock noise ceases to be of practical importance (normally less than a second) is very important and is normally referred to as the range of "short-term stability" of the slave clock. At these timescales, noise components injected after the VCO pass straight out unattenuated, while the noise components injected before the VCO are attenuated; however, they are also integrated thus changing their characteristics. The slope of the PSD is changed as it is multiplied by ω^{-2}. These latter noise components are normally called "frequency" noise components, while those injected after the VCO are normally called "phase" noise components. The noise can be modeled as a single noise source injected after the VCO as follows:

$$N(\omega)N^*(\omega) = N_f(\omega)N_f^*(\omega)\frac{\omega_{sc}^2}{\omega^2} + N_p(\omega)N_p^*(\omega)$$

The characteristics of the slave clock noise in the important region of short-term stability are summarized in Table A.1.

Table A.1

Types of Clock Noise

Type of Noise	*Slope of PSD*
White phase noise	ω^0
Flicker phase noise	ω^{-1}
Random walk phase noise or white frequency noise	ω^{-2}
Flicker frequency noise	ω^{-3}
Random walk frequency noise	ω^{-4}

A.4.2   Accumulation of Clock Noise

The input noise to a slave clock will generally come from internal noise of another slave clock, through which the slave receives its timing reference as described in Chapter 11. If all the slave clocks have the same characteristics, which is often the case in practice, then the output PSD can be expressed as:

$$O(\omega)O^*(\omega) = N(\omega)N^*(\omega)\left\{1 - [H(\omega)(H^*(\omega))]^n\right\}$$

where n is the number of slave clocks and $H(\omega)$ is the closed loop transfer function of the slave clock. The closed loop transfer function must always be a low-pass filter. When ω is below the filter cutoff frequency, $H(\omega)$ is close to unity and

$$O(\omega)O^*(\omega) = n\,N(\omega)N^*(\omega)\,[1 - H(\omega)H^*(\omega)]$$

When ω is below the filter cutoff frequency, $H(\omega)$ is less than unity and

$$O(\omega)O^*(\omega) = N(\omega)N^*(\omega)$$

This analysis applies to regenerators in a line system as well as to main slave clocks in a national synchronization network.

   If we assume that the noise on the synchronizing input of any one slave clock has the same basic characteristics as the internally generated noise, then for frequencies below ω_{sc}, all the noise components simply accumulate according to power law addition. For frequencies above ω_{sc}, the noise on the input is filtered; however, it is matched by the added internal noise.

A.4.3   Methods of Measuring Clock Noise

There are two basic tests for clock noise. The first measures the amount of phase variation generated by a clock, and therefore does not include any clock noise present on the synchronizing input. The second measures the absolute amount of phase variation on the output of the clock, and in the case of the slave clock this will include the effects of any noise on the synchronizing input. In the case of the master clock, each test will yield the same result. The first test can use the synchronizing input to the slave clock as a reference from which the phase variation can be calculated; however, the second test requires a separate reference clock to calculate the phase variation. Clearly, this reference clock could itself have significant phase variation, so it is not possible to measure "absolute" phase variation and only the phase variation between a clock and the best available reference clock. In the past, this has presented a real obstacle to clock noise measurement, as the best available clocks are based on the atomic resonance of a beam of cesium atoms. This type of clock is generally expensive and sensitive and so not well-suited to test gear. However, there is now a technology based on rubidium which, while slightly less accurate over the longer term (i.e., months), can be incorporated into relatively inexpensive test equipment. The test configuration for each test is shown in Figure A.7.

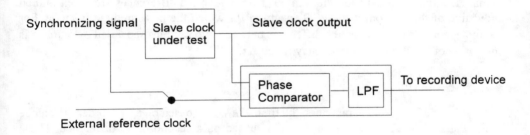

Figure A.7   Measurement configuration for clock noise.

The test equipment calculates the phase variation by measuring the difference in phase between the test signal and the reference. In practice, this difference will always be subject to a low-pass filter, even if this is simply the effect of sampling the phase variation signal as discussed above.

A.4.4   Measurement Parameters for Clock Noise

Over the years, many parameters have been proposed for clock noise measurement. The characteristics of noise are most often expressed as a PSD. This suggests that the most natural parameter would be the PSD measured by taking the square of the FFT of the sampled phase variation. However, this is not as simple as first appears. The phase variation of clock noise is a random signal, and therefore the calculated PSD is also random. The required characteristic is the underlying steady-state PSD and this must be estimated statistically.

Since it is always an underlying steadystate that is required, the problem of estimating the steadystate is common to all measurement parameters. This is achieved by averaging a number of successive measurements. However, to successfully and reliably carry out this averaging we must be sure that the averaging will converge on a stable result, the underlying steady state. There are two related difficulties with this. First, there must be some uncorrelated nature between the samples to be averaged. However, it is the very correlation in the noise signal that is the characteristic of the signal which is the subject of the measurement (the autocorrelation function of a noise signal is the Fourier transform pair of the PSD). The second difficulty is that clock noise often has no simple underlying steady state as the PSD rises to an infinity as the frequency approachs zero. In practical terms, a reliable measurement can only be made signal if it has a finite, well-behaved autocorrelation function. This, by definition, requires that the PSD is also finite and well-behaved at all frequencies, including the zero frequency.

Both of these problems can be overcome by differentiating the phase variation sufficiently to ensure that the PSD is finite as the frequency approaches zero. Each differentiation multiplies the PSD by ω^2, and this reduces any very long-term correlation of the noise and ensures an underlying steady state that can be estimated.

In practice, it is more useful to take differences between samples of the phase variation signal. The time gap between the samples from which the differences are taken is then indicative of the section of the PSD spectrum for which the measurement is made. For example, a measure based on the difference between two samples separated by a time τ is indicative of the PSD is the frequency range $1/\tau$.

A.4.4.1   TIE$_{rms}$ and A$_{dev}$

We start by defining two parameters; the first, TIE_{rms}, is based on a first-order difference while the second, Allan deviation (A_{dev}) is based on a second-order difference. These parameters are calculated as a "root mean square." Each difference, first or second order, is squared, then the squares are averaged over many samples, and finally a square root is taken of the average. Both are defined in absolute time and so must be scaled to phase variation ω_{nom}. For TIE_{rms}:

$$\omega_{nom}^2 TIE_{rms}^2(t,\tau) = E\{ [\, e(t) - e(t+\tau) \,]^2 \}$$

$$= E\{e^2(t)\} - 2\,E\{e(t)e(t+\tau)\} + E\{e^2(t+\tau)\}$$

$$= R(t,0) - 2R(t,\tau) + R(t+\tau,0)$$

$$\omega_{nom}^2 TIE_{rms}^2(\tau) = 2\,[\, R(0) - R(\tau) \,]$$

and for A_{dev}

$$\omega_{nom}^2 A_{dev}^2(t,\tau) = E\{ [\, e(t) - 2e(t+\tau) + e(t+2\tau) \,]^2 \}$$

$$
\begin{aligned}
= \; & \mathrm{E}\{e^2(t)\} - 4\,\mathrm{E}\{e^2(t+\tau)\} + \mathrm{E}\{e^2(t+2\tau)\} \\
& - 4\,\mathrm{E}\{e(t)e(t+\tau)\} - 4\,\mathrm{E}\{e(t+\tau)e(t+2\tau)\} + 2\,\mathrm{E}\{e(t)e(t+2\tau)\}
\end{aligned}
$$

$$
= R(t,0) - 4R(t+\tau,0) + R(t+2\tau,0) - 4R(t,\tau) - 4R(t+\tau,\tau) + 2R(t,2\tau)
$$

$$
\omega\,sunnom^2 Asundev^2(\tau) = 2\,[\,3R(0) - 4R(\tau) + R(2\tau)\,]
$$

$R(t,\tau)$ is the autocorrelation of the phase variation signal. If $R(t,\tau)$ is independent of t, which is the case if there is an underlying steady state, then the autocorrelation function is simply $R(\tau)$.

The characteristics of these two parameters can now be directly related to the PSD of the noise which allows the PSD of clock noise to be calculated using these parameters. If the autocorrelation function and the PSD are a Fourier transform pair, for $TIE^2{}_{\mathrm{rms}}$:

$$
\omega^2_{\mathrm{nom}} TIE^2_{\mathrm{rms}}(\tau) = 2 \int_{-\infty}^{\infty} O(\omega)O^*(\omega)\,[1 - e^{j\omega\tau}]\,d\omega
$$

$$
= 8 \int_{0}^{\infty} O(\omega)O^*(\omega)\,\sin^2(\tfrac{1}{2}\omega\tau)\,d\omega
$$

as $O(\omega)O^*(\omega)$ is an even function. Similarly, for A_{dev}

$$
\omega^2_{\mathrm{nom}} A^2_{\mathrm{dev}}(\tau) = 2 \int_{-\infty}^{\infty} O(\omega)O^*(\omega)\,[3 - 4e^{j\omega\tau} + e^{j2\omega\tau}]\,d\omega
$$

$$
= 32 \int_{0}^{\infty} O(\omega)O^*(\omega)\,\sin^4(\tfrac{1}{2}\omega\tau)\,d\omega
$$

When $O(\omega)O^*(\omega)$ is the form $\omega^{-\alpha}$, then

$$
\omega^2_{\mathrm{nom}} TIE^2_{\mathrm{rms}}(\tau) = \frac{8}{2^{\alpha-1}}\,\tau^{\alpha-1}\,f_{\mathrm{TIE}}(\alpha)
$$

$$
\omega^2_{\mathrm{nom}} A^2_{\mathrm{dev}}(\tau) = \frac{32}{2^{\alpha-1}}\,\tau^{\alpha-1}\,f_{A\mathrm{dev}}(\alpha)
$$

where $f_{\mathrm{TIE}}(\alpha)$ is finite if $0 < \alpha <= 2$ and $f_{A\mathrm{dev}}(\alpha)$ is finite if $0 < \alpha <= 4$. This indicates that each type of noise will give a different slope on a chart of TIE_{rms} and A_{dev} as long as it is within these limits. TIE_{rms} does not converge for frequency flicker noise and random walk frequency noise; however, A_{dev} does converge for these types of noise. Neither appear stable for white phase noise.

White phase noise (i.e., the case where $\alpha = 0$) is finite in most practical circumstances because the bandwidth of the noise is finite. It is limited either by low-pass filters, or more importantly, by the fact that most noise signals are sampled. This gives an automatic maximum frequency of the Nyquist frequency. However, the answer will be independent of τ and so will have the same slope as that of flicker phase noise, and therefore these parameters will not distinguish between these two.

A.4.4.2 T_{dev}

It is clearly desirable to have a parameter with a different slope for each type of noise. The parameter time deviation, T_{dev}, is an extension of the Allan deviation, which has now been defined as the primary parameter for measuring clock noise. This parameter gets around the difficulty of white phase noise by including sampling in the definition of the parameter, and also includes a small amount of averaging which has the affect of including a ω^{-2} term, the opposite of differentiating. This gives a stable convergence for white phase noise on a slope of $\tau^{-1/2}$. The definition of T_{dev} actually uses a summation and is

$$\omega_{nom}^2 T_{dev}^2(n\tau_0) = \frac{1}{6n^2} E\left\{ \left[\sum_{i=1}^{n} (e(t) - 2e(t + n\tau_0 + i\tau_0) + e(t + 2n\tau_0 + i\tau_0)) \right]^2 \right\}$$

It can be shown that the T_{dev} can be related to the PSD of the clock noise as follows:

$$= \omega_{nom}^2 T_{dev}^2(n\tau_0) = \frac{64}{6n^2} \int_0^{\infty} O(\omega)O^*(\omega) \frac{\sin^6(\frac{1}{2}\omega n\tau_0)}{\sin^2(\frac{1}{2}\omega \tau_0)} d\omega$$

The characteristics of all three of these parameters are summarized in Table A.2.

Table A.2
Characteristics of Different Clock Noise Measurement Parameters

Type of clock noise	PSD	TIE_{rms}	A_{dev}	T_{dev}
White phase noise	ω^0	τ^0*	τ^0*	$(n\tau_0)^{-1/2}$
Flicker phase noise	ω^{-1}	τ^0	τ^0	$(n\tau_0)^0$
White frequency noise	ω^{-2}	$\tau^{1/2}$	$\tau^{1/2}$	$(n\tau_0)^{1/2}$
Flicker frequency noise	ω^{-3}	—	τ^1	$(n\tau_0)^1$
Random walk frequency noise	ω^{-4}	—	$\tau^{3/2}$	$(n\tau_0)^{3/2}$

* Convergence only if noise bandwidth is limited.

A.5   JUSTIFICATION PHASE VARIATION

The characteristics of justification phase variation are highly complex. An initial view of its characteristics can be gained by carrying out a full simulation of all justification values. As described in Chapter 11, justification phase variation exists in the coding of a client layer signal in a server layer signal, and does not appear as a clock signal until it is desynchronized. This means the characteristics of the desynchronizer will affect the way in which justification phase variation appears on a clock signal.

When the justification phase variation does appear on a clock signal, it has a broad spectrum and so will have some jitter components and some wander components. In this section, we look at the analysis of the jitter aspects of justification phase variation; however, by varying the high-pass filter pole of the jitter measuring set, it is possible to create examples from the full spectrum of justification phase variation. The configuration used in the analysis is shown in Figure A.8.

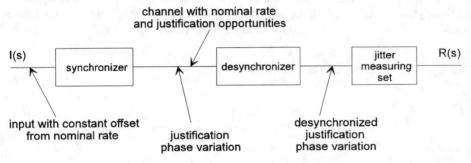

Figure A.8   Configuration for the analysis of justification phase variation.

The input to the synchronizer is assumed to be a clock signal with no phase variation other than a frequency offset. The frequency offset results in some justification opportunities being used and others not as the synchronizer buffer fills. The long-term proportion of justification opportunities used will be equal to the frequency offset. This will apply to positive, negative, and +0- justification schemes as well as quantization caused by ATM cell streams. The offset frequency can therefore be expressed as a proportion of the total available justification rate.

Figure 11.10 was calculated by directly simulating the effects of each value of justification, from zero to one, and so gives a full picture of its effects. We see from this the complex nature of justification jitter and how it does not yield easily to analytical solutions. However, what is hard to analyze, is seen directly from the simulation. The justification pattern is a complex but deterministic pattern of infinite depth, often called a "fractal."

This fractal is based on fractions; in particular, the integer value of the denominator of the fraction. For example, when the justification rate is around 1/3 or 2/3, there is a peak of justification jitter of 1/3 of the size of the phase quanta (note that in Figure 11.10, this is reduced by a factor of 4 by the desynchronizer and jitter measuring filters). Seeing this, the reason can be explained. When the justification rate is close to 1/3, every 1/3 justification opportunity is used for most of the time; however, every now and again a "reset" will occur in this pattern. If the justification rate is just above 1/3, sometimes the second justification opportunity is used rather than the third. This has the effect of creating a phase transient of 1/3 of a phase quanta. The closer the justification rate is to the rational fraction, the longer the time between the resets; indeed, the rate of resets can be calculated from a knowledge of the justification rate.

The same argument can be applied to any rational fraction justification rate. As well, the peaks in the fractal pattern have a width that will also, by definition, determine the total number of peaks in the pattern. In addition, it controls the size of the average peak.

From the understanding that these "resets" cause regular phase transients, it is possible to make a useful analysis of this pattern that gives the width and number of peaks and the probability of encountering a peak of a given size with an arbitrary justification rate.

A full analysis of phase transients was described above; however, in order to arrive at a final analytical answer, we take a simplified model of the jitter measuring — a single pole high-pass filter with pole ω_m. In a similar way, the desynchronizer is modeled by a single pole low-pass filter ω_d. If the height of each phase transient is A, and the gap between transients is T, then for the nth transient occurring at time $t = nT$,

$$R_n(s) = \frac{Ae^{j\omega nT}}{s} \frac{\omega_d}{s + \omega_d} \frac{s}{s + \omega_m}$$

$$r_n(t) = \frac{A\omega_d}{\omega_m - \omega_d} \left[e^{-\omega_d(t-nt)} - e^{-\omega_m(t-nt)} \right] u(t-nT)$$

A steady-state response for the time interval $nT <= t < (n+1)T$ can be found by adding the responses for all the previous transients (which form a geometric series),

$$r(t) = \sum_{i=n}^{-\infty} r_i(t)$$

$$r(t) = \frac{A\omega_d}{\omega_m - \omega_d} \left[\frac{e^{-\omega_d(t-nT)}}{1 - e^{-\omega_d T}} - \frac{e^{-\omega_m(t-nT)}}{1 - e^{-\omega_m T}} \right] u(t-nT)$$

This is the absolute response in the jitter measuring set, which is a combination of a static phase offset corresponding to the marginal frequency encoded in the phase transients, and an oscillating term with periodicity T. The static phase is the average over the interval $nT <= t < (n+1)T$,

$$r_{\text{static phase offset}} = \frac{1}{T} \int_{nT}^{(n+1)T} r(t)dt$$

$$r_{\text{static phase offset}} = \frac{A}{\omega_m T}$$

The oscillating term is the one of interest and we can calculate the peak to peak of this term. The minimum will occur at $t = nT$. As the shape of the oscillation is the sum of only two exponentials, there can be only one maximum when $r'(t) = 0$. From this,

$$r_{pk\text{-}pk} = \frac{A\omega_d}{\omega_m - \omega_d}\left[\frac{\left(\dfrac{\omega_d(1 - e^{-\omega_m T})}{\omega_d(1 - e^{-\omega_m T})}\right)^{\frac{\omega_d}{\omega_m - \omega_d}} - 1}{1 - e^{-\omega_d T}} - \frac{\left(\dfrac{\omega_d(1 - e^{-\omega_m T})}{\omega_d(1 - e^{-\omega_m T})}\right)^{\frac{\omega_m}{\omega_m - \omega_d}} - 1}{1 - e^{-\omega_m T}}\right]$$

The value of $r_{pk\text{-}pk}$ depends on the relative values $1/T$, ω_d, and ω_m. The above equation can be reduced to the following approximations when the conditions are met,

$$r_{pk\text{-}pk} \approx \frac{\omega_d T}{8} \quad \text{when } \omega_d T < \frac{\omega_d}{\omega_m}$$

$$r_{pk\text{-}pk} \approx \frac{\omega_d}{\omega_m} \quad \text{when } \omega_d T > \frac{\omega_d}{\omega_m} \text{ and } \omega_d < \omega_m$$

$$r_{pk\text{-}pk} \approx 1 \quad \text{when } \omega_d T > \frac{\omega_d}{\omega_m} \text{ and } \omega_d > \omega_m$$

The second and third cases correspond to the top of the peak, while the first is the slope of the peak. As T increases beyond the time constant of either of the filters, the transients are no longer averaged in the filters so each transient appears as a single isolated transient, and gives a maximum peak-to-peak reaction in the jitter measuring set. The width of the peak can therefore be determined by the point at which the top of the peak meets the side of the peak. If the width is expressed as a proportion of the maximum justification rate, then,

$$width = \frac{\omega_m T_{\text{jus}}}{8} \quad \text{when } \omega_d < \omega_m$$

$$width \approx f_m T_{\text{jus}} \text{ when } \omega_d < \omega_m$$

$$width = \frac{\omega_d T_{\text{jus}}}{8} \text{ when } \omega_d > \omega_m$$

$$width \approx f_d T_{\text{jus}} \text{ when } \omega_d > \omega_m$$

where T_{jus} is the time between justification opportunities. (Note in the case of ATM, T_{jus} is the inverse of the cell rate). A simple approximation to the number of peaks is simply the inverse of the width of the peaks, that is,

$$number \ of \ peaks \approx \frac{1}{width}$$

The smallest peaks must have denominators of the same order as the number of the peaks, and so relative to the highest peaks when the justification rate is nearly zero or nearly 1,

$$smallest \ relative \ peak \ height \approx width$$

and finally, the probability with an arbitrary justification rate of encountering a peak of a given height is determined by the number of peaks of roughly that height. Since the relative height is the inverse of the denominator, the number of peaks of a given height will be roughly equal to the inverse of the relative height, that is,

$$p(relative \ height) \approx \frac{width}{relative \ height}$$

This set of equations can be used for giving a good empirical picture of justification phase variation. Figure 11.10 showed the case of justification jitter with $T = 221$ μs, $\omega_d/2\pi =$ 10 Hz, and $\omega_m/2\pi = 20$ Hz. Figure A.9 and Figure A.10 illustrate the veracity of the equations by taking two different examples of justification phase variation from that shown in Figure 11.10. In Figure A.9, the value of the desynchronizer pole is reduced by a factor of ten which results in the height of the peaks being reduced by a factor of ten while the width and number of peaks has remained the same. In Figure A.10, the maximum justification rate increases by a factor of 78, but the filters are the same as those of Figure 11.10. In this case, the height of the peaks is the same; however the width decreases by a factor of 78, the number of peaks increases by a factor of 78, and the smallest peaks are 78 times smaller. (Note these examples correspond to the values of T corresponding to ATM cells in 2048 kbit/s and ATM cells in VC-4).

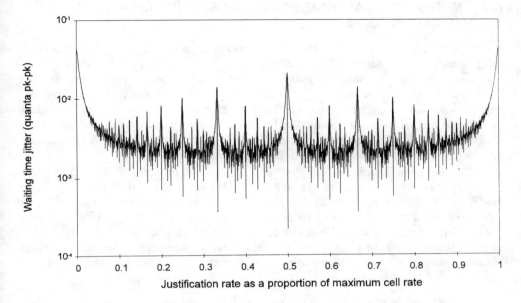

Figure A.9   Justification jitter with $T = 221$ μs, $\omega_d/2\pi = 1$ Hz, and $\omega_m/2\pi = 20$ Hz.

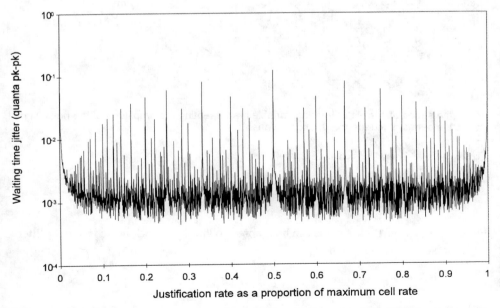

Figure A.10   Justification jitter with $T = 2.83$ μs, $\omega_d/2\pi = 10$ Hz, and $\omega_m/2\pi = 20$ Hz.

A.6   MODEL OF AN AU POINTER PROCESSOR SUBJECT TO CLOCK NOISE

It is possible to use the model of slave clock noise to estimate the pointer statistics of an AU pointer processor subject to slave clock noise.

The autocorrelation function gives the covariance of the random noise component of the phase variation signal with itself, but does not give the probability density function. The random noise comes from the combination of a very large number of individual noise-generating elements (generally, each electron in the conductor or semiconductor is the basic element!), and the sum of a large number of any probability density functions always tends to the Gaussian (or normal) distribution.

The random noise signal is sampled by the AU pointer processor every 125 μs (τ_{pp}), and every sample can be regarded as a random variable with a gaussian distribution. These random variables are all correlated with each other according to the autocorrelation function thus forming a correlated set of gaussian random variables, X_n. A second set of independent random variables, x_n, can be derived from this first set such that each member of the first set is equal to a linear sum of all the previous members of the second set. All members of the set have an identical probability density function. This is illustrated in Figure A.11.

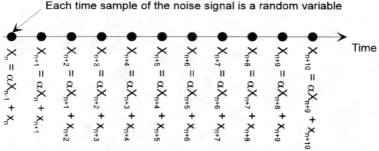

Each time sample of the noise signal is a random variable

X_i is the series of correlated random variables representing the amplitude of the noise signal.
x_i is the series of independent random variables.

Figure A.11   A random signal.

$$X_n = a_0 x_n + a_1 x_{n-1} + a_2 x_{n-2} + a_3 x_{n-3} + \ldots\ldots$$

For example, if the flicker noise can be ignored so the slave clock noise is simply filtered white noise, then $a_i = \alpha^i$ where

$$\alpha = e^{-\omega_{sc}\tau_{pp}}$$

and the variance of x_i

$$\sigma_x^2 = \frac{W\omega_{sc}}{4\omega_0^2}[1 - e^{-2\omega_{sc}\tau_{pp}}]$$

$$\approx \frac{N\omega_{sc}^2\tau_{pp}}{2\omega_0^2}$$

As the mean of the noise is zero, the gaussian probability density function is

$$P[x_i = x] = \frac{1}{(2\pi\sigma_x)^{1/2}} e^{-x^2/2\sigma_x^2} dx$$

To form a stochastic model, a difference equation is needed such that X_n is expressed as the linear sum a finite number of previous X_i and x_n, that is,

$$X_n = x_n + \sum_{p=1}^{k} b_p X_{n-p}$$

This assumes the slave clock noise is white noise, filtered by a linear digital filter, where b_p are the coefficients of the digital filter. In the case where the flicker noise can be ignored and the slave clock noise is simply white noise filtered by a single pole low pass filter:

$$X_n = x_n + \alpha X_{n-1}$$

This can now be used to form a Markov chain by using k state variables

$$X_{n_1} = x_n + \sum_{p=1}^{k} b_p X_{n-1_p}$$

$$X_{n_2} = X_{n-1_1}$$

$$X_{n_3} = X_{n-1_2}$$

.

.

.

$$X_{n_k} = X_{n-1_{k-1}}$$

The state transition matrix can now be formed as the probability of moving from any point in the k dimensional space of X_{n-1} to any point in the k dimensional space of X_n. Most of the probabilities are zero, as only one dimension, that is the value of X_{n1}, has a random element.

$$P[X_n = X_\text{bstate} \mid X_{n-1} = X_\text{astate}] = \quad \begin{array}{c} X_\text{bstate} \\ X_\text{astate} \left[\begin{array}{c} \text{transition probabilities} \end{array} \right] \end{array}$$

The probability used to describe the pointer adjustment statistics is the probability the number of τ_{pp} periods between one pointer adjustment and another is greater than or equal to n. If X_0 describes the state of the slave clock noise at the moment of a pointer adjustment, and n_{pa} is the number of τ_{pp} periods to the next pointer adjustment, then X_{01} is known precisely, as it is the value that generated the previous adjustment. When k is greater than 1, values are needed for X_{02}, X_{03}, etc. and these cannot be known with certainty. These must be within the previous position of the pointer processor buffer. However, the simplest approximation is to assume

$$X_{01} = X_{02} = X_{03} = \ldots = X_{0k}$$

and this gives the known starting state for the Markov chain we call $X_{startstate}$. We now need the possible trapping states for the Markov chain and this is if X_{n1} is outside the pointer processor buffer limits. As X_{n2} and so on are determined by a previous $X_{(n-1)1}$, $X_{(n-1)2}$ and so on which must be within the pointer processor buffer limits, assuming the Markov chain is not yet trapped. All the states where X_{n1} is greater than the upper limit of the buffer are $X_{upstates}$, and all the states where X_{n1} is less than the lower limit of the buffer are $X_{lostates}$. All other states are called $X_{bufstates}$.

$$P[X_n \text{ member } X_{lostates}] + P[X_n \text{ member } X_{bufstates}] + P[X_n \text{ member } X_{upstates}] = 1$$

The probability that the next pointer adjustment is after the first τ_{pp} period is

$$P[n_{pa} = 1] = P[X_1 \text{ member } X_{bufstates} \mid X_0 = X_{startstate}]$$

The probability that it is after the second τ_{pp} period can be found by summing the two step transition over all possible intermediate transition states

$$P[n_{pa} = 2] = \sum_{X_{1state} \text{ member } X_{bufstates}} P[X_2 \text{ member } X_{bufstates} \mid X_1 = X_{1state}] \, P[X_1 = X_{1state} \mid X_0 = X_{startstate}]$$

This process is repeated to obtain values for $n_{pa} = 1, 2, 4, 8, 16, \ldots$ and results in the curves for pointer statistics shown in Figure 11.29. Since the possible values of X_{np} are continuous, this summation is in fact an integration. However, the evaluation of the stochastic process has to be carried out numerically, as even the simplest models of slave clock noise lead to an analytically unsolvable integration, and all numerical integration techniques, in fact, use a summation.

References

The following gives a full analysis of clock noise measurement parameters:

Howe, D. A., Allan, D. W., Barnes, J. A., *Properties of Signal Sources and Measurement Methods*, Proceedings of the 35th Annual Symposium on Frequency Control, 1981.

Appendix B
Layer Designators and Diagrammatic Conventions

When discussing the SDH layers, the designations defined by CCITT have been used. This set of designations has been extended here to include other network layers relevant to the discussion of transmission networking. In particular, the PDH layers, the OSI layers, and certain auxiliary layers have been given similar succinct designations. These are all listed below for reference. The diagrammatic conventions have been introduced in the text, but are summarized in Figure B.1 for reference.

B.1 LAYER DESIGNATORS

OS-N	Optical section layer (signal rate index N = 1, 4, or 16)
RS-N	Regenerator section layer (signal rate index N = 1, 4, or 16)
MS-N	Multiplex section layer (signal rate index N = 1, 4, or 16)
S4	VC-4 path layer
S3	VC-3 path layer
S2	VC-2 path layer
S11	VC-11 path layer
S12	VC-12 path layer
S1	Equivalent to VC-11 or VC-12 layer generically
SD	Synchronization distribution layer
ECC(l)	Embedded communication channel (OSI Link layer)
ECC(t)	Embedded communication channel (OSI Transport layer)
P4s	140-Mbps synchronous PDH path layer (ETS 300 337)
P4a	140-Mbps clear channel path layer
P32	45-Mbps(DS3) path layer
P31	34-Mbps path layer

P22	8-Mbps path layer
P21	6-Mbps path layer
P12s	2-Mbps (G.704 structured) path layer (synchronous to network)
P12a	2-Mbps clear channel path layer (asynchronous to network)
P11s	1.5-Mbps (DS1) path layer (synchronous to network)
P11a	DS1 clear channel path layer (asynchronous to network)
P1	Equivalent to P11 or P12 generically (a or s)
P0	64-Kbps (DS0) circuit layer (implicitly synchronous to network)
Ex	Electrical Section (x = 11, 12, 22, 31, 32, 4) according to G.703.
VCh	ATM virtual channel layer
VP	ATM virtual path layer

Transport entities

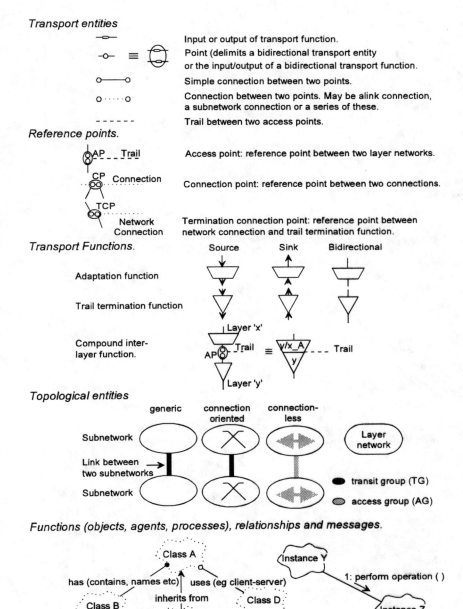

	Input or output of transport function.
	Point (delimits a bidirectional transport entity or the input/output of a bidirectional transport function.
	Simple connection between two points.
	Connection between two points. May be alink connection, a subnetwork connection or a series of these.
	Trail between two access points.

Reference points.

Access point: reference point between two layer networks.

Connection point: reference point between two connections.

Termination connection point: reference point between network connection and trail termination function.

Transport Functions.

Adaptation function

Trail termination function

Compound inter-layer function.

Topological entities

Subnetwork

Link between two subnetworks

Subnetwork

transit group (TG)

access group (AG)

Functions (objects, agents, processes), relationships and messages.

has (contains, names etc)   uses (eg client-server)

inherits from

1: perform operation ( )

Figure B-1   Diagrammatic conventions used in discussion of transmission network architecture.

Appendix C
List of Acronyms

ABR	available bit rate
ABT	ATM block transfer
ACSE	association control service element
ADM	add/drop multiplexer
ADPCM	adaptive differential pulse code modulation
AG	access group
AIS	alarm indication signal
AMI	alternate mark inversion
ANSI	American National Standards Institute
AP	access point
API	application programming interface
APLL	analog phase locked loop
APS	automatic protection switching
ASN1	abstract syntax notation no. 1
ASCII	American standard code for information interchange
ATDM	asynchronous time division multiplexing
ATM	asynchronous transfer mode
AU	administrative unit
AUG	administrative unit group
B3ZS	binary with three zeros suppression
B8ZS	binary with eight zeros suppression
BBER	background block error ratio
BER	binary error rate
BIP	bit interleaved parity
BIP-n	bit interleaved parity of depth n

BITS	building integrated timing supply
CAC	connection admission control
CAS	channel associated signaling
CBR	constant bit rate
CCITT	International Telephone and Telegraph Consultative Committee
CCS	common channel signaling
CDV	cell delay variation
CEPT	Committee European de Post et Telegraph
CI	concatenation indicator
CLP	cell loss priority
CMI	coded mark inversion
CMIP	common management information protocol
CMISE	common management information service element
CMTT	Mutual Television Committee
CP	connection point
CRC	cyclic redundancy check
CSMA/CD	carrier sense multiple access/collision detect
CTD	cell transfer delay
CTP	connection termination point
DAVIC	digital audio visual council
DBR	deterministic bit rate
DCC	data communications channel
DCN	data communications network
DCS	digital crossconnect system
DDF	digital distribution frame
DLC	digital loop carrier
DN	distinguished name
DPRing	dedicated protection ring
DQDB	distributed queue dual bus
DS	degraded second
DS1	primary rate digital signal
DSP	digital signal processing (implementation)
DSP	domain specific part (addressing)
DXC	digital crossconnect
EBU	European Broadcasting Union
ECC	embedded control channel
EEC	European Economic Community
EII	European information infrastructure
EOW	engineering order wire
E-R	entity-relationship
ERT	elementary regenerator section termination
ES	end system (in ISO messaging network)
ES	errored second
ESR	errored second ratio
ETSI	European telecommunications standards institute

FAW	frame alignment word
FDDI	fiber digital data interface
FDM	frequency division multiplex
FEC	forward error correction
FERF	far end receiver fail
GII	global information infrastructure
HDB3	high density bipolar with maximum of three zeros
HDTV	high-definition television
HEC	header error check
HO	higher order
HOP	higher order path
HVC	higher order virtual container
ICD	international code designator
I/D	increment/decrement
IDI	initial domain identifier
IDP	initial domain part
IEC	international exchange carriers
IEEE	Institution of Electrical and Electronic Engineers
IETF	internet engineering task force
IP	internetworking protocol
IS	intermediate system (in ISO messaging network)
ISO	International Standards Organization
LAN	local area network
LATA	local access and transit areas
LEC	local exchange carriers
LED	light emitting diode
LO	lower order
LOP	lower order path
LPF	low pass filter
LSS	link status signal
LTBER	long term bit error rate
LVC	lower order virtual containers
MAC	media access control
MAN	metropolitan area network
MBS	maximum burst size
MCR	minimum cell rate
MLM	multilongitudinal mode
MRTIE	maximum relative time interval error
MS	multiplex section
MSOH	multiplexer section overhead
MSPG	MS protection group
MTBF	mean time between failures
MTTR	mean time to restore
NDF	new data flag
NE	network element
NNI	network node interface

NPC	network parameter control
NPI	null pointer indicator
nrt-SBR	non-real-time statistical bit rate
NRZ	nonreturn to zero
NSAP	network service access point
OAM	operation administration and maintenance
OC-N	optical channel level N
ODP	open distributed processing
OH	overhead
OOF	out of frame
OS	optical section
OSI	open system interconnect
PCM	pulse code modulation
PCR	peak cell rate
PDH	plesiochronous digital hierarchy
PJE	pointer justification event
P-NNI	private network node interface
POH	path overhead
POP	point of presence
PS	protection switch
PSTN	public switched telephone network
PVC	permanent virtual connection
RBS	robbed bit signaling
RDN	relative distinguished name
RLT	regenerated line termination
RM	resource management
RMS	root mean square
ROSE	remote operations service element
RS	regenerator section
RS-DCC	regenerator section data communication channel
RS-EOW	regenerator section engineering order wire
RSOH	regenerator section overhead
rt-SBR	real time statistical bit rate
RTIE$_{rms}$	root mean square relative time interval error
RZ	return to zero
SAW	surface acoustic wave
SCR	sustained cell rate
SDCN	SDH data communication network
SDH	synchronous digital hierarchy
SE	service elements
SES	severely errored second
SESR	severely errored second ratio
SIE	short interruption event
SLM	single longitudinal mode
SMDS	switched multimegabit data service
SMS	SDH management subnetwork

SOH	section overhead
SONET	synchronous optical network
SPE	synchronous payload envelope
SPRing	shared protection ring
S-PVC	soft permanent virtual connection
STDM	synchronous time division multiplexing
STM-N	synchronous transport module level N
STS-N	synchronous transport signal level N
SVC	switched virtual connection
TCP	transmission control protocol
TCP	terminating connection points
TDM	time division multiplexing
TG	transit group
TMN	telecommunications management network
TP	termination point
TSI	time-slot interchange
TST	time-space-time
TTP	trail termination point
TTPId	trail termination point identifiers
TU	tributary unit
TUG	tributary unit groups
UBR	unspecified bit rate
UI	unit intervals
UNI	user-to-network interface
UPC	usage parameter control
UT	unavailable time
VBR	variable bit rate
VC	virtual containers (in SDH)
VCh	virtual channel (in ATM)
VP	virtual path
VT	virtual tributary
VTG	virtual tributary group
WDM	wavelength division multiplexing
WFQ	weighted fair queuing

Index

The Artech House Telecommunications Library

Vinton G. Cerf, Series Editor

For further information on these and other Artech House titles, including previously considered out-of-print books now available through our In-Print-Forever™ (IPF™) program, contact:

Artech House
685 Canton Street
Norwood, MA 02062
781-769-9750
Fax: 781-769-6334
Telex: 951-659
email: artech@artech-house.com

Artech House
46 Gillingham Street
London SW1V 1AH England
+44 (0) 171-973-8077
Fax: +44 (0) 171-630-0166
Telex: 951-659
email: artech-uk@artech-house.com

Find us on the World Wide Web at:
www.artech-house.com